Fluorine-containing
Amino Acids

Fluorine-containing Amino Acids

Synthesis and Properties

Edited by

V. P. KUKHAR' and V. A. SOLOSHONOK

National Academy of Sciences of Ukraine

JOHN WILEY & SONS

Chichester · New York · Brisbane · Toronto · Singapore

Copyright © 1995 by John Wiley & Sons Ltd,
Baffins Lane, Chichester,
West Sussex PO19 1UD, England

Telephone: National Chichester (01243) 779777
International +44 1243 779777

Other Wiley Editorial Offices

John Wiley & Sons, Inc., 605 Third Avenue,
New York, NY 10158-0012, USA

Jacaranda Wiley Ltd, 33 Park Road, Milton,
Queensland 4064, Australia

John Wiley & Sons (Canada) Ltd, 22 Worcester Road,
Rexdale, Ontario M9W 1L1, Canada

John Wiley & Sons (SEA) Pte Ltd, 37 Jalan Pemimpin #05-04,
Block B, Union Industrial Building, Singapore 2057

Library of Congress Cataloging-in-Publication Data

Fluorine-containing amino acids : synthesis and properties / V.P.
  Kukhar' and V. A. Soloshonok.
      p.   cm.
  Includes bibliographical references and index.
  ISBN 0-471-95203-6
  1. Amino acids—Synthesis.   2. Organofluorine compounds
  –Synthesis.   I. Kukhar', V. P. (Valerii) Pavlovich)   II. Soloshonok,
  V. A.
  QP521.F58   1994
  574.19'245—dc20   94-13954
                                                      CIP

British Library Cataloguing in Publication Data

A catalogue record for this book is available from the British Library

ISBN 0 471 95203 6

Typeset in 10/12pt Times by Alden Multimedia, Northampton
Printed and bound in Great Britain by Biddles Ltd, Guildford, Surrey

Contents

List of Contributors

Klaus Burger
Universität Leipzig, Institut für Organische Chemie, Talstr. 35, D-04103 Leipzig, Germany

Qing Dong
Department of Chemistry, State University of New York at Stony Brook, Stony Brook, NY 11794-3400, USA

S. V. Galushko
Institute of Biorganic Chemistry and Petrochemistry, National Academy of Sciences of Ukraine, Murmanskaya 1, 253660 Kiev-94, Ukraine

A. Gyenes
Department of Chemistry, University at Albany, Albany, NY 12222, USA

M. J. Jung
Marion Merrell Dow Research Institute, Cincinnati, OH, USA

Kenneth L. Kirk
Laboratory of Bioorganic Chemistry, National Institute of Diabetes and Digestive and Kidney Diseases, National Institutes of Health, Bethesda, MD 20892, USA

Valery P. Kukhar'
Institute of Bioorganic Chemistry and Petrochemistry, National Academy of Sciences of Ukraine, Murmanskaya 1, 253660 Kiev-94, Ukraine

Yasushi Matsumura
Asahi Glass Company Ltd, Research Center, Hazawa-cho, Kanagawa-ku, Yokohama 221, Japan

Toshifumi Miyazawa
Department of Chemistry, Faculty of Science, Konan University, Higashinada-ku, Kobe 658, Japan

Iwao Ojima
Department of Chemistry, State University of New York at Stony Brook, Stony Brook, NY 11794-3400, USA

Guiseppe Resnati
CNR, Centro Studio Sostanze Organiche Naturali, Dipartimento Chimica, Politecnico, Via Mancinelli 7, I-20131 Milan, Italy

Norbert Sewald
Universität Leipzig, Institut für Organische Chemie, Talstr. 35, D-04103 Leipzig, Germany

Hing L. Sham
Pharmaceutical Discovery Division, D-47D; AP9A, Abbott Laboratories, Abbott Park, IL 60064-3500, USA

Vadim A. Soloshonok
Institute of Bioorganic Chemistry and Petrochemistry, National Academy of Sciences of Ukraine, Murmanskaya 1, 253660 Kiev-94, Ukraine

Vladimír Tolman
Prague Institute of Chemical Technology, Department of Organic Chemistry, Vídeňská 1083, 142 20 Prague 4, Czech Republic

Masahiro Urushihara
Asahi Glass Company Ltd, Research Center, Hazawa-cho, Kanagawa-ku, Yokohama 221, Japan

J. T. Welch
Department of Chemistry, University at Albany, Albany, NY 12222, USA

Preface

The interest in amino acids (AAs) and their derivatives has existed for many years. More than 700 AAs have already been found in nature and their number is continually growing. Investigations into the biological properties of these AAs have been intensively prosecuted and have resulted in a dramatic acceleration of activity and interest in synthesis of unusual AAs directed, first of all, at the creation of new medicines and fine biochemicals. The synthesis of unusual AAs has continued to develop at a tremendous pace over the past 40 years, producing an incredible range of structurally exotic and novel compounds. The man-made area of fluorine-containing amino acids (FAAs) takes the most important place in the family of unusual AAs.

The upsurge in interest in FAA chemistry and biology has a number of origins. Among these, the similar geometry of FAAs to hydrocarbon patterns, the opposite polarization of C—H and C—F bonds and greater energy of the latter, have played a major part. Biologists and medicinal chemists have been quick to seize on the opportunities opened up by these unique basic physico-chemical properties of FAAs. It is an exciting story to see how chemists and biologists have inspired each other to build up the field of FAAs. It is a field to which both chemists and biologists have contributed significantly. The present brisk activity in the field, with the surprises that often emerge from research in this area, recently led Professor D. Seebach to coin a new term, *Flustrates* (*Flu*orine-containing sub*strates*). Being on the borderline between the organic chemistry and life science fields, FAAs possess enormous potential for future development. Suffice to say that the biochemistry of FAAs is still in its infancy. The origin of the biological activity of a number of FAAs has been studied, but these studies have not involved systematic investigations, but rather relatively random sampling. The application of FAAs in peptide design is also in the early stages of development.

The motives of authors of scientific books are as many and varied as the authors themselves. In our case we decided to edit this book mostly because we are in love with the subject. Also, we believe that the time is now ripe for a book in which the chemistry of FAAs and the existing knowledge of their biological properties are brought together and examined critically. We have attempted to produce a book as complete and comprehensive as the current state of knowledge and practical space limitations will permit. It was the enthusiastic consensus of the publisher, editors and contributors to achieve this goal. The individual authors have been given a great deal of freedom in preparing their contributions so that personal viewpoints could be presented and new, unpublished results could be disseminated.

Chapters 1–3 deal with a synthesis of FAAs by methods of classical amino acid chemistry, organofluorine chemistry and organometallic chemistry. Chapter 4 devoted to the synthesis of fluoroamino acids containing fluorine atom(s) in the β-position to the amino group. Among this type of FAAs the specific inhibitors of polyamine biosynthesis have been discovered. In the area of FAAs, which are mostly optically active, it is not possible to be successful without the preparation and investigation of individual stereoisomers (enantiomers) of these compounds. We have devoted four chapters to this aspect. Chemical and enzymatic asymmetric syntheses of FAAs are described in Chapters 5 and 6 and the enzymatic and chromatographic resolution of racemic FAAs in Chapters 7 and 8. In the final chapters, the biological properties of FAAs and their applications are reviewed. We hope that the contributions by ourselves and fifteen distinguished scientists collected in this volume will stimulate new ideas and initiate further research in all areas of FAAs chemistry and biology.

The chemistry of FAAs is only one part of a large field of biologically active organofluorine compounds. One can see growing interest in the chemistry and biology of fluorine-containing nucleosides, vitamins, steroids and carbohydrates and also new pharmaceuticals, pesticides and other biologically active substances. Obviously, we are still at the very beginning of crucial changes in scientific and industrial activities directed to the investigation, production and application of fine organofluorine compounds. The future of the field is assured by the results obtained so far and many new avenues of research await further interdisciplinary investigation.

The readership for which we have geared this book is research workers in the field of organic chemistry and organofluorine chemistry in general, medicinal and pharmaceutical chemists and others who are interested in the synthesis of organoelement analogues of natural compounds. The volume is designed as a source book and general reference and not as a textbook. It contains much that is of practical use to working chemists and biochemists in the identified fields.

We are grateful to all of the contributing authors for their dedication to the subject. Without their efforts this book could not have seen the light of day. We also thank Miss Jenny Cossham, Elspeth Tyler, Martin Röthlisberger and Irene Cooper (of John Wiley & Sons) for their pleasant cooperation and their toleration of several last-minute changes. Finally, the editors mutually acknowledge their unflagging support during the various stages of the enterprise.

V. P. Kukhar'	September	V. A. Soloshonok
Kiev	1993	Kiev
Ukraine		Ukraine

1 Syntheses of Fluorine-containing Amino Acids by Methods of Classical Amino Acid Chemistry

VLADIMÍR TOLMAN

Prague Institute of Chemical Technology, Department of Organic Chemistry, Vídeňská 1083, 142 20 Prague 4, Czech Republic

1.1 INTRODUCTION

The term 'methods of classical chemistry' should be understood only as a writer's licence, not as an explicit definition of any particular area created by exactly specified synthetic reactions. Of course, there is no doubt about the 'classical' character of most of the reactions reviewed in this chapter. Nevertheless, some cases *do* exist in which, for example, the synthesis is modified by 'modern' access to a 'classical' intermediate, or when a 'classical' starting compound is transformed into the product in a 'modern' way, but the *strategy* of the synthesis remains unchanged. The question necessarily arises of where the borders are. The answer may be a matter of discussion, but in this chapter the author will conform to the needs of practice. As a logical consequence, occasional overlapping with other chapters is inevitable.

Among the very numerous syntheses of fluorine-containing analogues of amino acids, those using the methods of classical amino acid chemistry are, on the whole, conceivably among the first pathways which the synthetic chemist takes into consideration and evaluates for their utility for the preparation of a desired fluorinated compound. This access is based on two premises: (1) stability and chemical inertness of the C—F bond(s); and (2) similar reactivities of the fluorinated and non-fluorinated reactants.

Provided that both of these hypotheses are valid, the concept of preparing fluoroamino acids by following the syntheses of their non-fluorinated counterparts represented a viable synthetic strategy and, as such, it was very frequently and successfully applied. However, this point of view is only a one-sided and mechanistic one, as it does not take into account the extreme electronegativity of fluorine (the standard electrode potential of F is 2.65 V, referred to the H^+/H couple as zero). This fact, in its consequences, is responsible for the limited applicability of the methods described in this chapter. The limitation is caused by:

Fluorine-containing Amino Acids: Synthesis and Properties   Edited by V. P. Kukhar' and V. A. Soloshonok
© 1995 John Wiley & Sons Ltd

(1) the change of polarity of the functional groups and bonds in the vicinity of fluorine atoms:
(2) lower reactivity of some fluorinated reactants, especially of polyfluorinated compounds;
(3) the formerly unexpected considerable lability of the C—F bonds in a β- and, to some extent, also in a γ-position to carbonyl.

Despite these limitations, many fluorinated amino acids have been successfully prepared by classical methods. The syntheses will be described in the following sections:

1.2   Amination and related reactions.
1.3   The Strecker synthesis.
1.4   The hydantoin synthesis.
1.5   The Erlenmeyer azlactone synthesis.
1.6   Syntheses using N-substituted aminomalonic esters and related compounds.
1.7   Syntheses involving other CH-acidic esters. The Curtius and Schmidt rearrangements.
1.8   Syntheses from 2-oxo acids and their derivatives.
1.9   Fluorinated amino acids from other amino acids.
1.10 Miscellaneous syntheses.

1.2 AMINATION AND RELATED REACTIONS

In this section, attention will be given to all reactions in which the amino group is introduced by direct reaction of ammonia or its equivalents with halo and hydroxy compounds (either free or as the sulfonate esters) and with compounds possessing carbon–carbon double bonds or an oxirane ring. As the ammonia equivalents we shall consider, for the purpose of this section, phthalimide and its potassium salt, azide ion and, in particular cases, also hexamethyldisilazane, dibenzylamine and trifluoroacetamide.

The simplest fluorinated amino acid, 3-fluoro-2-alanine (**1**) and its nearest homologue, 2-amino-4-fluorobutyric acid (**2**), were prepared in a two-step synthesis by bromination and ammonolysis of the appropriate alkanoic acids [1].

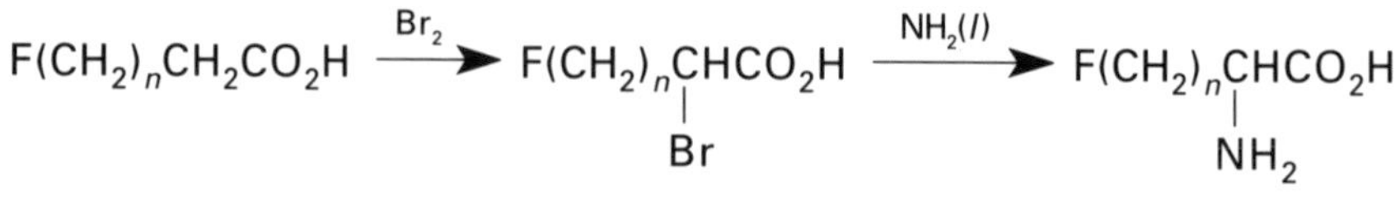

$$F(CH_2)_nCH_2CO_2H \xrightarrow{Br_2} F(CH_2)_n\underset{\underset{Br}{|}}{C}HCO_2H \xrightarrow{NH_3(l)} F(CH_2)_n\underset{\underset{NH_2}{|}}{C}HCO_2H$$

(1) $n = 1$
(2) $n = 2$

A series of homologous 2-amino-3-fluorocarboxylic acids were prepared similarly starting from the corresponding alk-2-enoic acids, which were first bromofluorinated with a mixture of N-bromoacetamide and hydrogen fluoride and then ammonolysed [2].

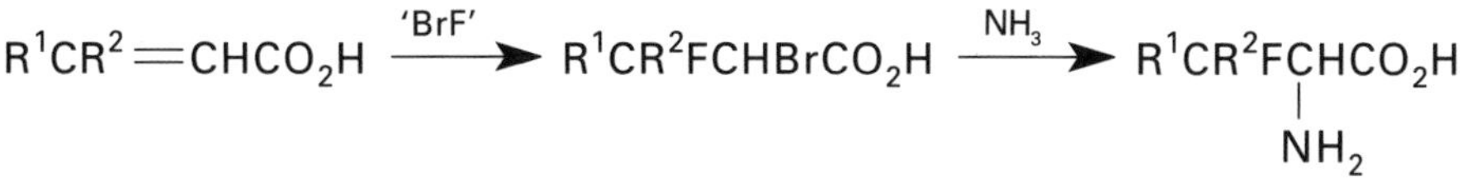

$$R^1CR^2\!\!=\!\!CHCO_2H \xrightarrow{\text{'BrF'}} R^1CR^2FCHBrCO_2H \xrightarrow{NH_3} R^1CR^2FCHCO_2H$$

with NH_2 on the final product

$$(1), (3)-(6)$$

(1) $R^1 = R^2 = H$ 3-fluoro-2-alanine
(3) $R^1 = Me, R^2 = H$ 2-amino-3-fluorobutyric acid
(4) $R^1 = Et, R^2 = H$ 3-fluoronorvaline
(5) $R^1 = Pr, R^2 = H$ 3-fluoronorleucine
(6) $R^1 = Bu, R^2 = H$ 2-amino-3-fluoroheptanoic acid

One branched-chain amino acid, 3-fluorovaline (7) ($R^1 = R^2 = Me$) has been also synthesized by this approach.

In addition to direct amination, the amino acid 1 has also been prepared recently from methyl 2-bromo-3-fluoropropionate by converting it under phase-transfer catalysis into the appropriate 2-azido ester, which was then hydrogenated under carefully controlled conditions to the methyl ester of 1. In order to avoid the simultaneous loss of fluorine during the reduction step, the hydrogenation was carried out in methanol–formic acid solution with saturation with hydrogen chloride [3].

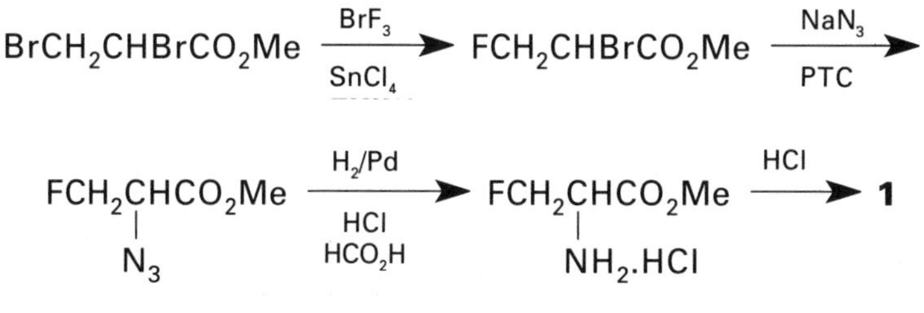

$$BrCH_2CHBrCO_2Me \xrightarrow[SnCl_4]{BrF_3} FCH_2CHBrCO_2Me \xrightarrow[PTC]{NaN_3}$$

$$FCH_2CHCO_2Me \text{ (with } N_3) \xrightarrow[HCl\ HCO_2H]{H_2/Pd} FCH_2CHCO_2Me \text{ (with } NH_2\!\cdot\!HCl) \xrightarrow{HCl} 1$$

The isomer of 1, 2-fluoro-3-alanine (8), resulted as the final product of two syntheses, both using potassium phthalimide as the aminating agent. Starting from diethyl fluoromalonate, amino acid 8 was prepared in four steps [4].

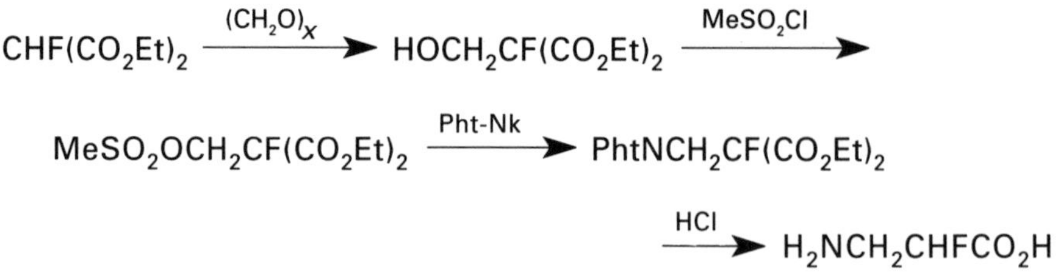

$$CHF(CO_2Et)_2 \xrightarrow{(CH_2O)_x} HOCH_2CF(CO_2Et)_2 \xrightarrow{MeSO_2Cl}$$

$$MeSO_2OCH_2CF(CO_2Et)_2 \xrightarrow{Pht-Nk} PhtNCH_2CF(CO_2Et)_2$$

$$\xrightarrow{HCl} H_2NCH_2CHFCO_2H$$

$$\textbf{(8)}$$

While this synthesis is a 'pure' substitution of the esterified hydroxy group, the other access to **8**, starting from ethyl 2-fluoro-3-hydroxypropionate, involves an elimination–addition mechanism [5].

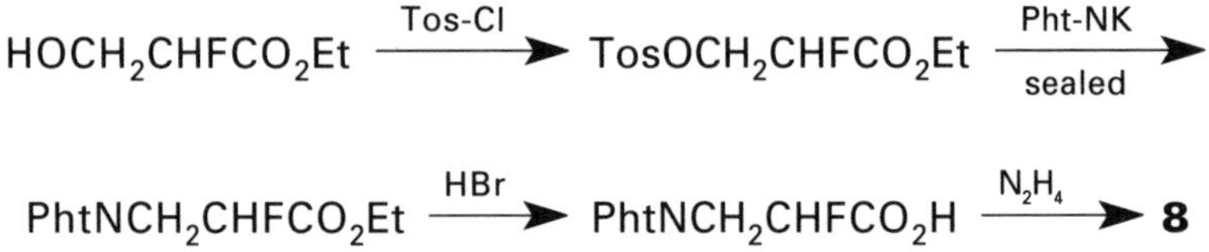

$$HOCH_2CHFCO_2Et \xrightarrow{Tos-Cl} TosOCH_2CHFCO_2Et \xrightarrow[\text{sealed}]{Pht-NK}$$

$$PhtNCH_2CHFCO_2Et \xrightarrow{HBr} PhtNCH_2CHFCO_2H \xrightarrow{N_2H_4} \textbf{8}$$

In the attempted preparation of 2-amino-4,4,4-trifluorobutyric acid (**9**) by reaction of either 2-bromo-4,4,4-trifluorobutyric acid or 4,4,4-trifluorocrotonic acid with ammonia, the isomeric 3-amino-4,4,4-trifluorobutyric acid (**10**) was the sole product. The strong electronegativity of the trifluoromethyl group was responsible for this unintentional result; however, this problem was successfully circumvented by reacting the bromo ester with sodium azide and hydrogenolysis of the 2-azido ester [6].

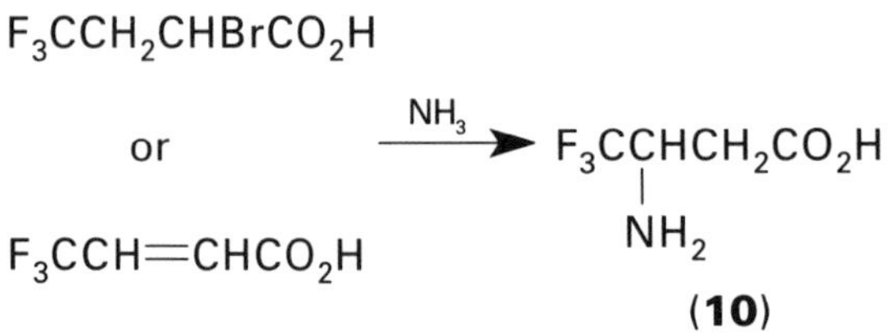

$$F_3CCH_2CHBrCO_2H$$

$$\text{or} \xrightarrow{NH_3} \underset{\underset{NH_2}{|}}{F_3CCHCH_2CO_2H}$$

$$F_3CCH{=}CHCO_2H$$

$$\textbf{(10)}$$

$$\underset{\underset{X}{|}}{F_3CCHCHBrCO_2Et} \xrightarrow{NaN_3} \underset{\underset{X}{|}\;\underset{N_3}{|}}{F_3CCHCHCO_2Et}$$

$$\xrightarrow[\text{2. } H_3O^+]{\text{1. } H_2/Pd} \underset{\underset{X}{|}\;\underset{NH_2}{|}}{F_3CCHCHCO_2H}$$

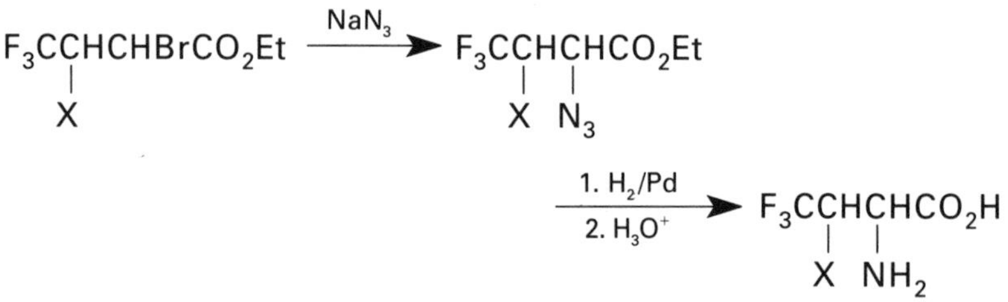

$$\textbf{(9)} \;\; X = H$$
$$\textbf{(11)} \;\; X = Me$$

In the same way, 4,4,4-trifluorovaline (**11**) was also prepared [7]. This methodology has been successfully applied also for the synthesis of 2-amino-4-chloro-4-fluorobutyric acid (**12**) and of 2-amino-4-chloro-4,4-difluorobutyric acid (**13**), the corresponding 2-chloro esters being the starting compounds [8].

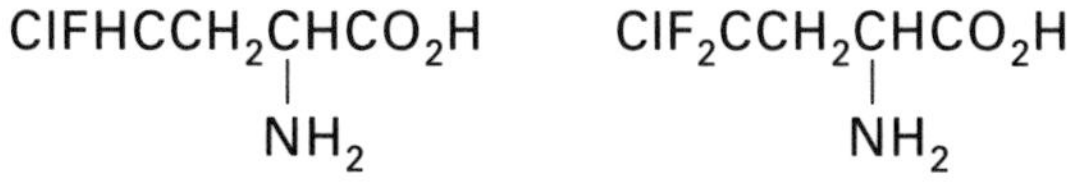

$$ClFHCCH_2\underset{\underset{NH_2}{|}}{C}HCO_2H \qquad ClF_2CCH_2\underset{\underset{NH_2}{|}}{C}HCO_2H$$

$$(\textbf{12}) \qquad\qquad (\textbf{13})$$

2-(Trifluoromethyl)acrylic acid has been found to add ammonia or hexamethyldisilazane fairly easily to give quantitatively [9] the 3-amino acid 3-amino-3',3',3'-trifluoroisobutyric acid (**14**); analogously, from methyl 3-methyl-4,4,4-trifluorocrotonate the methyl ester of 3-amino-3-methyl-4,4,4-trifluorobutyric acid (**15**) was prepared [10]. On the other hand, replacement of the 3-methyl group in the unsaturated ester by the second trifluoromethyl leads to the reversal of polarity of the double bond, so that the 2-amino acid 4,4,4,4',4',4'-hexafluorovaline (**16**) is the final product [11, 12, 13].

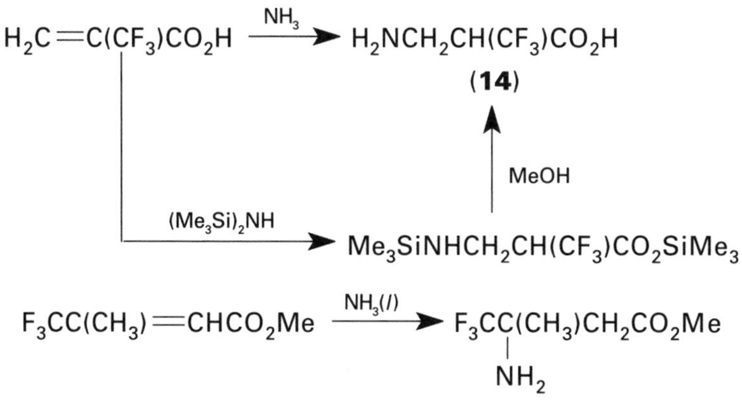

$$H_2C{=}C(CF_3)CO_2H \xrightarrow{\ NH_3\ } H_2NCH_2CH(CF_3)CO_2H$$

$$(\textbf{14})$$

$$\Big\uparrow MeOH$$

$$\xrightarrow{(Me_3Si)_2NH} Me_3SiNHCH_2CH(CF_3)CO_2SiMe_3$$

$$F_3CC(CH_3){=}CHCO_2Me \xrightarrow{\ NH_3(l)\ } F_3CC(CH_3)CH_2CO_2Me$$

$$\underset{\underset{NH_2}{|}}{}$$

$$(\textbf{15})$$

$$(F_3C)_2C{=}CXCO_2Et \xrightarrow[\text{2. HCl}]{\text{1. NH}_3(l)} (F_3C)_2CHCHCO_2H$$

$$\underset{\underset{NH_2}{|}}{}$$

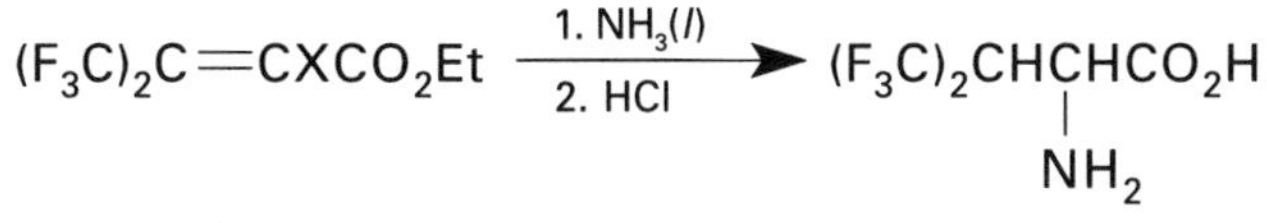

$$(\textbf{16})$$

X = H [11,12]
X = CO$_2$Et [13]

6 FLUORINE-CONTAINING AMINO ACIDS

The homologous 5,5,5,5',5',5'-hexafluoroleucine (17) resulted from a multi-step synthesis, involving the chain elongation of the starting 3-(trifluoromethyl)-4,4,4-trifluorobutyric acid [14].

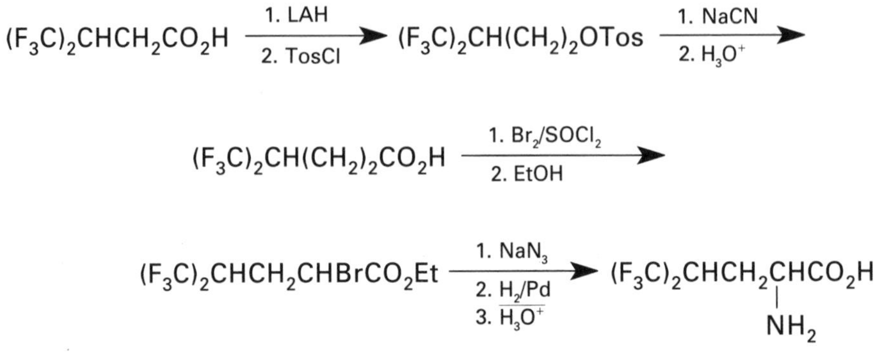

$$(F_3C)_2CHCH_2CO_2H \xrightarrow[\text{2. TosCl}]{\text{1. LAH}} (F_3C)_2CH(CH_2)_2OTos \xrightarrow[\text{2. H}_3\text{O}^+]{\text{1. NaCN}}$$

$$(F_3C)_2CH(CH_2)_2CO_2H \xrightarrow[\text{2. EtOH}]{\text{1. Br}_2/\text{SOCl}_2}$$

$$(F_3C)_2CHCH_2CHBrCO_2Et \xrightarrow[\substack{\text{2. H}_2/\text{Pd} \\ \text{3. H}_3\text{O}^+}]{\text{1. NaN}_3} (F_3C)_2CHCH_2CHCO_2H$$
$$\underset{NH_2}{|}$$

(17)

Ammonolysis of the 2-bromo acid was also the final step in the synthesis of two difluorinated amino acids. 4,4'-difluorovaline (18) and 5,5'-difluoroleucine (19). The toluenesulfonates of the starting 2-(alk-2'- or -3'-en-1'-yl)-propane-1,3-diols were converted into the unsaturated 1,3-difluorides, which on ozonolysis and subsequent bromination yielded the pertinent 2-bromo acids [1].

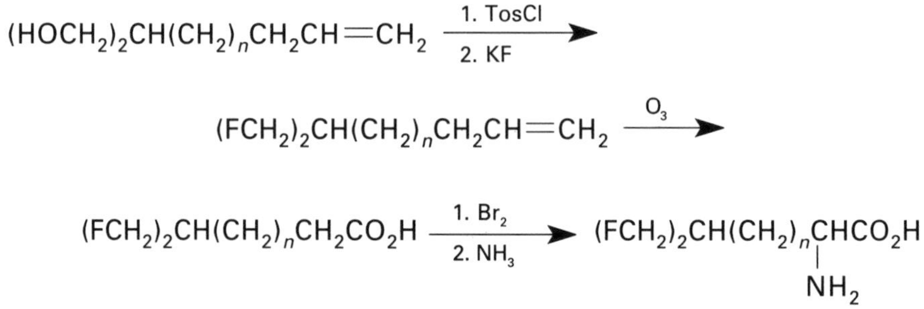

$$(HOCH_2)_2CH(CH_2)_nCH_2CH{=}CH_2 \xrightarrow[\text{2. KF}]{\text{1. TosCl}}$$

$$(FCH_2)_2CH(CH_2)_nCH_2CH{=}CH_2 \xrightarrow{\text{O}_3}$$

$$(FCH_2)_2CH(CH_2)_nCH_2CO_2H \xrightarrow[\text{2. NH}_3]{\text{1. Br}_2} (FCH_2)_2CH(CH_2)_nCHCO_2H$$
$$\underset{NH_2}{|}$$

(18) $n = 0$
(19) $n = 1$

An elegant access to amino acids containing a terminal perfluoroalkyl group was based on the free radical addition of perfluoroalkyl iodides to ethyl acrylate; the resulting fluorinated 2-iodo esters were converted into the amino acids using the azide route [15]:

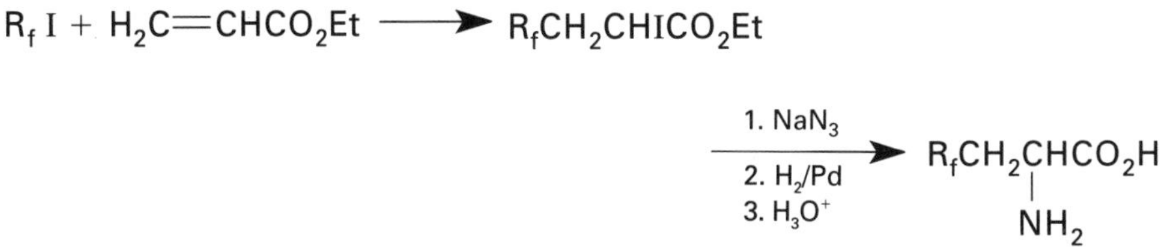

$$R_f I + H_2C{=}CHCO_2Et \longrightarrow R_fCH_2CHICO_2Et$$

$$\xrightarrow[\substack{2.\ H_2/Pd \\ 3.\ H_3O^+}]{1.\ NaN_3} R_fCH_2\underset{\underset{NH_2}{|}}{C}HCO_2H$$

(9)	**(20)**	**(21)**	**(22)**
$R_f = CF_3$	C_2F_5	C_3F_7	$(CF_3)_2CF$

Similar addition [16] of perfluoropropyl iodide to but-3-enoic acid and diethyl allylmalonate, respectively, followed by reductive removal of iodine from the primary adducts led to the heptafluoro derivatives of heptanoic and octanoic acids, which were then brominated and finally aminated to give 2-amino-5,5,6,6,7,7,7-heptafluoroheptanoic acid (23) and 2-amino-6,6,7,7,8,8,8-heptafluorooctanoic acid (24).

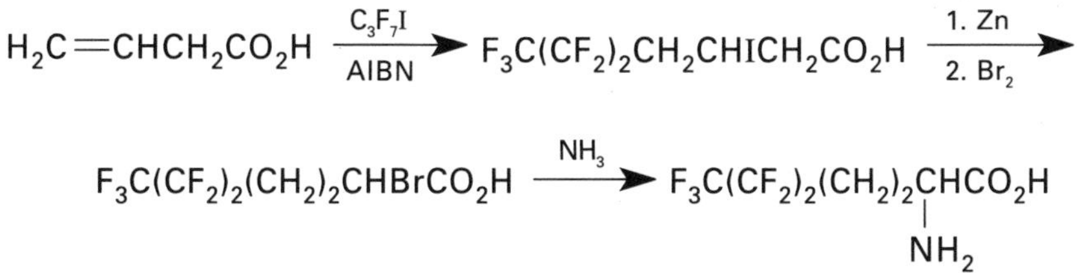

$$H_2C{=}CHCH_2CO_2H \xrightarrow[AIBN]{C_3F_7I} F_3C(CF_2)_2CH_2CHICH_2CO_2H \xrightarrow[2.\ Br_2]{1.\ Zn}$$

$$F_3C(CF_2)_2(CH_2)_2CHBrCO_2H \xrightarrow{NH_3} F_3C(CF_2)_2(CH_2)_2\underset{\underset{NH_2}{|}}{C}HCO_2H$$

$$(23)$$

$$H_2C{=}CHCH_2CH(CO_2Et)_2 \xrightarrow[AIBN]{C_3F_7I} F_3C(CF_2)_2CH_2CHICH_2CH(CO_2Et)_2$$

$$\xrightarrow[2.\ KOH]{1.\ Zn} F_3C(CF_2)_2(CH_2)_3CH_2CO_2H \xrightarrow[2.\ NH_3]{1.\ Br_2} F_3C(CF_2)_2(CH_2)_3\underset{\underset{NH_2}{|}}{C}HCO_2H$$

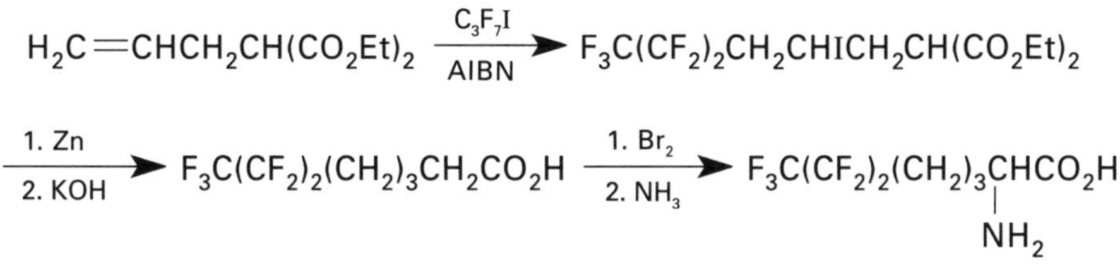

$$(24)$$

In the isoleucine series, Gershon *et al.* [17] used the anomalous result of bromofluorination of 4-methylpent-2-enoic acid in the synthesis of 4-fluoroisoleucine (25). The originally formed 3-carbocation rearranged by the methyl group transfer to the thermodynamically more stable 4-carbocation, which then added fluoride anion to give 2-bromo-4-fluoro-3-methylpentanoic acid; this on treatment with ammonia yielded 25.

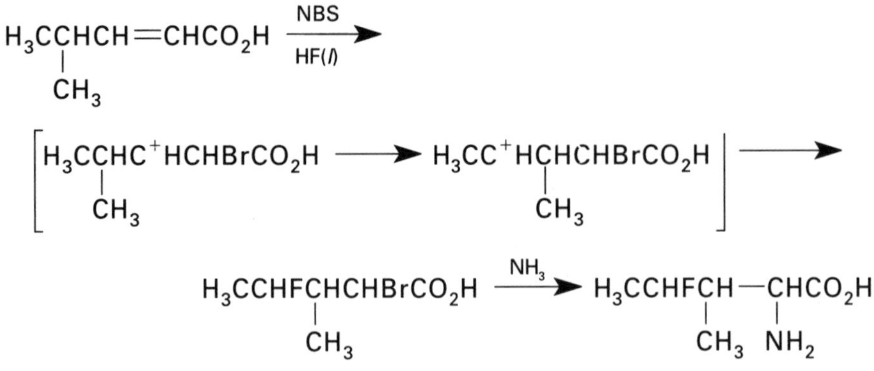

$$H_3CCHCH=CHCO_2H \xrightarrow[HF(\textit{l})]{NBS}$$
$$\underset{CH_3}{|}$$

$$\left[\underset{\underset{CH_3}{|}}{H_3CCHC^+HCHBrCO_2H} \longrightarrow \underset{\underset{CH_3}{|}}{H_3CC^+HCHCHBrCO_2H} \right] \longrightarrow$$

$$\underset{\underset{CH_3}{|}}{H_3CCHFCHCHBrCO_2H} \xrightarrow{NH_3} \underset{\underset{CH_3 \ NH_2}{|\quad|}}{H_3CCHFCH-CHCO_2H}$$

$$(\textbf{25})$$

In another study, the methyl ester of the same bromo acid was converted into **25** via the azide method; however, the yield was very poor [18]. On the other hand, a good result has been reported in the synthesis of 5,5,5,-trifluoroisoleucine (**26**) from the parent 3-methyl-5,5,5-trifluoropentanoic acid [19]:

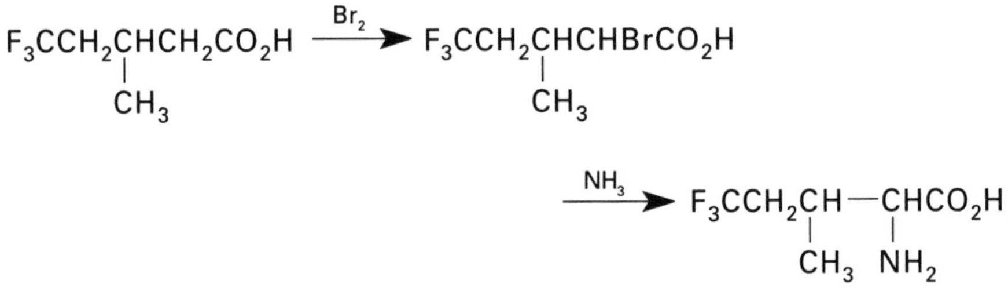

$$\underset{\underset{CH_3}{|}}{F_3CCH_2CHCH_2CO_2H} \xrightarrow{Br_2} \underset{\underset{CH_3}{|}}{F_3CCH_2CHCHBrCO_2H}$$

$$\xrightarrow{NH_3} \underset{\underset{CH_3 \ NH_2}{|\quad|}}{F_3CCH_2CH-CHCO_2H}$$

$$(\textbf{26})$$

4-Fluorothreonine (**27**) (probably the *allo* form) was prepared from 4-fluor-ocrotonic acid by adding the elements of hypobromous acid, following by ammonolysis of the 2-bromo-4-fluoro-3-hydroxybutyric acid so formed [1]. 4,4,4-Trifluorothreonine (**28**), also as the *allo* isomer, resulted from the stereo-selective opening of the *trans*-oxirane carboxylic ester by ammonia [20].

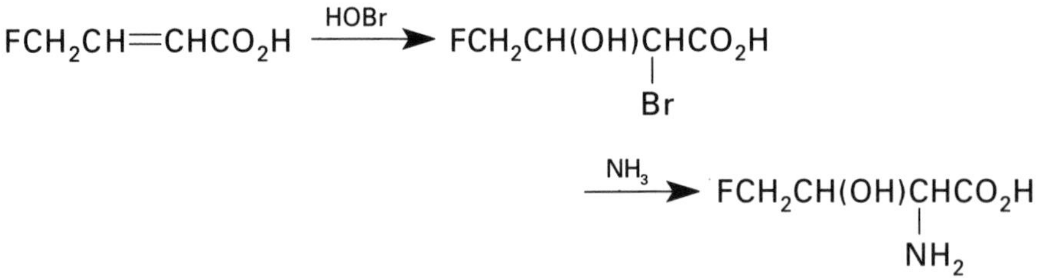

$$FCH_2CH=CHCO_2H \xrightarrow{HOBr} \underset{\underset{Br}{|}}{FCH_2CH(OH)CHCO_2H}$$

$$\xrightarrow{NH_3} \underset{\underset{NH_2}{|}}{FCH_2CH(OH)CHCO_2H}$$

$$(\textbf{27})$$

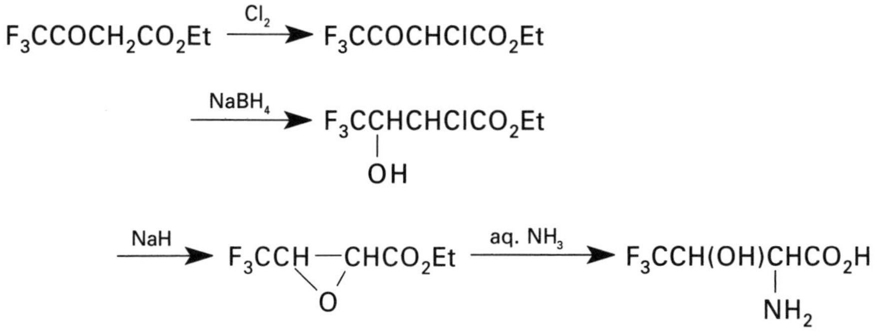

$$F_3CCOCH_2CO_2Et \xrightarrow{Cl_2} F_3CCOCHClCO_2Et$$

$$\xrightarrow{NaBH_4} F_3CC\underset{\underset{OH}{|}}{H}CHClCO_2Et$$

$$\xrightarrow{NaH} F_3CC\underset{\underset{O}{\diagdown\diagup}}{H}-CHCO_2Et \xrightarrow{aq.\ NH_3} F_3CCH(OH)C\underset{\underset{NH_2}{|}}{H}CO_2H$$

(28)

The simplest member of the monoaminodicarboxylic acid group, 3-fluoroaspartic acid (**29**), was the product of a three-step synthesis, starting with dibenzyl 2,3-difluoromaleate [21]. The high reactivity of fluorine in this ester permitted the successful introduction of one dibenzylamino group with the formation of an enamine ester, which was in turn reduced with sodium cyanoborohydride to dibenzyl 2-(N,N-dibenzylamino)-3-fluorosuccinate. Removal of all protecting groups by hydrogenolysis yielded **29**.

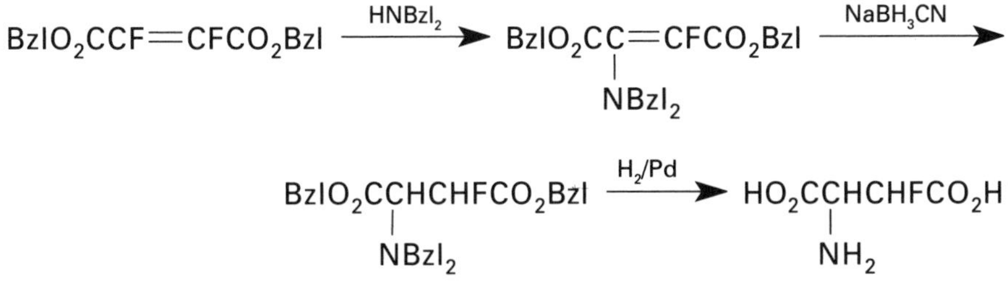

$$BzlO_2CCF\!=\!CFCO_2Bzl \xrightarrow{HNBzl_2} BzlO_2C\underset{\underset{NBzl_2}{|}}{C}\!=\!CFCO_2Bzl \xrightarrow{NaBH_3CN}$$

$$BzlO_2CC\underset{\underset{NBzl_2}{|}}{H}CHFCO_2Bzl \xrightarrow{H_2/Pd} HO_2CC\underset{\underset{NH_2}{|}}{H}CHFCO_2H$$

(29)

4,4-Difluoroglutamic acid (**30**) has recently been synthesised from methyl difluoroiodoacetate [22]. This ester is first converted into difluoroketene methyl triethylsilyl acetal, which adds on 3-acryloyl-4-phenyloxazolid-2-one to give the 5'-methyl ester of 3-(4',4'-difluoro-1'-glutaryl)-4-phenyloxazolid-2-one. By sequential treatment of this compound with dibutylboron triflate, N-bromosuccinimide and sodium azide, the 2'-azido derivative was prepared, which after hydrogenolysis and deprotection gave **30**.

$$\text{ICF}_2\text{CO}_2\text{Me} \xrightarrow[\text{2. Et}_3\text{SiCl}]{\text{1. Zn/MeCN}} \text{F}_2\text{C}{=}\text{C(OMe)} + \text{H}_2\text{C}{=}\text{CHCO}-\text{N}-\text{CO}$$

$$\underset{\text{OSiEt}_3}{|} \qquad\qquad\qquad \overset{\text{Ph}}{\wedge}\overset{|}{\underset{\wedge}{\text{O}}}$$

$$\longrightarrow \text{MeO}_2\text{CCF}_2\text{CH}_2\text{CH}_2\text{CO}-\text{N}-\text{CO} \qquad \begin{array}{l}\text{1. Bu}_2\text{BOTf}\\ \text{2. NBS}\\ \text{3. NaN}_3\end{array} \longrightarrow$$

$$\overset{\text{Ph}}{\wedge}\,\overset{|}{\text{O}}$$

$$\underset{\substack{|\\ \text{N}_3\\ \text{Ph}}}{\text{MeO}_2\text{CCF}_2\text{CH}_2\text{CHCO}}-\text{N}-\text{CO} \xrightarrow[\text{2. H}_2/\text{Pd}]{\text{1. LiOH}} \underset{\substack{|\\ \text{NH}_2}}{\text{HO}_2\text{CCF}_2\text{CH}_2\text{CHCO}_2\text{H}}$$

(30)

The initial step in the synthesis of 2-amino-6-fluoropimelic acid (**31**) was the condensation between diethyl fluoromalonate and ethyl 5-bromopentanoate, leading, after hydrolysis of the intermediate triester, to 2-fluoropimelic acid; this was then conventionally brominated and ammonolysed to yield **31** [23, 24].

$$\text{EtO}_2\text{C(CH}_2)_4\text{Br} + \text{HCF(CO}_2\text{Et})_2 \longrightarrow \text{EtO}_2\text{C(CH}_2)_4\text{CF(CO}_2\text{Et})_2$$

$$\xrightarrow{\text{HCl}} \text{HO}_2\text{C(CH}_2)_4\text{CHFCO}_2\text{H} \xrightarrow[\text{2. Br}_2]{\text{1. SOCl}_2}$$

$$\text{HO}_2\text{CCHBr(CH}_2)_3\text{CHFCO}_2\text{H} \xrightarrow{\text{aq. NH}_3} \underset{\substack{|\\ \text{NH}_2}}{\text{HO}_2\text{CCH(CH}_2)_3\text{CHFCO}_2\text{H}}$$

(31)

The same group [23, 24] elaborated a multi-step synthesis of 2,6-diamino-4-fluoropimelic acid (**32**). Diethyl 4-nitropimelate was first converted into 4-aminopimelic acid, which was thermally cyclized to 2-pyrrolidone-5-propionic acid. On treatment with sodium nitrate in aqueous HF, the lactam ring was reopened to give 4-fluoropimelic acid, albeit in a very low yield (6% in this single step). Bromination followed by amination was performed in the usual manner, potassium phthalimide being the aminating agent. The final deprotection of the 2,6-diphthalimido ester was achieved by hydrazinolysis. To improve the unsatisfactorily low yield of this reaction sequence, 4-fluor-opimelic acid was synthesized from 1,7-diacetoxyheptan-4-one [25]. This ketone was hydrogenated to the 4-hydroxy derivative, which on treatment with 2-chloro-1,1,2-trifluorotriethylamine (Yarovenko reagent, CTT) gave 1,7-diacetoxy-4-fluoroheptane. On oxidation of the free 1,7-diol with nitrogen tetroxide, 4-fluoropimelic acid was formed in a total yield of 20%.

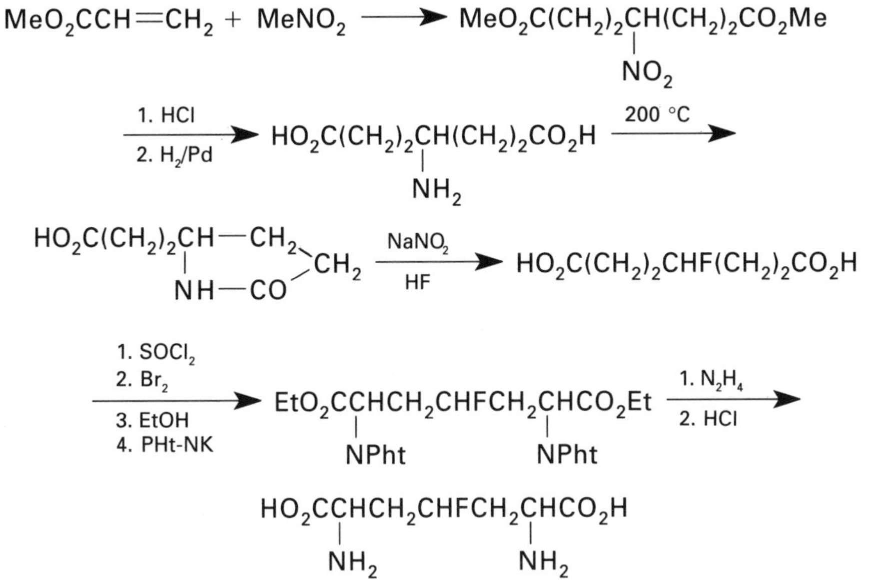

$$MeO_2CCH{=}CH_2 + MeNO_2 \longrightarrow MeO_2C(CH_2)_2\underset{\underset{NO_2}{|}}{CH}(CH_2)_2CO_2Me$$

$$\xrightarrow[\text{2. } H_2/Pd]{\text{1. HCl}} HO_2C(CH_2)_2\underset{\underset{NH_2}{|}}{CH}(CH_2)_2CO_2H \xrightarrow{\text{200 °C}}$$

$$HO_2C(CH_2)_2\underset{\underset{NH{-}CO}{|}}{CH}{-}\overset{}{\underset{}{CH_2}}{\big\rangle}CH_2 \xrightarrow[\text{HF}]{\text{NaNO}_2} HO_2C(CH_2)_2CHF(CH_2)_2CO_2H$$

$$\xrightarrow[\substack{\text{3. EtOH}\\ \text{4. PHt-NK}}]{\substack{\text{1. SOCl}_2\\ \text{2. Br}_2}} EtO_2CCH\underset{\underset{NPht}{|}}{}CH_2CHFCH_2\underset{\underset{NPht}{|}}{CH}CO_2Et \xrightarrow[\text{2. HCl}]{\text{1. } N_2H_4}$$

$$HO_2CCH\underset{\underset{NH_2}{|}}{}CH_2CHFCH_2\underset{\underset{NH_2}{|}}{CH}CO_2H$$

(32)

$$AcO(CH_2)_3CO(CH_2)_3OAc \xrightarrow[\text{2. CTT}]{\text{1. } H_2/Pt} AcO(CH_2)_3CHF(CH_2)_3OAc$$

$$\xrightarrow[\text{2. } N_2O_4]{\text{1. MeOH}} HO_2C(CH_2)_2CHF(CH_2)_2CO_2H$$

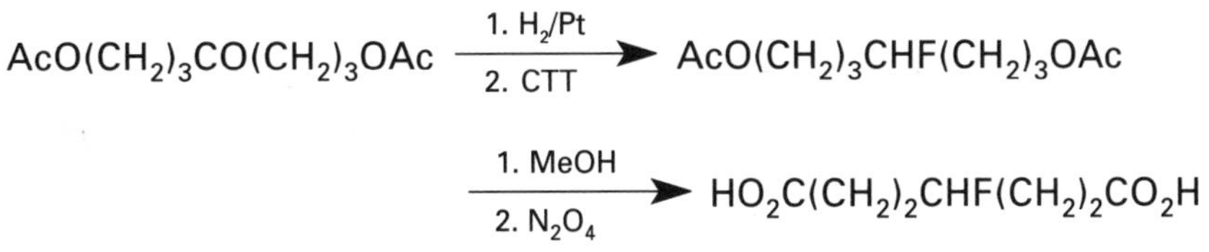

The key intermediates in the preparation of 4-fluoroornithine (**33**) and 5-fluorolysine (**34**) were the corresponding $(\omega - 1)$-fluoro-ω-phthalimi-doalkanoic acids [26]. Whereas in the synthesis of **34** this acid ($n = 3$) was successfully obtained from methyl hex-5-enoate through bromofluorination/ Gabriel reaction, the same approach failed when applied to methyl pent-4-enoate; 4-bromomethylbutyrolactone was the only product of the attempted bromofluorination. 4-Fluoro-5-phthalimidopentanoic acid was prepared by ozonolysis of 5-fluoro-6-phthalimidohex-1-ene [27]; the synthesis of both amino acids **33** and **34** was brought to the end by bromination at C-2, followed by Gabriel reaction and final deprotection.

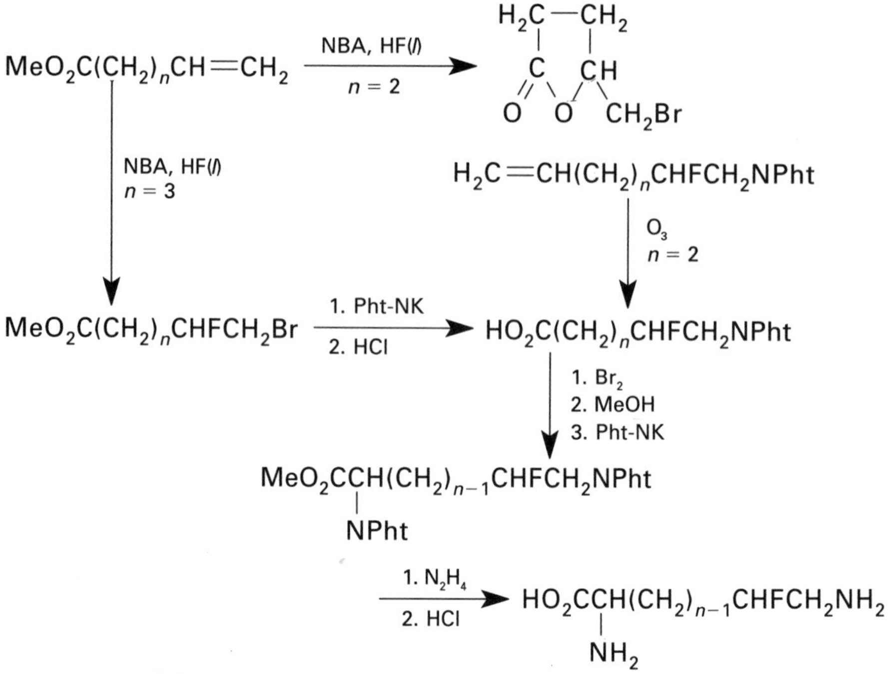

$$MeO_2C(CH_2)_nCH{=}CH_2 \xrightarrow[n=2]{NBA,\ HF(l)}$$

$$\begin{array}{c} H_2C{-}CH_2 \\ | \quad\ | \\ C \quad CH \\ /\!/\ \backslash\ /\ \backslash \\ O\ \ O\ \ CH_2Br \end{array}$$

$$\xrightarrow[n=3]{NBA,\ HF(l)}$$

$$H_2C{=}CH(CH_2)_nCHFCH_2NPht$$

$$\xrightarrow[n=2]{O_3}$$

$$MeO_2C(CH_2)_nCHFCH_2Br \xrightarrow[\text{2. HCl}]{\text{1. Pht-NK}} HO_2C(CH_2)_nCHFCH_2NPht$$

$$\xrightarrow[\substack{\text{1. Br}_2\\ \text{2. MeOH}\\ \text{3. Pht-NK}}]{}$$

$$\begin{array}{c} MeO_2CCH(CH_2)_{n-1}CHFCH_2NPht \\ | \\ NPht \end{array}$$

$$\xrightarrow[\text{2. HCl}]{\text{1. N}_2\text{H}_4} \begin{array}{c} HO_2CCH(CH_2)_{n-1}CHFCH_2NH_2 \\ | \\ NH_2 \end{array}$$

(**33**) $n = 2$
(**34**) $n = 3$

A few examples of unsaturated aminofluoro acids should also be mentioned here. 4-Amino-2-fluorocrotonic acid (**35**) was synthesized from ethyl 2-fluorocrotonate by allylic bromination and subsequent treatment of the 4-bromo ester with potassium phthalimide. Deprotection was accomplished by acid hydrolysis [28].

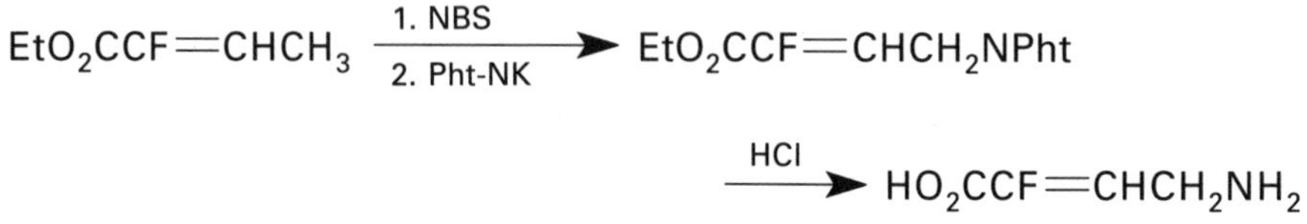

$$EtO_2CCF{=}CHCH_3 \xrightarrow[\text{2. Pht-NK}]{\text{1. NBS}} EtO_2CCF{=}CHCH_2NPht$$

$$\xrightarrow{HCl} HO_2CCF{=}CHCH_2NH_2$$

(**35**)

The opening reaction in the syntheses of the monoaminooxidase inhibitor 3-(fluoromethylene)-3-(3′-hydroxyphenyl)-2-alanine (**36**) [29] and of (*E*)-3-(fluoromethylene)glutamic acid (**37**) [30] is the base-promoted deconjugation of 2-bromo-2,3-unsaturated esters, which leads to intermediates possessing the *exo*-fluoromethylene group.

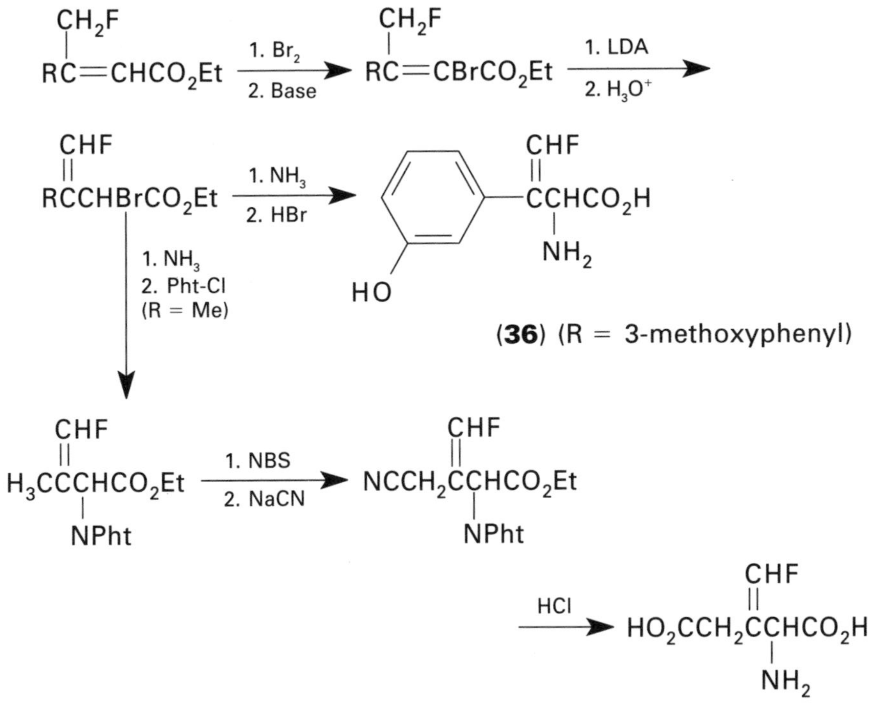

$$\underset{RC=CHCO_2Et}{\overset{\overset{\displaystyle CH_2F}{|}}{}} \xrightarrow[\text{2. Base}]{\text{1. Br}_2} \underset{RC=CBrCO_2Et}{\overset{\overset{\displaystyle CH_2F}{|}}{}} \xrightarrow[\text{2. H}_3O^+]{\text{1. LDA}}$$

$$\underset{RCCHBrCO_2Et}{\overset{\overset{\displaystyle CHF}{\|}}{}} \xrightarrow[\text{2. HBr}]{\text{1. NH}_3}$$

(36) (R = 3-methoxyphenyl)

with the central structure: HO-phenyl ring connected to $\overset{\overset{\displaystyle CHF}{\|}}{C}CHCO_2H$ with NH$_2$ substituent.

1. NH$_3$
2. Pht-Cl
(R = Me)

$$\underset{\underset{NPht}{|}}{\overset{\overset{\displaystyle CHF}{\|}}{H_3CCCHCO_2Et}} \xrightarrow[\text{2. NaCN}]{\text{1. NBS}} \underset{\underset{NPht}{|}}{\overset{\overset{\displaystyle CHF}{\|}}{NCCH_2CCHCO_2Et}}$$

$$\xrightarrow{\text{HCl}} \underset{\underset{NH_2}{|}}{\overset{\overset{\displaystyle CHF}{\|}}{HO_2CCH_2CCHCO_2H}}$$

(37)

The Mitsunobu reaction between a hydroxy ester and phthalimide in the presence of diethyl azodicarboxylate (DEAD) and triphenylphosphine was employed in the synthesis of two fluorovinyl analogues of the neurotransmitter 4-aminobutyric acid [31], namely 4-amino-6,6-difluorohex-5-enoic acid (38) and 4-amino-5,6,6-trifluorohex-5-enoic acid (39).

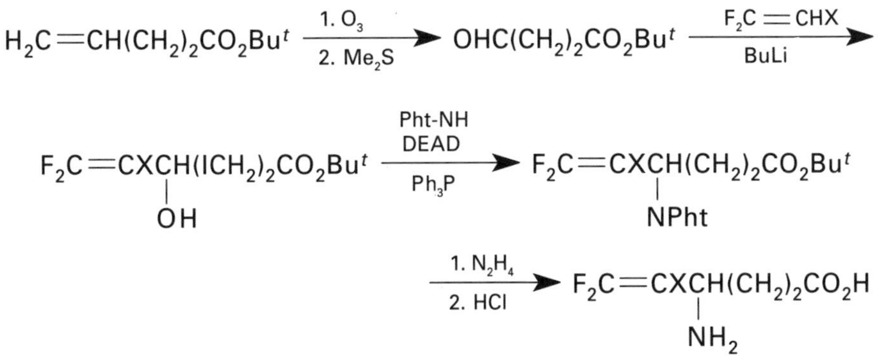

$$H_2C=CH(CH_2)_2CO_2Bu^t \xrightarrow[\text{2. Me}_2S]{\text{1. O}_3} OHC(CH_2)_2CO_2Bu^t \xrightarrow[\text{BuLi}]{F_2C=CHX}$$

$$\underset{\underset{OH}{|}}{F_2C=CXCH(ICH_2)_2CO_2Bu^t} \xrightarrow[\text{Ph}_3P]{\underset{\text{DEAD}}{\text{Pht-NH}}} \underset{\underset{NPht}{|}}{F_2C=CXCH(CH_2)_2CO_2Bu^t}$$

$$\xrightarrow[\text{2. HCl}]{\text{1. N}_2\text{H}_4} \underset{\underset{NH_2}{|}}{F_2C=CXCH(CH_2)_2CO_2H}$$

(38) X = H
(39) X = F

Three ring-fluorinated analogues of 3-phenyl-2-alanine and tyrosine were also prepared by the amination method. The requisite 2-halo-3-(fluoroaryl)-propionic acids are readily accessible by the Meerwein arylation of acrylic acid by fluorinated aryldiazonium halides. In the synthesis of 3-(pentafluorophenyl)-2-alanine (40) the azide route has been applied [32], whereas for the preparation of 2'-(trifluoromethyl)tyrosine (41) [34, 35] and 3'-(trifluoromethyl)tyrosine (42) [35] the direct action of ammonia on the halo acids was chosen.

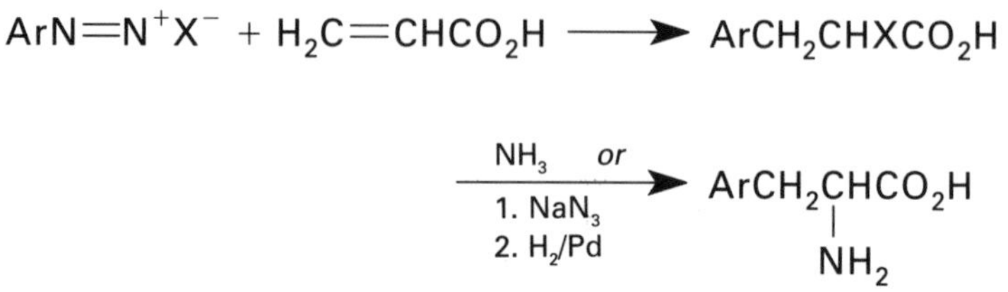

$$ArN{=}N^+X^- + H_2C{=}CHCO_2H \longrightarrow ArCH_2CHXCO_2H$$

$$\xrightarrow[\substack{1.\ NaN_3 \\ 2.\ H_2/Pd}]{NH_3 \quad or} ArCH_2\underset{\underset{NH_2}{|}}{C}HCO_2H$$

(40) Ar = C_6F_5
(41) Ar = 2-CF_3-4-OH-C_6H_3
(42) Ar = 3-CF_3-4-OH-C_6H_3

All three isomeric 2-(fluorophenyl)glycines 43–45 were prepared in a three-component reaction of the appropriate fluorobenzaldehyde, ammonia and bromoform [36] or chloroform under phase-transfer catalysis [37]. The suggested mechanism of this reaction involves an oxirane intermediate, which is cleaved by ammonia to give the amino acid.

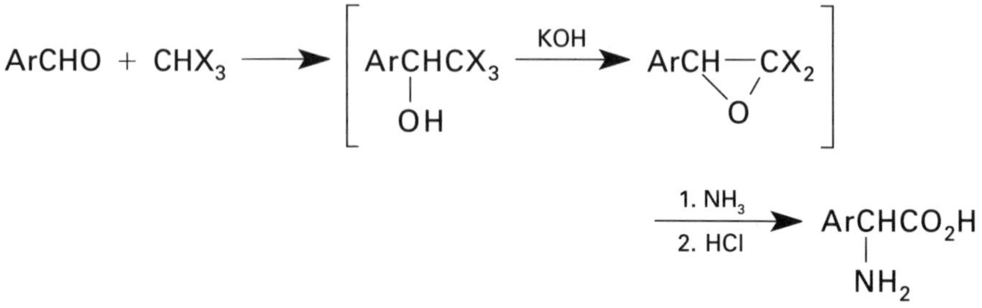

$$ArCHO + CHX_3 \longrightarrow \left[Ar\underset{\underset{OH}{|}}{C}HCX_3 \xrightarrow{KOH} ArCH{-}CX_2 \right]$$

$$\xrightarrow[\substack{2.\ HCl}]{1.\ NH_3} Ar\underset{\underset{NH_2}{|}}{C}HCO_2H$$

(43) Ar = 2-F-C_6H_4
(44) Ar = 3-F-C_6H_4
(45) Ar = 4-F-C_6H_4

In another synthesis of 45, trifluoroacetamide under phase-transfer conditions has been used as an effective aminating agent [38].

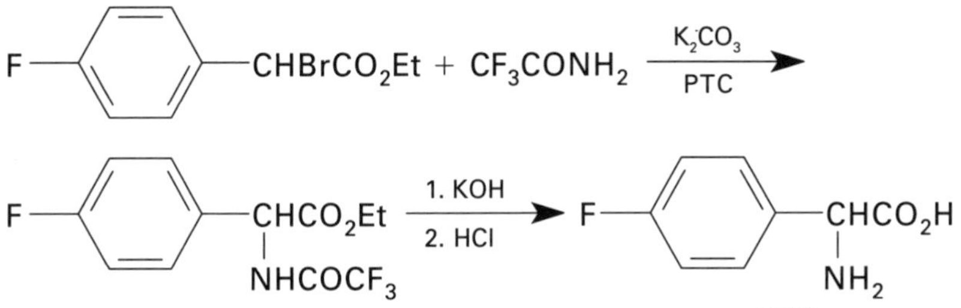

F—⟨benzene⟩—CHBrCO$_2$Et + CF$_3$CONH$_2$ $\xrightarrow[\text{PTC}]{\text{K}_2\text{CO}_3}$

F—⟨benzene⟩—CHCO$_2$Et $\xrightarrow[\text{2. HCl}]{\text{1. KOH}}$ F—⟨benzene⟩—CHCO$_2$H
                          | |
                        NHCOCF$_3$ NH$_2$

(45)

The stereospecific ammonolysis of the oxirane ring in ethyl *trans*-3-(4'-fluorophenyl)glycidate, followed by hydrolysis of the amino acid amide, gives high yield of *erythro*-3-(4'-fluorophenyl)isoserine (**46**). Alkaline hydrolysis is preferable, as it gives pure *erythro*-**46**; on the other hand, partial epimerization was observed after hydrolysis with hydrochloric acid [39, 40].

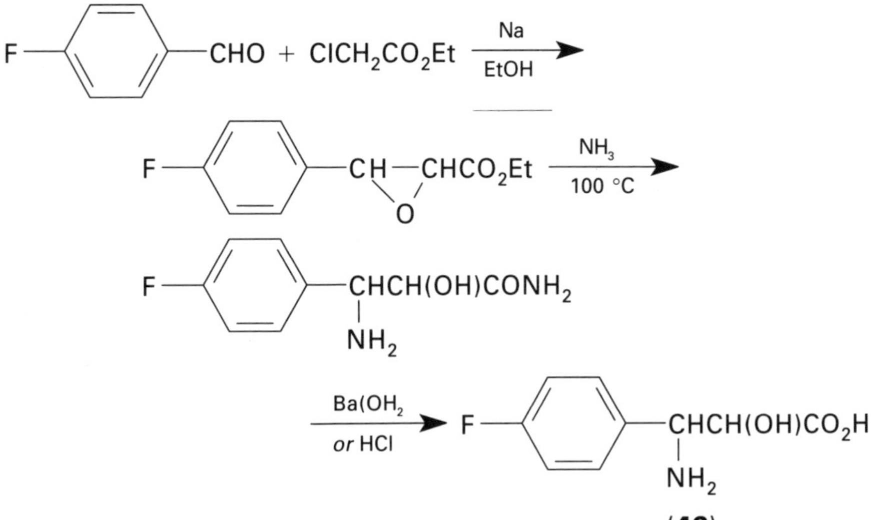

F—⟨benzene⟩—CHO + ClCH$_2$CO$_2$Et $\xrightarrow[\text{EtOH}]{\text{Na}}$

F—⟨benzene⟩—CH—CHCO$_2$Et $\xrightarrow[\text{100 °C}]{\text{NH}_3}$
                    \O/

F—⟨benzene⟩—CHCH(OH)CONH$_2$
              |
             NH$_2$

$\xrightarrow[\text{or HCl}]{\text{Ba(OH)}_2}$ F—⟨benzene⟩—CHCH(OH)CO$_2$H
                                                  |
                                                 NH$_2$

(46)

1.3 THE STRECKER SYNTHESIS

The discovery by Strecker, who in 1850 treated acetaldehyde successively with ammonia and hydrogen cyanide and after hydrolysis of the intermediate product isolated alanine [41], represented the first reported synthesis of an 2-amino acid and, as it was shown later, a generally useful method for the conversion of either aldehydes or ketones into 2-amino acids containing one carbon atom more than the parent oxo compound. The reaction proceeds via the formation of a 2-hydroxynitrile, which is converted into 2-aminonitrile by the action of ammonia or ammonium chloride. Final hydrolysis (preferably acidic) furnishes the 2-amino acid.

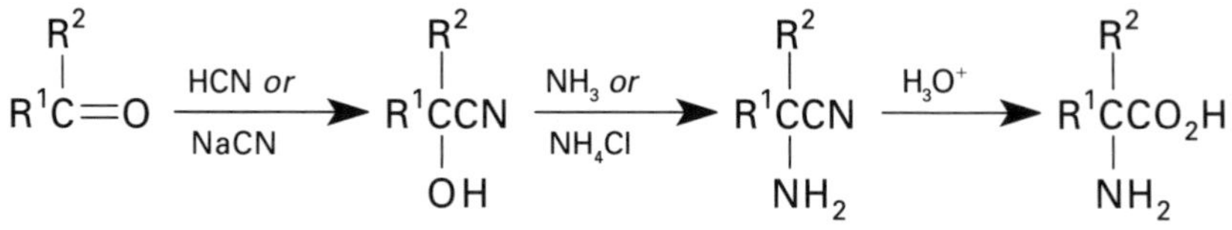

$$R^1\overset{R^2}{\underset{}{C}}=O \xrightarrow[\text{NaCN}]{\text{HCN or}} R^1\overset{R^2}{\underset{OH}{C}}CN \xrightarrow[\text{NH}_4\text{Cl}]{\text{NH}_3 \text{ or}} R^1\overset{R^2}{\underset{NH_2}{C}}CN \xrightarrow{H_3O^+} R^1\overset{R^2}{\underset{NH_2}{C}}CO_2H$$

Despite its versatility, the Strecker reaction found only limited application in the synthesis of fluorinated amino acids. In the following survey attention will be paid not only to syntheses realized according to Strecker's original scheme, but also to some modifications in which the intermediate hydroxy- or aminonitriles are prepared by a different route. In the normal procedure, the fluorinated aldehyde or ketone is converted into the aminonitrile in a single operation; the following amino acids were prepared in this way:

Amino acid	(26)	(47)	(48)	(49)	(50)
R^1	$CF_3CH_2C(CH_3)H$	CH_2F	CH_2F	CF_3	$CF_3(CH_2)_2$
R^2	H	CH_3	CH_2F	CH_3	H
Ref.	[19]	[42, 43]	[42]	[43]	[44]

26, 5,5,5-trifluoroisoleucine; 47, 2-amino-3-fluoroisobutyric acid; 48, 2-amino-3,3'-difluoroisobutyric acid; 49, 2-amino-3,3,3-trifluoroisobutyric acid; 50, 5,5,5-trifluoronorvaline.

When in the above scheme the sodium cyanide was used in combination with a chiral amine, e.g. (4S,5S)-(+)-5-amino-2,2-dimethyl-4-phenyl-1,3-dioxane (ADPD) [45], the 2S,4R- and 2S,4S- enantiomers of 5,5,5-trifluoroleucine (51) were prepared by this variation of the Strecker synthesis [46]. The reaction route started with ethyl 3-(trifluoromethyl)crotonate, which was reduced in two steps to 3-methyl-4,4,4-trifluorobutanol. This alcohol was then resolved through the ester with (−)-camphanic acid and the individual antipodes were oxidized by pyridinium chlorochromate (PCC) to give chiral 3-methyl-4,4,4-trifluorobutanal. Whereas on treating the R-(+)-enantiomer of this aldehyde with NaCN/ADPD (2S,4R)-51 was produced, the other antipode, (2S,4S)-51 arose from an analogous treatment of the racemic aldehyde.

$$H_3CC{=}CHCO_2Et \xrightarrow[\text{2. LAH}]{\text{1. } H_2/Pd} H_3CCHCH_2CH_2OH \xrightarrow[\text{2. PCC}]{\text{1. resolution}}$$
$$\quad\ \ |\qquad\qquad\qquad\qquad\qquad\qquad |$$
$$\quad\ \ CF_3\qquad\qquad\qquad\qquad\qquad CF_3$$

$$H_3CCHCH_2CHO \xrightarrow[\text{3. } HIO_4]{\begin{array}{l}\text{1. NaCN/ADPD}\\\text{2. HCl}\end{array}} H_3CCHCH_2CHCO_2H$$
$$\quad |\qquad\qquad\qquad\qquad\qquad\qquad\ |\qquad |$$
$$\quad CF_3\qquad\qquad\qquad\qquad\qquad\quad CF_3\ \ NH_2$$

(51)

Another modification of Strecker's route consisted in the cleavage of 2-cyano-3,3-dialkyloxiranes by hydrogen fluoride/pyridine (Olah's reagent) to give the requisite fluorinated 2-hydroxynitriles [47, 48]. For their successful conversion into the 2-aminonitriles by methanolic ammonia the presence of a dehydrating agent proved to be essential.

$$R^1R^2CO + ClCH_2CN \xrightarrow{t\text{-BuOK}} R^1R^2C{-}CHCN \xrightarrow[CH_2Cl_2]{HF/Py}$$
$$\qquad\qquad\qquad\qquad\qquad\qquad\ \backslash\,/$$
$$\qquad\qquad\qquad\qquad\qquad\qquad\ O$$

$$R^1R^2CFCHCN \xrightarrow[MgSO_4]{NH_3/MeOH} R^1R^2CFCHCN \xrightarrow{H_3O^+} R^1R^2CFCHCO_2H$$
$$\quad |\qquad\qquad\qquad\qquad\qquad\quad |\qquad\qquad\qquad\qquad\quad |$$
$$\quad OH\qquad\qquad\qquad\qquad\quad NH_2\qquad\qquad\qquad\qquad NH_2$$

(7), (52)–(56)

(7) $R^1 = R^2 = Me$ 3-fluorovaline
(52) $R^1 = Me, R^2 = Et$ 3-fluoroisoleucine
(53) $R^1 = R^2 = Et$ 3-ethyl-3-fluoronorvaline
(54) $R^1 + R^2 = {-}(CH_2)_4{-}$ 2-(1'-fluorocyclopentyl)glycine
(55) $R^1 + R^2 = {-}(CH_2)_5{-}$ 2-(1'-fluorocyclohexyl)glycine
(56) $R^1 = Me, R^2 = Ph$ 2-amino-3-fluoro-3-phenylbutyric acid

In the preparation of (E)-3,4-didehydro-2-(fluoromethyl)ornithine (57) the 2-aminonitrile was prepared at the beginning of the whole synthetic sequence. On reaction of 1-bromomagnesylprop-1-ene with fluoroacetonitrile an N-bromomagnesylimine was formed, which was immediately treated with NaCN/NH₄Cl to give (E)-2-amino-2-(fluoromethyl)pent-3-enenitrile. After protection of the amino group by phthaloylation, the desired second amino group was introduced at C-5 by Wohl–Ziegler bromination, followed by amination with potassium phthalimide. Acid hydrolysis terminated the synthesis [49].

$$FCH_2CN + BrMgCH=CHCH_3 \longrightarrow \left[BrMgN=\overset{\overset{\displaystyle CH_2F}{|}}{C}CH=CHCH_3 \right]$$

$$\xrightarrow[\text{NH}_4\text{Cl}]{\text{NaCN}} N\overset{\overset{\displaystyle CH_2F}{|}}{\underset{\underset{\displaystyle NH_2}{|}}{C}}CCH=CHCH_3 \xrightarrow[\text{3. Pht-NK}]{\substack{\text{1. Pht-Cl}\\ \text{2. NBS}}} N\overset{\overset{\displaystyle CH_2F}{|}}{\underset{\underset{\displaystyle NPht}{|}}{C}}CCH=CHCH_2NPht$$

$$\xrightarrow{\text{H}_3\text{O}^+} HO_2C\overset{\overset{\displaystyle CH_2F}{|}}{\underset{\underset{\displaystyle NH_2}{|}}{C}}CH=CHCH_2NH_2 \qquad (57)$$

An alternative to this synthesis of **57**, covered by a patent [50], differs from the preceding work only in that trifluoroacetylation was used for the protection of the 2-amino group and that the terminal amination was brought about with hexamethylenetetramine.

The recently developed multi-step synthesis of 5,5-difluorolysine (**58**) begins with an analogous reaction to that used in the preparation of 4,4-difluoroglutamic acid (**30**) (see Section 1.2), i.e. with the formation of difluoroketene methyl trimethylsilyl acetal. This intermediate underwent addition to acrolein diethyl acetal, giving methyl 2,2-difluoro-5-ethoxypent-4-enoate, which was further transformed in five steps into 5-benzoylamino-4,4-difluoro-1-ethoxypent-1-ene. From this enol ether, the required aldehyde was liberated by mild hydrolysis; the synthesis was finished by the original Strecker protocol, affording **58** in 13% overall yield [22].

$$ICF_2CO_2Me \xrightarrow[\text{Me}_3\text{SiCl}]{\text{Zn/MeCN}} \left[F_2C=\overset{\overset{\displaystyle C(OMe)}{|}}{\underset{\underset{\displaystyle OSiMe_3}{|}}{}} \right] \xrightarrow[\text{2. NaHCO}_3]{\text{1. CH}_2=\text{C(OEt}_2)}$$

$$MeO_2CCF_2CH_2CH=CHOEt \xrightarrow[\text{3. NaN}_3 \quad \text{4. Ph}_3\text{P} \quad \text{5. BzCl}]{\text{1. NaBH}_4 \quad \text{2. Tf}_2\text{O}}$$

$$BzNHCH_2CF_2CH_2CH=CHOEt \xrightarrow[\substack{\text{2. KCN, NH}_4\text{Cl}\\ \text{aq.NH}_3}]{\text{1. 1-M HCl}}$$

$$BzNHCH_2CF_2(CH_2)_2\overset{\overset{}{}}{\underset{\underset{\displaystyle NH_2}{|}}{C}}HCN \xrightarrow{\text{HCl}} H_2NCH_2CF_2(CH_2)_2\overset{}{\underset{\underset{\displaystyle NH_2}{|}}{C}}HCO_2H$$

$$(58)$$

The synthesis of 3,3,3-trifluoro-2-alanine (**59**), which involves N-substituted 2-aminonitrile as an intermediate, should also be included in this section. From N-benzoyl(or N-benzyloxycarbonyl)-1-chloro-2,2,2-trifluoroethylamine, the N-protected 2-amino-3,3,3-trifluoropropionitrile was prepared either through the corresponding trifluoracetaldimine [51] or directly by treatment with copper(I) cyanide [52]; acid hydrolysis then gave **59**.

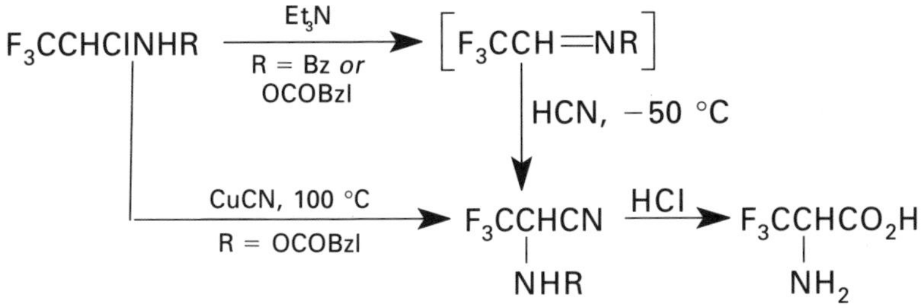

(59)

1.4 THE HYDANTOIN SYNTHESIS

Bücherer's hydantoin synthesis of 2-amino acids is very close to that of Strecker. It is based on the reaction of an oxo compound with a mixture of ammonium carbonate and alkali metal cyanide with the formation of 5-substituted 2,4-(3H, 5H)-imidazoline (hydantoin), which is in turn hydrolysed to the desired amino acid. The relationship between these two methods is well evidenced:

(a) identical starting material (aldehyde or ketone);
(b) identical final result, i.e. 2-amino acid with one more carbon atom;
(c) possible interconversion of the intermediates: the 2-amino(or 2-hydroxy)nitriles may be converted, by treatment with ammonium carbonate, into the corresponding hydantoins.

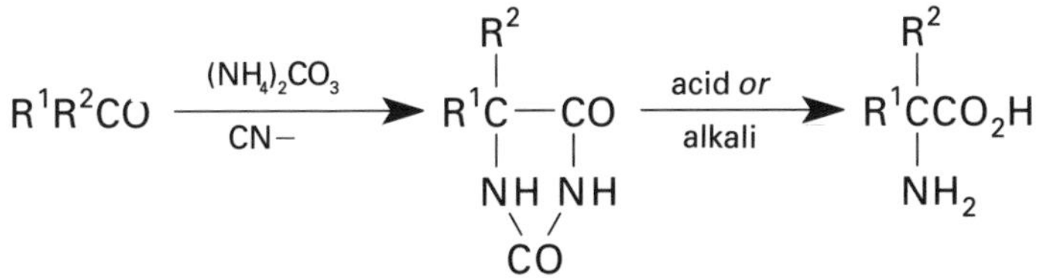

As with the Strecker method, the exploitation of the hydantoin synthesis in fluorinated amino acid chemistry is rather restricted; moreover, most of the examples are found only in the patent literature or are incomplete. According to the general scheme, 2-amino-3,3,3-trifluoroisobutyric acid (49) ($R^1 = CF_3$, $R^2 = Me$) was prepared via 5-methyl-5-(trifluoromethyl)hydantoin [53]. Similarly, 4,4,4-trifluorobutanal gave 5,5,5-trifluoronorvaline (50) in a substantially higher yield than under the Strecker synthesis conditions [44].

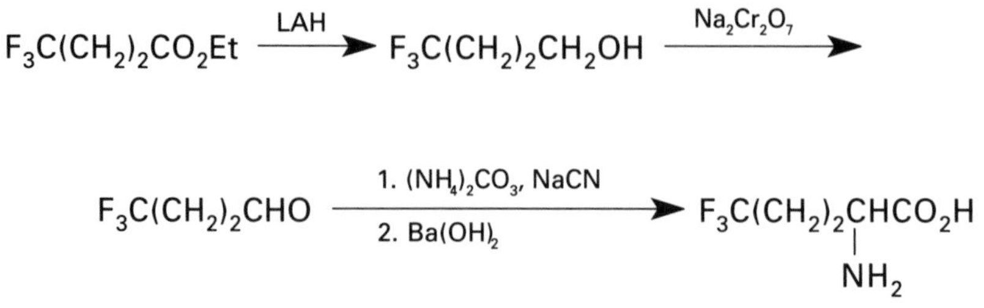

$$F_3C(CH_2)_2CO_2Et \xrightarrow{\text{LAH}} F_3C(CH_2)_2CH_2OH \xrightarrow{\text{Na}_2\text{Cr}_2\text{O}_7}$$

$$F_3C(CH_2)_2CHO \xrightarrow[\text{2. Ba(OH)}_2]{\text{1. (NH}_4)_2\text{CO}_3\text{, NaCN}} F_3C(CH_2)_2\underset{\underset{NH_2}{|}}{C}HCO_2H$$

(50)

In an analogous way, the hydantoin synthesis was successfully applied to 5-fluorolevulinic acid, from which 2-fluoromethylglutamic acid (60) was prepared [54]. The synthesis of 6,6,6-trifluoromethionine (61) has also been realized using the hydantoin method, the key steps being the preparation of trifluoromethanethiol and its addition to acrolein, which yielded the requisite 3-(trifluoromethylthio)propanal [55].

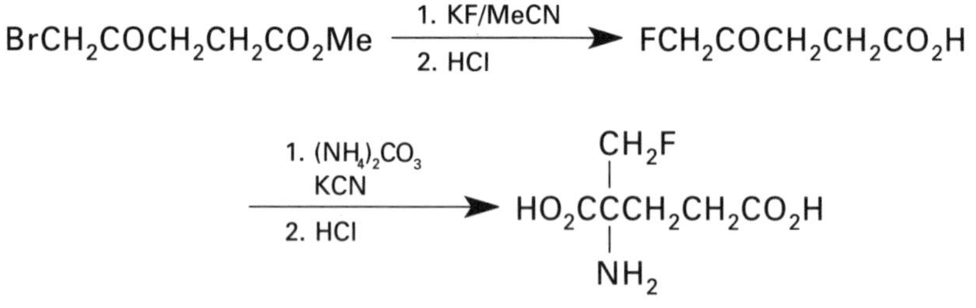

$$BrCH_2COCH_2CH_2CO_2Me \xrightarrow[\text{2. HCl}]{\text{1. KF/MeCN}} FCH_2COCH_2CH_2CO_2H$$

$$\xrightarrow[\text{2. HCl}]{\substack{\text{1. (NH}_4)_2\text{CO}_3 \\ \text{KCN}}} \underset{\underset{NH_2}{|}}{\overset{\overset{CH_2F}{|}}{HO_2CC}}CH_2CH_2CO_2H$$

(60)

$$HgF_2 + CS_2 \longrightarrow (F_3CS)_2Hg \xrightarrow{\text{HCl}} F_3CSH \xrightarrow{\text{H}_2\text{C}=\text{CHCHO}}$$

$$F_3CSCH_2CH_2CHO \xrightarrow[\text{2. NaOH \quad 3. HCl}]{\text{1. (NH}_4)_2\text{CO}_3\text{, NaCN}} F_3CSCH_2CH_2\underset{\underset{NH_2}{|}}{C}HCO_2H$$

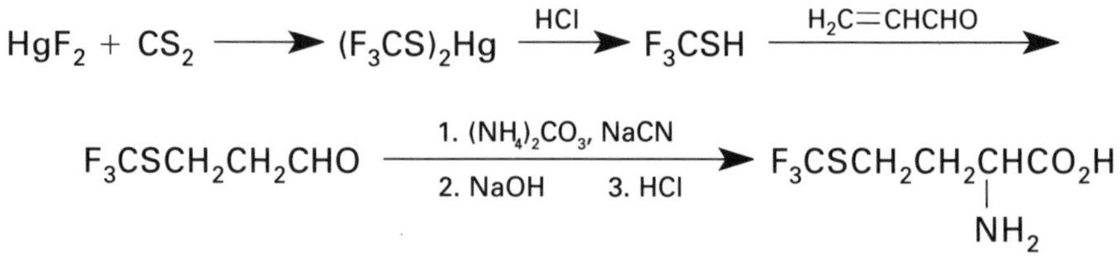

(61)

A few fluorinated aromatic amino acids were also obtained from the corresponding hydantoins. The formation of the 2-(3'-fluorophenyl)- and 2-(4'-

fluorophenyl)glycines **44** and **45** from the corresponding fluorobenzaldehydes was only briefly referred to, without any experimental data [56]. Several amino acids derived from 3-phenyl-2-alanine with a difluoromethyl or trifluoromethyl group at C-2 and with 1–3 hydroxyls on the benzene nucleus were synthesized by the hydantoin method from the suitably substituted acetophenones [57, 58].

$$Ar^1CHYCO \xrightarrow[\text{2. HBr } or \text{ HCl}]{\text{1. (NH}_4)_2CO_3, \text{ CN}^-} Ar^2CHYCCO_2H$$

with CXF$_2$ substituents above the respective carbons, and NH$_2$ on the product.

(62)–(66)

For Ar^1, R = Me; for Ar^2, R = H:
(**62**) Ar = 4-RO-C$_6$H$_4$, X = Y = H: 2-(difluoromethyl)tyrosine [57]
(**63**) Ar = 3-RO-C$_6$H$_4$, X = F, Y = H: 3-(3'-hydroxyphenyl)-2-(trifluoro-methyl)-2-alanine [58]
(**64**) Ar = 3,4-(RO)$_2$-C$_6$H$_3$, X = F, Y = H: 3-(3',4'-dihydroxyphenyl)-2-(tri-fluoromethyl)-2-alanine [58]
(**65**) Ar = 3,4-(RO)$_2$-C$_6$H$_3$, X = F, Y = Me: 3-(3',4'-dihydroxyphenyl)-3-methyl-2-(trifluoromethyl)-2-alanine [58]
(**66**) Ar = 3,4,5-(RO)$_3$-C$_6$H$_2$, X = F, Y = H: 3-(3',4',5'-trihydroxyphenyl)-2-(trifluoromethyl)-2-alanine [58]

To complete this short section, a brief mention should be made of an uncommon amino acid with a polyfluorinated cyclobutane ring [53], i.e. 2-(1'-methyl-2',2',3',3'-tetrafluorocyclobutyl)glycine (**67**).

$$\begin{array}{c} CH_3 \\ | \\ H_2C-C-CHO \\ | \quad | \\ F_2C-CF_2 \end{array} \xrightarrow[\text{2. 60\% H}_2SO_4]{\text{1. (NH}_4)_2CO_3, \text{ NaCN}} \begin{array}{c} CH_3 \\ | \\ H_2C-C-CHCO_2H \\ | \quad | \quad | \\ F_2C-CF_2 \; NH_2 \end{array}$$

(67)

1.5 THE ERLENMEYER AZLACTONE SYNTHESIS

The observation that oxo compounds condense under suitable conditions with *N*-acyl derivatives of glycine with the formation of 2,4-disubstituted 5-oxazolones is widely used in amino acid chemistry. The oxazolones, traditionally called azlactones, are converted into the desired amino acids in diverse ways. They are easily cleaved by a base or acid to *N*-acylamino-2,3-didehydro acids or their esters, which on reduction and hydrolysis give the amino acids (route 1). Reduction may be applied also to cleave the azlactone; according to the conditions used, it leads either to the saturated *N*-acylamino acid, which is finally hydrolysed (route 2) or, when carried out with a mixture of hydroiodic acid, red phosphorus and acetic anhydride, it furnishes the amino acid in one step, without any intermediate (route 3). As the reaction components, aromatic aldehydes and *N*-benzoylglycine are mostly, but not exclusively, used. Although several modifications of the original procedures also make possible the formation of azlactones derived from aryl alkyl ketones and from aliphatic oxo compounds, the aromatic amino acids substantially predominate among the products of the Erlenmeyer synthesis.

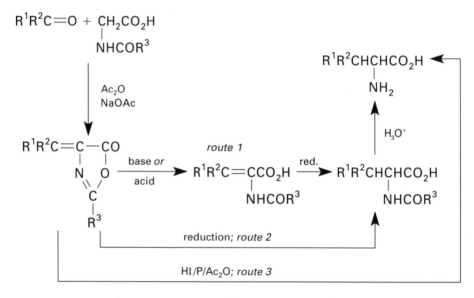

R^1 = aryl *or* alkyl; R^2 = H *or* alkyl; R^3 = Bz *or* Me

Many ring-substituted fluorinated analogues of 3-phenyl-2-alanine have been prepared by the azlactone method. All the three isomeric 3-(fluorophenyl)-2-alanines **68–70** were synthesized as early as in 1932 [59, 60] and later [61] using route 1 of the general scheme. Route 3 has also been applied in the preparation of these compounds [62] and in the synthesis of 3-(pentafluorophenyl)-2-alanine (**40**) [32]. Compound **40** was also successfully prepared by route 2, using zinc/acid for the reduction; by the same group, route 1 has been reported to fail [63, 64]. Route 2 was also applied in the synthesis of eight ring-fluorinated 3-phenyl-2-alanines [71–78) in unstated yields [65].

The synthesis of the three isomeric 3-(trifluoromethylphenyl)-2-alanines (**79–81**) from the respective trifluoromethylbenzaldehydes was accomplished by following route 1, modified in two respects: N-acetylglycine was used instead of the N-benzoyl derivative and the ring cleavage of the 2-methyla-zlactones was performed with aqueous acetone. Hydrogenation (H₂/Pd) followed by acid hydrolysis then gave **79–81**. N-Benzoylglycine was also tried as a starting reagent in this work; however, the benzoyl group was found to be unremovable in the last step [66].

$R^2 = H$, $R^3 = Bz$, R^1 = substituted phenyl, as indicated:

(40) 2,3,4,5,6-pentafluoro- [32, 63, 64]
(68) 2-fluoro- [60, 62]
(69) 3-fluoro- [59, 60, 62]
(70) 4-fluoro- [60, 61, 62]
(71) 2,4-difluoro- [65]
(72) 2,5-difluoro- [65]
(73) 2,6-difluoro- [65]
(74) 3,4-difluoro- [65]
(75) 3,5-difluoro- [65]
(76) 2,3,4,6-tetrafluoro- [65]
(77) 2,3,5,6-tetrafluoro- [65]
(78) 3,5-dichloro-2,4,6-trifluoro- [65]

$R^2 = H$, $R^3 = Me$, R^1 = substituted phenyl, as indicated:

(79) 2-(trifluoromethyl)- [66]
(80) 3-(trifluoromethyl)- [66]
(81) 4-(trifluoromethyl)- [66]

The applicability of the azlactone synthesis to the fluorinated ketones was confirmed by the successful synthesis of 3-(di- or trifluoromethyl)-3-phenyl-2-alanines and -tyrosines (**82–85**) from the fluorinated acetophenones [67].

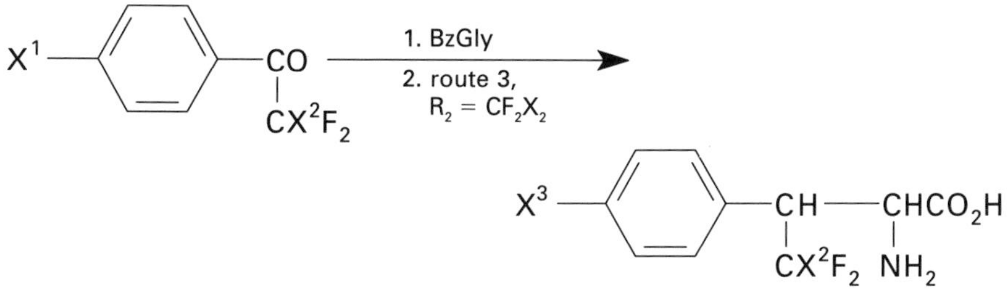

(82) $X^1 = X^2 = X^3 = H$
(83) $X^1 = X^3 = H$, $X^2 = F$
(84) $X^1 = MeO$, $X^2 = H$, $X^3 = HO$
(85) $X^1 = MeO$, $X^2 = F$, $X^3 = HO$

3'-Fluorotyrosine (86), a potent bactericidal agent, has been synthesized from 3-fluoro-4-methoxy(or 3-fluoro-4-ethoxy)benzaldehyde and N-benzoylglycine, following the general route 1 [59, 68] or route 3 [59, 69]; analogously, 3',5'-difluorotyrosine (87) was prepared from the difluorinated aldehyde [68]. Also, four O-fluoroalkyl ethers of tyrosine (88–91) resulted from the application of the azlactone synthesis to the corresponding 4-(fluoroalkoxy)benzaldehydes; both N-acetyl- and N-benzoylglycine were used in these reactions. Route 3 was successfully employed to prepare 89–91, but failed in the case of 88, where ether cleavage took place with the formation of tyrosine. Nevertheless, 88 was achieved by route 2 via hydrogenation of the azlactone [70, 71].

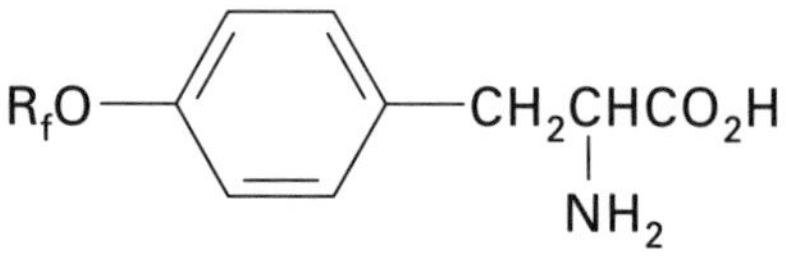

$$R_fO\text{---}\langle\text{benzene ring}\rangle\text{---}CH_2\underset{\underset{NH_2}{|}}{CH}CO_2H$$

(88) (89) (90) (91)

$R_f = CHF_2 \quad CF_3 \quad C_2F_5 \quad CF_3CHFCF_2$

A modification of the azlactone preparation, which consisted in the use of lead(II) acetate instead of the ineffective sodium salt [72], was essential for the successful synthesis of 5-(pentafluorophenyl)norvaline (92) from 3-(pentafluorophenyl)propanal [9]. The azlactone ring was opened by alcoholysis; route 2 was then terminated by hydrogenation and hydrolysis, affording 92 in high yield.

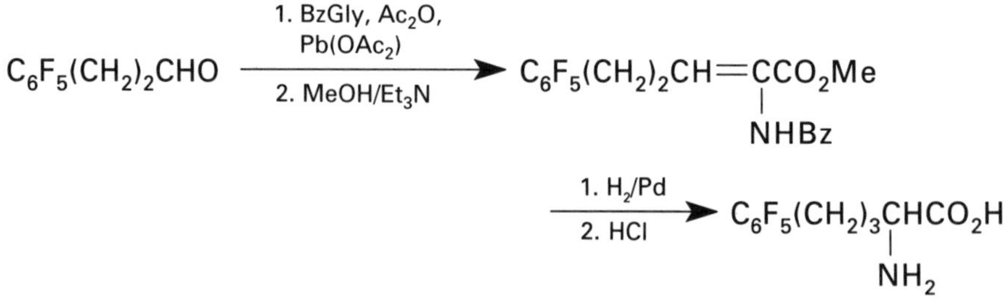

$$C_6F_5(CH_2)_2CHO \xrightarrow[\text{2. MeOH/Et}_3\text{N}]{\substack{\text{1. BzGly, Ac}_2\text{O,}\\ \text{Pb(OAc)}_2}} C_6F_5(CH_2)_2CH\!=\!\underset{\underset{NHBz}{|}}{C}CO_2Me$$

$$\xrightarrow[\text{2. HCl}]{\text{1. H}_2/\text{Pd}} C_6F_5(CH_2)_3\underset{\underset{NH_2}{|}}{CH}CO_2H$$

(92)

The fluoro derivatives of thyronine and its iodinated analogues with various combinations of the F and I substituents (93–95) were synthesized from the corresponding halogenated 4-(4'-methoxyphenoxy)benzaldehydes [59, 69, 73].

The aldehydes themselves are products of multi-step syntheses, whose description goes beyond the framework of this book.

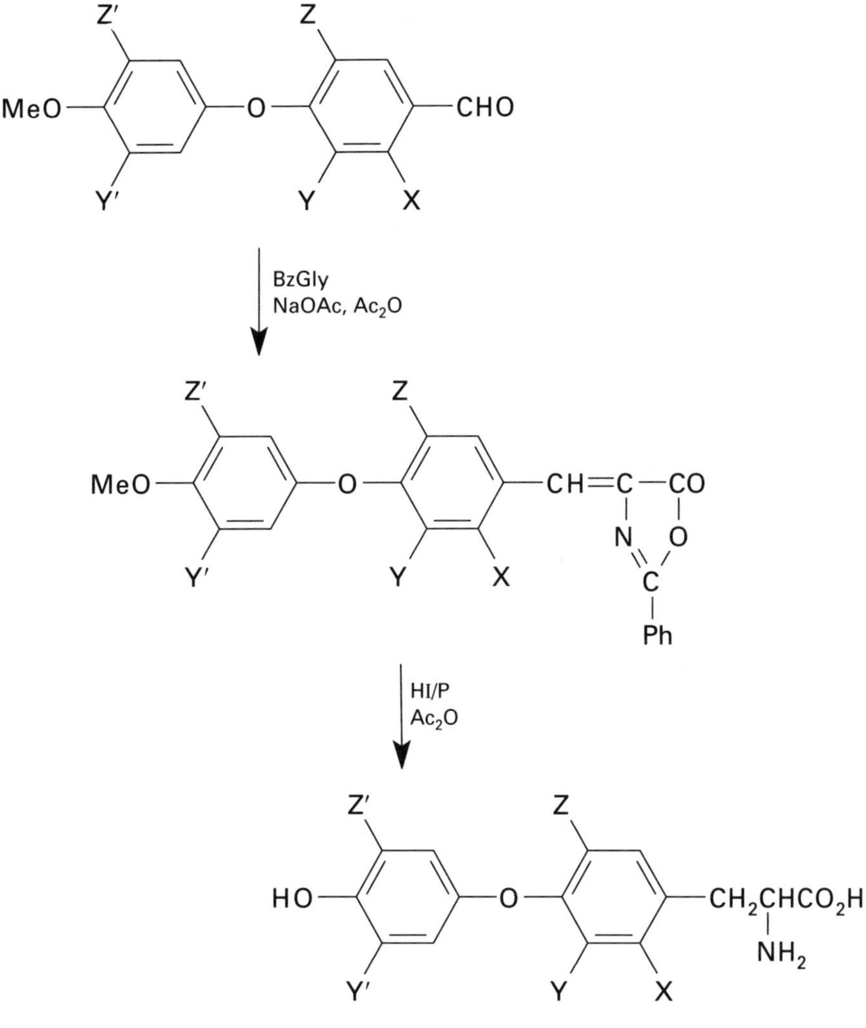

(93) X = F, Y = Y' = Z = Z' = H [59]
(94) X = Z' = H, Y = Z = I, Y' = F [73]
(95) X = H, Y = Z = I, Y' = Z' = F [69]

A heterocyclic amino acid, 4,5,6,7-tetrafluorotryptophan (96), has recently been prepared in a high yield from 3-formyl-4,5,6,7-tetrafluoroindole and N-benzoylglycine following route 2. In the same way as in the synthesis of 92, the azlactone was cleaved by methanol [9].

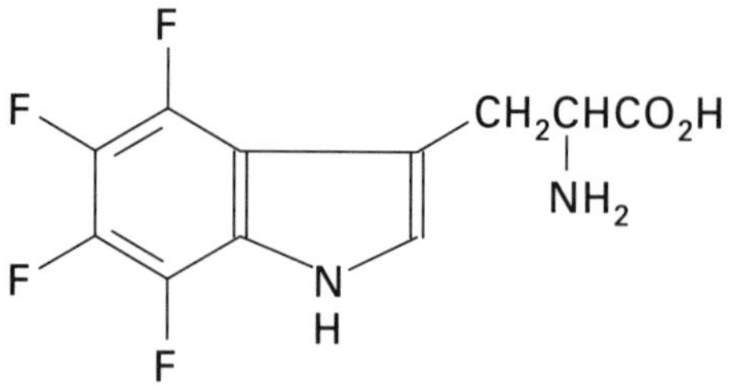

(96)

In the aliphatic series the azlactones were inaccessible by the original version of Erlenmeyer. However, the use of lead(II) acetate as catalyst [72] successfully circumvented this problem, as documented by the synthesis of 5,5,5-trifluoroleucine (**51**) and 6,6,6-trifluoronorleucine (**97**) from the respective fluorinated butanals [9]. In an another access to **51**, catalysis by zinc acetate was also applied [74–76].

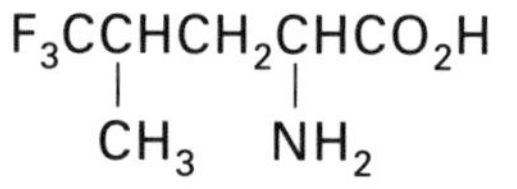

$$F_3CCHCH_2CHCO_2H \atop CH_3NH_2 \qquad\qquad F_3C(CH_2)_3CHCO_2H \atop NH_2$$

(51) 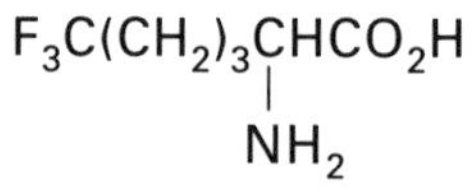   **(97)**

At this point, an alternative approach to fluorine-containing aliphatic azlactones should be mentioned, which is based on the Wittig reaction principle. Fluorinated oxo compounds react with 2-phenyl-4-triphenylphos-phoranylidene-5-oxazolone [77] to afford the desired azlactones [78]; however, their conversion into the fluorinated amino acids has not yet been reported.

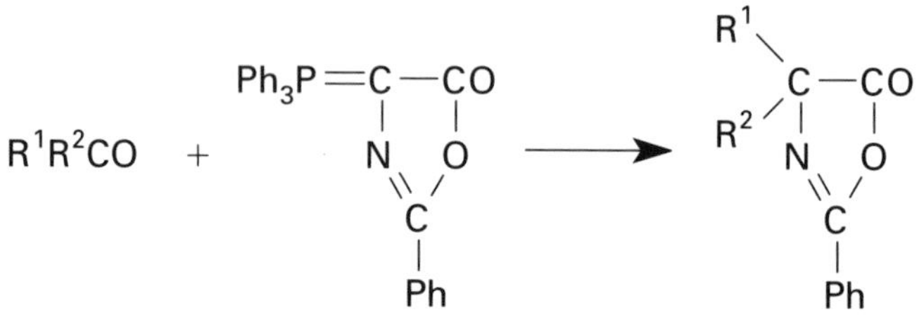

$R^1 = R^2 = CF_3$; $R^1 = CF_3$, $R^2 = CO_2Me$; $R^1 = H$, $R^2 = CHF_2(CF_2)_3$

An unusual synthesis of 3-fluoro-2-alanine (**1**), starting from an azlactone compound, 4-chloromethylene-2-phenyl-5-oxazolone, was reported by a Chinese group [61]. Fluorination with potassium fluoride furnished the requisite fluoromethylene-substituted azlactone, which was then converted, essentially by route 1, into **1** in a minute yield. No confirming reference on the use of this method has been found in the literature; further, its validity has been questioned [79].

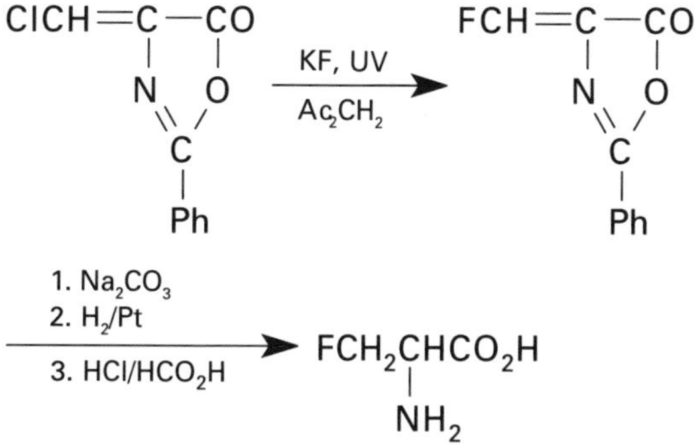

$$\text{ClCH}=\overset{|}{\text{C}}-\overset{|}{\text{CO}}\quad\xrightarrow[\text{Ac}_2\text{CH}_2]{\text{KF, UV}}\quad\text{FCH}=\overset{|}{\text{C}}-\overset{|}{\text{CO}}$$

$$\xrightarrow[\text{3. HCl/HCO}_2\text{H}]{\begin{array}{l}\text{1. Na}_2\text{CO}_3\\\text{2. H}_2\text{/Pt}\end{array}}\quad \text{FCH}_2\underset{\underset{\text{NH}_2}{|}}{\text{CHCO}_2\text{H}}$$

$$(\mathbf{1})$$

1.6 SYNTHESES USING N-SUBSTITUTED AMINO-MALONIC ESTERS AND RELATED COMPOUNDS

The N-protected aminomalonic esters are invaluable synthons in the preparation of fluorinated amino acids. Unlike the preceding methods, in the syntheses employing aminomalonates the basal unit of the 2-amino acids, i.e. the 2-aminoacetic acid moiety, is brought into the molecule *as a whole*. Diethyl N-acetamidomalonate is the most favoured reagent in these syntheses; the N-formyl derivative is used less frequently and other N-substituted aminomalonic esters occur only sporadically. The alkylation of the aminomalonate at C-2 is effected mostly by alkyl (or aralkyl) halides or sulfonate esters; the desired amino acids are then obtained by hydrolysis. Methods involving chemical transformations of the side-chain prior to hydrolysis, including those where fluorine or a fluorine-containing group are introduced, are also the subject of this section.

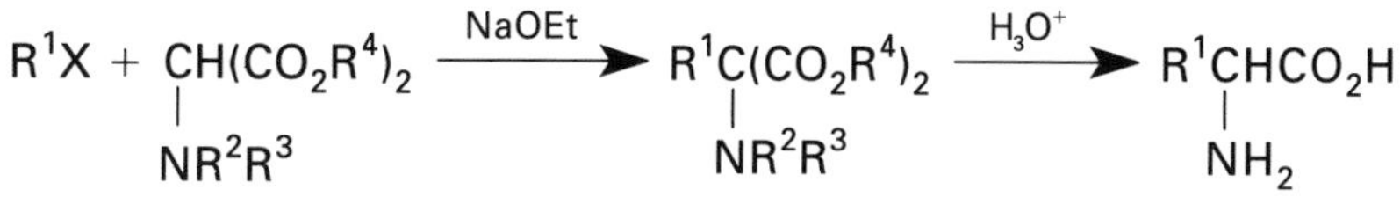

$$\text{R}^1\text{X} + \underset{\underset{\text{NR}^2\text{R}^3}{|}}{\text{CH(CO}_2\text{R}^4)_2} \xrightarrow{\text{NaOEt}} \underset{\underset{\text{NR}^2\text{R}^3}{|}}{\text{R}^1\text{C(CO}_2\text{R}^4)_2} \xrightarrow{\text{H}_3\text{O}^+} \underset{\underset{\text{NH}_2}{|}}{\text{R}^1\text{CHCO}_2\text{H}}$$

R^1 = alkyl, aralkyl
R^2 = H, R^3 = CHO, Ac, OCOBzl *or* R^2 + R^3 = Pht, =CHPh
R^4 = Et, Bzl
X = Br, I, OTos, OTf

A series of monofluorinated aliphatic amino acids were prepared by alkylation of diethyl acetamidomalonate with ω-fluoroalkyl bromides [53, 80, 81]. Difficulties were observed in the second step; to avoid loss of fluorine, hydrolysis by hydrofluoric acid [53, 80] or only a short treatment with hydrochloric acid [81] had to be applied. Nevertheless, the yields of the hydrolysis were low to moderate in all cases.

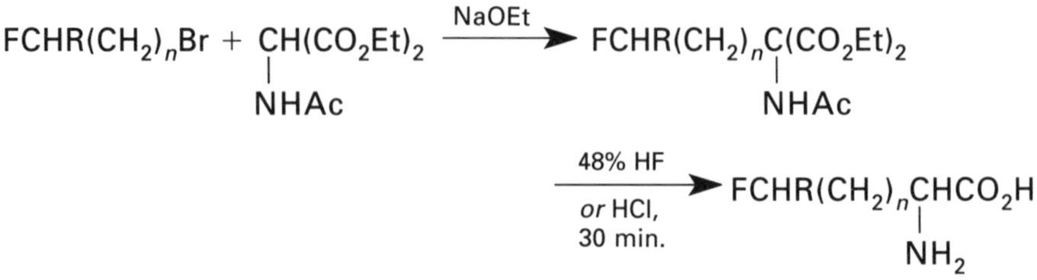

$$FCHR(CH_2)_nBr + \underset{\underset{NHAc}{|}}{CH(CO_2Et)_2} \xrightarrow{NaOEt} FCHR(CH_2)_n\underset{\underset{NHAc}{|}}{C(CO_2Et)_2}$$

$$\xrightarrow[\underset{30\ min.}{or\ HCl,}]{48\%\ HF} FCHR(CH_2)_n\underset{\underset{NH_2}{|}}{CHCO_2H}$$

$$(2),\ (98)-(100)$$

(2) $n = 1$, R = H   2-amino-4-fluorobutyric acid [53, 80]
(98) $n = 2$, R = H   5-fluoronorvaline [53, 80]
(99) $n = 2$, R = Me   5-fluoronorleucine [81]
(100) $n = 3$, R = H   6-fluoronorleucine [53, 80]

In contrast to the above, amino acids containing the trifluoromethyl group are resistant under the usual hydrolytic conditions; on the other hand, the presence of this group in the alkylating agent decreases its reactivity. Strong electrophiles such as the O-triflates must be used in cases when the distance between the CF_3 group and the alkylating centre is shorter than three carbon atoms [44, 82].

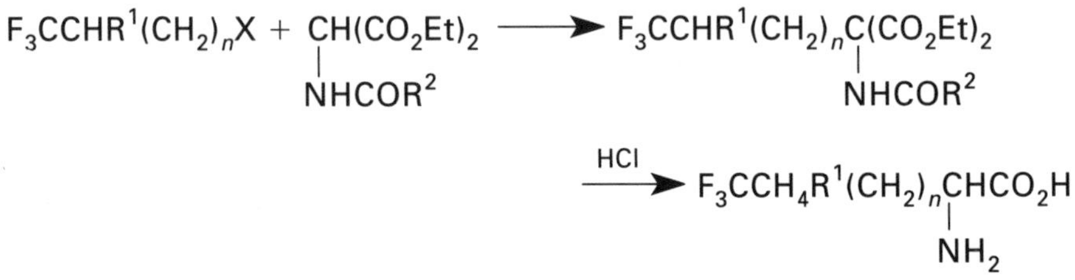

$$F_3CCHR^1(CH_2)_nX + \underset{\underset{NHCOR^2}{|}}{CH(CO_2Et)_2} \longrightarrow F_3CCHR^1(CH_2)_n\underset{\underset{NHCOR^2}{|}}{C(CO_2Et)_2}$$

$$\xrightarrow{HCl} F_3CCH_4R^1(CH_2)_n\underset{\underset{NH_2}{|}}{CHCO_2H}$$

$$(9),\ (50),\ (97),\ (101)$$

(9) $n = 0$, R^1 = H, R^2 = Me, X = OTf [82]
(50) $n = 1$, R^1 = H, R^2 = Me, X = OTf [82]
(97) $n = 2$, R^1 = H, R^2 = H or Me,/ X = I or OTos [44, 82]
(101) $n = 2$, R^1 = Me, R^2 = H, X = I [44]

In the synthesis of 4,4,4-trifluorovaline (**11**) the principle of Michael addition is applied. Diethyl acetamidomalonate adds quantitatively to phenyl 3,3,3-trifluoropropenylsulfone (available from 3,3,3-trifluoropropene in four steps); after reductive removal of the phenylsulfonyl group the synthesis is terminated by hydrolysis [83]:

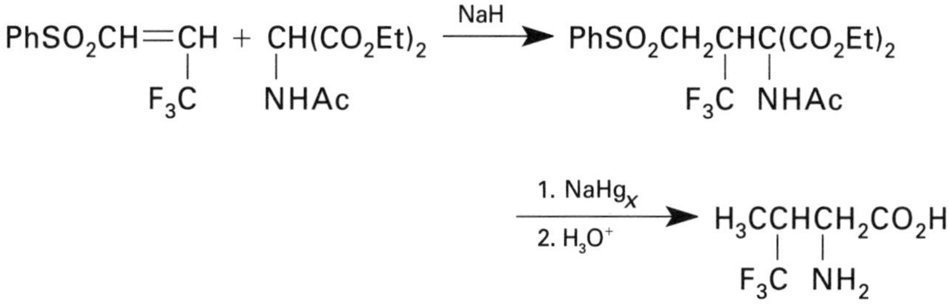

$$PhSO_2CH{=}CH + CH(CO_2Et)_2 \xrightarrow{\text{NaH}} PhSO_2CH_2CHC(CO_2Et)_2$$

with substituents: F_3C, $NHAc$ on left; F_3C, $NHAc$ on right

$$\xrightarrow[\text{2. H}_3\text{O}^+]{\text{1. NaHg}_x} H_3CCHCH_2CO_2H$$

with F_3C, NH_2

(11)

A few di- or polyfluorinated amino acids were achieved by insertion of fluorine atoms or a fluorinated group into the side-chain of substituted acet-amidomalonates. On treatment of diethyl 2-(3'-chlorobut-2'-enyl) acetamidomalonate with anhydrous hydrogen fluoride, the addition of HF occurred simultaneously with halogen exchange to give the *gem*-difluoro compound, which on hydrolysis afforded 5,5-difluoronorleucine (**102**) [84]. Synthesis of 5,5-difluorolysine (**58**) started with 2-acetamido-2-(3'-hydroxy-4'-phthalimidobutyl)malonate, which was first oxidized to the 3'-oxo ester and then fluorinated, using sulfur tetrafluoride, to diethyl 2-acetamido-2-(3',3'-difluoro-4'-phthalimidobutyl)malonate [85]. Acid hydrolysis of this ester yielded **58**.

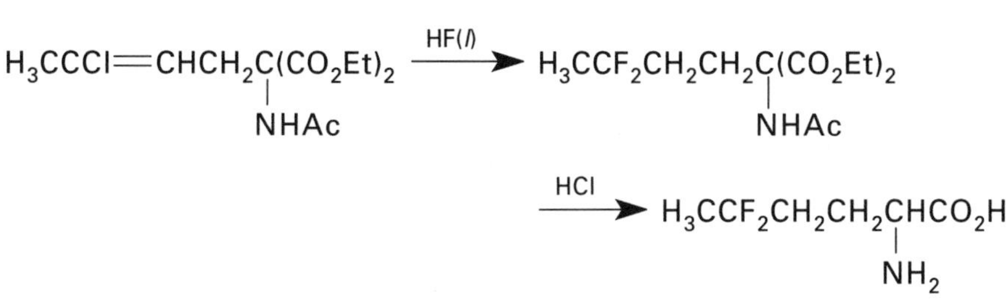

$$H_3CCCl{=}CHCH_2C(CO_2Et)_2 \xrightarrow{\text{HF}(l)} H_3CCF_2CH_2CH_2C(CO_2Et)_2$$

with $NHAc$ below on left; $NHAc$ below on right

$$\xrightarrow{\text{HCl}} H_3CCF_2CH_2CH_2CHCO_2H$$

with NH_2

(102)

$$\text{PhtNCH}_2\underset{\overset{|}{\text{OH}}}{\text{CH}}\text{CH}_2\underset{\overset{|}{\text{NHAc}}}{\text{C}}(\text{CO}_2\text{Et})_2 \xrightarrow[\text{DCCI}]{\text{DMSO}} \text{PhtNCH}_2\underset{\overset{||}{\text{O}}}{\text{C}}\text{CH}_2\text{CH}_2\underset{\overset{|}{\text{NHAc}}}{\text{C}}(\text{CO}_2\text{Et})_2$$

$$\xrightarrow{\text{SF}_4} \text{PhtNCH}_2\text{CF}_2\text{CH}_2\text{CH}_2\underset{\overset{|}{\text{NHAc}}}{\text{C}}(\text{CO}_2\text{Et})_2$$

$$\xrightarrow{\text{H}_3\text{O}^+} \text{H}_2\text{NCH}_2\text{CF}_2\text{CH}_2\text{CH}_2\underset{\overset{|}{\text{NH}_2}}{\text{CH}}\text{CO}_2\text{H}$$

(58)

The radical addition of perfluoropropyl iodide to the double bond of diethyl ω-alkenylacetamidomalonates opens up access to long-chain amino acids with a terminal heptafluoropropyl group (compare with the syntheses of **23** and **24** in Section 1.2). Preparations of 2-amino-4-hydroxy-6,6,7,7,8,8,8-heptafluorooctanoic acid **(103)** and of 2-acetylamino-8,8,9,9,10,10,10-heptafluorodecanoic acid **(104)** follow the routes shown [16].

$$\text{H}_2\text{C}=\text{CHCH}_2\underset{\overset{|}{\text{NHAc}}}{\text{C}}(\text{CO}_2\text{Et})_2 \xrightarrow[\text{AIBN}]{\text{C}_3\text{F}_7\text{I}} \text{F}_3\text{C}(\text{CF}_2)_2\text{CH}_2\text{CHICH}_2\underset{\overset{|}{\text{NHAc}}}{\text{C}}(\text{CO}_2\text{Et})_2$$

$$\xrightarrow[\text{2. AcOH}]{\text{1. HI}} \text{F}_3\text{C}(\text{CF}_2)_2\text{CH}_2\overset{\overset{\text{O}——\text{CO}}{|\qquad|}}{\text{CH}}\text{CH}_2\underset{\overset{|}{\text{NH}_2}}{\text{C}}\text{CO}_2\text{H}$$

$$\xrightarrow[\text{2. NaOH}]{\text{1. heat}} \text{F}_3\text{C}(\text{CF}_2)_2\text{CH}_2\underset{\overset{|}{\text{OH}}}{\text{CH}}\text{CH}_2\underset{\overset{|}{\text{NH}_2}}{\text{CH}}\text{CO}_2\text{H}$$

(103)

$$\text{H}_2\text{C}=\text{CH}(\text{CH}_2)_3\underset{\overset{|}{\text{NHAc}}}{\text{C}}(\text{CO}_2\text{Et})_2 \xrightarrow[\text{AIBN}]{\text{C}_3\text{F}_7\text{I}} \text{F}_3\text{C}(\text{CF}_2)_2\text{CH}_2\text{CHI}(\text{CH}_2)_3\underset{\overset{|}{\text{NHAc}}}{\text{C}}(\text{CO}_2\text{Et})_2$$

$$\xrightarrow[\text{2. H}_2/\text{Pt}]{\text{1. NaOH}} \text{F}_3\text{C}(\text{CF}_2)_2(\text{CH}_2)_5\underset{\overset{|}{\text{NHAc}}}{\text{C}}(\text{CO}_2\text{H})_2 \xrightarrow[-\text{CO}_2]{\text{heat}} \text{F}_3\text{C}(\text{CF}_2)_2(\text{CH}_2)_5\underset{\overset{|}{\text{NHAc}}}{\text{CH}}\text{CO}_2\text{H}$$

(104)

Two synthetic pathways to 4-fluoroglutamic acid **(105)** were also based on the acetamidomalonate approach. In 1960, Hudlický [86, 87] achieved the

Michael addition of diethyl acetamidomalonate on ethyl 2-fluoroacrylate. Hydrolysis of the addition product afforded the amino acid **105** in moderate yield. Owing to the availability of the starting compounds, this method may be considered preferable to other syntheses of **105**, including the 'reverse version' employing the Michael addition of diethyl fluoromalonate (see Section 1.7). The reaction conditions were recently optimized and the yields improved [88]. Another synthesis [89] of **105** started with ethyl 3-chlorolactate, which was condensed, with temporary protection of the hydroxy group, with diethyl acetamidomalonate to give the fully protected triester containing the 4-hydroxyglutamic acid skeleton. The released hydroxy group was then exchanged for fluorine using 2-chloro-1,1,2-trifluorotriethylamine to give the same fluorinated precursor of **105** as in the synthesis of Hudlický.

$$EtO_2CCF{=}CH_2 \ + \ \underset{\underset{NHAc}{|}}{CH(CO_2Et)_2} \ \xrightarrow{\text{NaOEt}} \ EtO_2CCHFCH_2\underset{\underset{NHAc}{|}}{C(CO_2Et)_2}$$

$$\xrightarrow{\text{HCl}} \ HO_2CCHFCH_2\underset{\underset{NH_2}{|}}{CHCO_2H} \qquad (\textbf{105})$$

1. CF_3CO_2H
2. CTT

$$EtO_2C\underset{\underset{OH}{|}}{CHCH_2Cl} \ \longrightarrow \ EtO_2C\underset{\underset{t\text{-BuO}}{|}}{CHCH_2Cl} \ \longrightarrow \ EtO_2C\underset{\underset{t\text{-BuO}}{|}}{CHCH_2}\underset{\underset{NHAc}{|}}{C(CO_2Et)_2}$$

Compound **105** so produced was formed as a mixture of diastereomers, whose separation was achieved either by ion-exchange chromatography [90] or by crystallization of their dialkyl ester hydrochlorides [91].

An amino acid richer by one carbon atom, namely 2-amino-5-fluorohex-4-enedioic acid (**106**) resulted from the acetamidomalonate alkylation by ethyl 4-bromo-2-fluorocrotonate [28].

$$EtO_2CCF{=}CHCH_2Br \ \longrightarrow \ EtO_2CCF{=}CHCH_2\underset{\underset{NHAc}{|}}{C(CO_2Et)_2}$$

$$\xrightarrow{\text{HCl}} \ HO_2CCF{=}CHCH_2\underset{\underset{NH_2}{|}}{CHCO_2H}$$

$$(\textbf{106})$$

The easily realizable alkylation of acetamidomalonic esters by benzyl halides was widely applied in the synthesis of many ring-fluorinated analogues of 3-phenyl-2-alanine, tyrosine and tryptophan. These reactions proceeded smoothly and in yields over 70% in most cases; the ensuing hydrolysis was carried out either by acid in one step or, less frequently, first by base to give the *N*-acylamino acid, which was then subjected to acid hydrolysis. Hydrogenolysis has also been applied for deprotection of a tribenzyl derivative.

All three monofluorinated 3-phenyl-2-alanines and some higher fluorinated analogues were prepared according to the following general scheme:

$$ArCH_2Hal + \underset{NYZ}{CH(CO_2X)_2} \longrightarrow \underset{NYZ}{ArCH_2C(CO_2X)_2} \longrightarrow \underset{NH_2}{ArCH_2CHCO_2H}$$

Hal = Cl, Br; Ar = substituted phenyl, as indicated:

(**40**) pentafluoro-; X = Et, Y = H, Z = Ac [32, 92]
(**68**) 2-fluoro-; X = Et, Y = H, Z = Ac [62, 93–95]
(**69**) 3-fluoro-; X = Et, Y = H, Z = Ac [62, 94, 95]
(**70**) 4-fluoro-; X = Et, Y = H, Z = Ac [62, 93–95]
      Y + Z = Pht [96]
(**72**) 2,5-difluoro-; X = Et, Y = H, Z = Ac [94, 95]
(**80**) 3-(trifluoromethyl)-; X = Et, Y = H, Z = Ac [94, 95]
      X = Bzl, Y =H, Z = OCOBzl [66]
(**107**) 2-chloro-5-(trifluoromethyl)-; X = Et, Y = H, Z = Ac [94, 95]
(**108**) 4-chloro-3-(trifluoromethyl)-; X = Et, Y = H, Z = Ac [94, 95]

It has been reported that the pathway successfully used for the preparation of **80** failed in an attempted synthesis of the 2′-isomer **79** [66].

The amino acid **70** has also been prepared by a combination of the acylaminomalonate synthesis with Schiemann reaction, thus avoiding the use of expensive 4-fluorobenzyl bromide. Diethyl 2-acetamido-2-(4′-nitrobenzyl) malonate was hydrogenated to the 4′-amino derivative, which was in turn converted into the diazonium fluoroborate. Thermal decomposition of this salt then gave the requisite 2-acetamido-2-(4′-fluorobenzyl)malonate [97].

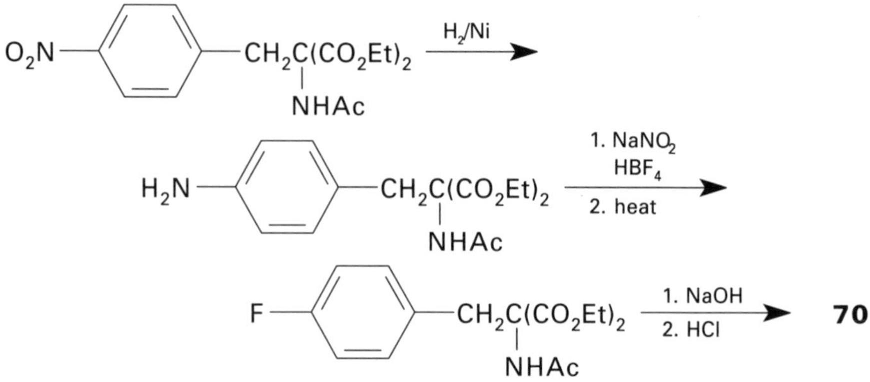

Four ring-fluorinated tyrosine analogues were also prepared according to the general scheme of the acetamidomalonic ester synthesis: 2'-fluorotyrosine (109) and its O-methyl ether (110) [98]. 3'-fluorotyrosine (86) [99] and 2',3',5',6'-tetra-fluorotyrosine (111) [32, 100]. Further, 3-(3',4'-dihydroxy-2'-fluorophenyl)-2-alanine (112) together with its O,O'-bis(methyl ether) (113) may be attached to this group [101]. In all these syntheses, diethyl acetamidomalonate is alkylated with the appropriate methoxybenzyl halide; if the ensuing hydrolysis is carried out with hydroiodic acid, the aminohydroxy acid is formed directly, whereas stepwise hydrolysis (first with hydrochloric and, then hydrobromic acid) allows isolation of the intermediate methoxy compounds 110 and 113.

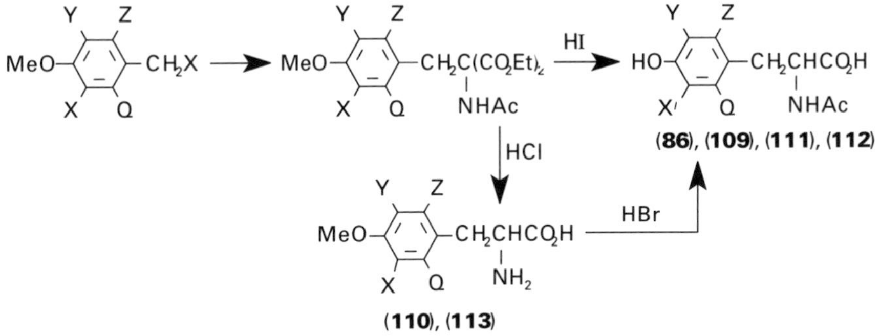

(86) Q = Y = Z = H, X = X' = F [99]
(109) Q = F, X = X' = Y = Z = H [98]
(110) Q = F, X = Y = Z = H [98]
(111) Q = X = X' = Y = Z = F [32, 100]
(112) Q = F, X = MeO, X' = HO, Y = Z = H [101]
(113) Q = F, X = MeO, Y = Z = H [101]

A few other analogues of tyrosine, in which the phenolic hydroxyl was substituted by a polyfluoroalkoxy- or polyfluoroalkylthio group, were prepared by alkylation of diethyl formamidomalonate with the corresponding 4-substituted benzyl bromides [70]. Prior to hydrolysis of the alkylation products, decarboxylation was effected separately by heating with aqueous dimethyl sulfoxide/NaCl.

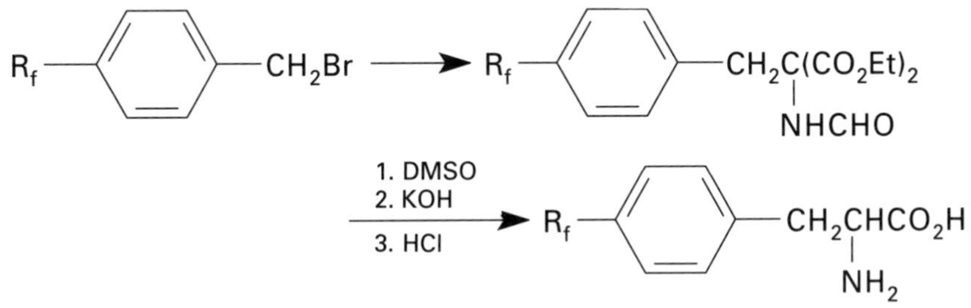

	(88)	(114)	(105)	(116)
R_f =	$OCHF_2$	OCH_2CF_3	$SCHF_2$	SCF_3

In the syntheses of various fluorinated analogues of tryptophan, the advantage was taken of the fact that each of the two starting compounds, either the acylaminomalonic ester and the fluorinated indole may undergo alkylation by a suitable derivative of the other. 5-Fluorotryptophan (117) and 6-fluorotryptophan (118) have been prepared from the corresponding fluoro-indoles, which were first converted into the Mannich bases, fluorinated 3-(dimethylaminomethyl)indoles (fluorogramines). The fluorogramines were used to alkylate acylaminomalonic esters, thus affording 5- or 6-fluoroskatyl-N-acylaminomalonates, which were finally hydrolysed to 117 or 118, respectively [102, 103]. In a similar manner, 4,5,6,7-tetrafluorotryptophan (96) was prepared from 3-piperidinomethyl-4,5,6,7-tetrafluoroindole or from its N-methyl methylsulfate; this quaternary salt was apparently a better alkylating agent than the simple tertiary amine [104, 105].

Both 117 and 118 and, in addition, 4-fluorotryptophan (119) were also synthesized by the inverse version of the above reaction, i.e. by the alkylation of the appropriate fluoroindole by diethyl 2-formamido-2-(piperidinomethyl) malonate [106].

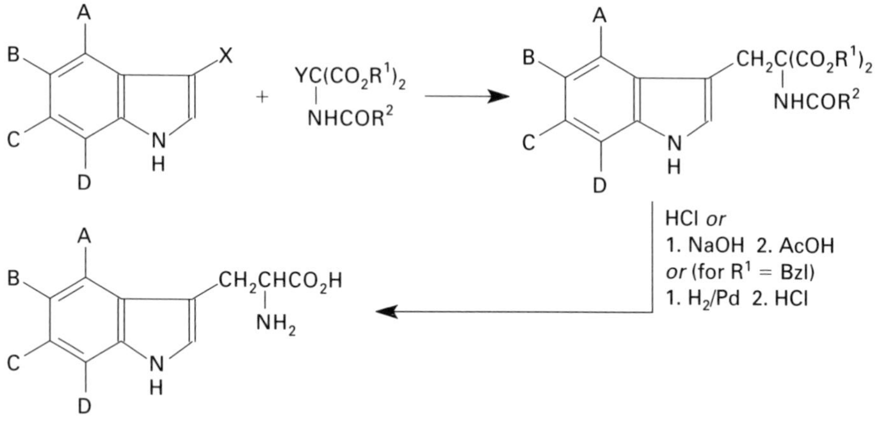

(96), (117)–(119)

(96) A = B = C = D = F, X = H$_2$C—N(piperidine) or H$_2$C—(Me)N$^+$(piperidine) . MeSO$_4^-$,

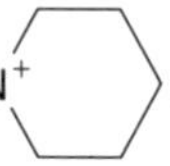

Y = H;
  (a) R^1 = Et, R^2 = H or Me [104]
  (b) R^1 = Bzl, R^2 = H [105]
(117) A = C = D = R^2 = H, B = F, R^1 = Et;
  (a) X = H$_2$C—NMe$_2$, Y = H [102]

  (b) X = H, Y = H$_2$C—N(piperidine) [106]

(118) A = B = D = H, C = F, R^1 = Et;
    (a) X = H$_2$C—NMe$_2$, Y = H, R^2 = H *or* Me [103]

    (b) X = H, Y = H$_2$C—N⟨ ⟩ , R^2 = H [106]

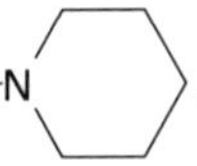

(119) A = F, B = C = D = X = R^2 = H, R^1 = Et. [106]

    Y = H$_2$C—N⟨ ⟩

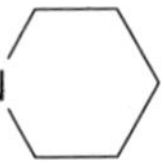

The amino acid **117** was also synthesized by the Fischer indole synthesis: the diester aldehyde resulting from the Michael addition of diethyl acetamidomalonate to acrolein was converted into the 4-fluorophenylhydrazone, which was then cyclized to the 5-fluoroskatylacetamidomalonic ester [107].

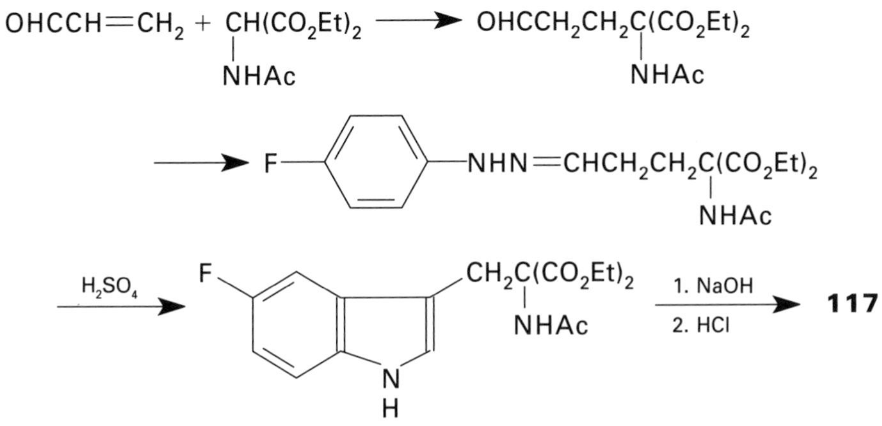

$$OHCCH=CH_2 + \underset{\underset{NHAc}{|}}{CH(CO_2Et)_2} \longrightarrow OHCCH_2CH_2\underset{\underset{NHAc}{|}}{C(CO_2Et)_2}$$

$$\longrightarrow F\text{—}⟨ ⟩\text{—}NHN=CHCH_2CH_2\underset{\underset{NHAc}{|}}{C(CO_2Et)_2}$$

A different approach to the amino group protection has been applied in the synthesis of 3,3,-difluoro-2-alanine (**120**). On condensing benzaldehyde with the free aminomalonic ester, the *N*-benzylidene derivative was formed quantitatively; its sodium salt was alkylated with chlorodifluoromethane and the intermediate hydrolysed to give **120** [108, 109].

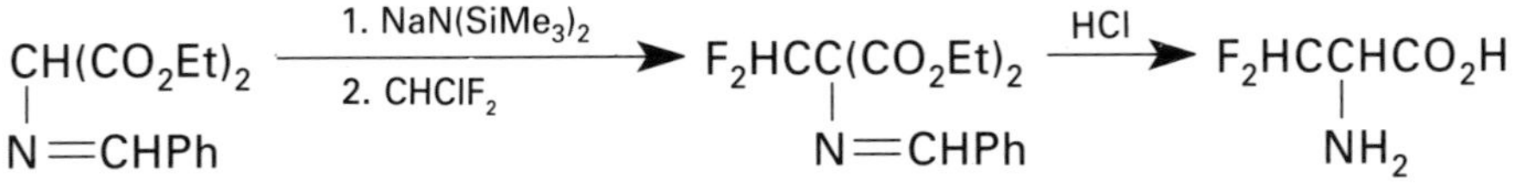

$$\underset{N=CHPh}{\overset{CH(CO_2Et)_2}{|}} \xrightarrow[\text{2. CHClF}_2]{\text{1. NaN(SiMe}_3)_2} \underset{N=CHPh}{\overset{F_2HCC(CO_2Et)_2}{|}} \xrightarrow{\text{HCl}} \underset{NH_2}{\overset{F_2HCCHCO_2H}{|}}$$

(**120**)

It has been shown and amply documented that for a successful alkylation of this type, the presence of two alkoxycarbonyl groups in the substrate to be alkylated is not essential. Many Schiff bases prepared from various 2-amino monocarboxylic esters undergo smooth alkylation at C-2 as well as do the acylaminomalonates. Effective deprotonation of the Schiff bases is essential; this is attained by the action of strong bases such as alkali metal hydrides or lithium organyls. Alkali metal hydroxides in the presence of phase-transfer catalysts are also successfully used for this purpose. In contrast to the acylaminomalonate syntheses, mild hydrolytic conditions are sufficient to effect deprotection of the alkylated intermediates. The use of diverse fluorohalomethanes as electrophiles in these reactions opens a way for the synthesis of 2-(fluorohalomethyl)amino acids from the parent amino acids. The applications of this modern variation on the classical theme are very numerous; aliphatic, aromatic and also heterocyclic fluorinated 2-amino acids were thus synthesized in a simple manner [110–113].

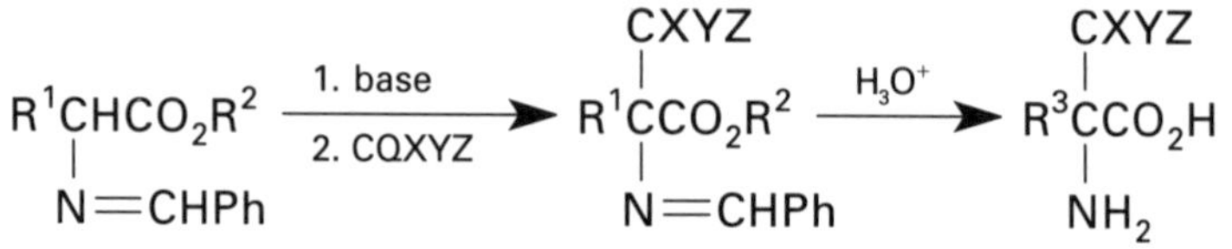

$$R^1CHCO_2R^2 \quad \xrightarrow[\text{2. CQXYZ}]{\text{1. base}} \quad R^1CCO_2R^2 \quad \xrightarrow{H_3O^+} \quad R^3CCO_2H$$

with substituents: $N{=}CHPh$ on starting material and middle; $CXYZ$ above middle, $CXYZ$ and NH_2 on product.

R^2 = Me or Et; Q = Cl (for **49** and **123**, Q = Br)

(a) aliphatic amino acids:

(**47**) $R^1 = R^3 =$ Me, X = Y = H, Z = F [110, 111]
(**49**) $R^1 = R^3 =$ Me, X = Y = Z = F [110]
(**121**) $R^1 = R^3 =$ Me, X = Cl, Y = H, Z = F [112]
(**122**) $R^1 = R^3 =$ Me, X = H, Y = Z = F [110, 111]
(**123**) $R^1 = R^3 =$ Me, X = Br, Y = Z = F [110]
(**124**) $R^1 =$ PhCH=N(CH$_2$)$_3$, $R^3 =$ H$_2$N(CH$_2$)$_3$, X = Y = H, Z = F [111]
(**125**) $R^1 =$ PhCH=N(CH$_2$)$_3$, $R^3 =$ H$_2$N(CH$_2$)$_3$,
    X = Cl, Y = H, Z = F [112]
(**126**) $R^1 =$ PhCH=N(CH$_2$)$_3$, $R^3 =$ H$_2$N(CH$_2$)$_3$,
    X = H, Y = Z = F [111]
(**127**) $R^1 =$ PhCH=N(CH$_2$)$_4$, $R^3 =$ H$_2$N(CH$_2$)$_4$,
    X = H, Y = Z = F [111]
(**128**) $R^1 =$ MeO$_2$C(CH$_2$)$_2$, $R^3 =$ HO$_2$C(CH$_2$)$_2$, X = H, Y = Z = F [113]
(**129**) $R^1 = R^3 =$ MeS(CH$_2$)$_2$, X = H, Y = Z = F [111]

(b) aromatic and heterocyclic amino acids:

(**62**) $R^1 = R^3 =$ 4-HO-C$_6$H$_4$CH$_2$, X = H, Y = Z = F [111]
(**130**) $R^1 = R^3 =$ PhCH$_2$, X = Y = H, Z = F [111]
(**131**) $R^1 = R^3 =$ PhCH$_2$, X = Cl, Y = H, Z = F [112]
(**132**) $R^1 = R^3 =$ PhCH$_2$, X = H, Y = Z = F [111]
(**133**) $R^1 =$ 3,4-(MeO)$_2$-C$_6$H$_3$CH$_2$, $R^3 =$ 3,4-(HO)$_2$-C$_6$H$_3$CH$_2$,
    X = H, Y = Z = F [111]
(**134**) $R^1 = R^3 =$ 3,4-(MeO)$_2$-C$_6$H$_3$CH$_2$, X = H, Y = Z = F [111]
(**135**) $R^1 =$ (1-Trt-5-imidazolyl)CH$_2$, $R^3 =$ (5-imidazolyl)CH$_2$,
    X = H, Y = Z = F [111]

It is worth noting that in the synthesis of **62** the usual protection of the phenolic hydroxyl may be omitted; on the other hand, the imidazole nitrogen has to be protected by a triphenylmethyl group in the preparation of 2-(difluoromethyl)histidine (**135**).

Another route to 2-(chlorofluoromethyl)ornithine (**125**) was begun with *N*-benzylidene-5,6-didehydronorleucine methyl ester, which was chloro-fluoromethylated at C-2 in the usual way. After changing the *N*-protecting group, the carbon chain was transformed by the reaction sequence shown in the following scheme; the final amination of the 5-hydroxy ester was achieved through Mitsunobu reaction [112]. In the synthesis of 2-(chloro-fluoromethyl)glutamic acid (**136**), methyl 2-(chlorofluoromethyl)-5,6-didehy-dronorleucinate was first converted into the *N-tert*-butyloxycarbonyl) derivative, the diastereoisomers of which were then separated by medium-pressure liquid chromatography (MPLC). Oxidation of the terminal double bond followed by hydrolysis afforded both diastereomeric forms of **136** [112, 114].

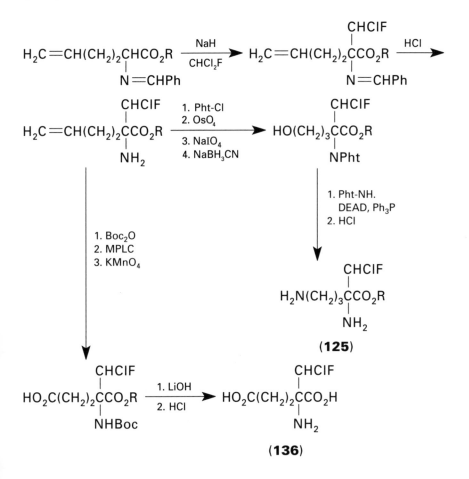

In the synthesis of 3-fluoro-3-phenyl-2-alanine (137), the Schiff base prepared from glycine benzyl ester and benzophenone has been alkylated by α,α-bromofluorotoluene and the resulting intermediate separated into the diastereomers. Deprotection by trimethylsilyl iodide and hydrolysis gave diastereomerically pure 137 [115]. 3-(4',5'-Dihydroxy-2'-fluorophenyl)-2-alanine (138) has also been synthesized analogously from a similar glycine-derived Schiff base [116].

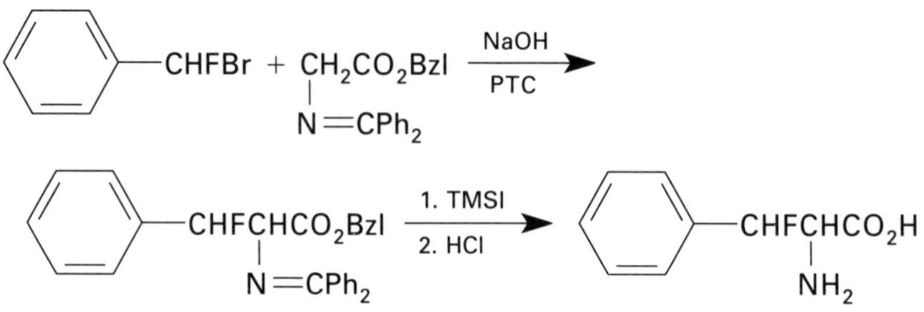

(137)

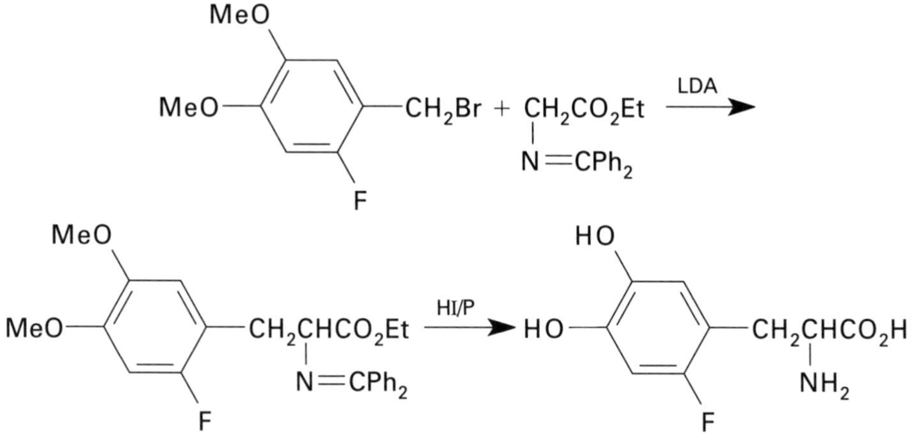

(138)

1.7 SYNTHESES INVOLVING OTHER CH-ACIDIC ESTERS. THE CURTIUS AND SCHMIDT REARRANGEMENTS

In addition to the N-acylaminomalonates and the Schiff bases discussed in the previous section, other CH-acids were also used in the syntheses of fluorinated amino acids. In these approaches, and also in the case of the glycine-derived Schiff bases, the advantage is often taken of the possible double alkylation by various electrophiles. Together with the ensuing chemical transformations of the newly introduced groups, this synthetic strategy gives rise to a wide variety of useful intermediates. The amino group is usually introduced towards the end of the reaction sequence through the Curtius or Schmidt rearrangement; less frequent are cases where the masked amino group is a structural part of the alkylating molecule.

In the following scheme the initial alkylation is carried out with either $CHCl_2F$ of $CHClF_2$, while CH_2ClF has been found unreactive under the conditions employed. Therefore, in the preparation of the 2-(fluoromethyl)amino acids **139–141**, alkylation by $CHCl_2F$ with ensuing reductive dechlorination of the chlorofluoromethylated intermediate was used instead. The CHClF group brings a second chiral centre into the molecule, giving rise to two diastereomeric forms; their chromatographic separation at the stage of the carbamate intermediate is demonstrated in the syntheses of **125** and **143**.

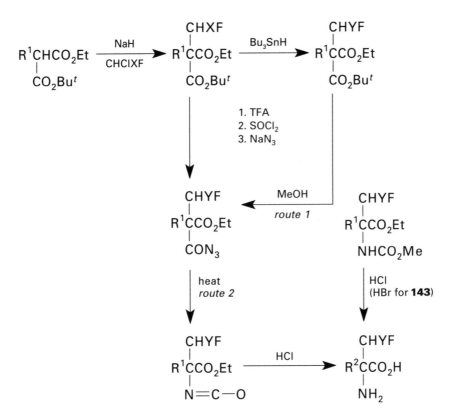

(125) $R^1 = PhtN(CH_2)_3$, $R^2 = H_2N(CH_2)_3$, $X = Y = Cl$; route 1 [114]
(126) $R^1 = PhtN(CH_2)_3$, $R^2 = H_2N(CH_2)_3$, $X = Y = F$; route 2 [117]
(132) $R^1 = R^2 = PhCH_2$, $X = Y = F$; route 2 [117]

(**139**) $R^1 = R^2 = H_2C=C(Me)$, $X = Cl$, $Y = H$; route 1 [112]
(**140**) $R^1 = R^2 = H_2C=CHCH(Me)$, $X = Cl$, $Y = H$; route 1 [112]
(**141**) $R^1 = MeO_2CCH_2$, $R^2 = HO_2CCH_2$, $X = Cl$, $Y = H$; route 1 [112]
(**142**) $R^1 = R^2 = 3\text{-}(MeO)\text{-}C_6H_4CH_2$. $X = Y = Cl$; route 1 [112]
(**143**) $R^1 = 3\text{-}(MeO)\text{-}C_6H_4CH_2$, $R^2 = 3\text{-}(HO)\text{-}C_6H_4CH_2$,
         $X = Y = Cl$; route 1 [114]

Synthesis of (*E*)-2-(difluoromethyl)-3,4-didehydroornithine (**144**) by this method involved a Wohl–Ziegler bromination of 2-allyl-2-(difluoromethyl) malonic ester, which afforded, under simultaneous allylic rearrangement, the 5-bromo derivative. Gabriel amination, followed by the usual transformation of the *tert*-butoxycarbonyl group into the carbamate and final hydrolysis, gave **144** [49].

5,5,5-Trifluoroisoleucine (**26**) has been synthesized by the modified Curtius method, employing Darapsky's procedure for the acyl azide preparation. 4,4,4-Trifluorobutan-2-one was condensed with ethyl cyanoacetate and the resulting unsaturated ester was hydrogenated to give ethyl 2-cyano-3-methyl-5,5,5-trifluoropentanoate. This compound was converted into the hydrazide, which on treatment with nitrous acid gave the desired azide. Route 1 was used to terminate the synthesis [19].

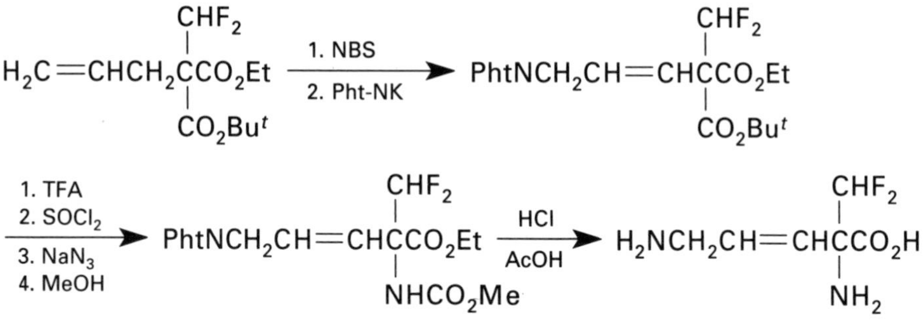

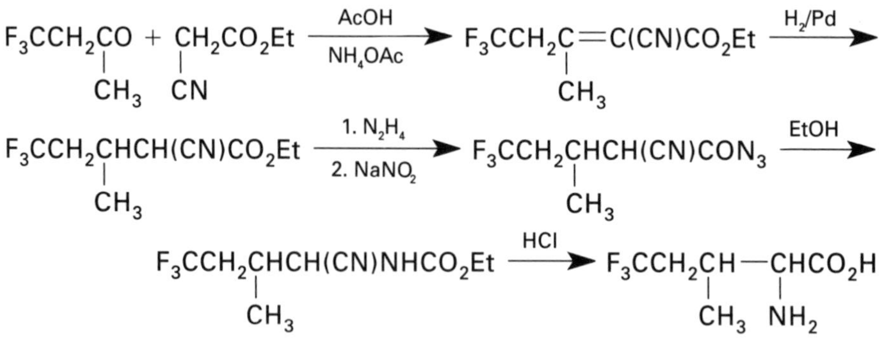

In the Curtius synthesis of 3,3,3-trifluoroalanine (59), the required isocyanate was prepared in a different way. Methyl 3,3,3,3',3',3'-hexa-fluoroisobutyrate was converted in two steps into methyl 2-fluorocarbonyl-3,3,3-trifluoropropionate, which in turn reacted with trimethylsilylazide (preferable to sodium azide in this case). The amino acid 59 was then obtained in high yield, following route 2 of the general scheme [118].

$$(F_3C)_2HCO_2Me \xrightarrow{BF_3.Et_3N} \underset{\overset{||}{CF_2}}{F_3CCCO_2Me} \xrightarrow[\text{2. heat}]{\text{1. AcOH}} \underset{\overset{|}{COF}}{F_3CCHCO_2Me}$$

$$\xrightarrow{Me_3SiN_3} \underset{\overset{|}{N=C=O}}{F_3CCHCO_2Me} \xrightarrow{HCl} \underset{\overset{|}{NH_2}}{F_3CCHCO_2H}$$

(59)

Applications of the Schmidt reaction in the syntheses of fluorinated amino acids are scarce in the literature. A synthetic pathway based on this principle was elaborated for 5-fluoroisoleucine (145), an amino acid accessible only with considerable difficulty. Alkylation of ethyl acetoacetate by 3-bromo-1-fluorobutane gave ethyl 2-acetyl-5-fluoro-3-methylpentanoate, which on treatment with hydrazoic acid yielded, after rearrangement, the ethyl ester of N-acetyl-5-fluoroisoleucine. Saponification then gave the free N-acetylamino acid as a mixture of the *threo* and *erythro* forms, separable by systematic crystallization. Hydrolysis was carried out either with hydrochloric acid or enzymatically (hog kidney acylase) and afforded the respective stereomers of 145; *threo*-D,L-, *threo*-L-, *threo*-D- and *erythro*-L-forms of 145 were prepared by this approach [119].

Another example of the Schmidt rearrangement is the synthesis of 3'-fluoro-tyrosine (86) [120]; ethyl acetoacetate was alkylated with 3-fluoro-4-meth-oxybenzyl chloride and the resulting keto ester was treated with hydrazoic acid. Hydrolysis with hydrobromic acid gave 86.

$$\underset{\overset{|}{CH_3}}{FCH_2CH_2CHBr} + \underset{\overset{|}{COCH_3}}{CH_2CO_2Et} \longrightarrow \underset{\overset{|}{H_3C}\;\overset{|}{COCH_3}}{FCH_2CH_2CHCHCO_2Et} \xrightarrow[\text{2. NaOH}]{\text{1. HN}_3}$$

$$\underset{\overset{|}{H_3C}\;\overset{|}{NHAc}}{FCH_2CH_2CHCHCO_2H} \xrightarrow[\text{2. HCl or acylase}]{\text{1. separation}} \underset{\overset{|}{H_3C}\;\overset{|}{NH_2}}{FCH_2CH_2CHCHCO_2H}$$

(145)

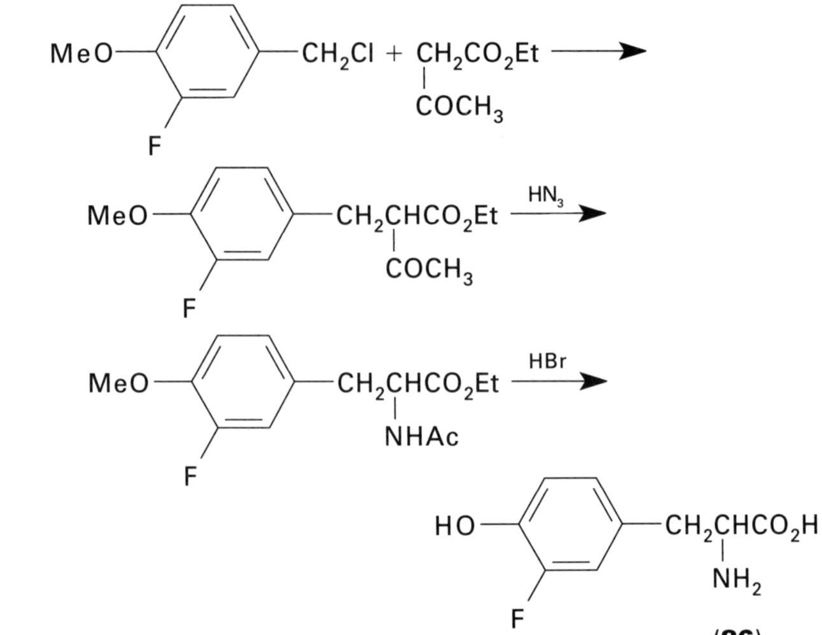

MeO—[benzene ring, F]—CH$_2$Cl + CH$_2$CO$_2$Et →
                                    |
                                    COCH$_3$

MeO—[benzene ring, F]—CH$_2$CHCO$_2$Et $\xrightarrow{HN_3}$
                          |
                          COCH$_3$

MeO—[benzene ring, F]—CH$_2$CHCO$_2$Et $\xrightarrow{HBr}$
                          |
                          NHAc

HO—[benzene ring, F]—CH$_2$CHCO$_2$H
                        |
                        NH$_2$

(86)

Derivatization of fluoromalonic esters was used several times in the synthesis of fluorinated amino acids. Michael addition of diethyl fluoromalonate to ethyl 2-acetamidoacrylate, followed by hydrolysis of the adduct led to 4-fluoroglutamic acid (105) [121]. An analogous reaction between dibenzyl fluoromalonate and methyl 2-phthalimidoacrylate gave, after hydrogenolytic removal of benzyl ester groups and reduction of the *gem*-dicarboxylic acid by borane/dimethyl sulphide, the methyl ester of the *N*-phthaloyl derivative of 5,5′-dihydroxy-4-fluoroleucine. The free amino acid 146 was liberated from this compound by hydrolysis [122].

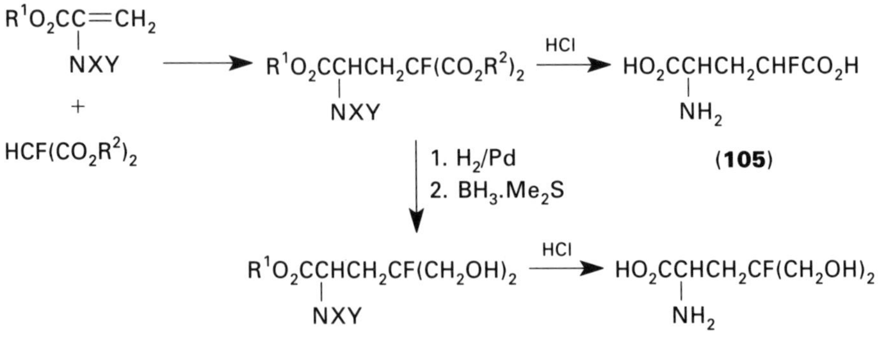

R^{1}O$_2$CC=CH$_2$
        |
        NXY   R^{1}O$_2$CCHCH$_2$CF(CO$_2$R^2)$_2$ $\xrightarrow{HCl}$ HO$_2$CCHCH$_2$CHFCO$_2$H
        +   → | |
                                  NXY   NH$_2$
HCF(CO$_2$R^2)$_2$
                                                                          (105)

                          $\downarrow$ 1. H$_2$/Pd
                              2. BH$_3$.Me$_2$S

              R^{1}O$_2$CCHCH$_2$CF(CH$_2$OH)$_2$ $\xrightarrow{HCl}$ HO$_2$CCHCH$_2$CF(CH$_2$OH)$_2$
                        | |
                        NXY   NH$_2$

                                                              (146)

(105) R^1 = R^2 = Et, X = H, Y = Ac [121]
(146) R^1 = Me, R^2 = Bzl, X + Y = Pht [122]

Similar addition of diethyl fluoromalonate to 2-nitrobut-1-ene afforded ethyl 2-ethoxycarbonyl-2-fluoro-4-nitrohexanoate. Hydrolysis with simultaneous decarboxylation, followed by reduction of the nitro group, furnished 4-amino-2-fluorohexanoic acid (**147**). If the reduction is carried out prior to hydrolysis, the synthesis stops at the stage of 4-amino-2-carboxy-2-fluorohexanoic acid (**148**) which is resistant towards attempted decarboxylation [123].

Diethyl fluoromalonate anion also cleaves the aziridine ring, as shown by the synthesis of 2-fluoro-4-aminobutyric acid (**149**). In the reaction with 1-(4'-nitrobenzoyl)aziridine, ethyl 2-ethoxycarbonyl-2-fluoro-4-(4'-nitrobenzoyl-amino)butyrate is formed, which is hydrolysed to **149** by acid [123].

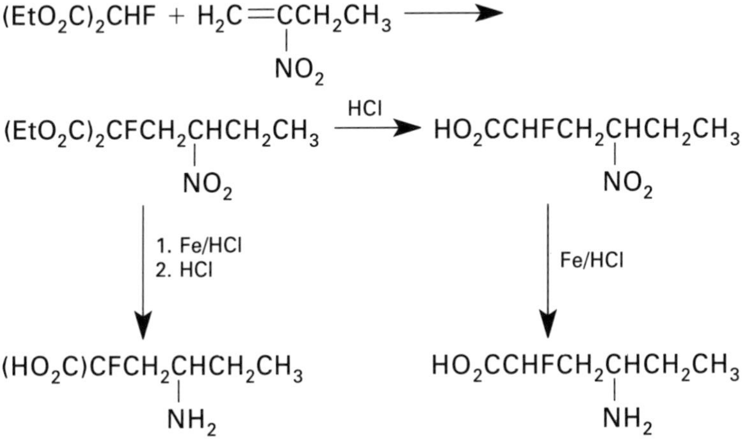

$$(EtO_2C)_2CHF \ + \ H_2C{=}\underset{\underset{NO_2}{|}}{C}CH_2CH_3 \longrightarrow$$

$$(EtO_2C)_2\underset{\underset{NO_2}{|}}{CFCH_2CHCH_2CH_3} \xrightarrow{HCl} HO_2C\underset{\underset{NO_2}{|}}{CHFCH_2CHCH_2CH_3}$$

1. Fe/HCl
2. HCl

Fe/HCl

$$(HO_2C)\underset{\underset{NH_2}{|}}{CFCH_2CHCH_2CH_3}$$

(148)

$$HO_2C\underset{\underset{NH_2}{|}}{CHFCH_2CHCH_2CH_3}$$

(147)

$$(EtO_2C)_2CHF \ + \ \underset{H_2C}{\overset{H_2C}{\diagdown}}NCO{-}\bigcirc{-}NO_2 \xrightarrow{NaOEt}$$

$$(EtO_2C)_2CF(CH_2)_2NHCO{-}\bigcirc{-}NO_2$$

$$\xrightarrow{HCl} HO_2CCHFCH_2CH_2NH_2$$

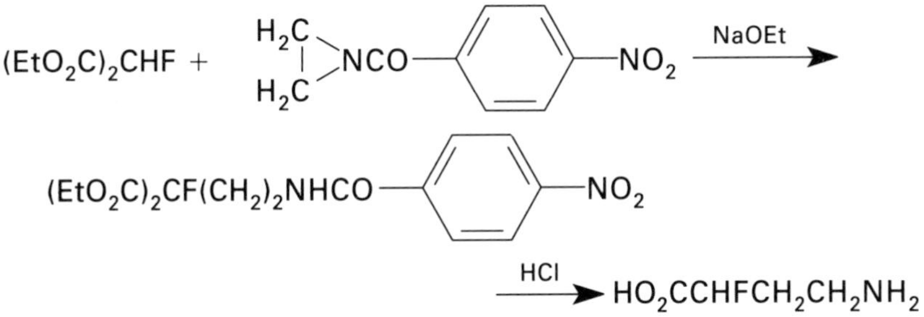

(149)

1.8 SYNTHESES FROM 2-OXO ACIDS AND THEIR DERIVATIVES

2-Oxo acids may be converted into the corresponding 2-amino acids either directly, i.e. by reductive amination, or by reduction of their nitrogen-containing derivatives, such as oximes and oxime ethers and also of the phenylhydrazones. To achieve reductive amination the enzymatic approach is often used, which is not a subject of this chapter. The non-enzymatic method, applied several times in the synthesis of fluorinated amino acids, consists in the reduction of a 2-oxo acid alkali metal salt with a borohydride in the presence of ammonia or an ammonium salt. High yields were reported when this method was used in a simple synthesis of a broad-spectrum antimicrobial 3-fluoro-2-alanine-2-d (1-2-d) from fluoropyruvic acid [124].

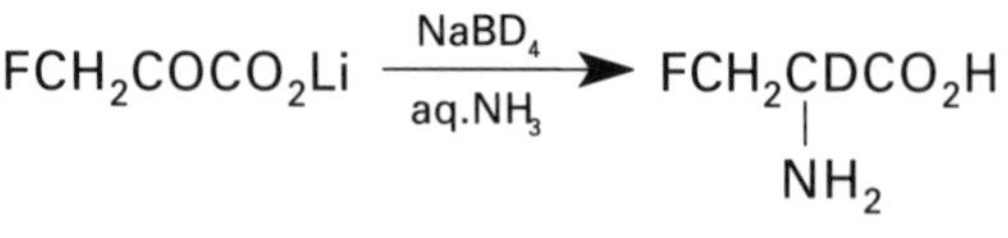

$$FCH_2COCO_2Li \xrightarrow[\text{aq.NH}_3]{\text{NaBD}_4} FCH_2\underset{\underset{NH_2}{|}}{C}DCO_2H$$

(1–2–d)

In an analogous way, *erythro*-3-fluoro-3-phenyl-2-alanine (137) was prepared from 3-fluoro-3-phenylpyruvic acid. Two reagents were compared, the sodium borohydride/ammonia system being much more effective [125, 126].

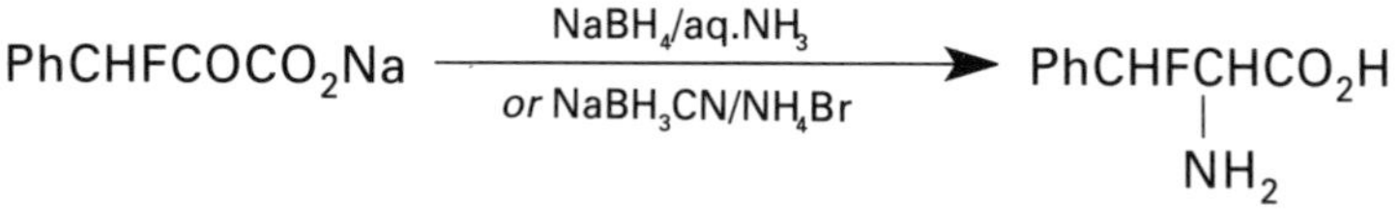

$$PhCHFCOCO_2Na \xrightarrow[\text{or NaBH}_3\text{CN/NH}_4\text{Br}]{\text{NaBH}_4/\text{aq.NH}_3} PhCHF\underset{\underset{NH_2}{|}}{C}HCO_2H$$

(137)

The oxime reduction was used in the synthesis of 5,5,5-trifluoroleucine (51) from ethyl 3-(trifluoromethyl)crotonate [127]. The starting ester was first hydrogenated to ethyl 3-(trifluoromethyl)butyrate; condensation with diethyl oxalate, followed by acidolysis of the 2-oxo ester, gave 2-oxo-4-(trifluoromethyl)pentanoic acid, isolated as the oxime. Catalytic hydrogenation of this oxime yielded 51.

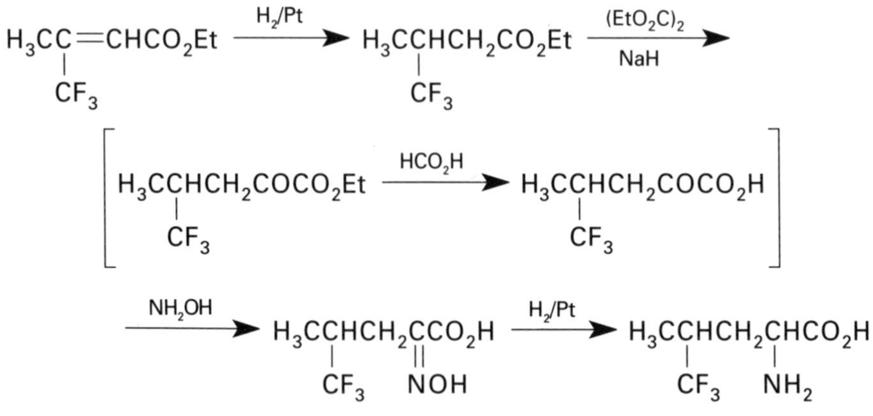

$$H_3C\underset{\underset{CF_3}{|}}{C}C=CHCO_2Et \xrightarrow{H_2/Pt} H_3C\underset{\underset{CF_3}{|}}{C}HCH_2CO_2Et \xrightarrow[\text{NaH}]{(EtO_2C)_2}$$

$$\left[H_3C\underset{\underset{CF_3}{|}}{C}HCH_2COCO_2Et \xrightarrow{HCO_2H} H_3C\underset{\underset{CF_3}{|}}{C}HCH_2COCO_2H \right]$$

$$\xrightarrow{NH_2OH} H_3C\underset{\underset{CF_3}{|}}{C}HCH_2\underset{\underset{NOH}{||}}{C}CO_2H \xrightarrow{H_2/Pt} H_3C\underset{\underset{CF_3}{|}}{C}HCH_2\underset{\underset{NH_2}{|}}{C}HCO_2H$$

(51)

4,4,4-Trifluorothreonine (28) has been synthesized from ethyl 4,4,4-tri-fluoroacetoacetate by two related sequences. On treatment of the starting ester with nitrous acid, followed by O-methylation, ethyl 2-(methoxyimino)-3-oxo-4,4,4-trifluoropropionate was formed, which was first reduced by borohydride to the 3-hydroxy ester and then, by zinc in acidic medium, to a mixture of both diastereomers of 28, containing 60% of the *threo* isomer. When the final reduction was applied to free oximino acid instead of the oximino ester, the product composition changed to only 32% of the *threo* form [128].

In another approach to 28, the Japp–Klingemann reaction was applied to the same starting oxo ester, giving rise to the 3-oxo-2-phenylhydrazono derivative. The conversion of this compound to 28 of undetermined steric composition proceeded in a manner similar to that described above [20]. The diastereomers of 28 may be separated chromatographically or by crystallization; interestingly, the *allo* (i.e. the *erythro*) form may be epimerized to the other isomer through the oxazolidone [128].

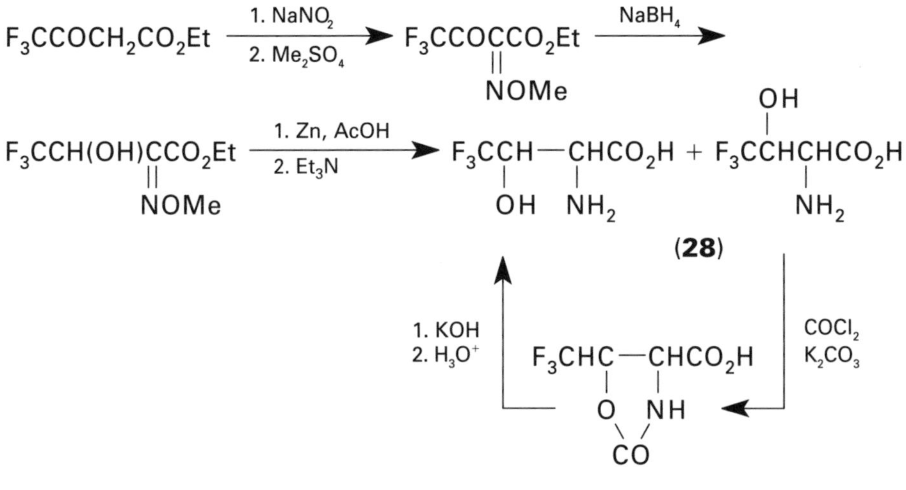

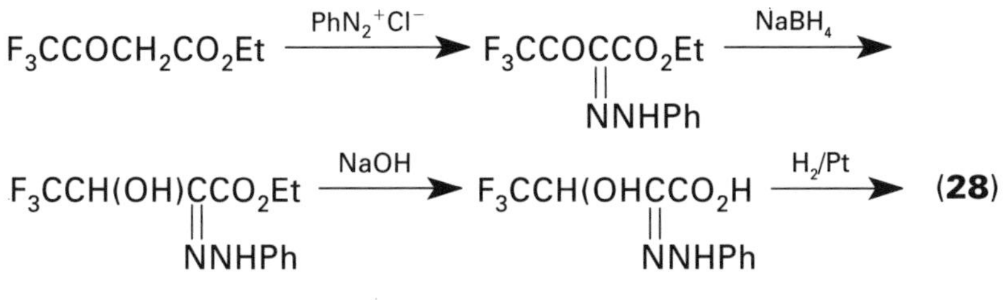

The synthesis of 3,3-difluoroaspartic acid (150) started with di-tert-butyl 2-oxosuccinate, which was fluorinated to the 3,3-difluoro-2-oxo ester hemiketal. Reaction with O-methylhydroxylamine converted it into the 2-methoxyimino derivative, which on reduction with amalgamated aluminium and subsequent removal of the ester groups gave **150**. Successful resolution of the racemate through the brucine salt was also reported [129].

$$Bu^tO_2CCOCH_2CO_2Bu^t \xrightarrow[\text{EtOH}]{FClO_3} Bu^tO_2C\overset{\overset{\displaystyle OH}{|}}{\underset{\underset{\displaystyle OEt}{|}}{C}}CF_2CO_2Bu^t \xrightarrow{NH_2OMe}$$

$$Bu^tO_2C\overset{\overset{\displaystyle}{\|}}{\underset{\underset{\displaystyle NOMe}{}}{C}}CF_2CO_2Bu^t \xrightarrow[\text{2. TFA}]{\text{1. AlHg}_x} HO_2C\overset{}{\underset{\underset{\displaystyle NH_2}{|}}{C}}HCF_2CO_2H$$

(150)

1.9 FLUORINATED AMINO ACIDS FROM OTHER AMINO ACIDS

In this section, several syntheses of fluorinated amino acid are associated not on the basis of their contingent methodical affinity, but with preferential emphasis on the chemical nature of the starting compounds. These are in all cases other amino acids, no matter whether they contain fluorine or not. Some related syntheses, namely those involving fluorohalomethylations of the amino acid-derived Schiff bases, have been discussed in Section 1.6.

The fluorinated ornithines **33** and **124** have been converted into 4-fluoro-arginine (**151**) [26] and 2-(difluoromethyl)arginine (**152**) [111], respectively, by treatment with S-alkylisothiouronium salts. Also, 4-fluorocitrulline (**153**) has been prepared from **33** by the action of potassium cyanate [26]. In these reactions, except for **152**, the 2-amino group was temporarily protected by formation of a copper(II) complex.

$$HO_2C\overset{\overset{\displaystyle Q}{|}}{\underset{\underset{\displaystyle NH_2}{|}}{C}}CH_2CHXCH_2NH_2 \xrightarrow[\text{3. H}_2\text{S}]{\substack{\text{1. CuCO}_3 \\ \text{2. KCNO}}} HO_2C\overset{\overset{\displaystyle Q}{|}}{\underset{\underset{\displaystyle NH_2}{|}}{C}}CH_2CHXCH_2NHCONH_2$$

(153)

1. CuCO3
2. RSC(=NH)NH₂.HY
3. H₂S

$$HO_2C\overset{\overset{\displaystyle Q}{|}}{\underset{\underset{\displaystyle NH_2}{|}}{C}}CH_2CHXCH_2NHC\overset{\overset{\displaystyle}{\|}}{\underset{\underset{\displaystyle NH}{}}{N}}H_2$$

(151), (152)

(**151**) Q = H, X = F, R = Me, Y = Cl [26]

(**152**) Q = CHF₂, X = H, R = Et, Y = Br [111]

(**153**) Q = H, X = F [26]

The diastereomers of 3-fluoroaspartic acid (**29**) were made accessible from 3,3-difluoroaspartiic acid (**150**) by an elimination–reduction pathway. On treatment with 1,5-diazabicyclo[4.3.0]non-5-ene, one molecule of hydrogen fluoride was split off from the di-*tert*-butyl ester of **150**, giving quantitatively the monofluorinated enamino ester with a Z/E ratio of 20:1 (the same enamine with a Z/E ratio of 1:1 is also available by reaction of di-*tert*-butyl 3-fluoro-2-oxosuccinate with ammonium acetate). Cyanoborohydride reduction then afforded a mixture of *erythro*- and *threo*- **29** *tert*-butyl esters, which was efficiently separated by chromatographic mens. The predominant *erythro* isomer was then deprotected by trifluoroacetic acid to yield free *erythro*- **29** [130].

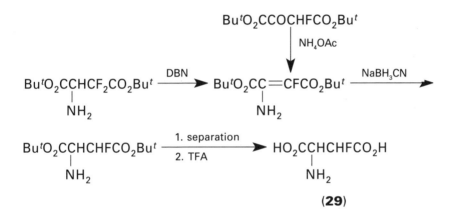

(**29**)

A stereospecific synthesis of all four enantiomers of 4-fluoroglutamic acid (**105**) from the sterically relevant 4-hydroxyprolines has been accomplished by a four-step reaction sequence. The starting amino acid, after being protected at nitrogen and carboxyl, was fluorodehydroxylated by either diethylaminosulfur trifluoride (DAST) or 2-chloro-1,1,2-trifluorotriethylamine to give the respective 4-fluoroproline derivative. The reaction proceeded with inversion at C-4, the DAST reagent being superior to CTT because of its higher stereoselectivity. Ruthenium tetraoxide oxidation then converted the fluoroproline derivative into protected 4-fluoropyroglutamic acid, which was finally hydrolysed to **105** [131, 132].

A related route, starting from L-pyroglutamic acid, has been devised [133]. The protected lactam was hydroxylated at C-4 by 2-toluenesulfonyl-3-phenyloxaziridine (TSPA) [134] to give the derivative of (2*S*,4*R*)-4-hydroxy-pyroglutamic acid. Treatment with DAST then furnished the 4-fluoro lactam with 2*S*, 4*S* configuration; at this stage the experiments were interrupted.

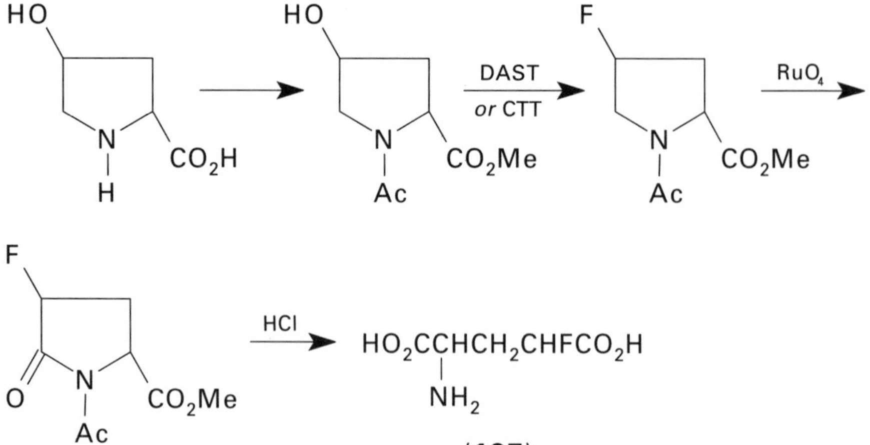

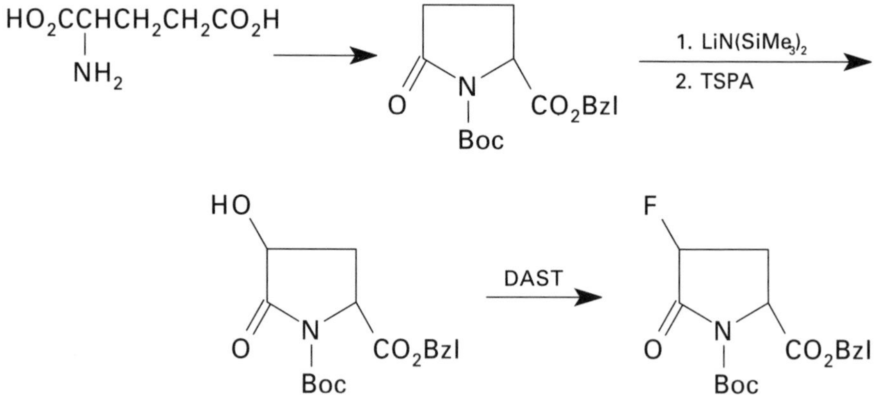

The ω-monoamides of fluorine-containing dicarboxylic acids, i.e. the fluoro derivatives of asparagine and glutamine, have been prepared from the parent amino acids. The acids were selectively esterified to the ω-methyl esters using methanol/thionyl chloride according to Boissonas [135]. erythro-3-Fluoroasparagine (154) [130] and 3,3,-difluoroasparagine (155) [129] resulted from direct ammonolysis of the unprotected esters, whereas in the case of 4-fluoroglutamine (156) temporary protection of the amino group was essential. It was accomplished either by dithiocarbamate formation [28] or more conveniently with the easily removable tert-butyloxycarbonyl group. Both erythro and threo forms of 156 were prepared using this second route [136]. When the amino protection was omitted, ring closure occurred, which led to the 4-fluoropyroglutamic acid **157** as the sole product. Thus *cis*-**157** was prepared from *threo*-**105**, while cyclization of *erythro*-**105** yielded *trans*-**157** [91].

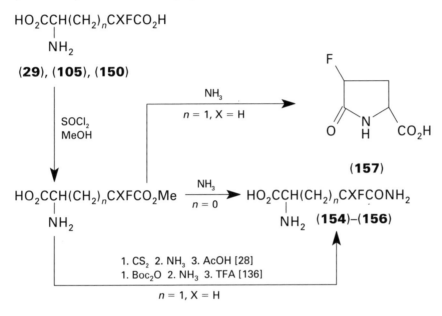

$$HO_2CCH(CH_2)_nCXFCO_2H$$
$$| \atop NH_2$$

(29), (105), (150)

SOCl$_2$
MeOH

NH$_3$
$n = 1, X = H$

(157)

$$HO_2CCH(CH_2)_nCXFCO_2Me \quad \xrightarrow[n=0]{NH_3} \quad HO_2CCH(CH_2)_nCXFCONH_2$$
$$| \atop NH_2 \qquad\qquad\qquad\qquad\qquad\qquad | \atop NH_2 \; \textbf{(154)–(156)}$$

1. CS$_2$ 2. NH$_3$ 3. AcOH [28]
1. Boc$_2$O 2. NH$_3$ 3. TFA [136]

$n = 1, X = H$

(29) and **(154)** $n = 0$, X $= $ H [130]
(150) and **(155)** $n = 0$, X $= $ F [129]
(105) and **(156)** $n = 1$, X $= $ H [28, 136]

The conversion of **105** into the lactam **157** was also achieved by pyrolysis at 170 °C [90] or by distillation of diethyl 4-fluoroglutamate, followed by saponification of the **157** ethyl ester formed [28].

Similar formation of the lactam ring is extremely easy in the case of glutamic acid analogues with a polyfluoromethyl substituent at C-2. Whereas 2-(difluoromethyl)pyroglutamic acid (**158**) is formed quantitatively from **128** by boiling its aqueous solution [113], ring closure of the trifluoromethyl compound to 2-(trifluoromethyl)pyroglutamic acid (**159**) occurs spontaneously within a few hours at ambient temperature [137].

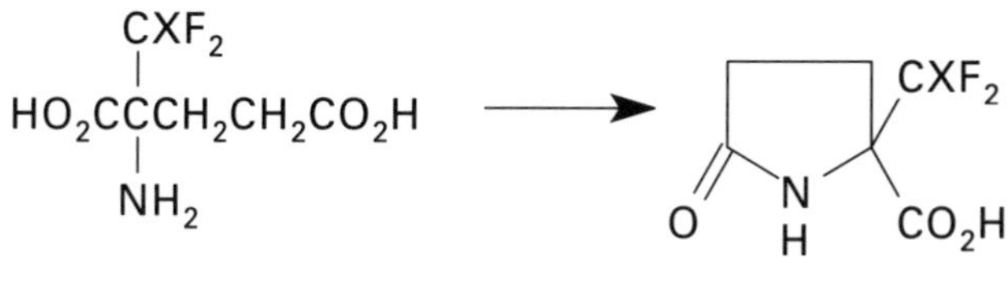

$$\begin{array}{c} CXF_2 \\ | \\ HO_2CCCH_2CH_2CO_2H \\ | \\ NH_2 \end{array}$$

(158) X $= $ H
(159) X $= $ F

4-Fluoroisoglutamine (160) has been prepared from the acid 105 by two methods. Dialkyl 4-fluoroglutamate was converted into 4-fluoropyroglutamamide, which was then cleaved by HCl/alcohol to the alkyl ester of 160. Mild acid or alkaline hydrolysis then afforded 160 [28, 136]. In a simpler alternative, 105 was treated with diphosgene to give the N-carboxanhydride, which on ammonolysis furnished 160 in good overall yield [136]. Both diastereomers of 160 have been prepared by both routes.

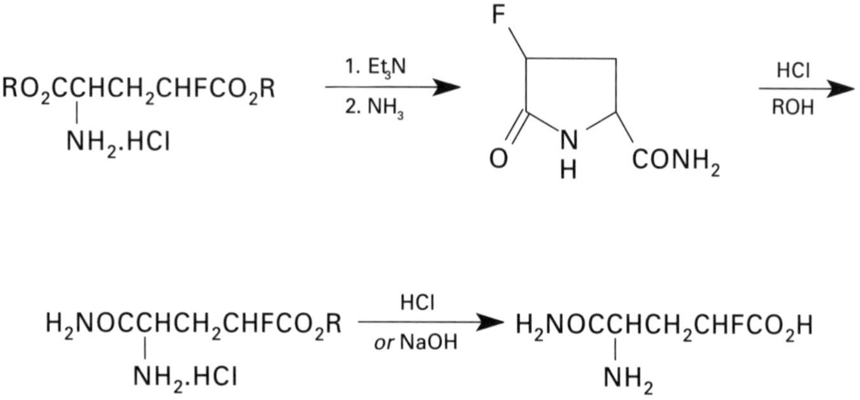

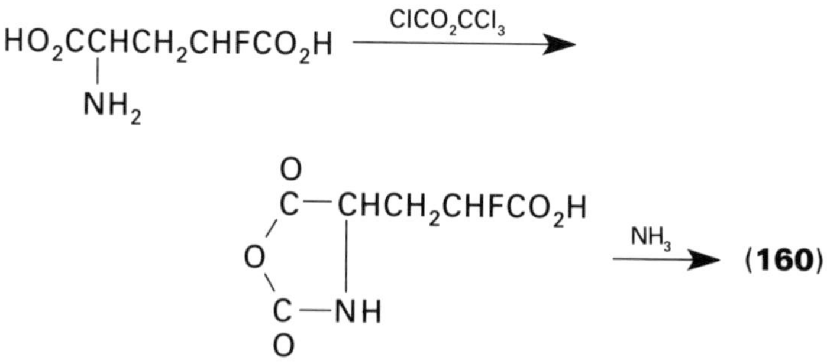

Reduction of the ester group in 4-fluoroglutamic acid 5-methyl ester, effected by sodium borohydride in water and without protection of the functional groups, resulted in moderate yields of 4-fluoro-5-hydroxynorvaline (161) [28]. As a minor byproduct, 4-fluoroproline (162) was identified [28] and later isolated by chromatography [138].

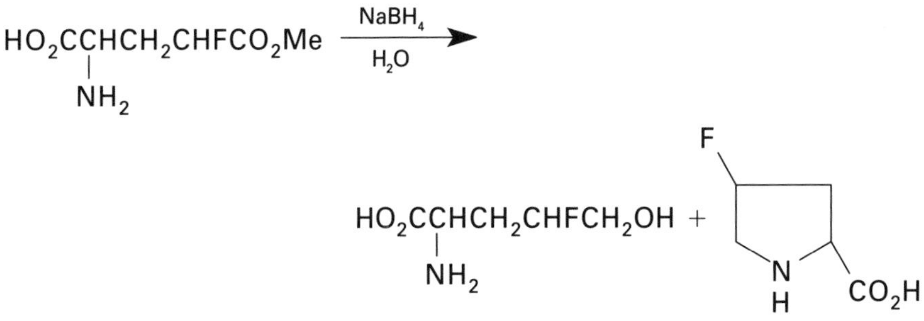

$$HO_2CCHCH_2CHFCO_2Me \xrightarrow[\text{H}_2\text{O}]{\text{NaBH}_4}$$

$$\underset{NH_2}{|}$$

$$HO_2CCHCH_2CHFCH_2OH \quad +$$

$$\underset{NH_2}{|}$$

(161)   **(162)**

The methyl–benzyl thioether interchange [139] has been applied in the conversion of 2-(difluoromethyl)methionine **(129)** into S-benzyl-2-(difluoromethyl)homocysteine **(163)**. The reaction proceeded via the sulfonium intermediate, from which the methyl group is cleaved off preferentially [111].

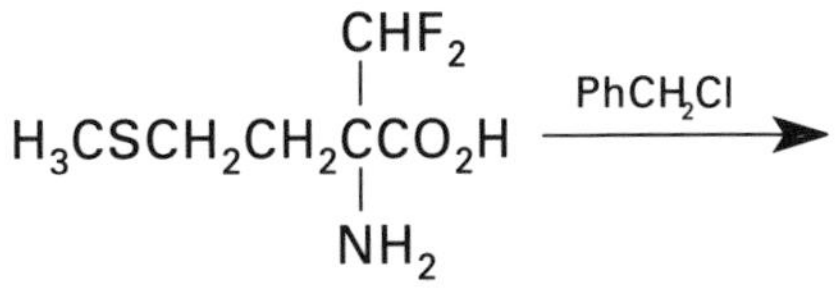

$$H_3CSCH_2CH_2\overset{\overset{\displaystyle CHF_2}{|}}{\underset{\underset{\displaystyle NH_2}{|}}{C}}CO_2H \xrightarrow{PhCH_2Cl}$$

(129)

$$\left[C_6H_5CH_2\overset{Cl^-}{\underset{\underset{\displaystyle CH_3}{|}}{S^+}}CH_2CH_2\overset{\overset{\displaystyle CHF_2}{|}}{\underset{\underset{\displaystyle NH_2}{|}}{C}}CO_2H \right] \xrightarrow{HCl} C_6H_5CH_2SCH_2CH_2\overset{\overset{\displaystyle CHF_2}{|}}{\underset{\underset{\displaystyle NH_2}{|}}{C}}CO_2H$$

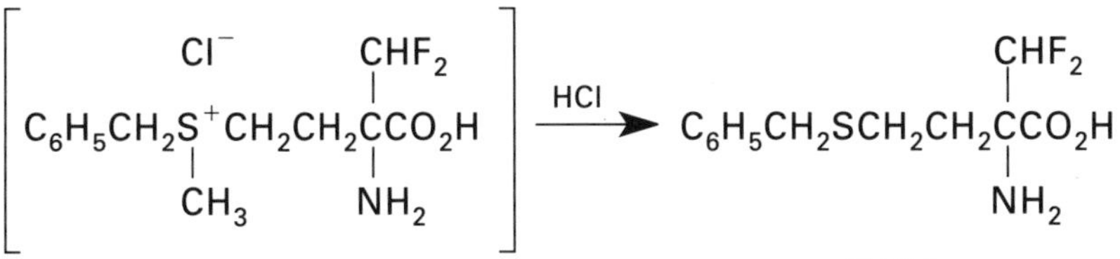

(163)

Three 4'-substituted derivatives of 3-(3'-fluorophenyl)-2-alanine were prepared from the parent amino acid **69** within one reaction sequence. Nitration of unprotected **69** gave 3-(3'-fluoro-4'-nitrophenyl)-2-alanine **(164)**, which was in turn reduced by tin metal to 3-(4'-amino-3'-fluorophenyl)-2-alanine **(165)**. Final treatment of **165** with nitrous acid yielded 3'-fluorotyrosine **(86)** [59].

Direct halogenation of **86** afforded 3'-bromo-5'-fluorotyrosine (**166**) [120, 140] and 3'-fluoro-5'-iodotyrosine (**167**), respectively [68, 99, 120, 140].

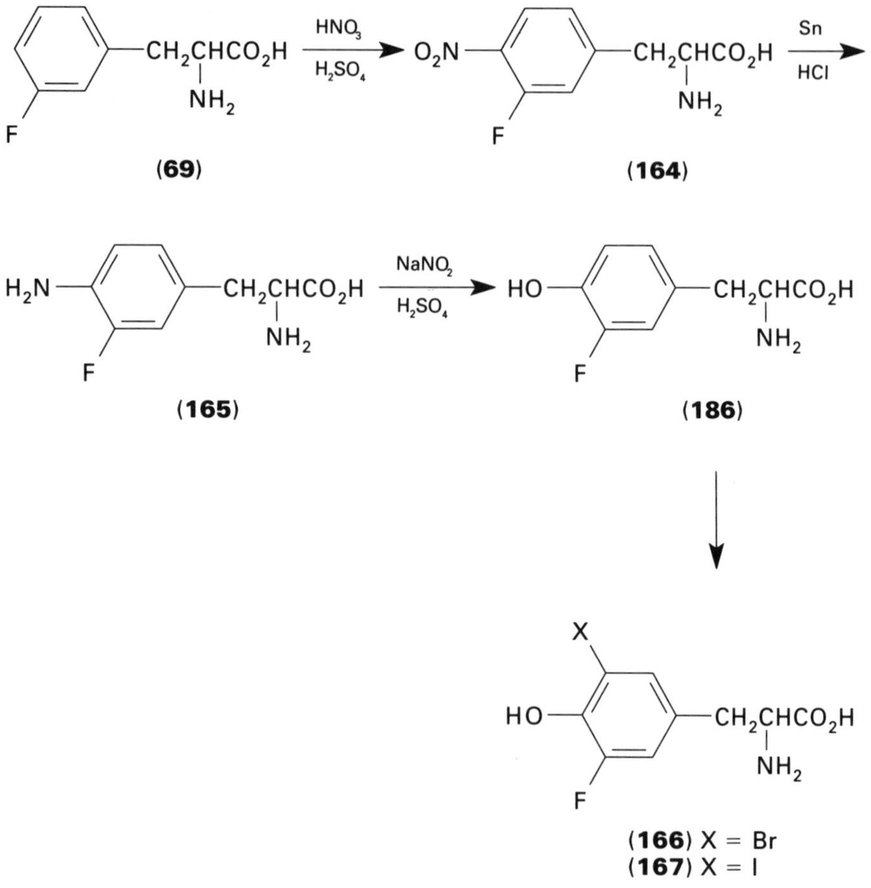

(**69**)　　(**164**)

(**165**)　　(**186**)

(**166**) X = Br
(**167**) X = I

(**166**) X = Br
(**167**) X = I

The tyrosine derivative **167** was used as the starting compound in the synthesis of 3'-fluoro-5'-iodothyronine (**168**) and its *O*-methyl ether (**169**). The ethyl ester of *N*-acetyl-**167** was condensed with bis(4-methoxyphenyl)iodonium sulfate (BMIS) to give the fully protected intermediate, which was hydrolysed with either HBr or HCl to afford **168** or **169**, respectively [99].

Other thyronine derivatives were prepared either by further iodination or by reductive deiodination of fluoroiodothyronines **94** and **95**, mentioned in Section 1.5 [69, 73].

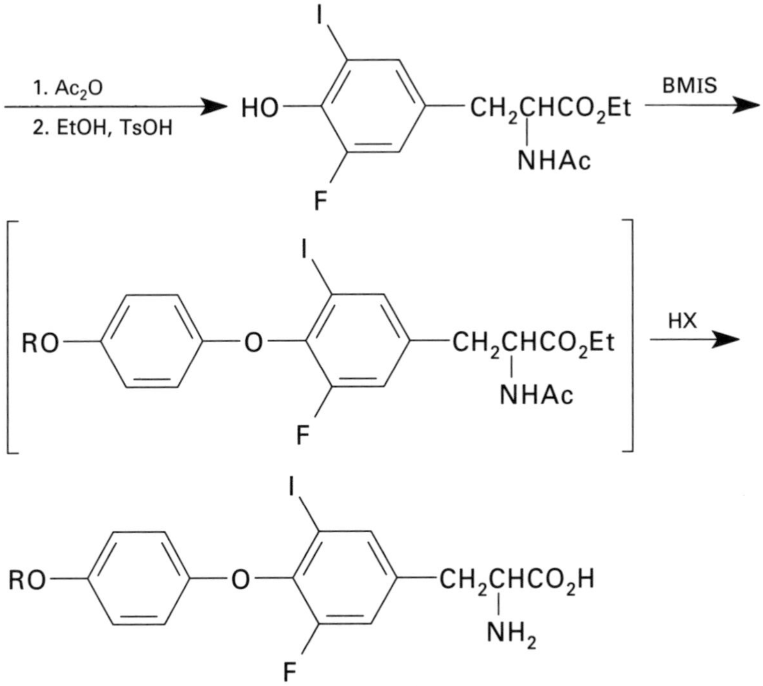

(168) R = H
(169) R = Me

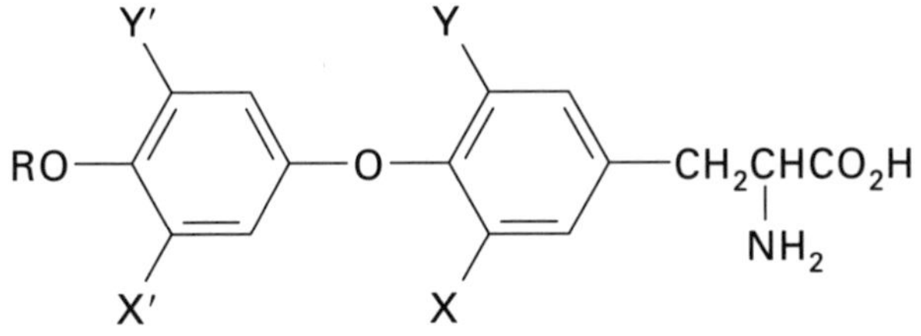

(170)–(172)

(170) X = Y = Y′ = H, X′ = F (from **94** by H$_2$/Pd) [73]
(171) X = Y = Y′ = I, X′ = F (from **94** by I$_2$/KI) [73]
(172) X = Y = H, X′ = Y′ = F (from **95** by H$_2$/Pd) [69]

L-Glutamic acid, a widely used chiral synthon, has been chosen as the starting compound for the synthesis of (S)-4-amino-5-fluoropentanoic acid (**173**). After ring closure and esterification, the ethyl ester of pyrrolid-5-one-2-carboxylic acid was reduced to the 2-hydroxymethyl derivative, which was then converted within two steps into 2-fluoromethylpyrrolid-5-one. Acid hydrolysis of this lactam yielded **173** [141].

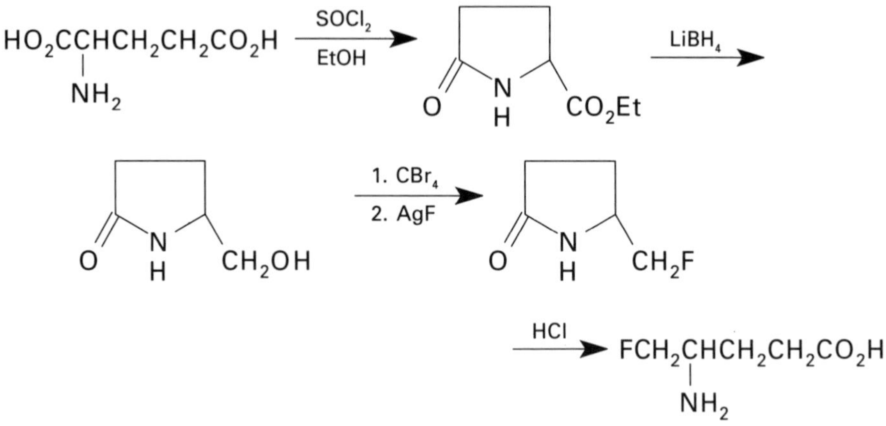

$HO_2CCHCH_2CH_2CO_2H$ (with NH_2) $\xrightarrow[\text{EtOH}]{SOCl_2}$ [pyrrolidin-2-one with CO_2Et] $\xrightarrow{LiBH_4}$

[pyrrolidin-2-one with CH_2OH] $\xrightarrow[\text{2. AgF}]{\text{1. }CBr_4}$ [pyrrolidin-2-one with CH_2F]

$\xrightarrow{HCl}$ $FCH_2CHCH_2CH_2CO_2H$ (with NH_2)

(173)

1.10 MISCELLANEOUS SYNTHESES

A few synthetic routes to fluorinated amino acids, different in principle from the syntheses described in the foregoing sections, and hardly incorporatable in any of them, are described in this separate section.

3-(Fluoroaryl)-3-alanines are attainable by the Rodionov reaction, which consists in condensation of aromatic aldehydes with malonic acid in the presence of ammonium acetate. The desired 3-amino acids are prepared in good yields in a single operation [61, 142, 143].

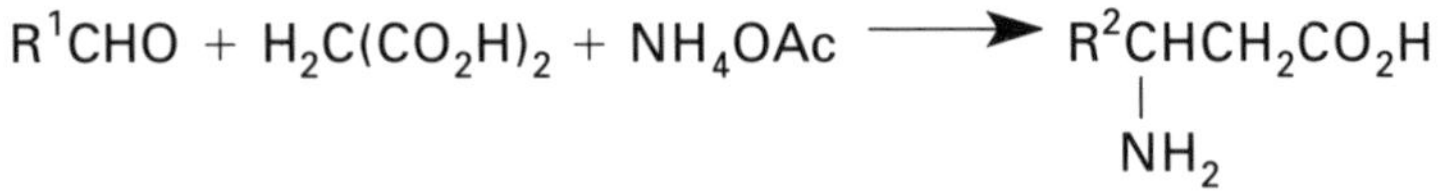

$R^1CHO + H_2C(CO_2H)_2 + NH_4OAc \longrightarrow R^2CHCH_2CO_2H$ (with NH_2)

(174)–(178)

(**174**) $R^1 = R^2 = $ 4-F-C_6H_4 [61, 142]
(**175**) $R^1 = R^2 = $ 3-F-4-(MeO)-C_6H_3 [142]
(**176**) $R^1 = $ 3-F-4-(MeO)-C_6H_3, $R^2 = $ 3-F-4-(HO)-C_6H_3 (from **175** by HBr) [142]
(**177**) $R^1 = R^2 = $ 2-(F_2HCO)-C_6H_4 [143]
(**178**) $R^1 = R^2 = $ 4-(F_2HCO)-C_6H_4 [143]

Another condensation of aromatic aldehydes, with unprotected glycine in alkaline medium, gives rise to the 3-aryl derivatives of serine. 3-(2'-Fluorophenyl)serine (**179**) and 3-(4'-fluorophenyl)serine (**180**) have been prepared in this way [39, 61, 144], while attempts to prepare the 3′-fluoro isomer have been reported to fail [144]. It has been found that 4-fluoro-benzaldehyde and free glycine gave the *threo* diastereomer of **180**, whereas in the reaction with glycine ethyl ester *erythro*-**180** was formed after hydrolysis [39].

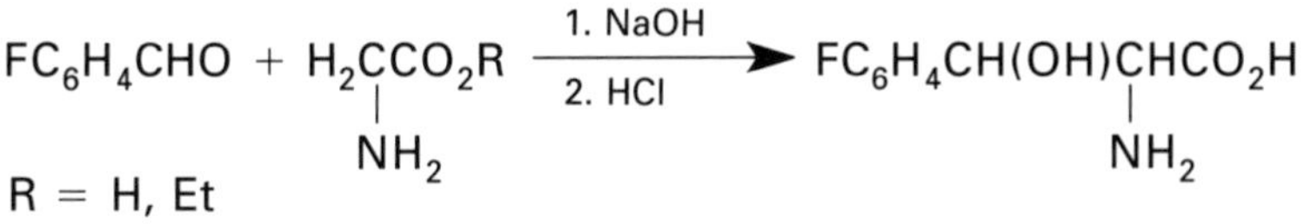

$$FC_6H_4CHO + \underset{\underset{NH_2}{|}}{H_2CCO_2R} \xrightarrow[\text{2. HCl}]{\text{1. NaOH}} FC_6H_4CH(OH)\underset{\underset{NH_2}{|}}{CHCO_2H}$$

R = H, Et

(179), (180)

In the synthesis of the isomeric 3-(4′-fluorophenyl)isoserine (**46**), 4-fluoro-phenylnitromethane was condensed with ethyl glyoxylate ethyl hemiacetal with the formation of ethyl 3-(4′-fluorophenyl)-2-hydroxy-3-nitropropionate. Hydrogenation of this ester, followed by hydrolysis, furnished the *threo* diastereomer of **46** [39].

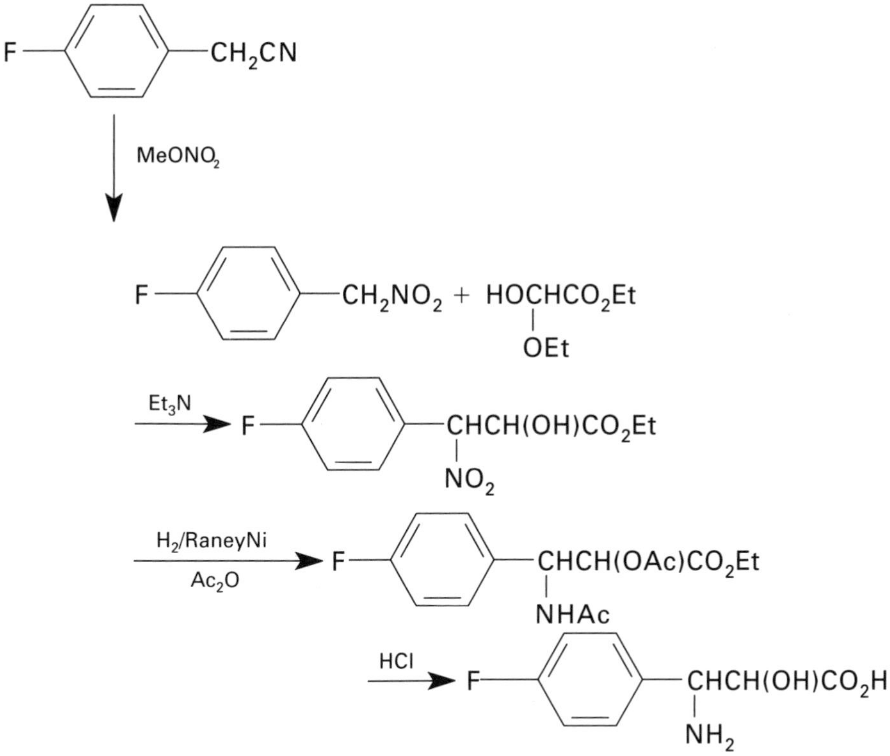

(46)

S-Benzyl-3-(4′-fluorophenyl)cysteine (**181**) and *S*-ethyl-3-(4′-fluorophenyl) cysteine (**182**) were prepared by addition of the appropriate thiol to ethyl 2-benzamido-4′-fluorocinnamate, followed by hydrolytic removal of the protecting groups. Compound **181** was debenzylated by the action of sodium in liquid ammonia, affording 3-(4′-fluorophenyl)cysteine (**183**) in unstated yield [61].

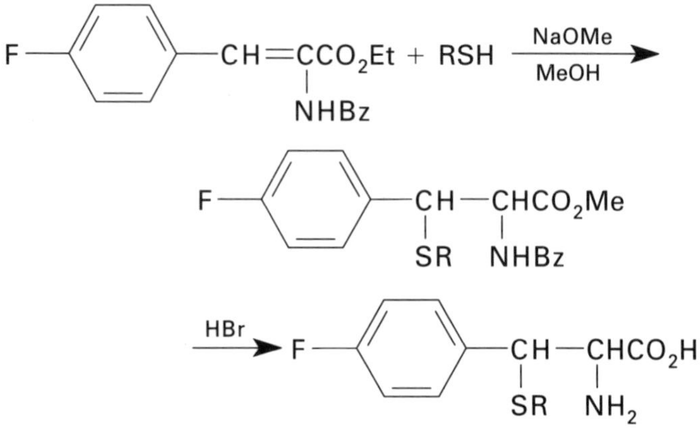

(**181**) R = Bzl
(**182**) R = Et

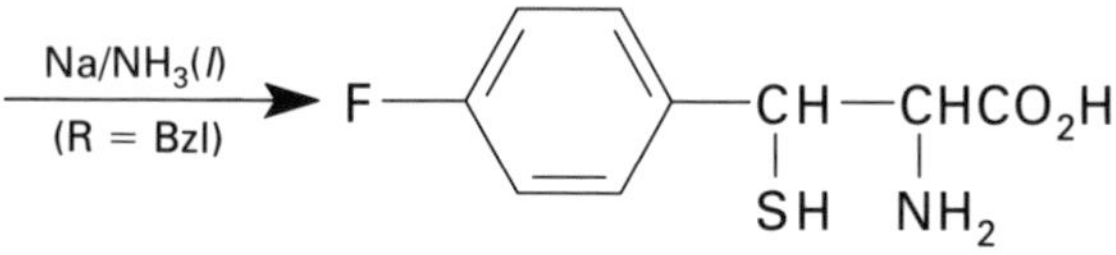

(**183**)

Reaction of diethyl 2-fluoro-3-oxosuccinate with aldehydes, suggested as a general route to 2,3-didehydro-2-fluoro carboxylic esters [145], has been adapted for the synthesis of 4-amino-2-fluorocrotonic acid (**35**) and its homologue, 5-amino-2-fluoro-2-pentenoic acid (**184**). On condensation of the oxo ester sodium enolate with either phthalimidoacetaldehyde or 3-phthalimido-propanal, the ethyl 2-fluoro-ω-phthalimidoalk-2-enoates were formed, which were in turn hydrolysed to the desired amino acids [146].

$$\mathrm{PhtN(CH_2)_{\it n}CHO + EtO_2C\underset{\displaystyle ONa}{\underset{|}{C}}{=}CFCO_2Et \longrightarrow}$$

$$\mathrm{PhtN(CH_2)_{\it n}CH{=}CFCO_2Et \xrightarrow{HCl} H_2N(CH_2)_{\it n}CH{=}CFCO_2H}$$

(**35**) $n = 1$
(**184**) $n = 2$

Two syntheses of 4,4,4-trifluorothreonine (**28**) should also be included in this section. Both diastereomers of **28** were prepared by a reaction sequence starting with Claisen acylation of N,N-dibenzylglycine ethyl ester by ethyl trifluoroacetate. The resulting 3-oxo ester was reduced by borohydride to ethyl 2-(N,N-dibenzylamino)-3-hydroxy-4,4,4-trifluorobutyrate of *threo* configuration, which on hydrogenolytic debenzylation afforded the ethyl ester of *threo*-**28**. Saponification of the N-protected ester yielded, however, a mixture of both *threo* and *allo* forms of N,N-dibenzyl-4,4,4-trifluorothreonine; after being separated by chromatographic means, both forms were debenzylated to *threo*-**28** and *allo*-**28**, respectively [128].

On the other hand, only the *allo* isomer of **28** was produced from the lithium derivative of ethyl 3-hydroxy-4,4,4-trifluorobutyrate, which on reaction with bis-*tert*-butyl azodicarboxylate (BTBAD), followed by removal of the protecting groups, gave 2-hydrazino-3-hydroxy-4,4,4-triifluorobutyric acid. (According to the authors' claim, this acid was the first fluorinated 2-hydrazino acid known.) Hydrogenolysis of the N—N bond then furnished *allo*-**28** [147].

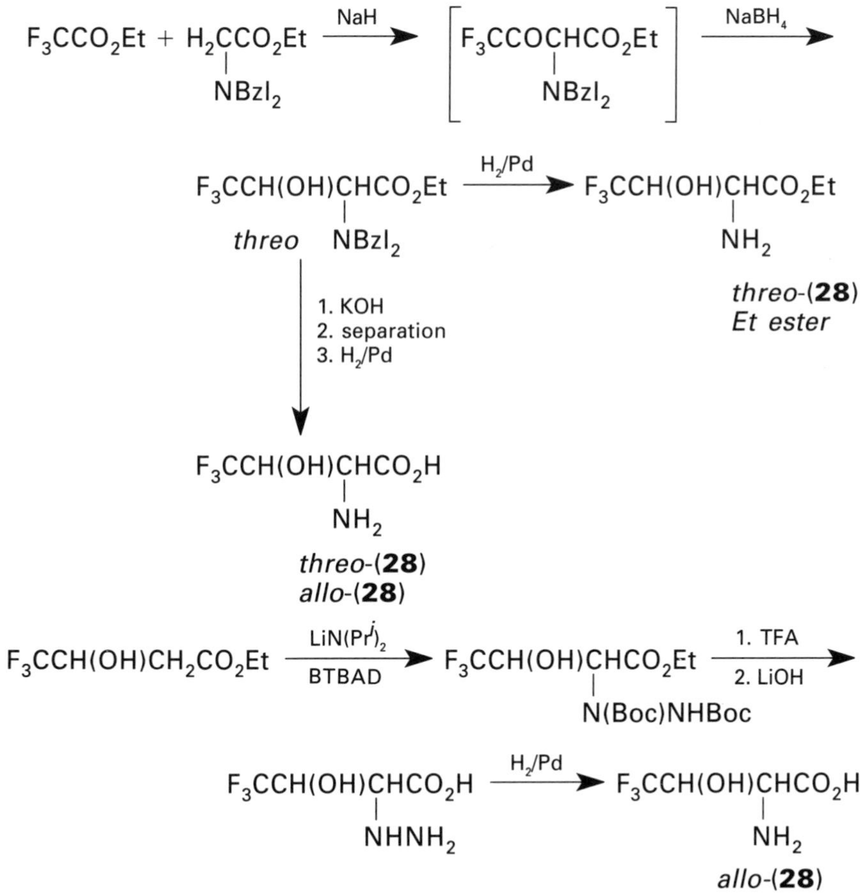

As far as could be established, the Beckmann rearrangement has been applied only once in the synthesis of fluorine-containing amino acids. In the preparation of 6-amino-4-(trifluoromethyl)hexanoic acid (185) 4-(trifluoromethyl)phenol was transformed into 4-(trifluoromethyl)cyclohexanone, the oxime of which was rearranged to 4-(trifluoromethyl)caprolactam. Acid hydrolysis of the lactam gave 185 [148].

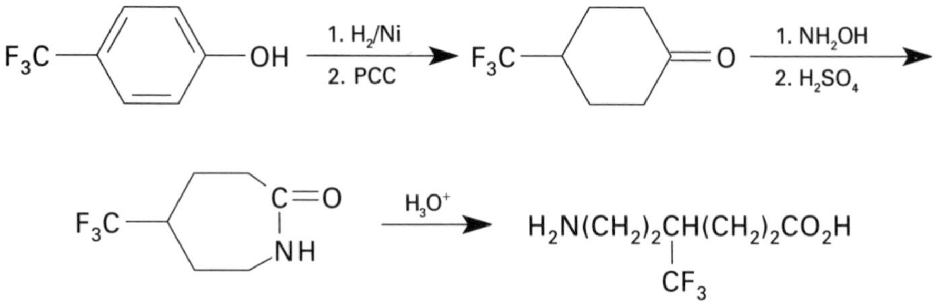

(185)

Finally, attention should be paid to the unique application of a Friedel–Crafts-type reaction in amino acid chemistry. On alkylation of fluorobenzene with 2-benzamido-2-hydroxyacetic acid, N-benzoyl-2-(4'-fluorophenyl)glycine was formed in high yield, which on hydrolysis furnished 2-(4'-fluorophenyl)glycine (45) [149].

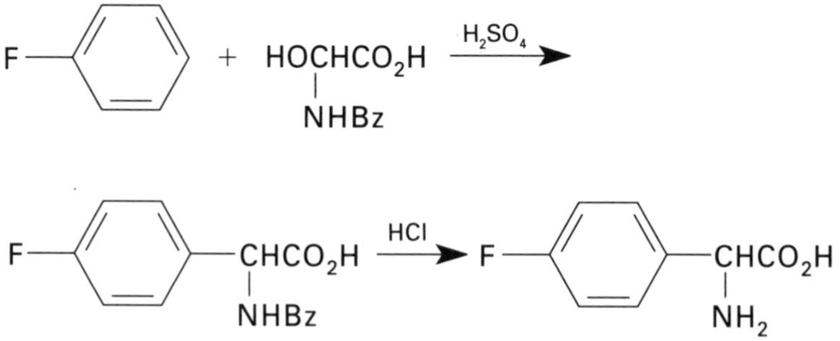

(45)

Table 1.1   Fluorine-containing amino acids prepared by classical synthetic methods

Amino acid	No.	Section	Total yield (%)/ No. of steps	Reference
2-Alanine				
3,3-difluoro-	**120**	1.6	36/2	108,109
3-fluoro-	**1**	1.2	23/2	1
		1.2	13/2	2
		1.2	—/4	3
		1.5	2.7/4	61
3-fluoro-2-*d*	**1-2-*d***	1.8	70/1	124
3,3,3-trifluoro-	**59**	1.3	52/3	51
		1.3	60/2	52
		1.7	47/3	118
3-Alanine				
2-fluoro-	**8**	1.2	6.4/4	4
		1.2	62/4	5
2-Aminobut-3-enoic acid				
2-(fluoromethyl)-3-methyl-	**139**	1.7	—/4	112
2-Aminobutyric acid				
4-chloro-4,4-difluoro-	**13**	1.2	59/3	8
4-chloro-4-fluoro-	**12**	1.2	57/3	8
3-fluoro-	**3**	1.2	25/2	2
4-fluoro-	**2**	1.2	57/2	1
		1.6	—/2	53, 80
3-fluoro-3-phenyl-	**56**	1.3	46/2	47, 48
4,4,4-trifluoro-	**9**	1.2	15/4	6
		1.2	17–29/4	15
		1.6	38/2	82
3-Aminobutyric acid				
3-methyl-4,4,4-trifluoro-,				
Me ester	**15**	1.2	61/3	10
4,4,4-trifluoro-	**10**	1.2	39/1	6
4-Aminobutyric acid				
2-fluoro-	**149**	1.7	31/2	123
4-Aminocrotonic acid				
2-fluoro-	**35**	1.2	23/3	28
		1.10	14/2	146
2-Aminodecanoic acid				
N-acetyl-8,8,9,9,10,10,10-	**104**	1.6	23/4	16
heptafluoro-				
2-Aminoheptanoic acid				
3-fluoro-	**6**	1.2	3/2	2
5,5,6,6,7,7,7-heptafluoro-	**23**	1.2	19/4	16
4-Aminohexanoic acid				
2-carboxy-2-fluoro-	**148**	1.7	45/2	123
2-fluoro-·	**147**	1.7	31/3	123
6-Aminohexanoic acid				
4-(trifluoromethyl)-	**185**	1.10	6.7/5	148

(*continued overleaf*)

Table 1.1 (*continued*)

Amino acid	No.	Section	Total yield (%)/ No. of steps	Reference
2-Aminohex-4-enedioic acid				
5-fluoro-	**106**	1.6	44/2	28
4-Aminohex-5-enoic acid				
6,6-difluoro-	**38**	1.2	3.6/5	31
5,6,6-trifluoro-	**39**	1.2	8/5	31
2-Aminoisobutyric acid				
3-bromo-3,3-difluoro-	**123**	1.6	—/2	110
3-chloro-3-fluoro-	**121**	1.6	—/3	112
3,3-difluoro-	**122**	1.6	60/3	110, 111
3,3′-difluoro-	**48**	1.3	16/2	42
3-fluoro-	**47**	1.3	Low/2	42
		1.3	64/2	43
		1.6	50/2	110, 111
3,3,3-trifluoro-	**49**	1.3	20/2	43
		1.4	13/2	53
		1.6	—/2	110
3-Aminoisobutyric acid				
3′,3′,3′-trifluoro-	**14**	1.2	100/1	9
2-Aminooctanoic acid				
6,6,7,7,8,8,8-heptafluoro-	**24**	1.2	22/5	16
6,6,7,7,8,8,8-heptafluoro-				
-4-hydroxy-	**103**	1.6	63/5	16
4-Aminopentanoic acid				
5-fluoro-(*S*)	**173**	1.9	44/5	141
5-Aminopent-2-enoic acid				
2-fluoro-	**184**	1.10	23/2	146
2-Aminopent-4-enoic acid				
2-(fluoromethyl)-3-methyl-	**140**	1.7	—/7	112
2-Aminopimelic acid				
6-fluoro-	**31**	1.2	14/4	23, 24
Arginine				
2-(difluoromethyl)-	**152**	1.9	50/1	110
4-fluoro-	**151**	1.9	37/3	26
Asparagine				
3,3-difluoro-	**155**	1.9	61/2	129
3-fluoro-(*erythro*)	**154**	1.9	75/2	130
Aspartic acid				
3,3-difluoro-	**150**	1.8	22/4	129
3-fluoro-(*threo*)	**29**	1.2	27/3	21
3-fluoro-(*erythro*)		1.9	40/3	130
2-(fluoromethyl)-	**141**	1.7	—/7	112
Citrulline				
4-fluoro-	**153**	1.9	50/3	26
Cysteine				
S-benzyl-3-(4′-fluorophenyl)-	**181**	1.9	79/2	61
S-ethyl-3-(4′-fluorophenyl)-	**182**	1.9	36/2	61
3-(4′-fluorophenyl)-	**183**	1.9	—/3	61

Table 1.1 (*continued*)

Amino acid	No.	Section	Total yield (%)/ No. of steps	Reference
2,6-Diaminopimelic acid				
4-fluoro-	**32**	1.2	0.16/8	23, 24
		1.2	4.1/7	25
Glutamic acid				
2-(chlorofluoromethyl)-	**136**	1.6	5.4/6	112
2-(chlorofluoromethyl)- (*RR, SS*)		1.6	1.2/6	114
2-(chlorofluoromethyl)- (*RS, SR*)		1.6	1.7/6	114
4,4-difluoro-	**30**	1.2	19/6	22
2-(difluoromethyl)-	**128**	1.6	14/2	113
4-fluoro-	**105**	1.6	34/2	86, 87
		1.6	70/3	88
		1.6	6.8/5	89
		1.7	56/2	121
4-fluoro- (L-*threo*)	**105**	1.9	14/5	131, 132
4-fluoro- (D-*threo*)		1.9	4.5/5	131, 132
4-fluoro- (L-*erythro*)		1.9	2.2/5	131, 132
4-fluoro- (D-*erythro*)		1.9	18/5	131, 132
2-(fluoromethyl)-	**60**	1.4	13/4	54
3-(fluoromethylene)-	**37**	1.2	1.2/10	30
Glutamine				
4-fluoro-	**156**	1.9	36/3	28
4-fluoro- (*erythro*)		1.9	46/4	136
4-fluoro- (*threo*)		1.9	38/4	136
Glycine				
2-(2'-fluorophenyl)-	**43**	1.2	37/1	36
2-(3'-fluorophenyl)-	**44**	1.2	57/1	36
		1.2	59/1	37
		1.4	—/—	35
2-(4'-fluorophenyl)-	**45**	1.2	54/1	36
		1.2	50/1	37
		1.2	72/2	38
		1.4	—/—	56
		1.10	66/2	149
2-(1'-fluorocyclohexyl)-	**55**	1.3	50/2	47, 48
2-(1'-fluorocyclopentyl)-	**54**	1.3	47/2	47, 48
2-(1'-methyl-2',2',3',3'- tetrafluorocyclobutyl)-	**67**	1.4	33/2	53
Histidine				
2-(difluoromethyl)-	**135**	1.6	8.1/4	111
Homocysteine				
S-benzyl-2-(difluoromethyl)-	**163**	1.9	39/1	111
Isoglutamine				
4-fluoro-	**160**	1.9	31/4	28
4-fluoro- (*threo*)		1.9	59/2	136
		1.9	46/4	136, acc. to 28

(*continued overleaf*)

Table 1.1 (*continued*)

Amino acid	No.	Section	Total yield (%)/ No. of steps	Reference
Isoglutamine (*continued*)				
4-fluoro- (*erythro*)	**160**	1.9	62/2	136
		1.9	35/4	136, acc. to 28
Isoleucine				
3-fluoro-	**52**	1.3	57/3	47
4-fluoro-	**25**	1.2	4.7/2	17
5-fluoro- (DL-*threo*)	**145**	1.7	3.2/3	119
5-fluoro- (L-*threo*)		1.7	2.8/4	119
5-fluoro- (D-*threo*)		1.7	0.85/5	119
5-fluoro- (L-*erythro*)		1.7	0.52/4	119
5,5,5-trifluoro-	**26**	1.2	42/2	19
		1.3	21/3	19
		1.7	18/6	19
Leucine				
5,5'-difluoro-	**19**	1.2	2.7/6	1
5,5'-dihydroxy-4-fluoro-	**146**	1.7	8.3/4	122
4,5,5,5,5',5',5'-heptafluoro-	**22**	1.2	17–29/3	15
5,5,5,5',5',5'-hexafluoro-	**17**	1.2	0.5/9	14
5,5,5,-trifluoro-	**51**	1.5	53/2	9
		1.5	61/4	74–76
		1.8	29/5	127
5,5,5,-trifluoro- (2*S*, 4*S*)	**51**	1.3	4.5/6	46
5,5,5,-trifluoro- (2*S*, 4*R*)		1.3	2.2/7	46
Lysine				
5,5-difluoro-	**58**	1.3	13/10	22
		1.6	32/4	85
2-(difluoromethyl)-	**127**	1.6	40/4	111
5-fluoro-	**34**	1.2	14/6	26
Methionine				
2-(difluoromethyl)-	**129**	1.6	44/4	111
6,6,6-trifluoro-	**61**	1.4	4.6/5	55
Norleucine				
5,5-difluoro-	**102**	1.6	40/2	84
3-fluoro-	**5**	1.2	2.3/2	2
5-fluoro-	**99**	1.6	16/2	81
6-fluoro-	**100**	1.6	50/2	53
		1.6	47/1	80
4,4,5,5,6,6,6-heptafluoro-	**21**	1.2	17–29/3	15
5-methyl-6,6,6-trifluoro-	**101**	1.6	0.34/2	44
6,6,6-trifluoro-	**97**	1.5	36/4	9
		1.6	23/3	44
		1.6	49/2	82

Table 1.1 (*continued*)

Amino acid	No.	Section	Total yield (%)/ No. of steps	Reference
Norvaline				
3-ethyl-3-fluoro-	**53**	1.3	55/3	48
3-fluoro-	**4**	1.2	3/2	2
5-fluoro-	**98**	1.6	17/2	53, 80
4-fluoro-5-hydroxy-	**161**	1.9	45/1	28
4,4,5,5,5-pentafluoro-	**20**	1.2	17–29/4	15
5-(pentafluorophenyl)-	**92**	1.5	24/4	9
5,5,5-trifluoro-	**50**	1.3	4.6/2	44
		1.4	13/4	44
		1.6	46/2	82
Ornithine				
2-(chlorofluoromethyl)-	**125**	1.6	Low/2	112
		1.6	6/7	112
2-(chlorofluoromethyl)-(*RR, SS*)		1.7	3.8/8	114
2-(chlorofluoromethyl)-(*RS, SR*)		1.7	3.6/8	114
3,4-didehydro-2-(difluoro- methyl)-	**144**	1.7	1.4/9	49
3,4-didehydro-2-(fluoromethyl)-	**57**	1.3	14/5	49
		1.3	—/5	50
2-(difluoromethyl)-	**126**	1.6	48/4	111
		1.7	46/6	111
4-fluoro-	**33**	1.2	11/6	26, 27
2-(fluoromethyl)-	**124**	1.6	74/3	111
3-Phenyl-2-alanine				
4′-amino-3′-fluoro-	**165**	1.9	42/2	59
2-(chlorofluoromethyl)-	**131**	1.6	23/2	112
2-(chlorofluoromethyl)- 3′-hydroxy- (*RR, SS*)	**143**	1.7	5.2/7	114
2-(chlorofluoromethyl)- 3′-hydroxy- (*RS, SR*)		1.7	5.8/7	114
2-(chlorofluoromethyl)- 3′-methoxy-	**142**	1.7	18/6	112
2′-chloro-5′-(trifluoromethyl)-	**107**	1.6	—/2	94, 95
4′-chloro-3′-(trifluoromethyl)-	**108**	1.6	—/2	94, 95
3′,5′-dichloro-2′,4′,6′-trifluoro-	**78**	1.5	—/3	65
2′,4′-difluoro-	**71**	1.5	—/3	65
2′,5′-difluoro-	**72**	1.5	—/3	65
		1.6	—/2	94, 95
2′,6′-difluoro-	**73**	1.5	—/3	65
3′,4′-difluoro-	**74**	1.5	—/3	65
3′,5′-difluoro-	**75**	1.5	—/3	65
2-(difluoromethyl)-	**132**	1.6	31/4	111
		1.7	68/6	117
3-(difluoromethyl)-	**82**	1.5	61/1	67
2-(difluoromethyl)-3′,4′- dihydroxy-	**133**	1.6	55/4	111

(*continued overleaf*)

Table 1.1 (*continued*)

Amino acid	No.	Section	Total yield (%)/ No. of steps	Reference
3-Phenyl-2-alanine (*continued*)				
2-(difluoromethyl)-3′,4′- dimethoxy-	**134**	1.6	34/4	111
4′-(difluoromethylthio)-	**115**	1.6	19/2	70
3′,4′-dihydroxy-2′-fluoro-	**112**	1.6	33/4	101
4′,5′-dihydroxy-2′-fluoro-	**138**	1.6	55/2	116
3′,4′-dihydroxy-3-methyl- 2-(trifluoromethyl)-	**65**	1.4	—/2	58
3′,4′-dihydroxy-2- (trifluoromethyl)-	**64**	1.4	—/2	58
2′-fluoro-	**68**	1.5	12/3	60
		1.5	12/5	62
		1.6	39/4	62
		1.6	44/3	93
		1.6	—/2	94, 95
3-fluoro- (*erythro*)	**137**	1.6	14/3	115
		1.8	52/1	125, 126
3′-fluoro-	**69**	1.5	78/1	59
		1.5	24/3	60
		1.5	13/5	62
		1.6	43/4	62
		1.6	—/2	94, 95
4′-fluoro-	**70**	1.5	17/3	60
		1.5	51/2	61
		1.5	16/4	62
		1.6	40/4	62
		1.6	46/3	93
		1.6	—/2	94, 95
		1.6	50/2	96
		1.6	50/4	97
2′-fluoro-3′,4′-dimethoxy-	**113**	1.6	35/3	101
2′-fluoro-4′-methoxy-	**110**	1.6	33/3	98
2-(fluoromethyl)-	**130**	1.6	75/2	111
3-(fluoromethylene)-3′-hydroxy-	**36**	1.2	32/8	29
3′-fluoro-4′-nitro-	**164**	1.9	69/1	59
3′-hydroxy-2-(trifluoromethyl)-	**63**	1.4	—/2	58
2′,3′,4′,5′,6′-pentafluoro-	**40**	1.2	—/4	32
		1.5	32/2	32
		1.5	77/3	63, 64
		1.6	52/3	32
		1.6	32/2	92
2′,3′,4′,6′-tetrafluoro-	**76**	1.5	—/3	65
2′,3′,5′,6′-tetrafluoro-	**77**	1.5	—/3	65
2′-(trifluoromethyl)-	**79**	1.5	6.5/4	66
3-(trifluoromethyl)-	**83**	1.5	34/2	67
3′-(trifluoromethyl)-	**80**	1.5	28/4	66
		1.6	27/2	94, 95
		1.6	26/4	66

Table 1.1 (*continued*)

Amino acid	No.	Section	Total yield (%)/ No. of steps	Reference
3-Phenyl-2-alanine (*continued*)				
4'-(trifluoromethyl)-	**81**	1.5	19/4	66
4'-(trifluoromethylthio)-	**116**	1.6	23/2	70
2-(trifluoromethyl)-3',4',5'-				
trihydroxy-	**66**	1.4	—/2	58
3-Phenyl-3-alanine				
2'-(difluoromethoxy)-	**177**	1.10	40/1	142
4'-(difluoromethoxy)-	**178**	1.10	52/1	142
4'-fluoro-	**174**	1.10	20/1	61
		1.10	63/1	142
3'-fluoro-4'-hydroxy-	**176**	1.10	32/2	142
3'-fluoro-4'-methoxy-	**175**	1.10	58/1	142
3-Phenylisoserine				
4'-fluoro-	**46**	1.2	56/3	40
4'-fluoro- (*threo*)		1.10	16/4	39
4'-fluoro (*erythro*)		1.2	52/3	39
3-Phenylserine				
2'-fluoro-	**179**	1.10	—/1	144
4'-fluoro-	**180**	1.10	47/1	61
		1.10	—/1	144
4'-fluoro- (*threo*)		1.10	45/1	39
4'-fluoro- (*erythro*)		1.10	43/2	39
Proline				
4-fluoro-	**161**	1.9	—/1	28, 138
Pyroglutamic acid				
2-(difluoromethyl)-	**158**	1.9	100/1	113
4-fluoro-	**157**	1.9	55/2	28
		1.9	—/1	90
4-fluoro- (*cis*)		1.9	72/2	91
4-fluoro- (*trans*)		1.9	76/2	91
2-(trifluoromethyl)-	**159**	1.9	70/1	137
Threonine				
4-fluoro- (*allo?*)	**27**	1.2	4.7/2	1
4,4,4-trifluoro-	**28**	1.8	4.8/4	20
		1.8	40/4	128
4,4,4-trifluoro- (*allo*)		1.2	1.4/4	20
		1.10	—/4	128
		1.10	26/3	147
4,4,4-trifluoro-, Et ester (*threo*)		1.10	59/3	128
Thyronine				
3'',5''-difluoro-	**172**	1.9	—/1	69
3'',5''-difluoro-3',5'-diiodo-	**95**	1.5	37/2	69
3',5'-diiodo-3''-fluoro-	**94**	1.5	40/2	73
2'-fluoro-	**93**	1.5	4.9/4	59
3''-fluoro-	**152**	1.9	—/1	73
3'-fluoro-5'-iodo-	**168**	1.9	13/4	99
3'-fluoro-5'-iodo-*O*-methyl-	**169**	1.9	32/4	99
3''-fluoro-3',5',5''-triiodo-	**171**	1.9	—/1	73

(*continued overleaf*)

Table 1.1 (*continued*)

Amino acid	No.	Section	Total yield (%)/ No. of steps	Reference
Tryptophan				
4-fluoro-	**119**	1.6	65/2	106
5-fluoro-	**117**	1.6	—/3	102
		1.6	70/2	106
		1.6	41/4	107
6-fluoro-	**118**	1.6	62/3	103
		1.6	76/2	106
4,5,6,7-tetrafluoro-	**96**	1.5	50/4	9
		1.6	53/2	104
		1.6	53/3	105
Tyrosine				
3′-bromo-5′-fluoro-	**166**	1.9	78/1	120
		1.9	76/1	140
3′-5′-difluoro-	**87**	1.5	32/3	68
2-(difluoromethyl)-	**62**	1.4	—/3	57
		1.6	—/4	111
3-(difluoromethyl)-	**84**	1.5	—/2	67
O-(difluoromethyl)-	**88**	1.5	—/3	71
2′-fluoro-	**109**	1.6	27/4	98
3′-fluoro-	**86**	1.5	Route 1, 41/3	59
		1.5	Route 3, 57/2	59
		1.5	49/2	68
		1.5	21/5	69
		1.6	63/2	99
		1.7	37/3	120
		1.9	16/3	59
3′-fluoro-5′-iodo-	**167**	1.9	47/1	68
		1.9	56/1	99
		1.9	60/1	120
		1.9	76/1	140
O-(2,2,3,4,4,4-hexafluoropropyl)-	**91**	1.5	—/2	71
O-(pentafluoroethyl)-	**90**	1.5	43/2	70, 71
2′,3′,5′,6′-tetrafluoro-	**111**	1.6	55/3	32, 100
O-(2,2,2-trifluoroethyl)-	**114**	1.6	31/2	70
2′-(trifluoromethyl)-	**41**	1.2	17/3	33, 34
3-(trifluoromethyl)-	**85**	1.5	—/2	67
3′-(trifluoromethyl)-	**42**	1.2	41/4	35
0-(trifluoromethyl)-	**89**	1.5	—/2	71
Valine				
4,4′-difluoro-	**18**	1.2	8/4	1
3-fluoro-	**7**	1.2	39/2	2
		1.3	58/3	47
		1.3	55/3	48
4,4,4,4′,4′,4′-hexafluoro-	**16**	1.2	53/5	11
		1.2	46/3	12
		1.2	22/3	13
4,4,4-trifluoro-	**11**	1.2	3.5/5	7
		1.6	12/7	83

1.11 REFERENCES

1. H. Lettré and U. Wölcke, *Justus Liebigs Ann. Chem.*, **708**, 75 (1967).
2. H. Gershon, M. W. McNeil and E. D. Bergmann, *J. Med. Chem.*, **16**, 1407 (1973).
3. I. I. Gerus, Yu. D. Yagupol'skii and V. P. Kukhar', *6th All-Union Conference on the Chemistry of Fluorinated Organic Compounds, Novosibirsk, 1990, Abstracts*, p. 245.
4. E. D. Bergmann and S. Cohen, *J. Chem. Soc.*, 4669 (1961).
5. V. Tolman and K. Vereš, *Collect. Czech. Chem. Commun.*, **29**, 234 (1964).
6. H. M. Walborsky and M. E. Baum, *J. Org. Chem.*, **21**, 538 (1956).
7. D. F. Loncrini and H. M. Walborsky, *J. Med. Chem.*, **7**, 369 (1964).
8. F. Heinzer and P. Martin, *Helv. Chim. Acta*, **64**, 1379 (1981).
9. I. Ojima, K. Kato and K. Nakanashi, *J. Org. Chem.*, **54**, 4511 (1989).
10. I. L. Knunyants and Yu. A. Cheburkov, *Izv. Akad. Nauk SSSR. Ser. Khim.*, 1057 (1961); *Chem. Abstr.*, **55**, 27046i (1961).
11. I. L. Knunyants and Yu. A. Cheburkov, *Izv. Akad. Nauk SSSR, Ser. Khim.*, 2162 (1960); *Chem. Abstr.*, **55**, 16412i (1961).
12. W. H. Vine, K. Hsieh and G. R. Marshall, *J. Med. Chem.*, **24**, 1043 (1981).
13. T. -Y. Chen, N. P. Gambaryan and I. L. Knunyants, *Dokl. Akad. Nauk SSSR*, **133**, 1113 (1960); *Chem. Abstr.*, **54**, 24385g (1960).
14. J. Lazar and W. A. Sheppard, *J. Med. Chem.*, **11**, 138 (1968).
15. Y. Maki and K. Inukai, *J. Synth. Org. Chem. Jpn.*, **34**, 722 (1976); *Chem. Abstr.*, **87**, 68600x (1977).
16. N. O. Brace, *J. Org. Chem.*, **32**, 430 (1967).
17. H. Gershon, L. Shanks and D. D. Clarke, *J. Pharm. Sci.*, **67**, 715 (1978).
18. D. Butina and M. Hudlický, *J. Fluorine Chem.*, **16**, 301 (1980).
19. N. Muller, *J. Fluorine Chem.*, **36**, 163 (1987).
20. H. M. Walborsky and M. E. Baum, *J. Am. Chem. Soc.*, **80**, 187 (1958).
21. M. Hudlický, *J. Fluorine Chem.*, **40**, 99 (1988).
22. O. Kitagawa, A. Hashimoto, Y. Kobayashi, *et al.*, *Chem. Lett.*, 1307 (1990).
23. B. Cavalleri, E. Bellasio and E. Testa, *Gazz. Chim. Ital.*, **96**, 227 (1966).
24. B. Cavalleri, E. Bellasio and E. Testa, *Gazz. Chim. Ital.*, **96**, 253 (1966).
25. J. Hanuš, V. Tolman and K. Vereš, *Collect. Czech. Chem. Commun.*, **38**, 1212 (1973).
26. V. Tolman and J. Beneš, *J. Fluorine Chem.*, **7**, 397 (1976).
27. V. Tolman and K. Vereš, *Collect. Czech. Chem. Commun.*, **37**, 2962 (1972).
28. V. Tolman and K. Vereš, *Collect. Czech. Chem. Commun.*, **32**, 4460 (1967).
29. I. McDonald, M. Lacoste, P. Bey, *et al.*, *J. Am. Chem. Soc.*, **106**, 3354 (1984).
30. I. McDonald, M. Palfteyman, M. Junge, *et al.*, *Tetrahedron Lett.*, 4091 (1985).
31. M. Kolb, J. Barth, J. -G. Heydth, *et al.*, *J. Med. Chem.*, **30**, 267 (1987).
32. R. Filler, N. R. Ayyangar, W. Gustowski, *et al.*, *J. Org. Chem.*, **34**, 534 (1969).
33. R. Filler and H. Novar, *Chem. Ind. (London)*, 468 (1960).
34. R. Filler and H. Novar, *J. Org. Chem.*, **26**, 2707 (1961).
35. R. Filler, B. T. Khan and C. W. McMullen, *J. Org. Chem.*, **27**, 4660 (1962).
36. E. L. Campere, Jr, and D. A. Weinstein, *Synthesis*, 852 (1977).
37. D. Landini, F. Montanari and F. Rolla, *Synthesis*, 26 (1979).
38. D. Landini and M. Penso, *J. Org. Chem.*, **56**, 420 (1961).
39. N. Blažević and F. Zymalkowski, *Arch. Pharm.*, **308**, 541 (1975).
40. E. Kamandi, A. W. Frahm and F. Zymalkowski, *Arch. Pharm.*, **308**, 135 (1975).
41. A. Strecker, *Justus Liebigs Ann. Chem.*, **75**, 27 (1850).
42. E. D. Bergmann and A. Shani, *J. Chem. Soc.*, 3462 (1963).
43. H. N. Christensen and D. L. Oxender, *Biochim. Biophys. Acta*, **74**, 386 (1963).

44. H. M. Walborsky, M. Baum and D. F. Loncrini, *J. Am. Chem. Soc.*, **77**, 3637 (1955).
45. K. Weinges, G. Graab, D. Nagel, *et al.*, *Chem. Ber.*, **104**, 3594 (1971).
46. K. Weinges and E. Kramm, *Justus Liebigs Ann. Chem.*, 90 (1985).
47. A. I. Ayi, M. Remli and R. Guedj, *Tetrahedron Lett.*, **22**, 1505 (1981).
48. A. I. Ayi and R. Guedj, *J. Fluorine Chem.*, **24**, 137 (1984).
49. P. Bey, G. Gerhart, V. Van Dorsselaer, *et al.*, *J. Med. Chem.*, **26**, 1551 (1983).
50. F. Gerhart, *Eur. Pat. Appl.*, 46710; *Chem. Abstr.*, **97**, 38498j (1982).
51. F. Weygand, W. Steglich and F. Fraunberger, *Angew. Chem.*, **71**, 822 (1967); *Agnew. Chem., Int. Ed. Engl.* **6**, 808 (1967).
52. A. Uskert, Á. Néder and E. Kasztreiner, *Magy. Kem. Foly.*, **79**, 333 (1973); *Chem. Abstr.*, **79**, 79147r (1973).
53. J. F. Lontz and M. S. Raasch, *US Pat.*, 2662915; *Chem. Abstr.*, **48**, 12795a (1954).
54. D. Kuo and R. R. Rando, *Biochemistry*, **20**, 506 (1981).
55. R. L. Dannley and R. G. Taborsky, *J. Org. Chem.*, **22**, 1275 (1957).
56. A. H. Neims, D. C. De Luca and L. Hellerman, *Biochemistry*, **5**, 203 (1966).
57. J. Kollonitsch, A. A. Patchett and S. Marburg, *Eur. Pat. Appl.* 7600; *Chem. Abstr.*, **93**, 114978a (1980).
58. M. Sletzinger and W. A. Gaines, *US Pat.*, 3046300; *Chem. Abstr.*, **57**, 16740h (1962).
59. G. Schiemann and W. Winkelmüller, *J. Prakt. Chem.*, **135**, 101 (1932).
60. G. Schiemann and W. Roselius, *Chem. Ber.*, **65**, 1439 (1932).
61. C. -Y. Yüan, C. -N. Chang and I. -F. Yeh, *Yao Hsüeh Hsüeh Pao*, 7, 237 (1959); *Chem. Abstr.*, **54**, 12096i (1960).
62. E. L. Bennett and C. Niemann, *J. Am. Chem. Soc.*, **72**, 1800 (1950).
63. G. V. Shishkin and V. P. Mamaev, *Izv. Akad. Nauk SSSR, Ser. Khim.*, 934 (1965); *Chem. Abstr.*, **63**, 5734g (1965).
64. G. V. Shishkin and V. P. Mamaev, *Zh. Obsch. Khim.*, **36**, 660 (1966); *Chem. Abstr.*, **65**, 9010h (1966).
65. A. T. Prudchenko, *Izv. Sib. Otd. Akad. Nauk SSSR, Ser. Khim. Nauk.*, (14), 95 (1970); *Chem. Abstr.*, **75**, 6265k (1971).
66. R. Filler and H. Novar, *J. Org. Chem.*, **25**, 733 (1960).
67. M. T. Kolycheva, Yu. L. Yagupol'skii, L. M. Zaitsev, *et al.*, *Khim.-Farm. Zh.*, **22**, 159 (1988); *Chem. Abstr.*, **109**, 213r (1988).
68. C. English, Jr, J. E. Mead and C. Niemann, *J. Am. Chem. Soc.*, **62**, 350 (1940).
69. C. Niemann, A. A. Benson and J. F. Mead, *J. Am. Chem. Soc.*, **63**, 2204 (1941).
70. M. T. Kolycheva, Yu. L. Yagupol'skii, I. I. Gerus, *et al.*, *Zh. Org. Khim.*, **25**, 1306 (1989); *Chem. Abstr.*, **112**, 179728z (1970).
71. Yu. L. Yagupol'skii, *Aromatic and Heterocyclic Compounds wiith Fluorine-containing Substituents*, Naukovaya Dumka, Kiev, 1987, p. 320.
72. E. Baltazzi and R. Robinson, *Chem. Ind. (London).* 191 (1954).
73. C. Niemann, J. F. Mead and A. A. Benson, *J. Am. Chem. Soc.*, **63**, 609 (1941).
74. Sagami Chemical Research Centre, *Jpn. Kokai*, 81164173; *Chem. Abstr.*, **96**, 142836b (1982).
75. Sagami Chemical Research Centre, *Jpn. Kokai*, 81161340; *Chem. Abstr.*, **96**, 180787a (1982).
76. Sagami Chemical Research Centre, *Jpn. Kokai*, 81164151; *Chem. Abstr.*, **96**, 181623f (1982).
77. V. S. Brovarets, O. P. Lobanov, A. A. Kisilenko, *et al.*, *Zh. Obsch. Khim.*, **56**, 1492 (1986); *Chem. Abstr.*, **107**, 96787p (1987).

78. V. A. Soloshonok, V. S. Brovarets and Yu. L. Yagupol'skii, *Zh. Org. Khim.*, **23**, 2475 (1987); *Chem. Abstr.*, **109**, 110292y (1988).
79. D. F. Loncrini and R. Filler, *Adv. Fluorine Chem.*, **6**, 43 (1970).
80. M. S. Raasch, *J. Org. Chem.*, **23**, 1567 (1958).
81. M. Hudlický and B. Kakáč, *Collect. Czech. Chem. Commun.*, **31**, 1101 (1966).
82. T. Tsushima, K. Kawada, S. Ishihara, *et al.*, *Tetrahedron*, **44**, 5375 (1988).
83. T. Taguchi, G. Tomizawa, M. Nakajima, *et al.*, *Chem. Pharm. Bull.*, **33**, 4077 (1985).
84. M. Hudlický, *Collect. Czech. Chem. Commun.*, **32**, 453 (1967).
85. F. N. Shirota, H. T. Nagasawa and J. A. Elberling, *J. Med. Chem.*, **20**, 1623 (1977).
86. M. Hudlický, *Tetrahedron Lett.*, 21 (1960).
87. M. Hudlický, *Collect. Czech. Chem. Commun.*, **26**, 1414 (1961).
88. V. Tolman, *J. Fluorine Chem.*, **60**, 179 (1993).
89. E. D. Bergmann and L. Chun-Hsu, *Synthesis*, 44 (1973).
90. J. C. Unkeless and P. Goldman, *Mol. Pharmacol.*, **7**, 293 (1971).
91. V. Tolman, V. Vlasáková and J. Němeček, *J. Fluorine Chem.*, **60**, 185 (1993).
92. R. Filler and W. Gustowski, *Nature (London)*, **205**, 1105 (1965).
93. M. D. Armstrong and J. D. Lewis, *J. Biol. Chem.*, **188**, 91 (1951).
94. Y. Maki, *Jpn. Kokai*, 77 14 741; *Chem. Abstr.*, **87**, P85237b (1977).
95. Y. Maki, S. Fujii and K. Inukai, *J. Synth. Org. Chem. Jpn.*, **35**, 421 (1977); *Chem. Abstr.*, **87**, 118052t (1977).
96. E. D. Bergmann, *J. Am. Chem. Soc.*, **74**, 4947 (1952).
97. T. Okuda and S. Tatsumi, *J. Biochem. (Tokyo)*, **44**, 631 (1957).
98. E. L. Bennett and C. Niemann, *J. Am. Chem. Soc.*, **72**, 1806 (1950).
99. A. Dibbo, L. Stephenson, T. Walker, *et al.*, *J. Chem. Soc.*, 2645, (1961).
100. R. Filler and H. H. Kang, *Chem. Commun.*, 627 (1965).
101. C. Kaiser and A. Burger, *J. Am. Chem. Soc.*, **79**, 4365 (1957).
102. E. Hoffmann, R. Ikan and A. B. Galun, *J. Heterocycl. Chem.*, **2**, 298 (1965).
103. E. D. Bergmann and E. Hoffmann, *J. Chem. Soc.*, 2827 (1962).
104. T. D. Petrova, T. I. Savchenko, T. F. Ardyukova, *et al.*, *Khim. Geterotsikl. Soedin.*, **7**, 213 (1971); *Chem. Abstr.*, **75**, 48812s (1971).
105. H. H. Rajh, J. H. Uizetter, L. W. Westerhaus, *et al.*, *Int. J. Pept. Protein Res.*, **14**, 68 (1979).
106. M. Bentov and C. Roffman, *Isr. J. Chem.*, **7**, 835 (1969).
107. H. Rinderknecht and C. Niemann, *J. Am. Chem. Soc.*, **72**, 2296 (1950).
108. T. Tsushima and K. Kawada, *Tetrahedron Lett.*, **26**, 2445 (1985).
109. T. Tsushima and K. Kawada, *Jpn. Pat.*, 60 172 194; *Chem. Abstr.*, **104**, 110175j (1986).
110. P. Bey and J. -P. Vevert, *Tetrahedron Lett.*, 1215 (1978).
111. P. Bey, J. -P. Vevert, V. Van Dorsselaer, *et al.*, *J. Org. Chem.*, **44**, 2732 (1979).
112. P. Bey, J. B. Ducep and D. Schirlin, *Tetrahedron Lett.*, **25**, 5657 (1984).
113. T. Tsushima, K. Kawada, O. Shiratori, *et al.*, *Heterocycles*, **23**, 45 (1985).
114. D. Schirlin, J. B. Ducep, S. Baltzer, *et al.*, *J. Chem. Soc., Perkin Trans. II.* 1053 (1992).
115. M. J. O'Donnell, C. L. Barney and J. R. McCarthy, *Tetrahedron Lett.*, **26**, 3067 (1985).
116. J. R. Grierson and M. J. Adam, *J. Labelled Compd. Radiopharm.*, **23**, 1019 (1986).
117. P. Bey and D. Schirlin, *Tetrahedron Lett.*, **52**, 5225 (1978).
118. Yu. L. Yagupol'skii, V. A. Soloshonok and V. P. Kukhar', *Zh. Org. Khim.*, **22**, 517 (1986); *Chem. Abstr.*, **105**, 115390f (1986).

119. M. Hudlický, V. Jelínek, K. Eisler, *et al., Collect. Czech. Chem. Commun.*, **35**, 498 (1970).
120. K. Kraft, *Chem. Ber.*, **84**, 150 (1951).
121. R. L. Buchanan, F. H. Dean and F. L. M. Pattison, *Can. J. Chem.*, **40**, 1571 (1966).
122. J. Dubois, C. Fourès, S. Bory, *et al., Tetrahedron*, **47**, 1001 (1991).
123. R. L. Buchanan and F. L. M. Pattison, *Can. J. Chem.*, **43**, 3466 (1965).
124. V. -H. Dolling, A. W. Douglas, E. J. J. Grabowski, *et al., J. Org. Chem.*, **43**, 1634 (1978).
125. S. Misaki and M. Suefuji, *Tetrahedron Lett.*, **21**, 3593 (1980).
126. T. Tsushima, K. Kawada, J. Nishikawa, *et al., J. Org. Chem.*, **49**, 1163 (1984).
127. O. M. Rennert and H. S. Anker, *Biochemistry*, **2**, 471 (1963).
128. C. Scholastico, E. Conca, L. Prati, *et al., Synthesis*, 850 (1985).
129. J. J. M. Hageman, M. J. Wanner, G. -J. Koomen, *et al., J. Med. Chem.*, **20**, 1677 (1977).
130. M. J. Wanner, J. J. M. Hageman, G. -J. Koomen, *et al., J. Med. Chem.*, **23**, 85 (1980).
131. M. Hudlický and J. Merola, *Tetrahedron Lett.*, **31**, 7403 (1990).
132. M. Hudlický, *J. Fluorine Chem.*, **60**, 193 (1993).
133. A. G. Avent, A. N. Bowler, P. M. Doyle, *et al., Tetrahedron Lett.*, **33**, 1509 (1992).
134. F. A. Davis, R. H. Jenkins, S. B. Awad, *et al., J. Am. Chem. Soc.*, **104**, 5412 (1982).
135. R. A. Boissonas, *Ger. Pat.*, 1 080 113; *Chem. Abstr.*, **56**, 4859 (1962).
136. V. Tolman, paper presented at the ACS 10th Winter Fluorine Conference. St Petersburg, FL, 1991; *Tetrahedron Lett.*, in preparation.
137. K. Burger and K. Gaa, *Chem.-Ztg.*, **114**, 101 (1990).
138. V. Tolman, unpublished results.
139. C. A. Dekker and J. S. Fruton, *J. Biol. Chem.*, **173**, 471 (1948).
140. K. Kraft and F. Denzel, *Ger. Pat.*, 895 292; *Chem. Abstr.*, **52**, 10184d (1958).
141. R. B. Silverman and M. A. Levy, *J. Org. Chem.*, **45**, 815 (1980).
142. V. P. Mamaev, *Zh. Obsch. Khim.*, **27**, 1290 (1957); *Chem. Abstr.*, **52**, 2748c (1958).
143. M. T. Kolycheva, I. I. Gerus and V. P. Kukhar', *Zh. Org. Khim.*, **25**, 2367 (1989); *Chem. Abstr.*, **113**, 23212m (1990).
144. E. J. Edmond, C. M. Volkmann and E. Beerstecher, Jr, *Proc. Soc. Exp. Biol. Med.*, **92**, 80 (1956); *Chem. Abstr.*, **50**, 12174c (1956).
145. E. D. Bergmann, I. Shahak, E. Sal'i, *et al., J. Chem. Soc. C*, 1232 (1968).
146. E. D. Bergmann and A. Cohen, *Tetrahedron Lett.*, 2085 (1965).
147. G. Guanti, L. Banfi and E. Narisano, *Tetrahedron*, **44**, 5553 (1988).
148. I. M. Zalesskaya, A. N. Blakitnyi, E. P. Saenko, *et al., Zh. Org. Khim.*, **16**, 1194 (1980); *Chem. Abstr.*, **93**, 167704g (1980).
149. W. -D. Sprung, M. Kobow and E. Schultz, *Pharmazie*, **44**, 540 (1989).

2 Preparation of Fluorine-containing Amino Acids by Methods of Organofluorine Chemistry

VALERY P. KUKHAR'

Institute of Bioorganic Chemistry and Petrochemistry, National Academy of Sciences of Ukraine, Murmanskaya 1, 253660 Kiev-94, Ukraine

2.1 INTRODUCTION

After traditional methods of amino acid synthesis, the direct introduction of fluorine or fluorine-containing groups into amino acids or their precursors by the methods of organofluorine chemistry is a second general route to synthesize fluorine-containing amino acids in racemic or enantiomeric states. Generally it is very attractive to use available amino acids as starting materials for transformations into desired derivatives of amino acids. Probably this methodology can also be applied to modifications of more complex compounds such as peptides or other amino acid conjugates (including natural substances).

For more than 50 years of the development of organofluorine chemistry a number of methods were elaborated to construct desired molecules with fluorine atoms or fluorine-containing groups based on specific reactions of fluorination reagents or peculiar properties of polyfluorine-containing compounds. Some of them were used in syntheses of fluorine-containing amino acids.

2.2 DIRECT FLUORINATION OF AMINO ACIDS BY FLUORINE

The direct substitution of hydrogen atoms with fluorine by the action of molecular fluorine is an attractive and the most promising method for the synthesis of fluorine-containing amino acids. This method provides the shortest route, by means of which it is possible to obtain enantiomeric fluorine-containing amino acids with retention of their optical activity, especially in the case of aromatic amino acids. On the other hand, there are important disadvantages of these methods. First there is low regioselectivity owing to a number of reactive C—H bonds in the molecules of amino acids,

Fluorine-containing Amino Acids: Synthesis and Properties Edited by V. P. Kukhar' and V. A. Soloshonok
© 1995 John Wiley & Sons Ltd

other functional groups and the high reactivity of fluorine, resulting in the formation of reaction product mixtures that hinder the resolution and isolation of the target substances. The high energy of the direct fluorination reaction results in destruction of C—C and other bonds in a molecule, so it is necessary to use a mixture of fluorine with inert diluents or solvents. Further, the use of fluorine and its equivalents needs the use of special equipment, which must be stable under the fluorination process conditions.

Nevertheless, there are a number of examples of the successful direct fluorination of amino acids predominantly for the preparation of ^{18}F-labelled amino acids. Many efforts have been concentrated on the synthesis of labelled 6-fluorodihydroxyphenylalanine (fluoro-DOPA) and related fluorinated DOPA derivatives. For example, the reaction of fluorine-18 with 3′,4′-dihydroxyphenylalanine (DOPA) in liquid hydrogen fluoride gave 2′-, 5′- and 6′-[^{18}F]fluoro-3′,4′-dihydroxyphenylalanines in a 35:5:60 ratio and a low isolated yield of desired labelled 6′-fluoro-DOPA (ca 3%) [1,2]. Application of boron trifluoride as a Lewis acid resulted in an improved radiochemical yield of up to 9% [3].

DOPA

The reaction of fluorine with protected DOPA (with an *N*-acetyl protecting group) yields a 1:1 mixture of 2′- and 6′-fluoro-3′,4′-dihydroxyphenylalanine, from which 6′-fluoroDOPA can be obtained by HPLC [4]. 3′-*O*-Methyl-6′-fluoro-DOPA has been obtained by this method in 9–15% yield [5]. The protection of the catechol ring is desirable in direct fluorination reactions. For this purpose *O*-acetyl and *O*-methyl protection of hydroxyl catechol groups was used [6].

2-F and 6-F

The synthesis of ^{18}F-labelled fluoro-DOPA using fluorine-18 is much more effective when carried out via fluorodemetallation reaction of the corresponding trimethylsilyl derivative [7].

The interaction of tyrosine with fluorine-18 in hydrogen fluoride yielded exclusively 3-fluorotyrosine in 46% yield, whereas the same reaction of O,N-diacetyl-L-tyrosine methyl ester gave a mixture of 2'- and 3'-fluorotyrosines in 26% yield and a 40:60 ratio. The reaction of O-acetyltyrosine without protection at amino and acid groups with acetyl [^{18}F]hypofluorite resulted in an 83:17 ratio of 2'- and 3'-isomers and a 20% total yield [8].

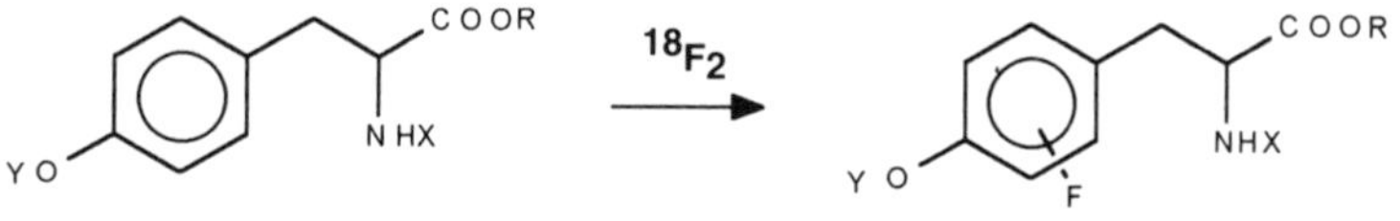

Y = H or Ac , R = H or Me   2'- and 3'-F-isomers

X = H or Ac

(S)-Fluorophenylalanine with an ^{18}F label in the benzene ring was also prepared by the action of fluorine in hydrogen fluoride on (S)-phenylalanine; the resulting mixture of labelled (S)-2'-, -3'- and -4'-fluorophenylalanines was separated by HPLC [9].

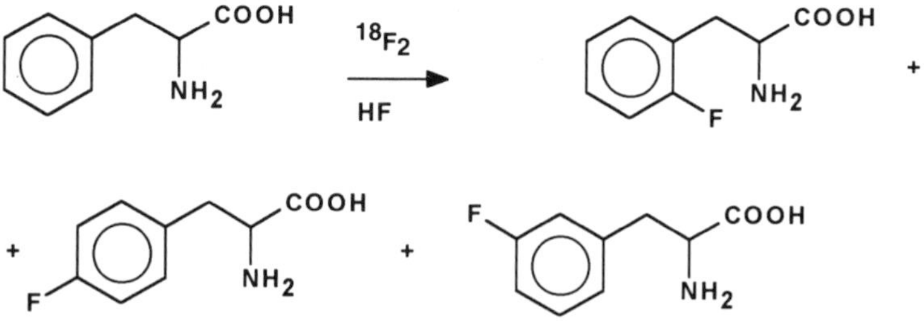

Probably the direct fluorination of the aromatic ring could be applied to obtain peptides with aromatic amino acid residues. For instance, it has been shown recently that the action of fluorine and sodium acetate in $CFCl_3$ at 20 °C on (O-Bzl-N-Boc)-Tyr-D-Ala-L-Phe-Gly-NH$_2$ peptide **2** gave the corresponding 3'-fluorotyrosine-containing peptide **1** in 65% yield [10].

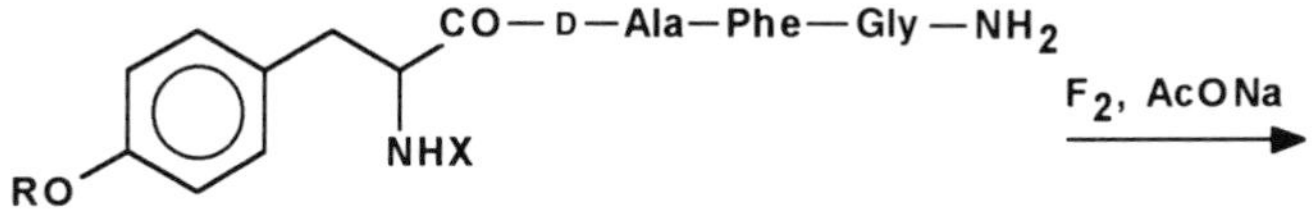

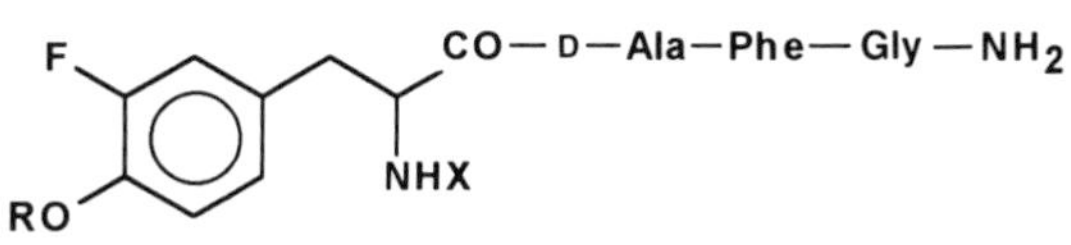

Electrophilic substitution in the aromatic ring by the action of fluorine was used for the syntheses of 2'-fluoro-DOPA [11], 3'-O-methyl-6'-fluoro-DOPA [12], 2'-fluorophenylalanine [13], 2'-fluoromethyl-6'-fluoro-DOPA [14], 2'- and 3'-fluorotyrosines [8, 13] and 5-hydroxy-6-fluoro- and 5-hydroxy-4-fluoro-tryptophans [15]. A number of cases of the application of fluorine, and also other fluorination reagents, for the preparation of ^{18}F-labelled fluorine-containing amino acids have been reviewed recently [16].

There are some examples of the utilization of direct fluorination with fluorine to prepare fluorinated precursors of fluorine-containing amino acids. For instance, in the synthesis of 3-fluorophenylalanine or 3-fluoro-p-nitrophenylalanine the first step is fluorination of phenylpyruvic acid or p-nitrophenylpyruvic acid methyl esters (3) with a 10% fluorine–nitrogen mixture, leading to 3-fluoro derivative (4) formation. The following asymmetric reductive amination with diborane gives the desired *erythro*-3-fluorophenylalanines [17–20].

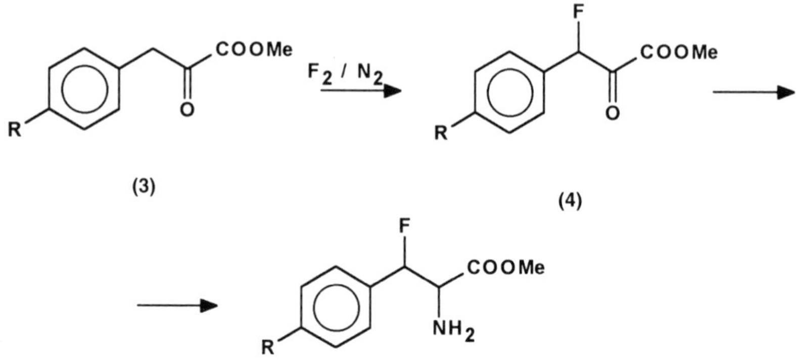

(3)   (4)

erythro,   R = H, NO$_2$

Molecular fluorine was used for desulfurization of the SH group in cysteine, which resulted in 3-fluoroalanine (5) [and 3,3-difluoroalanine (6) as a minor product] [21].

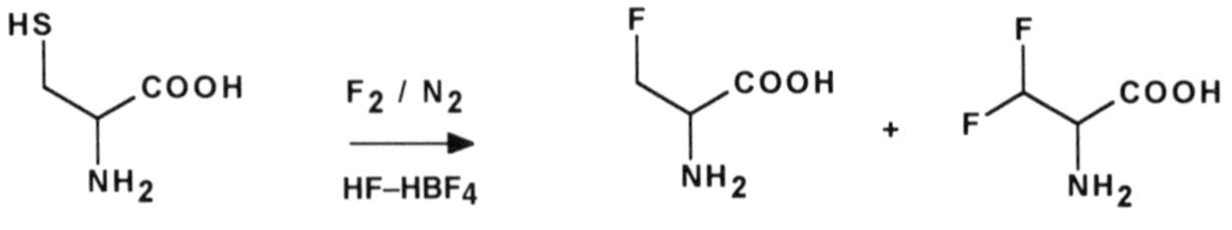

(5)   (6)

Utilization of the direct fluorination methods to obtain aliphatic fluorine-containing amino acids is represented by only a few examples. This is evidently connected both with the fact that molecules of aliphatic amino acids are far less stable towards the action of fluorine or its equivalents than aromatic amino acids and with a number of reactive C—H bonds causing the very low regioselectivity of these processes.

For the preparation of fluorinated aromatic amino acids it is preferable to use fluorine auxiliaries such as xenon difluoride (XeF_2) and acetyl hypofluorite (AcOF). The reaction of 3'-O-methyl-3',4'-dihydroxyphenylalanine with xenon difluoride and hydrogen fluoride as a catalyst was found to be a reasonable method for synthesizing 6'-fluoro-3',4'-dihydroxyphenylalanine, although for preparation of ^{18}F-labelled 6'-fluoro-DOPA this method gave poor radio-chemical results owing to exchange of ^{18}F in xenon difluoride with unlabelled fluorine in hydrogen fluoride [22, 23].

Xenon difluoride was used to synthesize the monofluoro ketone peptide isosteres **7** for the investigation of the interaction between these compounds and serine proteinases. The mild reaction of xenon difluoride with an O-silylated vinyl ether fragment offers an opportunity to introduce a fluorine atom into a required position in a chain in the presence of other reactive groups [24].

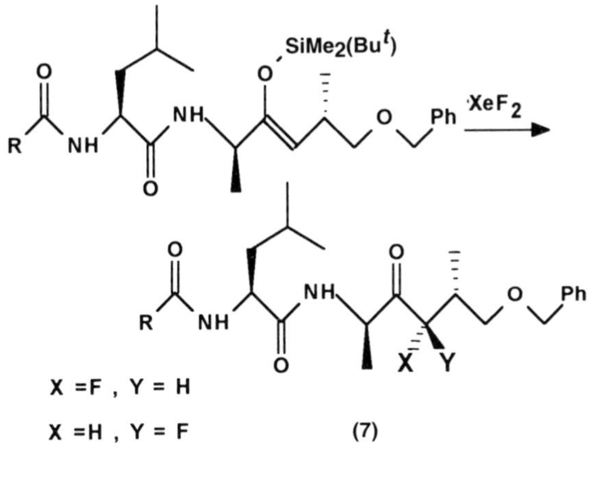

The difference in reactivity of methionine C—H bonds enabled a preparative method to be developed for the synthesis of S-fluoromethylmethionine and S-fluoromethylmethionylglycine dipeptide (**8**) using xenon difluoride as a fluorination reagent [25].

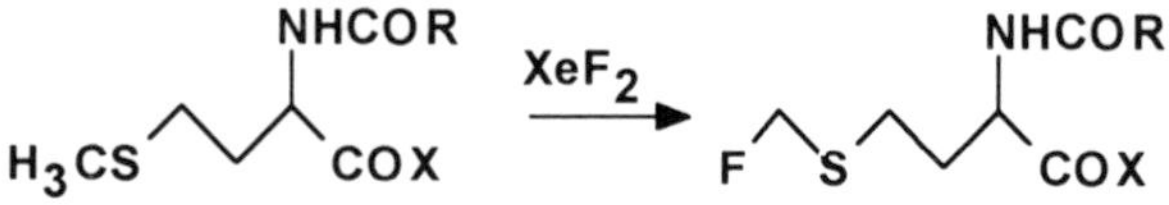

$$R = CF_3 \ , \ X = C_6H_4NO_2\text{-}p$$

(8)

$$R = PhCH_2O \ , \ X = NHCH_2COOC_2H_5$$

$$R = CF_3 \ , \ X = OCH_3$$

Acetyl hypofluorite was used as a fluorination reagent in the synthesis of labelled 2'- and 6'-fluoro-3',4'-dihydroxyphenylalanine. The acetyl hypofluorite required for the reaction is obtained, as a rule, by passing molecular fluorine into sodium acetate solution in a neutral solvent, e.g. fluorotrichloromethane [26].

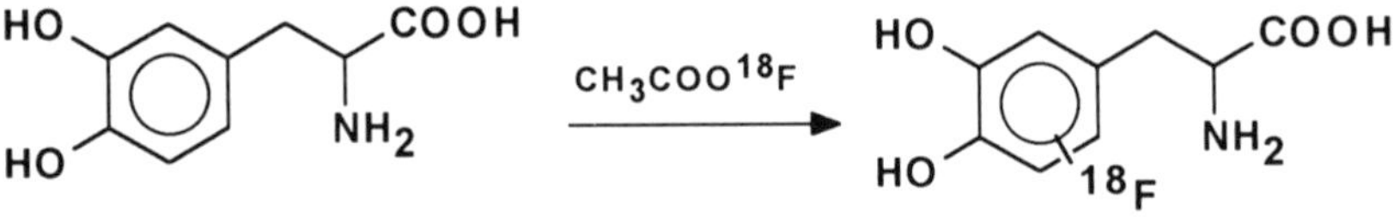

The application of the fluorination method to 3'-O-methyl-3'-4'-dihydroxy-phenylalanine ethyl ester resulted in a mixture of 2',5'-difluoro- and 6'-fluoro-DOPA, from which the desired 6'-fluoro-DOPA was isolated by HPLC in low yield [27]. Fully protected DOPA (to prevent oxidative degradation of the catechol ring) gave a slightly higher yield of the desired compound [28]. A more successful regioselective method of introducing an ^{18}F label by this route is to use a combination of acetyl hypofluorite with a fluorodemetallation reaction of the corresponding mercury(II) precursor 9 [29–31]. Thus, treatment of protected (S)- or (R)-3',4'-dihydroxyphenylalanine derivatives with mercury(II) trifluoroacetate and subsequent action of acetyl hypofluorite afford exclusively the corresponding homochiral 6'-difluoro-3',4'-dihydroxy-phenylalanines in 20% yield [30, 32].

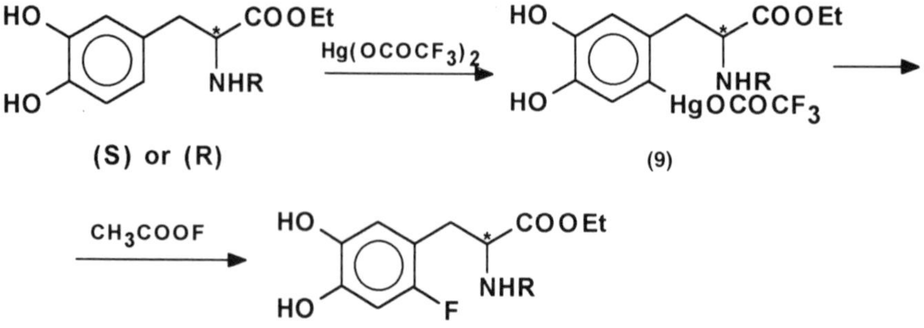

Combination of mercuration with fluorination has shown good results in the synthesis of labelled 4'-[^{18}F]fluoro-m-tyrosine [33–35]. This method was also applied to the preparation of 2'-fluoro-5'-hydroxyphenylserine (10) [36].

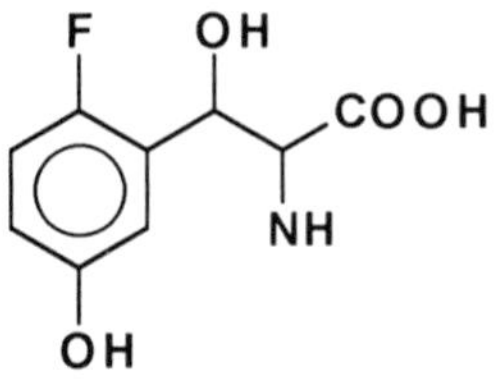

(10)

2.3 TRIFLUOROMETHYL HYPOFLUORITE

The photochemical fluorination reaction with trifluoromethyl hypofluorite is a selective method for the introduction of fluorine atoms into organic compounds. In the case of fluorination of amino acids as a starting material, this reaction is carried out under UV irradiation in a solution of hydrogen fluoride or trifluoroacetic acid to protect the amino group by protonation with a strong acid.

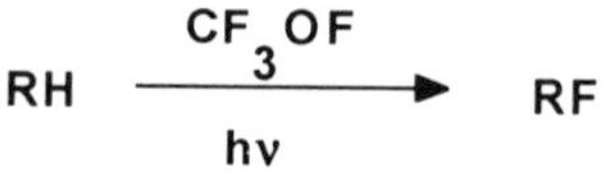

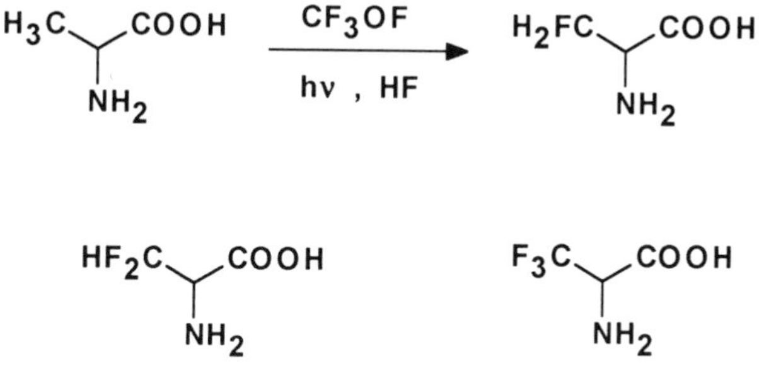

Monofluoroalanine [37–39], difluoroalanine and trifluoroalanine [40], γ-fluoromethyl-γ-aminobutyric acid (11) [41], β-fluoroaspartic acid (12) [40] and α-fluoromethylglutamic acid (13) [42, 43] were synthesized in this way directly from non-fluorinated amino acids.

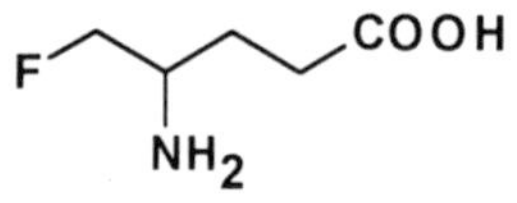

(11) 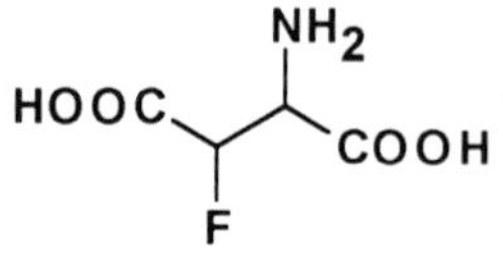   (12)

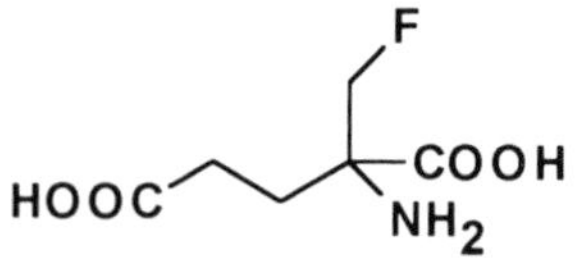

(13)

In the fluorination of lysine, the reaction proceeds with loss of regioselectivity and 5-fluorolysine (**14**) was the only reaction product [44].

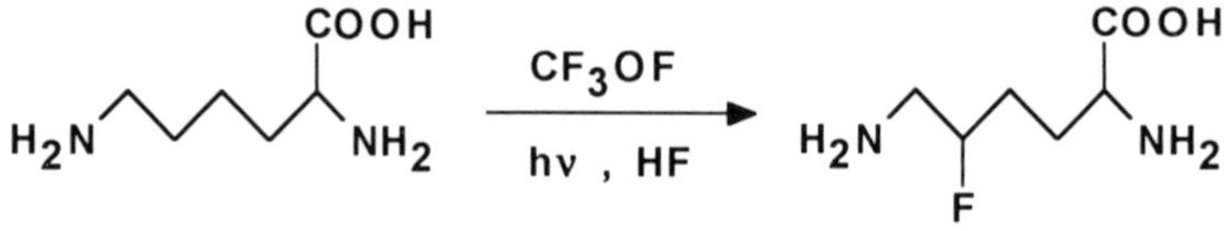

(14)

Fluorination of L-isoleucine by this method yielded *trans*-3-methyl-L-proline
(**15**) [42]. Obviously after fluorination of the δ-position of the carbon chain the
elimination of hydrogen fluoride occurs with the formation of an unsaturated
non-fluorine-containing amino acid and the subsequent heterocyclization of
the latter gives 3-methylproline (**15**).

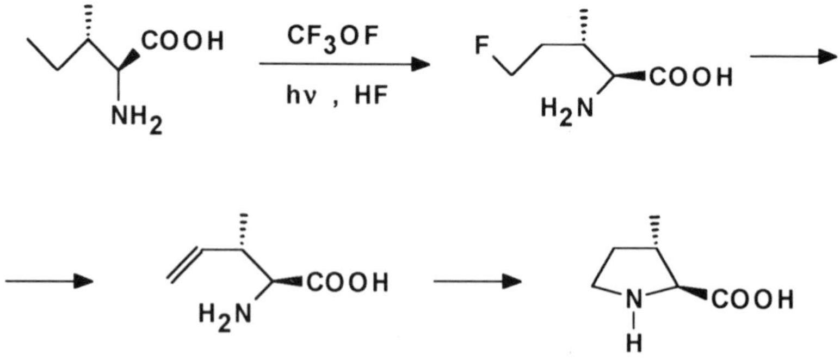

(15)

Trifluoromethyl hypofluorite was used as a reagent for the fluorodesulfurization process of the C—S bond in SH-containing amino acids [45].
Thus, the reaction of trifluoromethyl hypofluorite with penicillamine (**16**)
proceeds in hydrogen fluoride solution at $-78\,°C$ to give 3-fluorovaline (**17**).

Trifluoromethyl hypofluorite was found to be capable of being replaced by chlorine of N-chlorosuccinimide because the hydrogen fluoride solvent is a source of fluorine in the reaction of C—F bond formation.

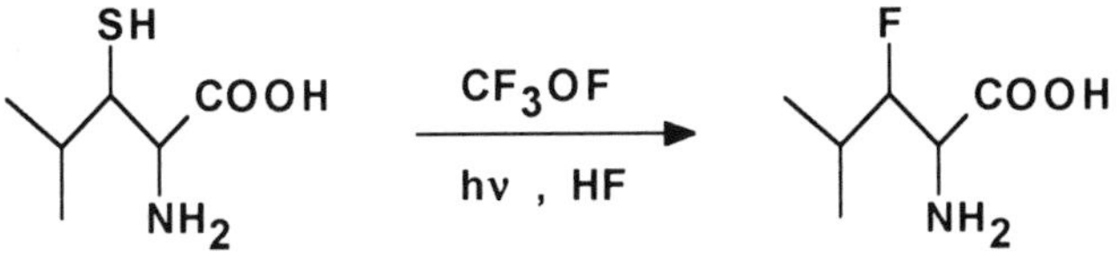

(16)   (17)

Analogously 3-mercaptoalanine (18) could be converted into mono- and difluoroalanine (19 and 20) [45].

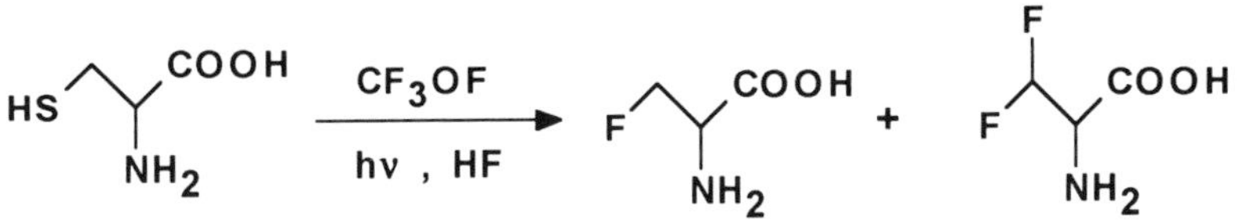

(18)   (19)   (20)

Apparently, fluorodesulfurization with trifluoromethyl hypofluorite is relatively unsuitable for wide application to fluorine-containing amino acid synthesis because it needs mercaptoamino acids as starting materials and the yields of the final products are low. Trifluoromethyl hypofluorite fluorodesulfurization can also be realized by the action of fluorine-diluted helium in hydrogen fluoride.

2.4 PERCHLORYL FLUORIDE

Another very popular reagent for the replacement of a C—H bond with a C—F bond, perchloryl fluoride, was firstly applied to convert amino acids in the synthesis of 4-fluoroglutamic acid (21) [46–49]. Fluorination is directed at the γ-position of 2-acetamido-2,4-dicarboxyglutarate (22) and subsequent hydrolysis and decarboxylation yield 4-fluoroglutamic acid (21).

(22)

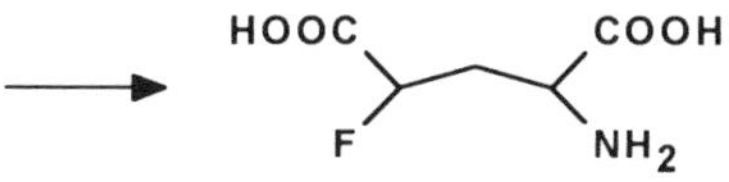

(21)

The action of perchloryl fluoride on di-*tert*-butyl-2-ketosuccinate in ethanol results in the formation of the difluorohemiacetal **23**, the subsequent ox-imination and reduction of which yield β,β-difluoroaspartic acid (**24**) [50]. Further transformations lead to difluoroasparagine (**25**) and β-fluoroaspartic acid (**26**) [50, 51].

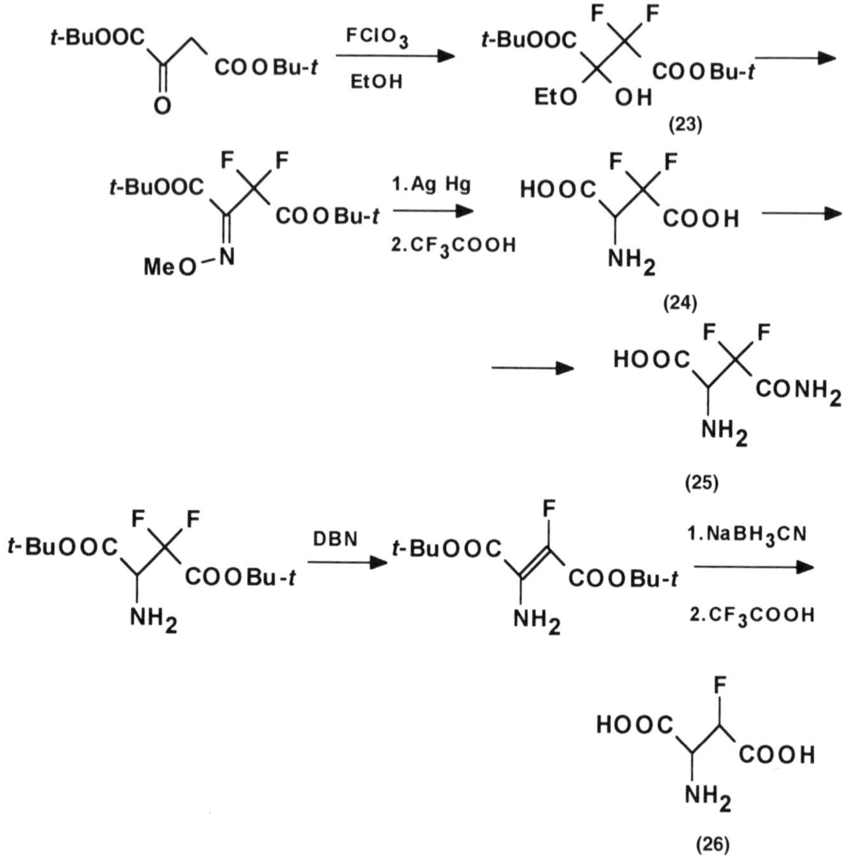

2.5 SULFUR TETRAFLUORIDE AND DIETHYLAMINOSULFUR TRIFLUORIDE

Sulfur tetrafluoride is one of the most useful reagents for the transformation of carbonyl groups into geminal difluoromethylene groups. This reagent has also been applied to the preparation of fluorine-containing amino acids from cor-responding synthons. Thus, the action of sulfur tetrafluoride in liquid hydrogen fluoride gives the possibility of converting amino acids bearing keto carbonyl groups into *gem*-difluoro derivatives, e.g. to obtain 5,5-difluorolysine (**27**) [52].

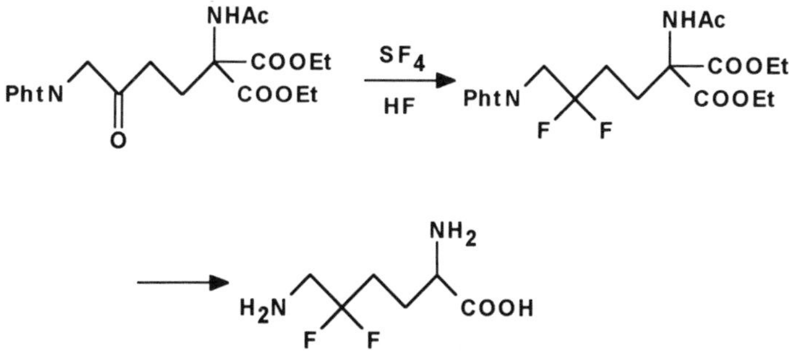

(27)

The decreased activity of the mixture of SF_4 and HF towards an amide carbonyl group in the presence of a ketone group allows sulfur tetrafluoride to be used to transform the diketopiperazine derivative of 4-keto-L-proline (28) into 4,4-difluoroproline (29) [53].

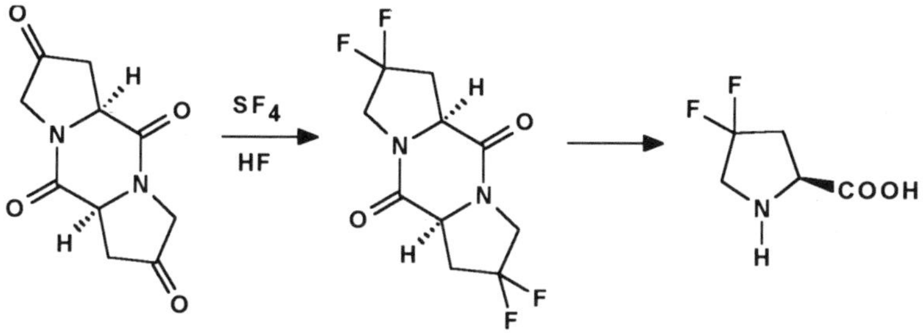

(28)   (29)

Only recently has sulfur tetrafluoride been applied to replace hydroxyl groups with fluorine atoms. The method has been applied successfully in syntheses of β-monofluoro derivatives of amino acids. Since sulfur tetrafluoride reacts with amino and carboxyl groups, liquid hydrogen fluoride, which converts an amino group into the ammonium group (NH_4^+), which is inert to the action of sulfur tetrafluoride, was used as a solvent in the reaction. In order to prevent fluorination of carboxy to trifluoromethyl groups, the reaction was performed at low temperature [54–56].

In the case of β-hydroxy-α-amino acids, the fluorination must be carried out at low reactant concentrations, in addition to the conditions above. Interaction of sulfur tetrafluoride with serine is complicated by the formation of a sulfenium salt (30). Further reaction of 30 leads to the formation of the initial amino acids and monofluoroalanine (19) in the ratio 1:1 [57]. When the reaction is carried out at high dilution, the formation of 30 is prevented, which makes it possible to obtain monofluoroalanine (19) with a very low serine content (the ratio is 99:1).

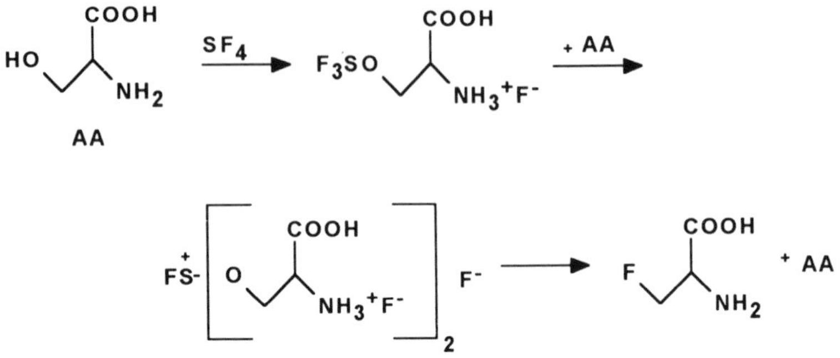

The reaction of fluorodehydroxylation with sulfur tetrafluoride was used for the synthesis of 3-fluoro-D-alanine (51% yield) [52, 56], 3-fluoro-L-2-amino-butyric acid (85%) and β-fluoro-D,L-phenylalanine (65%) from the corresponding hydroxy-containing derivatives [54].

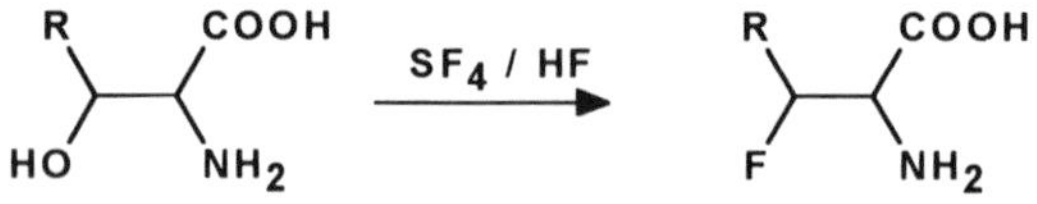

By this method 3-hydroxyaspartic acid dimethyl ester was converted into 3-fluoroaspartic acid dimethyl ester in 71% yield, in comparison with a 42% yield of this compound obtained from free 3-hydroxyaspartic acid [55]. The reaction occurred with inversion of configuration due to a carbonium ion being formed after the C—OH bond had been broken. A higher percentage of inversion was found for *threo* isomers [56]. Fluorination of *erythro*- and *threo*-3-hydroxyglutamic acid (31) with sulfur tetrafluoride gives only fluorinated lactams (32) [58]. Acid hydrolysis of 32 was found to take place only under severe conditions and was accompanied by elimination of hydrogen fluoride, as a result of which the keto acid 33 was isolated as the main reaction product [58].

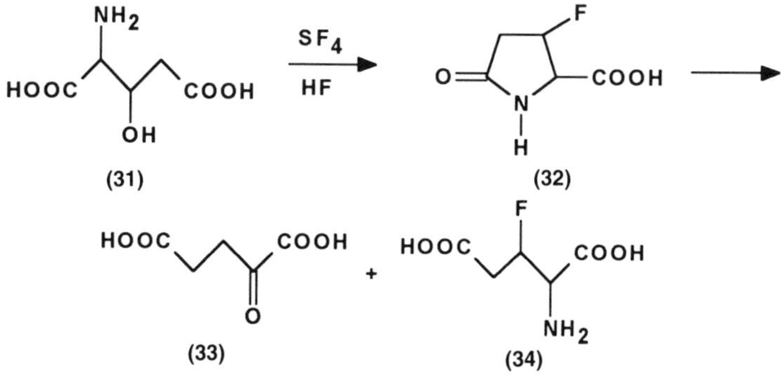

(31)   (32)

(33)   (34)

The application of N-acetyl-3-hydroxyglutamic acid (35) as a starting material in this reaction resulted in the formation of N-acetyl-3-fluoroglutamic acid (36). When the acetyl protecting group in 36 was removed with acylase, L-*threo* and L-*erythro* isomers of 3-fluoroglutamic acid (34) were obtained in a pure state [58].

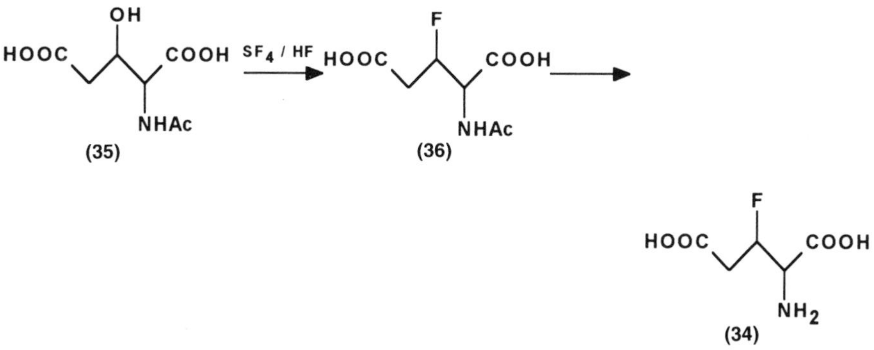

(35)   (36)

(34)

The advantage of the fluorodehydroxylation method is that it is based on readily available initial compounds. It is noted that the initial α-hydroxymethyl derivatives of amino acids are formed readily on treatment of copper complexes of the corresponding amino acids with formaldehyde. Fluorodehydroxylation makes it possible to obtain α-fluoromethyl derivatives of aliphatic, aromatic and heterocyclic amino acids [59]. Further, optical activity is not lost in the fluorination of pure enantiomers of β-hydroxy-α-amino acids since the α-chiral carbon atom is not involved in the reaction. A disadvantage of the method is the incomplete conversion of the initial amino acids in many cases, which complicates the isolation of the final products and the formation of fluoro-substituted amino acids as hydrofluorides.

The fluorodehydroxylation of protected 4-hydroxylysine (37) by sulfur tetra-fluoride in liquid hydrogen fluoride proved to be a more efficient method for obtaining 4-fluorolysine than direct fluorination of lysine with trifluoromethyl hypofluorite discussed above [60].

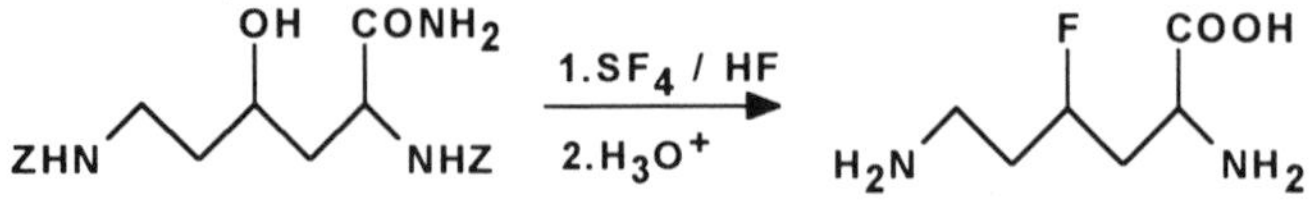

(37)   (38)

Diethylaminosulfur trifluoride (DAST) is a very popular reagent for the replacement of a hydroxyl or carbonyl oxygen with fluorine under very mild conditions [61]. There are some advantages of DAST in comparison with sulfur tetrafluoride, e.g. ease of application. Sulfur tetrafluoride is a gas of boiling point $-40\,^\circ$C and a toxicity comparable to that of phosgene, and hence must be handled with caution. Most reactions with SF_4 require special stainless-steel vessels since fluorination, as a rule, is carried out at elevated temperatures. DAST mimics the reactions of sulfur tetrafluoride, but avoiding the problems associated with the application of the latter. It is a commercially available liquid which can be used at ambient temperature with standard chemical laboratory equipment. In many cases the application of DAST gives the possibility of avoiding the using of hydrogen fluoride, which could induce the rearrangement of sensitive alcohols.

Diethylaminosulfur trifluoride has been used in many cases for the synthesis of fluorine-containing amino acids from corresponding oxygen synthons. For example, the replacement of the hydroxyl group in *threo*-β-phenylserine with fluorine using DAST leads to stereoisomers of β-fluoro-β-phenylalanines (39) [62, 63].

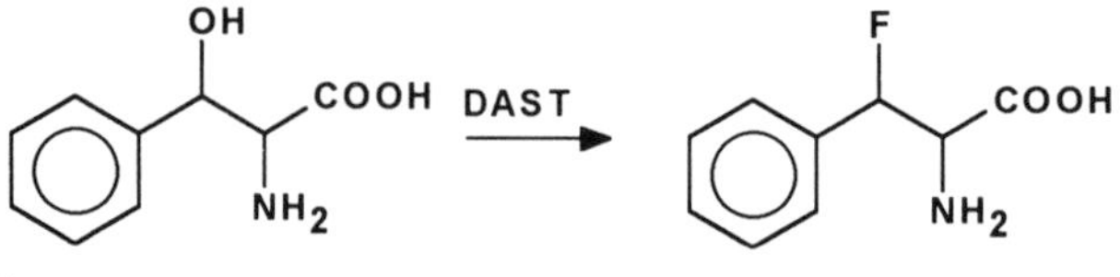

(39)

Optically pure (2*S*,3*S*)-4-fluorothreonine (40) was synthesized via replacement of the hydroxy group (4*S*,5*S*)-oxazolone (41) [64].

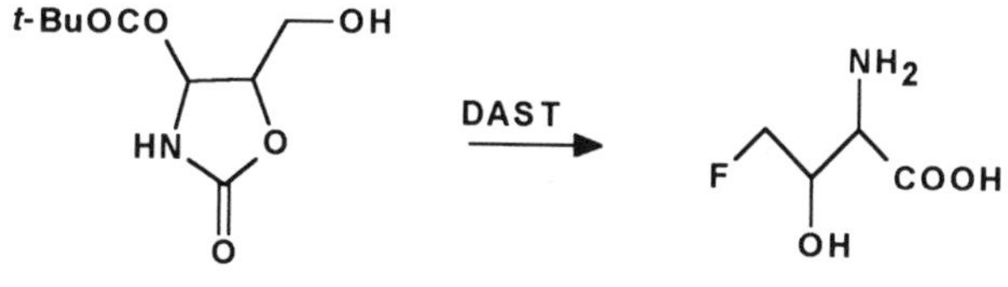

(41) (40)

The possibility of converting hydroxyproline (42) into glutamic acid has been used recently for the synthesis of optically pure (2S,4S)-4-fluoroglutamic acid (43) [65, 66]. Treatment of 42 with DAST or 2-chloro-1,1,2-trifluorotriethylamine gives a 3:2 mixture of *trans*- and *cis*-methyl N-acetyl-4-fluoro-L-prolinate (44) in 50% yield. Subsequent oxidation with ruthenium tetraoxide and hydrolysis convert 4-fluoro-L-pyrrolidine-5-one-2-carboxylate into (+)-L-*threo*-4-fluoroglutamic acid (43) [(+)-(2S,4S)-4-fluoroglutamic acid] in 77% yield.

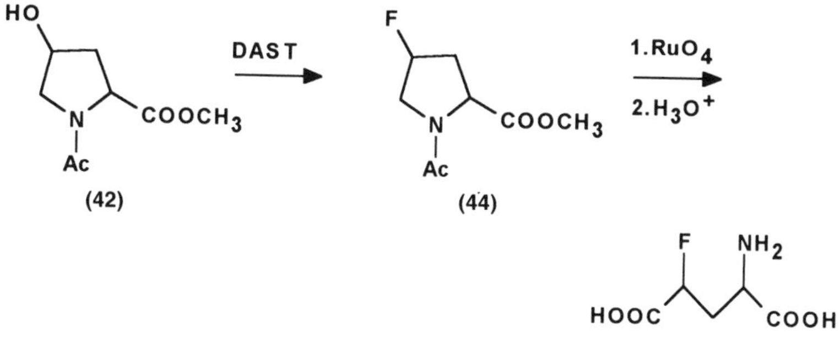

(42) (44) (43)

Analogously, from *cis*-4-hydroxy-D-proline methyl N-acetyl-*trans*-4-fluoro-D-prolinate could be obtained in 80% yield and 20% of its *cis* isomer in 40% total yield; oxidation of this mixture and subsequent hydrolysis led to (−)-(2R,4S)-fluoroglutamic acid.

Groth and Schollkopf [67] used DAST for the stereoselective synthesis of 3-fluorovaline and 3-fluorophenylalanine in high optical yield. Compound 46 on treatment with diethylaminosulfur trifluoride yielded fluoro derivatives of dihydropyrazines (47), subsequent acid hydrolysis of which afforded (R)-3-fluorovalinate and a mixture of diastereomers of 3-fluorophenylalanine methyl esters (48). Some epimerization observed in the reaction indicates a carbenium ion-pair mechanism for the fluorodehydroxylation reaction with DAST.

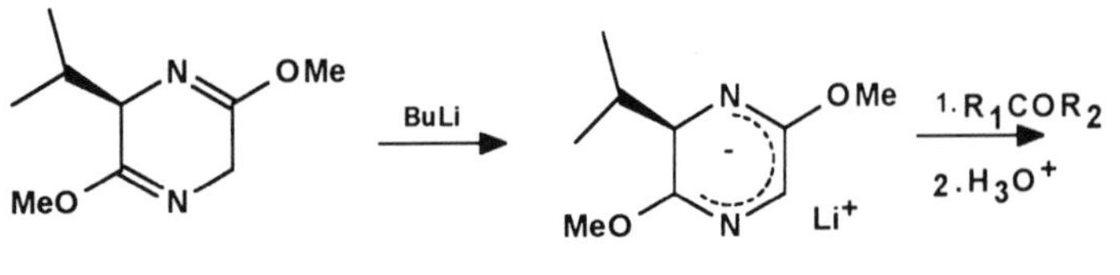

(45)

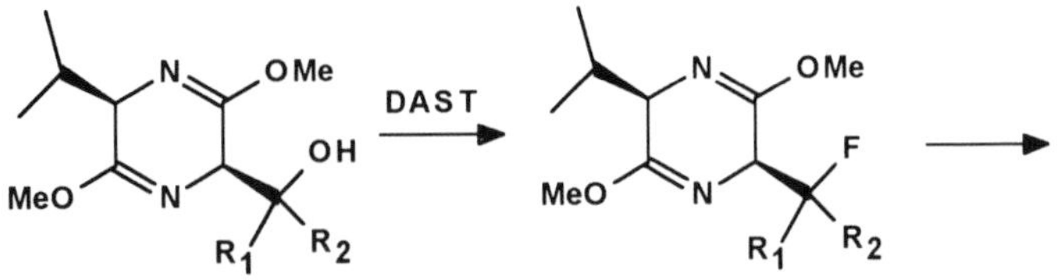

(46)   (47)

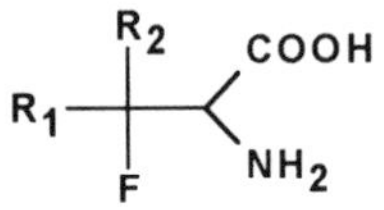

(48)

Alkylation of the anion **45** with carbonyl compounds (e.g. acetone, benz-aldehyde) takes place with different optical yields, but not less than 85%. In the case of **46** with $R_1 = H$ and $R_2 = Ph$ or $R_1 = Me$ and $R_2 = Ph$ the ratios of *erythro* and *threo* isomers are 6:1 and 12:1, respectively. Further conversion of hydroxy derivatives **46** into the fluorine-substituted products **47** proceeds with moderate yields (ca 45%) owing to the formation of side-produces, i.e. 3-isopropylidenedihydropyrazine (**49**) or pyrazine derivatives (**50**).

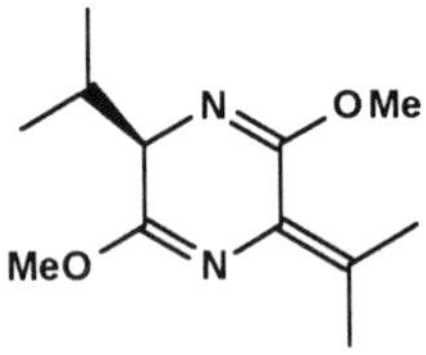

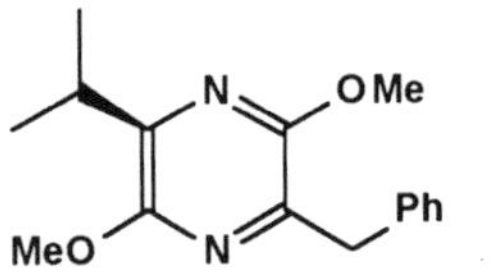

(49)   (50)

In compound **46a** ($R = R_1 = H$), the hydroxy group could not be replaced with fluorine using DAST. In order to obtain monofluoroalanine, the anion **45** was treated with fluoroiodomethane, giving the corresponding mono-fluoromethyl derivative **47** in 80% yield; the optical yield did not exceed 45%.

N-Protected 4-amino-2-(hydroxymethyl)but-2-enoic acid (**51**) was converted into 4-amino-2-(fluoromethyl)but-2-enoic acid (**52**) with DAST at $-78\,^{\circ}C$ in 55% yield [68].

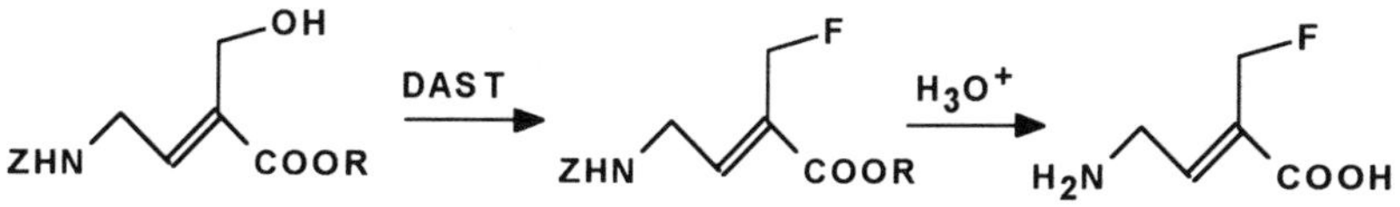

(51)　　　　　　　　　　　　　　　　　　　(52)

The use of DAST for fluorination of N,N-dibenzyl derivatives of β-hydroxy α-amino acid esters (53) leads to α-fluoro-β-amino derivatives (56) [69].

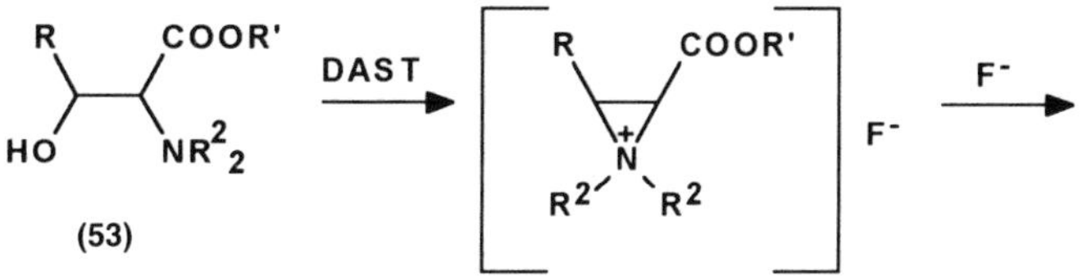

(53)　　　　　　　　　　　　　　　　　(54)

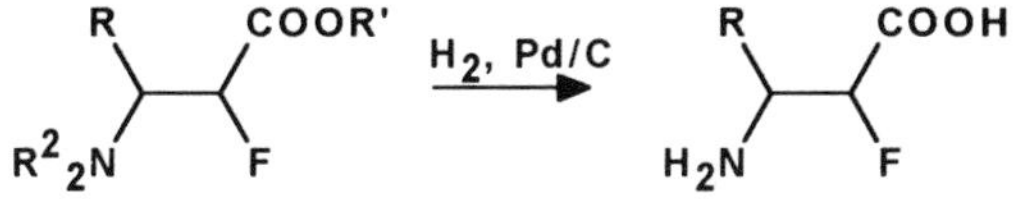

(55)　　　　　　　　　　　　　　(56)

The formation of the β-amino acids 56 from their α-amino derivatives 53 can be explained by the intermediate formation of the aziridinium ion 54, which is attacked by fluoride anion at the α-carbon atom, affording the substituted β-amino acids 56 in 90% yields. An advantage of this method is the high stereospecificity. Thus double inversion of the configuration at the α-carbon atom (the formation of aziridinium ion and ring opening by fluoride anion) results in the formation of corresponding optically active α-fluoro derivatives from optically active L- or D-amino acids. Further, in the case of β-hydroxy-α-amino acids having two asymmetric centres (α- and β-carbon atoms) the initial *erythro* or *threo* configuration is retained during their conversion into α-fluoro-β-amino derivatives [69].

The decrease in the nucleophilicity of the nitrogen atom when an amino group is protected by employing 4,5-diphenyl-4-oxazoline-2-one (57) rules out the possibility of the isomerization of α-amino-β-hydroxy acids via the aziridinium ion 54. The application of DAST in this instance makes it possible to obtain β-fluoro derivatives 58 of various natural amino acids under mild conditions and in high yields [70].

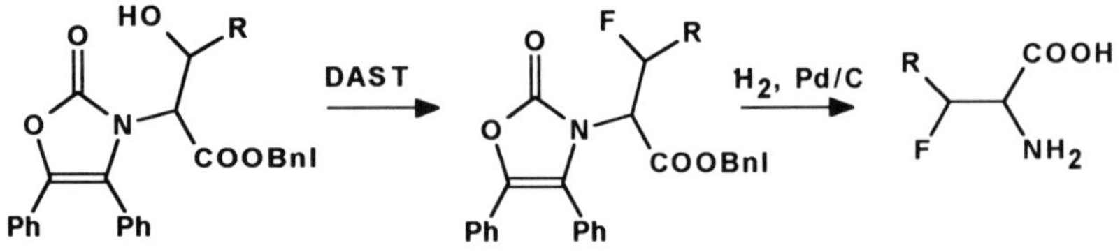

(57)   (58)

The reaction of DAST with the hindered hydroxymethyl group of the tryptophan derivative **59** provided only a trace of the desired 2-fluoromethyl compound **60** and the oxazoline **61** as the predominant product. To synthesize the fluorine-containing compound **60** the treatment of **62** with DAST was applied, which yielded the fluoromethyl compound **63** in 50–60% yield, ring opening of which gave the desired monofluorine derivative in 95–100% yield [71].

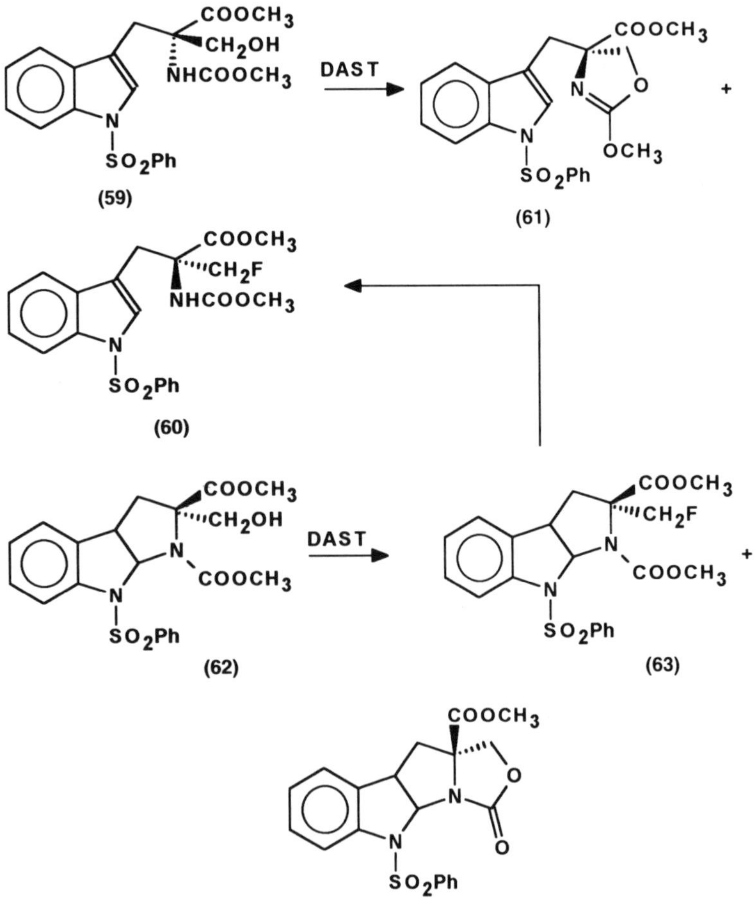

DAST has recently used in the synthesis of N-acetyl-9'-deoxy-9-fluoro-neuraminic acid [72].

2.6 HYDROGEN FLUORIDE AND METAL FLUORIDES

Hydrogen fluoride is often used as a reagent in the syntheses of organofluorine compounds, and has also been applied to the preparation of fluorine-containing amino acids. The action of hydrogen fluoride on the acet-amidomalonic acid ester **64** leads to the simultaneous addition of hydrogen fluoride to the C=C double bond and replacement of the chlorine atom by fluorine. After hydrolysis of the intermediate product formed (**65**), a 50% yield of 5,5-difluoronorleucine (**66**) is obtained [73].

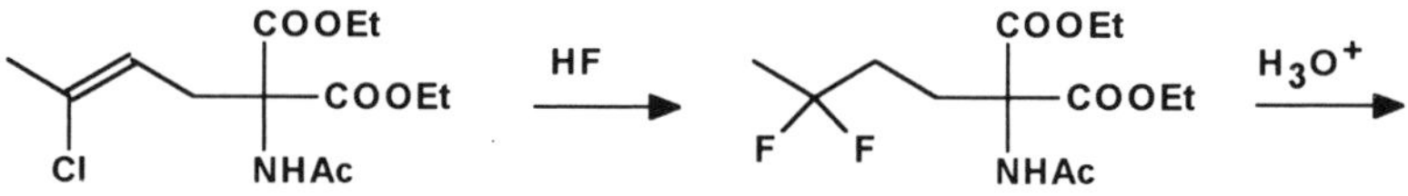

The combined addition of fluorine and bromine to 5-vinylpyrrolid-2-one (**67**) gives a mixture of isomeric 5-(bromofluoroethyl)pyrrolidinones (**68** and **69**) which is readily separated by chromatography. Dehydrobromination of **68** and **69** and subsequent hydrolysis lead to the formation of γ-amino-γ-fluoro-vinylbutyric acid (**70** and **71**) [74].

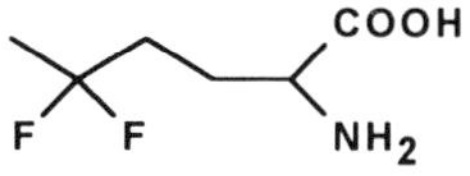

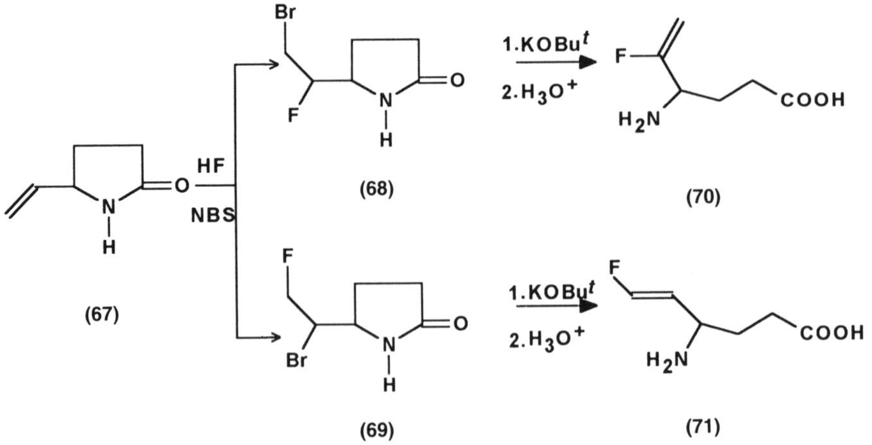

The problem of using of the cheapest fluorinating agent, anhydrous hydrogen fluoride, connected with its corrosive nature and low boiling point has been overcome by the application of polypyridinium hydrogen fluoride (Olah reagent). It is a composition of 30% (w/w) pyridine and 70% (w/w) hydrogen fluoride and is a commercially available, stable liquid which has been used in a number of fluorination reactions [75]. Polypyridinium hydrogen fluoride solution is much more convenient than hydrogen fluoride because it permits reaction to be effected at atmospheric pressure.

The procedure based on opening of the aziridine [76–80] or azirine [81] ring using Olah reagent can be regarded as one of the most modern and general methods for the synthesis of β-fluoro derivatives of amino acids (mono- and difluorinated). The starting material for the synthesis of monofluorinated compounds was acrylic esters (72), which were converted by bromination into dibromo derivatives (73), and aziridines (74) were obtained by subsequent treatment with ammonia. The aziridines 74 afford the corresponding amino acids in high yields on treatment with Olah reagent. The use of alkylamines instead of ammonia in this reaction permits the synthesis of N-substituted derivatives of fluorine-containing amino acids (75).

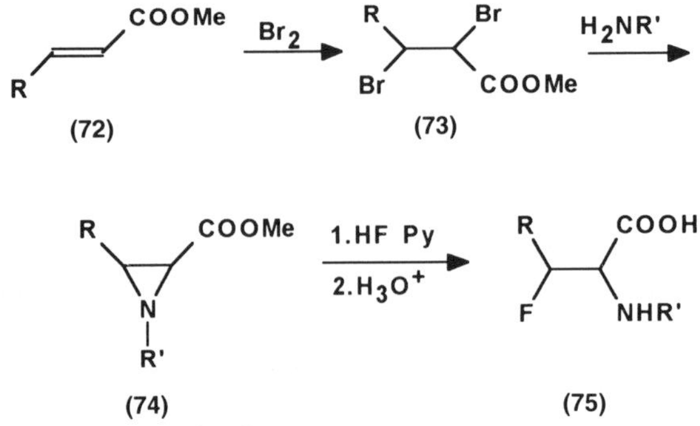

A number of alkyl and aryl derivatives of both β-substituted acrylic acid esters and amines have been introduced into this reaction. An advantages of this method is the regioselective manner of the ring opening—fluorine is directed to the β-position of the amino acid obtained; bromination of crotonic acid ester and subsequent treatment with ammonia and Olah reagent were shown to give β-fluoroalanine in the *threo* configuration as a major product [20]. The mechanism of the ring-opening aziridine reaction is suggested to consist in the protonation of ring-nitrogen and subsequent formation of a carbonium ion at C-2. Fluoride is directed to the most substituted ring carbon or to the benzylic carbon in accordance with an S_N1-type process. The formation of the similar aziridinium cation **76** as an intermediate was suggested to occur in the above-described reaction of DAST with *N,N*-dibenzyl derivatives of β-hydroxy-α-amino acids, which led to α-fluoro-β-amino acids [69]. In the last case the fluoride anion is directed to the α-carbon carbonium centre formed on aziridinium cation ring opening.

The treatment of the azirinecarboxylate **78**, formed n thermal decomposition of the azide **77**, with Olah reagent is a convenient method for synthesizing β,β-difluoro derivatives of amino acids in 37–42% yields [18, 81, 82–86].

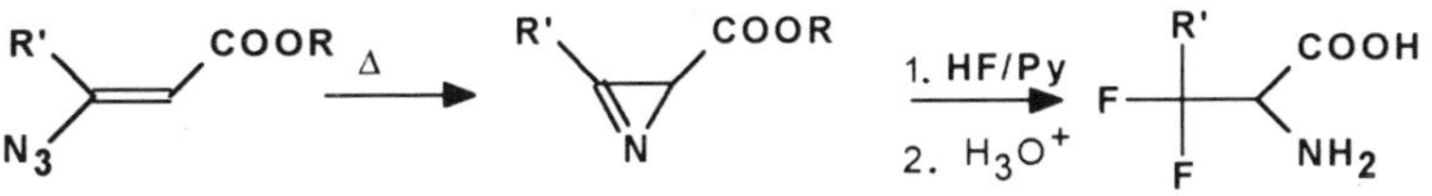

(77)   (78)   (79)

Substituted azirines react with Olah reagent more easily than aziridines. Presumably, the formation of the fluoroaziridine **76** is the first step of the reaction. After hydrolysis of the reaction mixture, the subsequent two competing ring-opening pathways yield β,β-difluoroamino acids (**79**) and α-keto acids (**80**) owing to the formation of β-amino-α,β-difluoro compounds (**81**) as intermediate products.

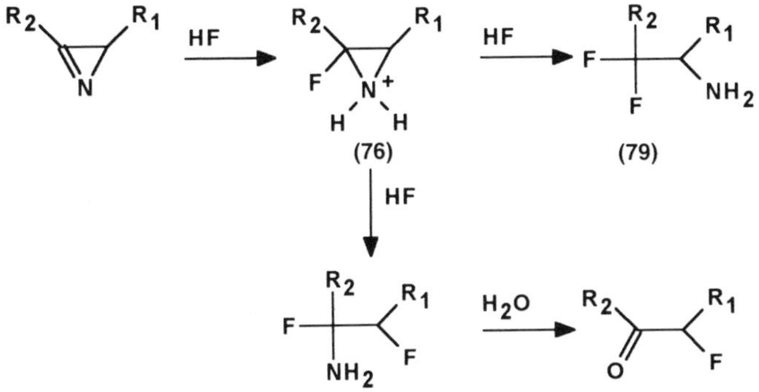

Generally, the yields of fluorine-containing amino acids in this reaction are lower compared with the reaction of aziridine. Closely related to the aziridine ring-opening reaction is the ring-opening reaction of epoxides bearing a cyano group with Olah reagent [83, 87, 88]. By this method it is possible to obtain 3-fluorine-containing amino acids in 60% yields.

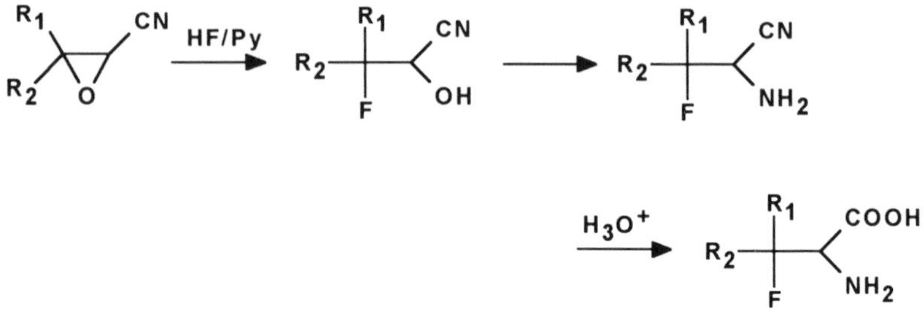

The method of synthesis of fluorine-containing amino acids involving replacement of a halogen atom with fluorine under the action of hydrogen fluoride or metal fluorides has not found wide application, apparently owing to the low availability of the initial compounds.

The replacement of a bromine atom by fluorine in 5-bromo-methylpyrrolidin-2-one (82) on treatment with silver fluoride in anhydrous acetonitrile and subsequent acid hydrolysis afforded γ-fluoromethyl-γ-amino-butyric acid (83), which proved to be an effective inactivator of γ-aminobutyric acid transaminase [89, 90].

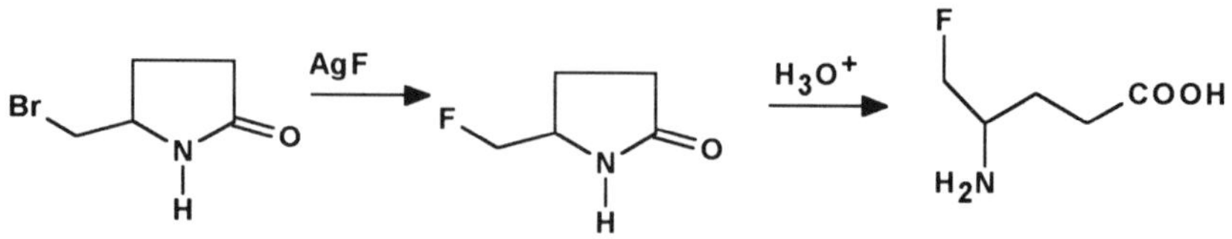

(82)   (83)

3-Fluoroalanine (19) was first obtained by replacement of the chlorine atom with fluorine in the treatment of 4-chloromethylenephenyloxazolin-2-one (84) with potassium fluoride and subsequent hydrolysis and reduction of the fluoro derivative (85) [91].

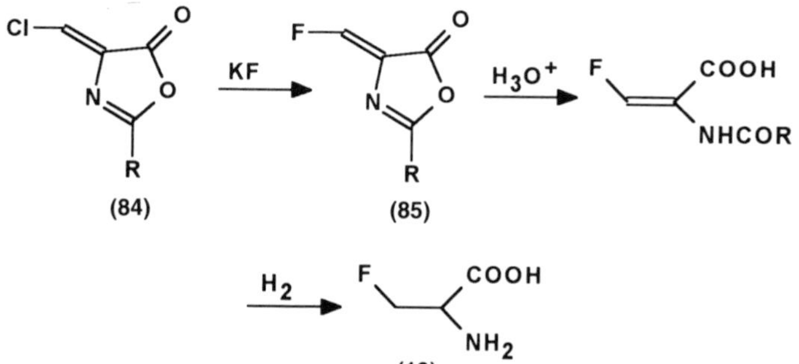

(84)   (85)

(19)

Another method for substitution by a fluorine atom consists in the treatment of *O*-tosyl derivatives of hydroxyamino acids (86) with anhydrous potassium fluoride. The procedure led to the formation of *cis*- and *trans*-4-fluoroprolines (87) from hydroxyproline and hydroxy-*allo*-proline [92, 93].

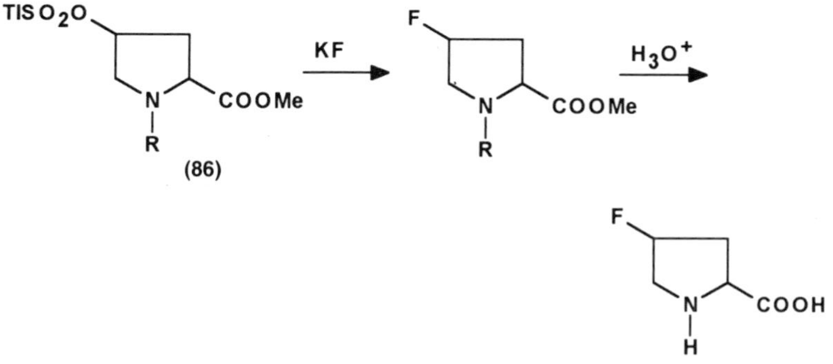

(86)

(87)

The employment of antimony trifluoride as a fluorinating agent made it possible to synthesize 3-fluoroglutamic acid (88) from diethyl 2-acetamido-3-ketoglutarate (89) [94]. The diethyl ester of the keto acid 89 was reduced and converted by subsequent chlorination into the chlorine derivative 90, from which the amino acid 88 was obtained by treatment with antimony trifluoride and subsequent hydrolysis.

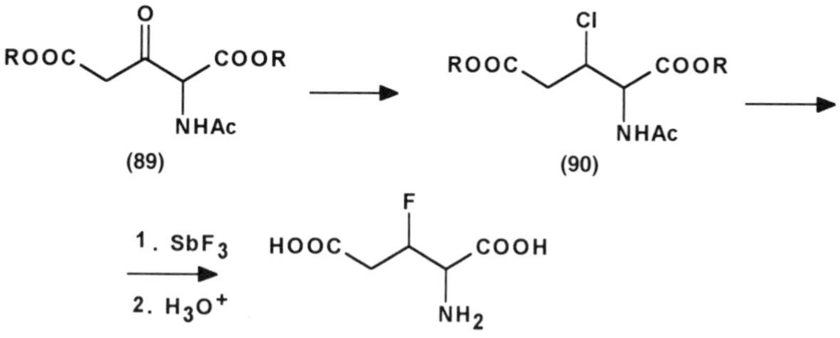

(89)   (90)

(88)

2.7 YAROVENKO REAGENT

The first reagent for the substitution of hydroxyl group with fluorine was Yarovenko reagent or 2-chloro-1,1,2-trifluoroethyldiethylamine, easily obtained by addition of diethylamine to chlorotrifluoroethylene [95]. A major advantage of this reagent is attack only towards hydroxyl groups without involving other functional groups such as carbonyl or ester. By this method 4-fluoroglutamic acid was synthesized from a hydroxyl-containing precursor [96].

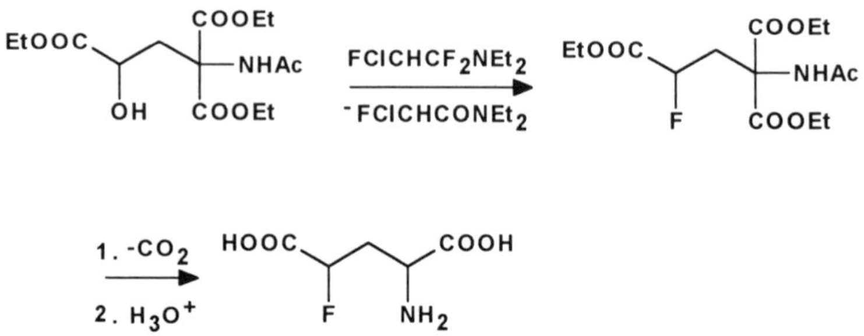

2.8 BALZ–SCHIEMANN AND RELATED REACTIONS

Photoreaction between diazo compounds and tetrafluoroboric acid is closely related to the Balz–Schiemann reaction and was shown to be the route to obtaining fluorine-containing aromatic amino acids. $3'$-Fluoro-L-tyrosine and $3',5'$-difluoro-L-tyrosine were synthesized by this method [97].

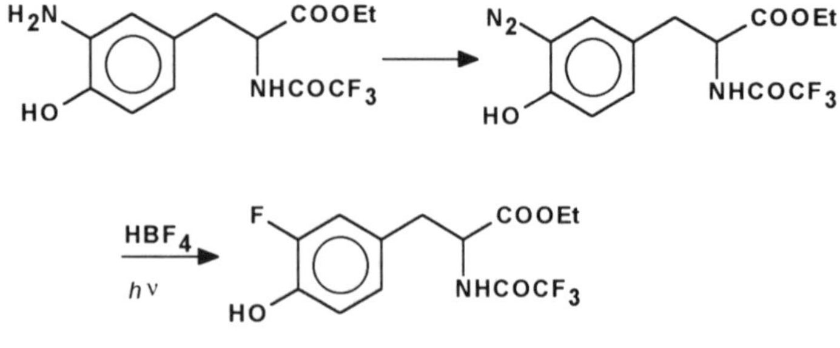

Similarly, $2'$-fluoro- and $4'$-fluorohistidines were synthesized by such methodology [98, 99]. Isolation of the intermediate diazonium fluoroborate was shown to be unnecessary.

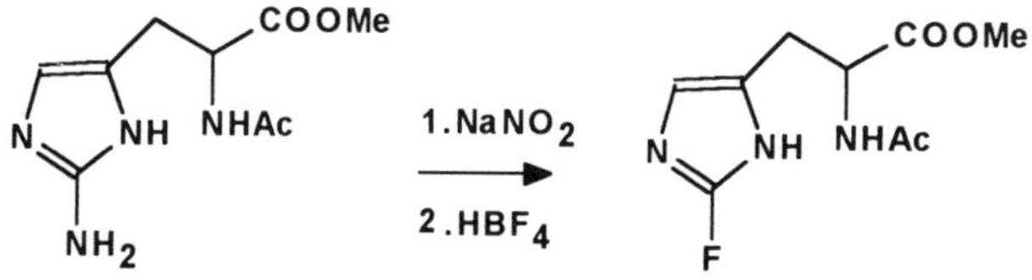

2.9 SYNTHESIS OF PERFLUOROALKYLAMINO ACIDS FROM DERIVATIVES OF FLUORINATED ALDEHYDES AND KETONES

The high reactivity of the C=N double bond in imines of perfluoroalkyl aldehydes and ketones induced by strong electronegative effects of perfluoroalkyl substituents opened up the possibility of developing a general method for the synthesis of α-perfluoroalkylglycines [100]. For example, trifluoroalanine (93) could be obtained by addition of vinyl Grignard reagent to N-acyltrifluoroacetaldimine (91) and subsequent oxidation of the unsaturated compound 92 with potassium permanganate [101, 102].

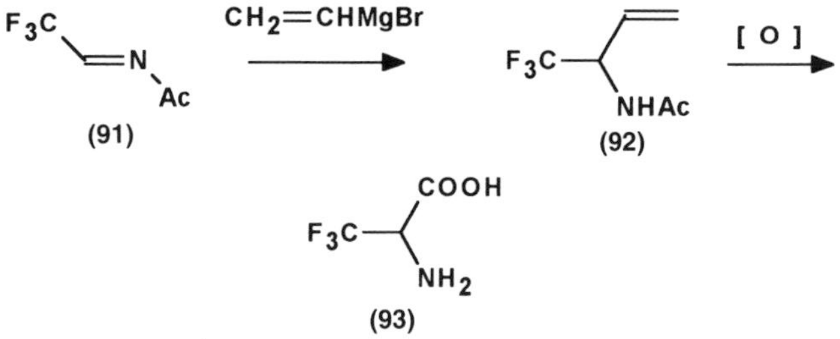

This method can be used to synthesize a number of α-perfluoroalkyl-substituted amino acids when other perfluoroalkylazomethynes, $R_fCH=NAc$, are introduced into the reaction [103, 104].

N-Acyltrifluoroacetaldimine (91) reacts with anhydrous hydrogen cyanide to yield 2-acylamino-3,.3,3-trifluoropropionitrile (94), which was hydrolysed to 3,3,3-trifluoroalanine (93) [105].

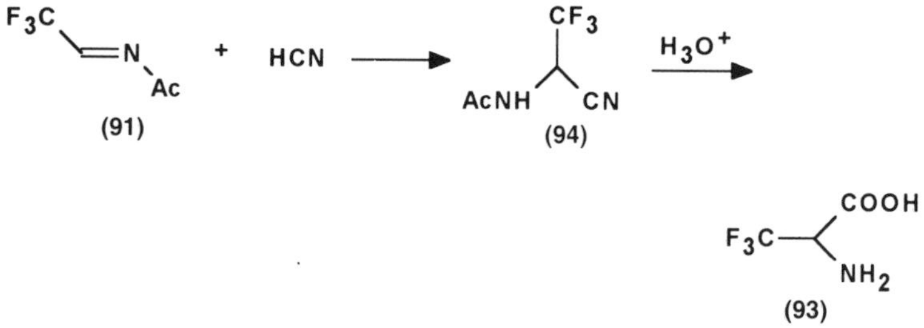

Sulphones (95) or halo derivatives (96) can be used as equivalents to the imine 91 in the above reaction, but the presence of a base (triethylamine) is essential.

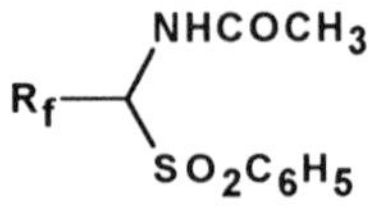

(95) 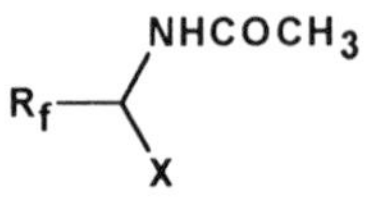   (96)

Trifluoroalanine (93) has also been obtained by replacement of the chlorine atom by CN by treatment of the chloro derivative 96 with CuCN [106] and subsequent hydrolysis of the corresponding cyanamide.

A method based on alkylation of the N-methoxycarbonylimine of methyl trifluoropyruvate (97) for the synthesis of β-fluoro-substituted amino acids which are the most difficult to obtain, namely α-trifluoromethylamino acids (98), has been proposed [107, 108].

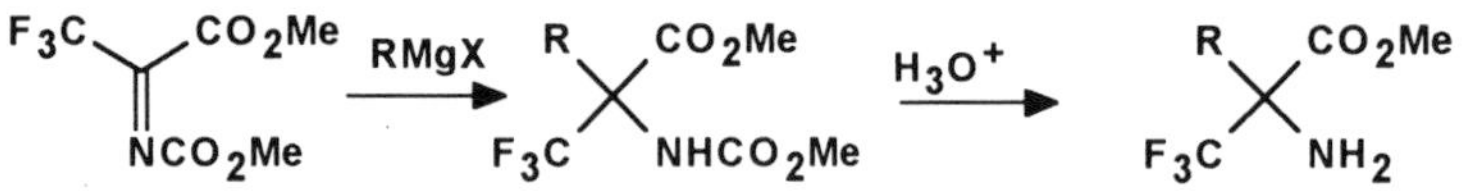

(97)   (98)

Because of the high electrophilicity of the imine 97, the regioselective alkylation of N-methylcarbonylimine with organomagnesium compounds can be achieved only by mixing the reactants at liquid nitrogen temperature and subsequent heating of the reaction mixture at 20 °C. The use of less reactive organocadmium compounds permits alkylation of the imine at 4 °C. The yields of amino esters 98 are almost independent of the metal employed (60–90%).

In the synthesis of α-trifluoromethylvaline (100) by alkylation of N-meth-oxycarbonylimine (97) with diisopropylcadmium or isopropylmagnesium bromide, the reaction is complicated by reduction of the imine 97 to the methyl ester of N-methoxycarbonyl-3,3,3-trifluoroalanine (99) [109].

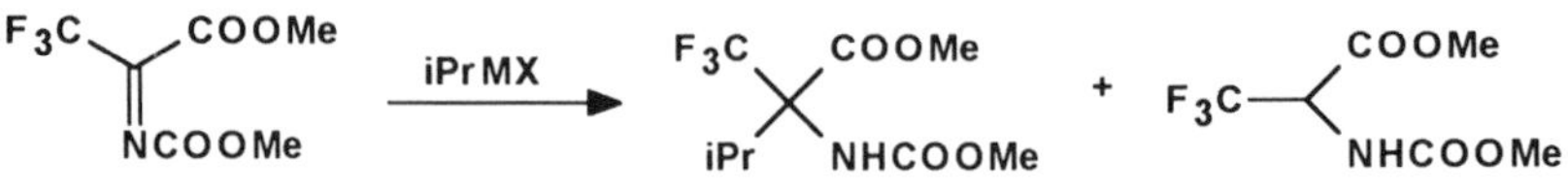

(97)   (100)   (99)

In the case of isopropylmagnesium bromide, the ratio of the yield of 100 to that of 99 is 1.5:1, whereas the use of diisopropylcadmium makes it possible to obtain predominantly the valine derivative 100. The amido esters 100 and 99 proved to be resistant to acid hydrolysis. To obtain hydrochlorides of α-tri-fluoromethylamino acids (98) it is necessary to heat the latter in a mixture of concentrated hydrochloric and formic acid at 100 °C for 40 h.

The possibility has been also demonstrated of synthesizing α-trifluoromethylamino acids (102) by treating the diacetyl derivative 101, which can be regarded as equivalent to the N-acetylimine 97, with 2 mol of organomagnesium reagent [108].

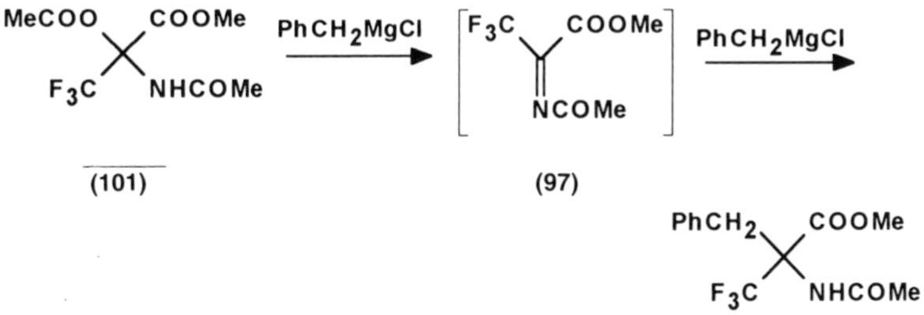

MeCOO, COOMe PhCH₂MgCl [F₃C, COOMe] PhCH₂MgCl
F₃C   NHCOMe   →   [NCOMe] →

(101)   (97)

PhCH₂, COOMe
F₃C   NHCOMe

(102)

The use of N-alkylimines of methyl trifluoropyruvate in reactions with organometallic compounds for synthesizing N-alkyl substituted α-trifluoromethylamino acids failed. However, the N-phenylimine 103 was alkylated with dimethylcadmium to form the methyl ester of N-phenyl-α-trifluoromethylalanine (104) [108].

F₃C, COOMe Me₂Cd Me, COOMe
    NPh   →   F₃C NHPh

(103)   (104)

The advantages of this method are the availability of the reagent employed and the high yields of α-trifluoromethylamino acids obtained. However, the disadvantage is that its application to the synthesis of heterocyclic or polyfunctional natural amino acids is a problem.

One of the prospects for further application of imines of methyl trifluoropyruvate in the synthesis of α-trifluoromethyl-substituted amino acids involves the use of these compounds in C-alkylation and cycloaddition reactions. Thus, the high electrophilicity of the N-acylimines 105 makes it possible to synthesize the derivatives of 2-aryl- and 2-hetaryl-3,3,3-trifluoroalanines (106) by regioselective alkylation of N,N-dimethylaniline and a series of heterocycles [110, 111].

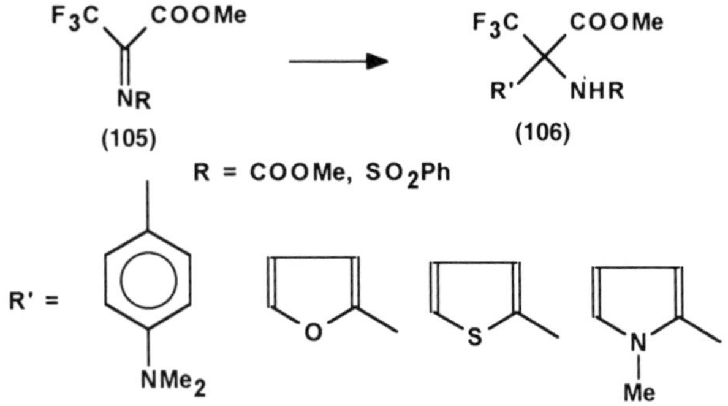

$R = COOMe, SO_2Ph$

Further, the introduction of the N-trifluoroacetylimine of methyl trifluoropyruvate (107) into the reaction with diazo compounds and alkenes made it possible to obtain derivatives of α-trifluoromethylcoronamic acid (108) and α-trifluoromethyl-γ,δ-dehydroleucine (109) [112].

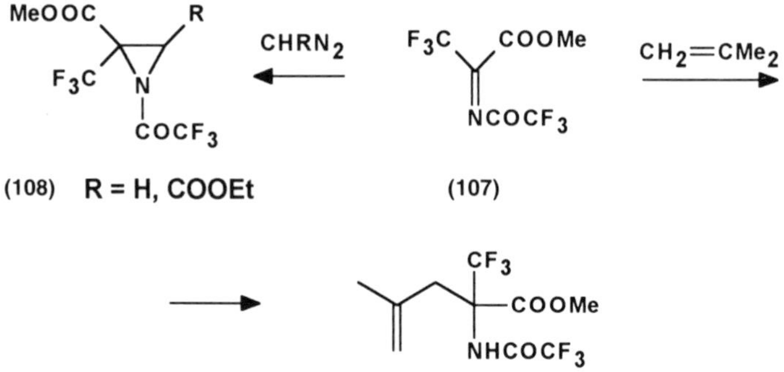

(108)   R = H, COOEt   (107)

(109)

Hexafluoroacetone reacts with benzamide to give the bis(trifluoromethyl)amino ketone 110, which can be transformed into the oxazole 111. The latter yielded trifluoroalanine (93) on reaction with potassium hydroxide in ethanol in the presence of hexamethyldisilane [102].

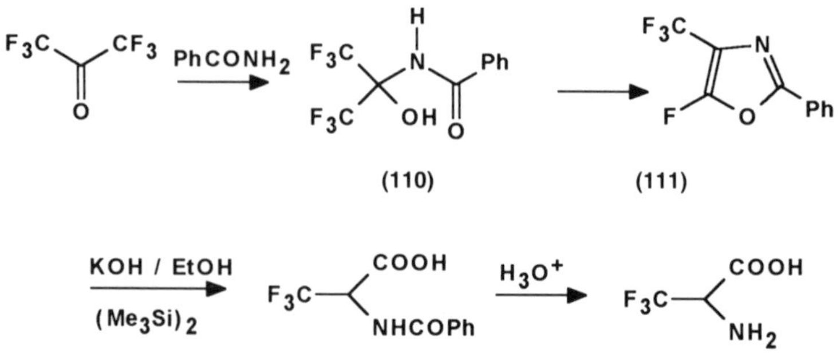

(110)   (111)

(93)

2.10 RING-OPENING OF FLUORINE-CONTAINING HETEROCYCLES

The majority of methods considered in this section are feasible because of the specific effect of fluorine or perfluoroalkyl groups. In the case of non-fluorinated compounds these reactions are inapplicable or lead to products other than amino acids. Syntheses of this type give rise to the possibility of obtaining individual β-fluoro derivatives of amino acids, but in many instances the approaches are general and can be used to prepare a series of fluorinated amino acids.

First we can mention the opening of aziridine and azirine rings as described earlier. The opening of three-membered oxirane and azirine rings containing a perfluoroalkyl group has also been used to synthesize β-fluoro-containing amino acids. Thus, the interaction of perfluoroisobutene oxide (112) with ammonia affords the amide of 3,3,3,3',3',3'-hexafluoro-2-aminoisobutyric acid (113) [113].

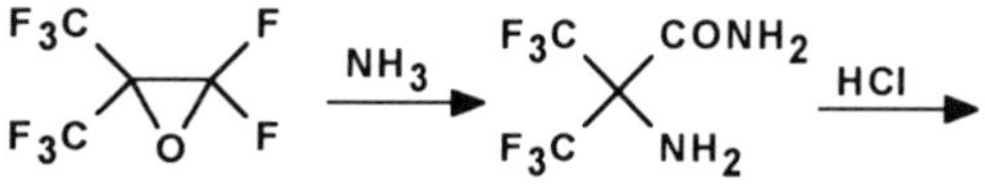

(112)   (113)

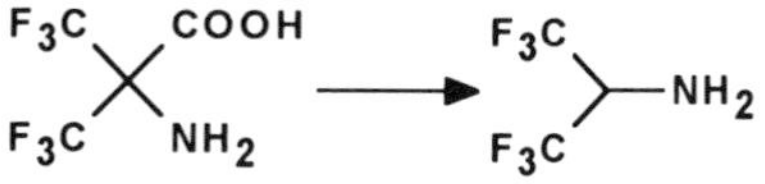

(114)   (115)

However, the attempt to hydrolyse the amide 113 with dilute hydrochloric acid causes decarboxylation of the amino acid 114 with formation of hexafluoroisopropylamine (115). Nevertheless, by means of alkaline hydrolysis it has been possible to isolate the sodium salt of hexafluoro-2-aminoisobutyric acid (114), which, however, also is decarboxylated on acidification [114].

Other derivatives of hexafluoro-α-aminoisobutyric acid (dialkylamides and the ethyl esters) were obtained by hydrolysis of the corresponding azirines 116, which are readily formed on treatment of dialkylamino or ethoxy derivatives of perfluoroisobutene (117) with sodium azide [115, 116].

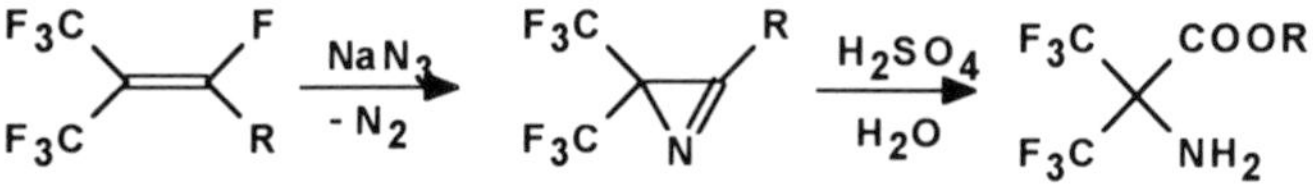

(117)   (116)   (118)

It is noteworthy that derivatives of hexafluoro-α-aminoisobutyric acid (118) do not form stable salts even with H_2SO_4, which has been explained by the electron acceptor influence of the two trifluoromethyl groups in the α-position to the amino group.

A series of methods based on the formation and opening of the oxazole ring have been proposed for the synthesis of 3,3,3-trifluoroalanine (93). The disadvantage of these methods is their multi-stage character, which reduces appreciably the yield of the desired product. Trifluoroalanine was first obtained by acid hydrolysis of the oxazolines 119 formed by [1 + 4] cycloaddition of N-acyltrifluoroacetaldimines (91) to isonitriles [100]. If isonitriles obtained from N-formylamino acids are used in this reaction, it is possible to synthesize trifluoroalanine-containing peptides (120) [100]. It is noteworthy that alkaline hydrolysis of trifluoroalanine derivatives takes place with conversion of the trifluoromethyl group into a carboxy group.

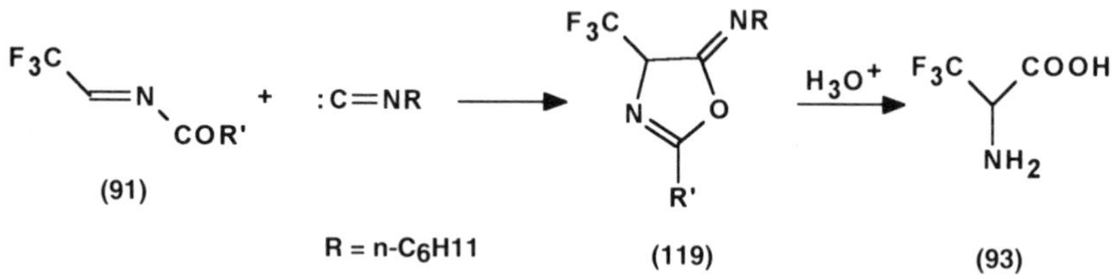

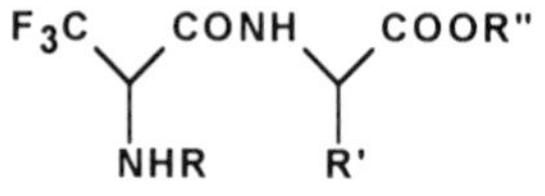

Like N-acyltrifluoroacetaldimines (91), hexafluoroacetone N-benzoylimine (121) readily enters into [1 + 4] cycloaddition reactions with isonitriles and orthoformic acid esters. Hydrolysis of the oxazolines 122 and 123 formed yielded derivatives of unstable 3,3,3,3′,3′,3′-hexafluoro-2-aminoisobutyric acid (118) [117, 118].

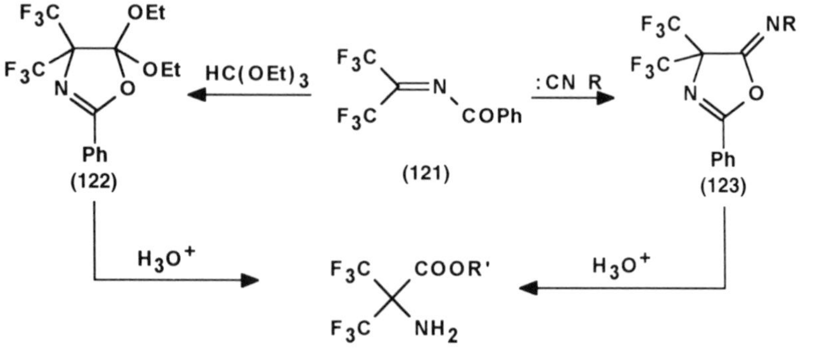

Another approach to the synthesis of trifluoroalanine involves the cyclization of *N*-trifluoroacetyl derivatives of amino acids to oxazolines (**124**), their subsequent conversion into oxazoles (**125**) and isomerization of the latter by treatment with 4-dimethylaminopyridine to oxazolines (**126**), brief heating of which at 240 °C affords oxazoles (**127**). Acid hydrolysis of **127** yields trifluoroalanine [119, 120].

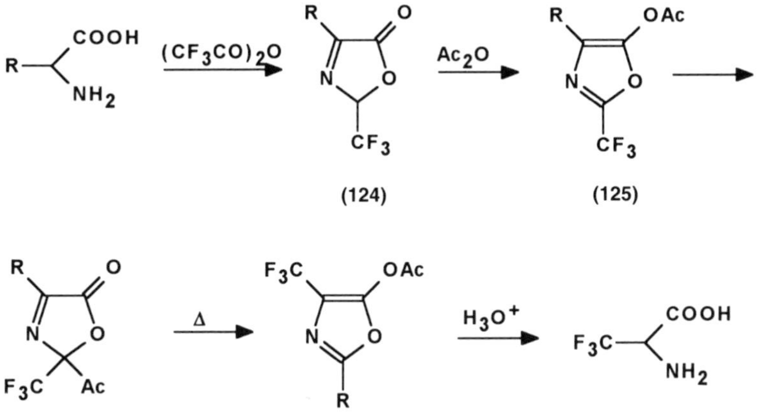

An interesting method for the synthesis of trifluoroalanine from hexafluoroacetone-*N*-benzoylimine (**128**) has been proposed [121]. The latter was transformed by treatment with tin(II) chloride into the oxazole **129**, hydrolysis of which afforded trifluoroalanine in overall 20% yield.

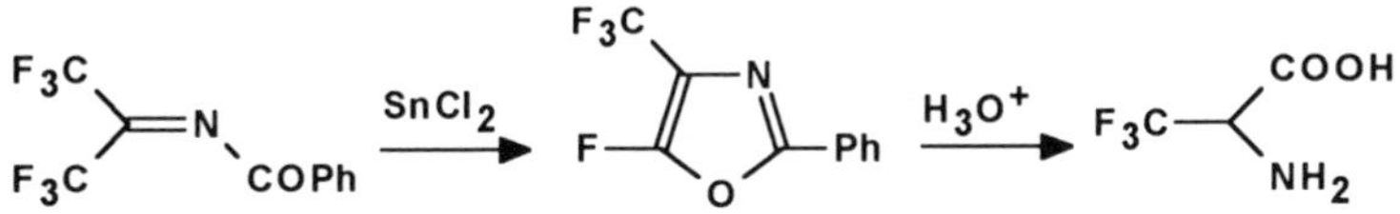

The introduction of hexafluoroacetone-*N*-acylimines (**128**) into [4 + 2] cycloaddition reactions with ketene and subsequent hydrolysis of the oxazolinones obtained (**130**) leads to β-amino-β,β-bis(trifluoromethyl)propionic acid (**131**) [118].

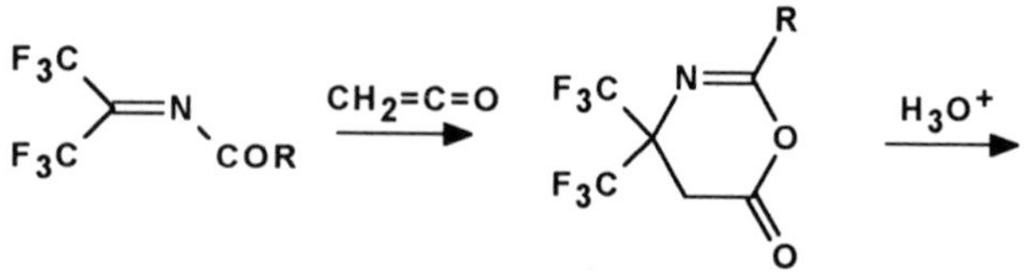

(128)   (130)

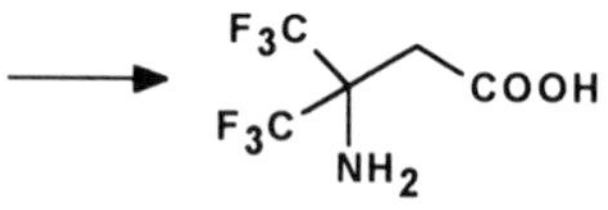

(131)

A β-alanine derivative has also been obtained by the reaction of ketene with hexafluoroacetone-N-toluene-p-sulfonylimine and subsequent hydrolysis of the propiolactam [122].

2.11 ADDITION OF AMMONIA TO UNSATURATED POLYFLUOROCARBOXYLIC ACIDS

Addition of ammonia or some of its derivatives to fluorine-containing acrylic acid (the double bond of which is strongly activated by two trifluoromethyl groups) as the final stage in the construction of fluorine-containing amino acids is used mainly to obtain derivatives of hexafluorovaline and α-trifluoromethyl-β-alanine. Hexafluorovaline (132) was synthesized for the first time in a preparative yield of 80–90% by addition of ammonia in diethyl ether at $-80\,^{\circ}$C to ethyl β,β-bis(trifluoromethyl)acrylate [123].

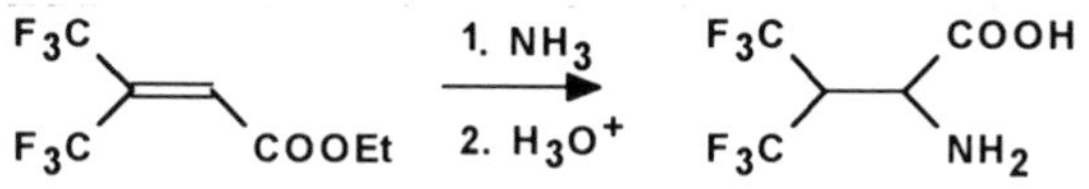

(133)   (132)

By either the addition of benzylamine to the ester 133 in a solution of methyl L-lactate or the reaction of the N-(α-phenylethyl)amide of β,β-bis(trifluoromethyl)acrylic acid with α-phenylethylamine, the asymmetric synthesis of hexafluorovaline (132) may be realized in low optical yield [124]. It is possible to use a secondary amine, hydroxylamine and derivatives of hydrazine and aziridine in addition to ammonia in the condensation with 133 [125–128].

Attempts to obtain hexafluorovalinamide (135) by the reaction of the ester 133 and ammonia failed. Compound 135 was successfully synthesized only by the reaction of ammonia with β,β-bis(trifluoromethyl)acrylic acid chloride (134) [123].

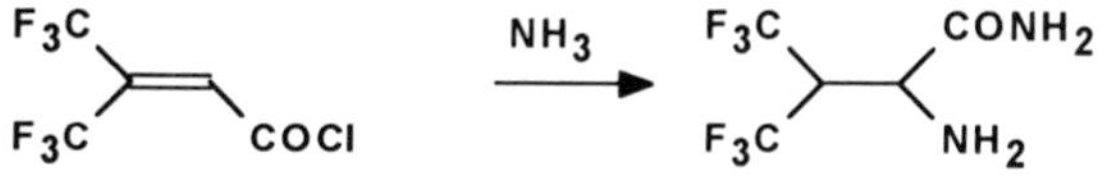

(134)   (135)

Instead of ethyl β,β-bis(trifluoromethyl)acrylate (133), it is possible to use diethyl bis(trifluoromethyl)methylidene malonate (136) in the synthesis of hexafluorovaline [129]. The addition of ammonia to the diester 136 occurs regioselectively at the α-carbon atom to give the amino diester 137, hydrolysis of which gives hexafluorovaline (132).

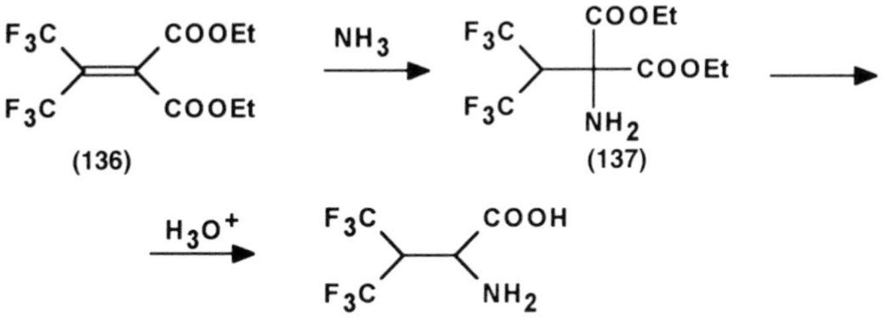

(136)   (137)

(132)

It must be emphasized that the effect of two trifluoromethyl groups on polarization of the C=C bond in the diester 136 proved to be stronger than that of two ethoxycarbonyl groups, whereas in ethyl β-trifluoromethylacrylate (138) the reverse polarization of the C=C bond occurred.

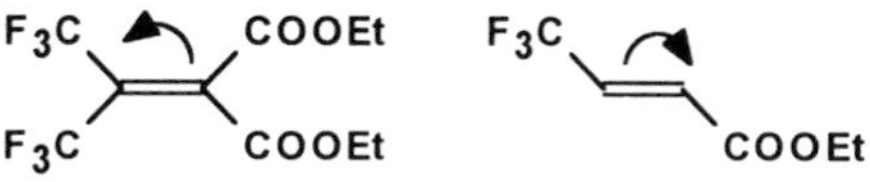

(136)   (138)

α-Trifluoroacrylic acid adds ammonia, benzylamine or hexamethyldisilazane very easily at 0–5 °C to yield α-trifluoromethyl-β-alanine or its derivatives [130, 131].

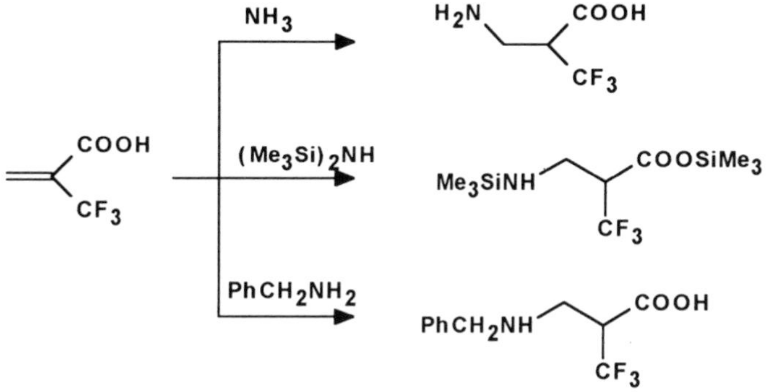

The asymmetric synthesis of N,N-diethyl-α-trifluoromethyl-β-alanine ethyl ester (140) with 35% enantiomeric excess has been achieved via condensation of ethyl α-trifluoromethylacrylate (139) with diethylamine in benzene in the presence of lipase modified with lipophilic fluorine-containing compounds [132].

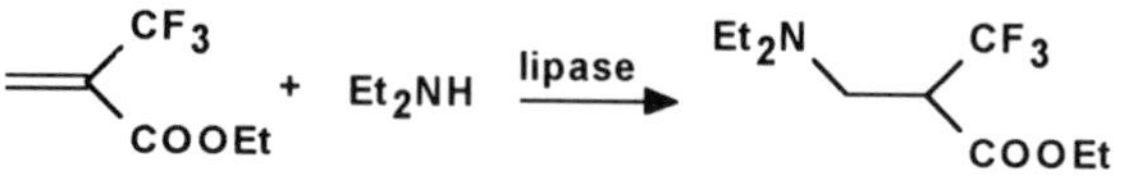

(139)   (140)

2.12 POLYFLUOROALKYLATION OF AMINO ACIDS

One of the most attractive routes to fluorine-containing amino acids involves the introduction of polyfluoroalkyl groups directly into the amino acids molecule. Very often this approach allows fluorine-containing amino acids to be obtained in the shortest time and in good yields. At the same time, if the polyfluoroalkylation conditions are sufficiently mild, no racemization takes place and enantiomers of fluorine-containing amino acids may be obtained by this method.

The result of the photochemical trifluoromethylation of (S)-N-acylhistidine methyl esters with iodotrifluoromethane is a mixture of $2'$- and $4'$-trifluoromethylhistidines. $4'$-Trifluoromethyl derivatives (32–34% yields) prevail over $2'$-isomers (21–27% yields), and small amounts of the $2',4'$-bis(trifluoromethyl) derivative of N-trifluoroacetylhistidine are also formed. Pure isomers of trifluoromethylhistidine were readily obtained by silica gel chromatography and acid hydrolysis [133].

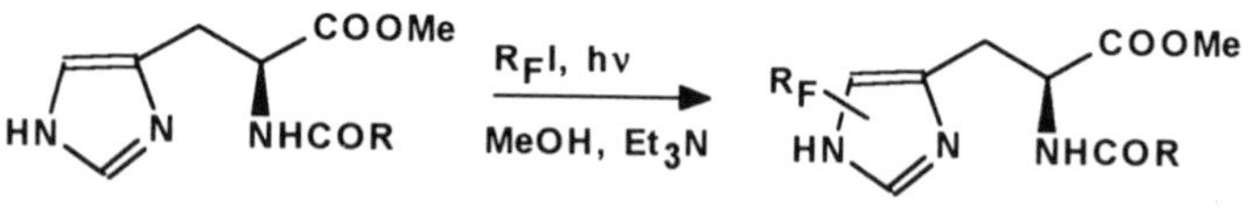

$$R_F = CF_3, \; C_2F_5$$

$$R = CH_3, \; CF_3$$

Similar results are obtained in the photochemical pentafluoroethylation of (S)-N-trifluoroacetylhistidine methyl ester with iodopentafluoroethane [134]. Polyfluoroalkylation of mercapto and phenol groups in amino acid molecules is widely used for the syntheses of fluorine-containing amino acids. (S,R)-N-Acetyl-S-trifluoromethylhomocysteine (142) was prepared in 61% yield by photochemical trifluoromethylation of (S,R)-N-acetylhomocysteine, prepared by in $situ$ alkaline hydrolysis of the thiolactone 141 [135].

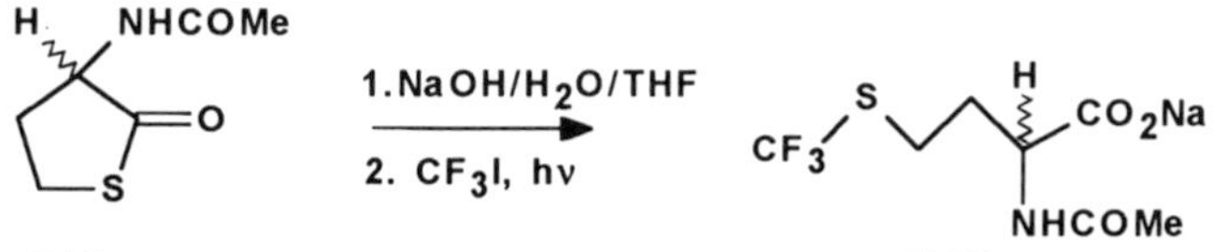

(141)　　　　　　　　　　　　　　　　　(142)

Photochemical polyfluoroalkylation of free mercapto-containing amino acids in liquid ammonia allows optically pure amino acids with polyfluoroalkylthio groups to be prepared, e.g. S-trifluoromethyl-L-homocysteine, -L-cysteine and -D-penicillamine in 60–80% yields [136]. In this reaction, unprotected mercaptoamino acids can be used because trifluoromethylation of the mercapto group is preferred over that of other nucleophilic functional groups such as free amino and carboxylate groups.

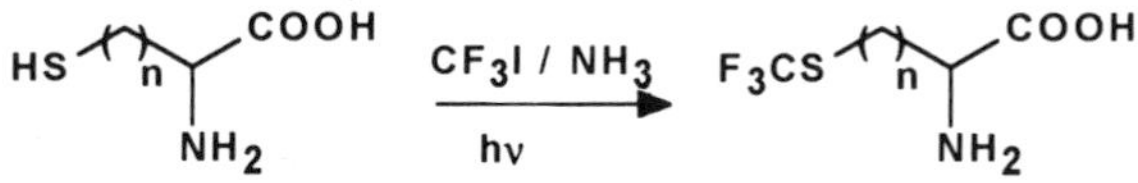

The trifluoromethylation of mercapto-containing amino acids can be carried out with trifluoromethylating agents other than iodotrifluoromethane, e.g. N-nitroso-N-trifluoromethylphenylsulfamide (143). Thus, the S-trifluoromethylcysteine derivative 145 is formed after UV irradiation of the N-trifluoroacetylcystine methyl ester 144 with the sulfamide 143 [137].

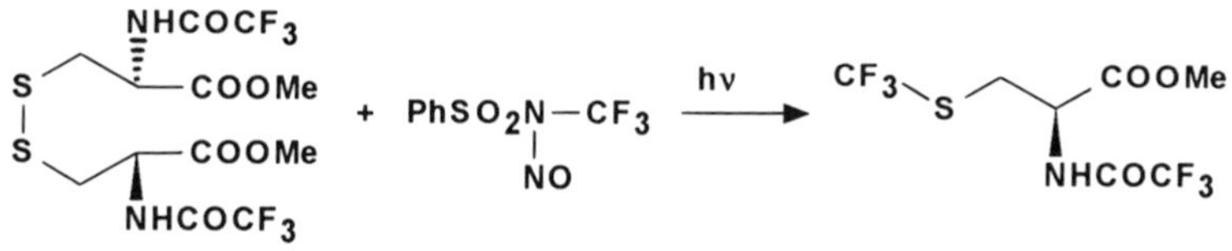

(144)　　　　　　　　(143)　　　　　　　　(145)

Simple and readily available reagents such as Freons and polyfluoroalkenes have also been successfully used to synthesize fluorine-containing amino acids. Thus, (S, R)-N-acetyl- [135] and free (S)-homocysteine [138] react with difluorocarbene formed from difluorochloromethane (Freon-11) under alkaline conditions and the corresponding S-difluoromethyl derivatives of homocysteine are formed. Difluoromethylation of (S)-homocysteine proceeds without racemization [138].

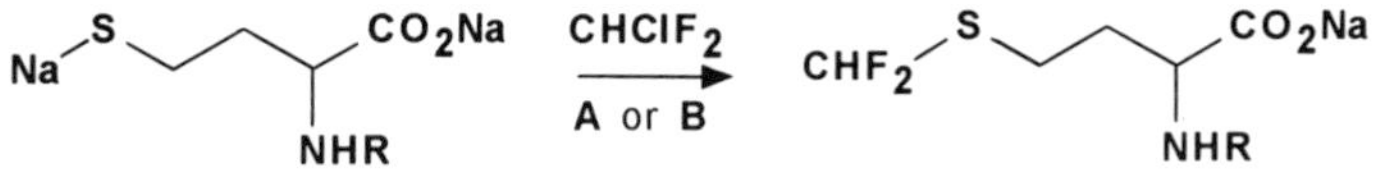

R = COCH$_3$, A = NaOH/H$_2$O/THF [135]

R = H, B = KOBu-t/EtOH [138]

S-(1',1',2'-Trifluoro-2'-chloroethyl)-containing (S)cysteine was obtained by nucleophilic addition of chlorotrifluoroethylene to the mercapto group of N-Boc-protected and free (S)-cysteine under alkaline conditions [139].

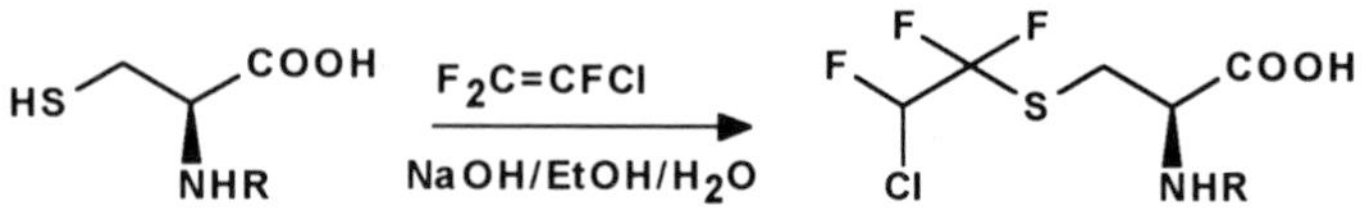

R = H, Boc

The procedure for the polyfluoroalkylation of phenols has been applied to the synthesis of O-polyfluoroalkylated (S)-tyrosine derivatives [140]. Thus, treatment of N-Boc or N-Cbz-(S)-tyrosine with chlorodifluoromethane in the presence of base affords the corresponding optically active N-protected O-difluoromethyltyrosine derivatives.

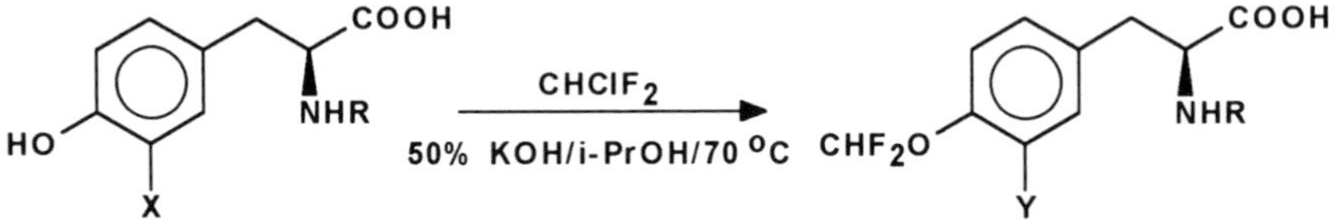

R = Boc, Cbz ; X = Y = H
R = Boc ; X = OH ; Y = CHF$_2$O

Free O-difluoromethyl-(S)-tyrosine was isolated by the action of trifluoro-acetic acid on the N-Boc derivative. In a similar way, difluoromethylation of N-Boc-(S, R)-3',4'-dihydroxyphenylalanine affords 3',4'-bis(difluoromethoxy)-(S,R)-phenylalanine [140].

The interaction of N-Boc- or N-Cbz-(S)-tyrosine with hexafluoropropylene in aqueous triethylamine at 70 °C provides the corresponding (S)-4'-(1,2,2,3,3,3-hexafluoropropoxy)phenylalanine. On the other hand, the 4'-poly-haloethoxy derivative of N-Boc-(S)-tyrosine has been prepared by treatment of N-Boc-(S)-tyrosine with Halotan ($CF_3CHBrCl$) in 50% aqueous potassium hydroxide solution containing a phase-transfer catalyst [140].

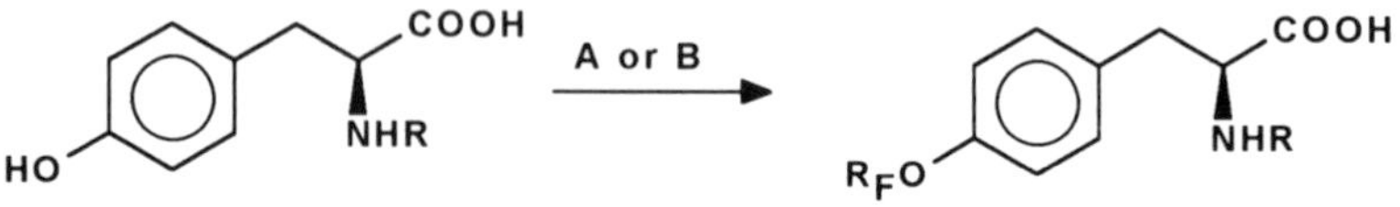

A: $CF_3CF=CF_2/H_2O/Et_3N/70^oC$; R = Boc, Cbz ; R_F = CF_3CHFCF_2

B: $CF_3CHBrCl/50\%$ $KOH/Bu_4N^+Br^-/90^oC$; R = Boc ; R_F = $CHBrClCF_2$

In contrast to N-Boc-(S)-tyrosine, N-Boc-(R)-4-hydroxyphenylglycine was completely racemized under the difluoromethylation conditions whereas a hexafluoropropoxy-containing amino acid was formed in optically active form [141].

A: $CHClF_2$/50% KOH/i-$PrOH$/70°C;; R_F = CHF_2

B: $CF_3CF=CF_2$/H_2O/Et_3N/70°C ; R_F = CF_3CHFCF_2

The different stability of tyrosine and phenylglycine derivatives to racemization under difluoromethylation conditions can be explained by the greater mobility of the α-proton in the latter case.

2.13 REFERENCES

1. G. Firnau, R. Chiracal, and E. S. Garnett, *J. Nucl. Med.*, **25**, 1228 (1984).
2. G. Firnau, R. Chiracal, and E. S. Garnett, *J. Labelled Compd. Radiopharm.*, **23**, 1106 (1986).
3. Firnau, R. Chiracal, S. Sood, C. Nachmias and G. Schrobilgen, *Appl. Radiat. Isot.* **37**, 669 (1986); *Chem Abstr.*, **105**, 205455k (1986).
4. M. J. Adam, T. J. Rith, J. R. Grierson, B. Abeysekera and B. D. Pate, *J. Nucl. Med.*, **27**, 1462 (1986); *Chem Abstr.*, **106**, 214330p (1987).
5. T. Nozaki and Y. Tanaka, *Int. J. Appl. Radiat. Isot.*, **18**, 111 (1967); *Chem. Abstr.*, **67**, 11283d (1967).
6. Y. Adam, J. R. Grierson, T. J. Ruth and S. Jivan, *Appl. Radiat. Isot.*, **37**, 877 (1986); *Chem. Abstr.*, **106**, 19012j (1987); M. J. Adam and S. Jivan, *J. Labelled Compd. Radiopharm.*, **31**, 39 (1992); *Chem. Abstr.*, **116**, 152319n (1992).
7. M. Diksic and S. Farrokhzad, *J. Nucl. Med.*, **26**, 1314 (1985); *Chem. Abstr.*, **105**, 153500z (1986).
8. R. Chiracal, K. L. Brown, G. Firnau and E. S. Garnett, *J. Fluorine Chem.*, **37**, 267 (1987).
9. H. H. Coenen, W. Bodsch and K. Takahashi, *et al. Nucl. Med., Suppl.*, **22**, 600 (1986).
10. D. Hebel, K. L. Kirk, L. A. Cohen and V. M. Labroo, *Tetrahedron Lett.*, **31**, 619 (1990).
11. M. G. Straatmann and M. J. Welch, *J. Nucl. Med.*, **18**, 151 (1977).
12. T. Nozaki and Y. Tanaka, *Int. J. Appl. Radiat. Isot.*, **18**, 111 (1967).
13. M. Murakami, K. Takahashi, Y. Kondo, S. Mizusawa, H. Nakamichi, H. Sasaki, E. Hagami, H. Iida, I. Kanno, S. Miura and K. Uemura, *J. Labelled Compd. Radiopharm.*, **25**, 773 (1988).
14. R. Chiracal, G. Firnau and E. S. Garnett, *J. Labelled Compd. Radiopharm.*, **26**, 228 (1989).
15. R. Chiracal, B. G. Sayer, G. Firnau and E. S. Garnett, *J. labelled Compd. Radiopharm.*, **25**, 63 (1988).
16. M. R. Kilburu (Ed.), *Fluorine-18 Labelling of Radiopharmaceuticals*, National Academy Press, Washington, DC, 1990.

17. T. Tshushima, J. Nishikawa, T. Sato, H. Tanida, K. Tori, T. Tsuji, S. Misaki and M. Suefuji, *Tetrahedron Lett.*, **21**, 3593 (1980).
18. T. Tsushima, K. Kawada, J. Nishikawa, T. Sato, K. Tori, T. Tsuji and S. Misaki, *J. Org. Chem.*, **49**, 1163 (1984).
19. T. Tsushima, K. Kawada and T. Tsuji, *J. Org. Chem.*, **47**, 1107 (1982).
20. T. Tsushima, K. Kawada, T. Tsuji and K. Tawara, *J. Med. Chem.*, **28**, 253 (1985).
21. J. Kollonitsch, S. Marburg and L. M. Perkins, *J. Org. Chem.*, **41**, 3107 (1976).
22. D. Block, H. H. Coenen and G. Stoecklin, *J. Labelled Compd. Radiopharm.*, **25**, 201 (1988).
23. G. Firnau, R. Chiracal and S. Sood, *Can. J. Chem.*, **58**, 1449, (1980).
24. G. S. Garrett, T. J. Emge, S. C. Lee, E. M. Fisher, K. Dyehouse and J. M. McIver, *J. Org. Chem.*, **56**, 4823 (1991).
25. A. F. Jansen, P. M. C. Wang and A. E. Lemire, *J. Fluorine Chem.*, **22**, 557 (1983).
26. M. J. Adam, T. J. Ruth, J. R. Grierson, B. Abeysekers and B. D. Pate, *J. Nucl. Med.* **27**, 1462 (1986).
27. R. Chiracal, G. Firnau, J. Couse and E. S. Garnett, *Int. J. Appl. Radiat. Isot.*, **35**, 651 (1984).
28. M. J. Adam, J. R. Grierson, T. J. Ruth and S. Jivan, *Appl. Radiat. Isot.*, **37**, 877 (1986).
29. M. J. Adam and S. Jivan, *Appl. Radiat. Isot.*, **39**, 1203 (1988).
30. A. Luxen, J. R. Barrio, J. T. Bida and N. Satyamurthy, *J. Labelled Comp. Radiopharm.*, **23**, 1066 (1986).
31. A. Luxen, J. T. Bida, M. E. Phelps and J. R. Barrio, *J. Nucl. Med.*, **28**, 624 (1987).
32. A. Luxen and J. R. Barrio, *Tetrahedron Lett.*, 1501 (1988).
33. J. R. Barrio, M. M. Perlmutter, A. Luxen *et al.*, *J. Nucl. Med.*, **30**, 752 (1989).
34. O. T. De Jesus, J. J. Sunderland, C. -A. Chen *et al.*, *J. Nucl. Med.*, **30**, 930 (1989).
35. D. L. Gildersleeve, M. E. Van Dort, K. S. Rosenspire, S. Toorongian and D. M. Wieland, *J. Nucl. Med.*, **30**, 752 (1989).
36. M. Van der Ley, *J. Labelled Comp. Radiopharm.*, **20**, 453 (1983).
37. J. Kollonitsch and L. Barash, *J. Am. Chem. Soc.*, **98**, 5591 (1976).
38. J. Kollonitsch, *US Pat.* 3 956 367 (1974); *Chem. Abstr.*, **85**, 160526 (1976).
39. F. M. Kahan, *US Pat.* 4 028 405 (1976); *Chem. Abstr.*, **87**, 68655 (1977).
40. J. Kollonitsch, *US Pat.* 4 030 994 (1976); *Chem. Abstr.*, **87**, 83904 (1977).
41. J. Kollonitsch, *US Pat.* 4 431 817 (1978); *Chem. Abstr.*, **90**, 186348 (1979).
42. J. Kollonitsch, L. Barash and G. A. Dolduras, *J. Am. Chem. Soc.*, **92**, 7494 (1970).
43. J. Kollonitsch, *Dutch Pat.* 7 514 240 (1975); *Chem Abstr.*, **86**, 44236 (1976).
44. P. E. Curtley and R. F. Hirschmann, *US Pat.* 4 427 661 (1983); *Chem. Abstr.*, **100**, 192286 (1984).
45. J. Kollonitsch, S. Marburg and L. M. Perkins, *J. Org. Chem.*, **41**, 3107 (1976).
46. V. Tolman and K. Veres, *Tetrahedron Lett.*, 3909 (1966).
47. V. Tolman and K. Veres, *Collect. Czech. Chem. Commun.*, **32**, 4460 (1967).
48. V. Tolman, and K. Veres, *Tetrahedron Lett.*, 1967 (1964).
49. L. V. Alekseeva, B. N. Lundin and N. L. Burde, *Zh. Obshch. Khim.*, **37**, 1754 (1967).
50. J. J. M. Hageman, M. J. Wanner, G. Koomen and U. K. Pandit, *J. Med. Chem.*, **20**, 1677 (1977).
51. M. J. Wanner, J. J. M. Hageman, G. Koomen and U. K. Pandit, *J. Med. Chem.*, **23**, 85 (1980).
52. F. N. Shirota, H. T. Nagasawa and J. A. Elberling, *J. Med. Chem.*, **20**, 1623 (1977).
53. F. N. Shirota, H. T. Nagasawa and J. A. Elberling, *J. Med. Chem.*, **20**, 1176 (1977).

54. J. Kollonitsch, S. Marburg and L. M. Perkins, *J. Org. Chem.*, **40**, 3808 (1975).
55. J. Kollonitsch, S. Marburg and L. M. Perkins, *J. Org. Chem.*, **44**, 771 (1979).
56. P. J. Reider, R. S. E. Conn, P. Davis, V. J. Grenda, A. J. Zambito and E. J. J. Grabowski, *J. Org. Chem.*, **52**, 3326 (1987).
57. A. W. Douglas and P. J. Reider, *Tetrahedron Lett.*, **25**, 2851, (1984).
58. A. Vidal-Cros, M. Gaudry and A. Maraquet, *J. Org. Chem.*, **50**, 3163 (1985).
59. J. Kollonitsch, A. A. Patchett and S. Marburg, *US Pat.* 4 325 961 (1978); *Chem. Abstr.*, **91**, 21 109 (1979).
60. P. E. and R. F. Hirshmann, *US Pat.* 4 427 661 (1900); *Chem. Abstr.*, **100**, 192286 (1984).
61. M. Hudlicky, *Org. React.*, **35**, 513 (1988).
62. T. Tsushima, T. Sato and T. Tsuji, *Tetrahedron Lett.*, **21**, 3591 (1980).
63. T. Tsushima, J. Nishikawa, T. Sato *et al.*, *Tetrahedron Lett.*, **21**, 3593 (1980).
64. C. Scolastico, E. Conca, L. Prati *et al.*, *Synthesis*, 850 (1985).
65. M. Hudlicky and J. S. Merola, *Tetrahedron Lett.*, **31**, 7403 (1990).
66. A. G. Avent, A. N. Bowler, P. M. Doyle, C. M. Marchand and D. W. Young, *Tetrahedron Lett.*, **33**, 1509 (1992).
67. U. Groth and U. Schollkopf, *Synthesis*, 673 (1983).
68. R. B. Silverman, S. C. Durkee and B. J. Invergo, *J. Med. Chem.*, **29**, 764 (1986).
69. L. Somekh and A. Shanzer, *J. Am. Chem. Soc.*, **104**, 5836 (1982).
70. S. V. Pansare and J. C. Vederas, *J. Org. Chem.*, **52**, 4804 (1987).
71. D. E. Zembover, J. A. Gilbert and M. M. Ames, *J. Med. Chem.*, **36**, 305 (1993).
72. M. Sharma, C. R. Petrie and W. Korytnyk, *Carbohydr. Res.*, **175**, 25 (1988).
73. M. Hudlicky, *Collect. Czech. Chem. Commun.*, **32**, 453 (1967).
74. M. Kolb, J. Barth, J. -G. Heydt *et al.*, *J. Med. Chem.*, **30**, 267 (1987).
75. J. A. Wilkinson, *Chem. Rev.*, **92**, 505 (1992).
76. A. I. Ayi and R. Guedj, *J. Chem. Soc., Perkin Trans. 1*, 2045 (1983).
77. T. N. Wade, F. Graymard and R. Guedj, *Fr. Pat.* 2 449 675 (1980); *Chem. Abstr.*, **95**, 81527 (1981).
78. T. N. Wade, F. Graymard and R. Guedj, *Tetrahedron Lett.*, **20**, 2681 (1979).
79. A. Barama, R. Condom and R. Guedj, *J. Fluorine Chem.*, **16**, 183 (1980).
80. M. L. M. Alkaniz, N. Patino, R. Condom *et al.*, *J. Fluorine Chem.*, **35**, 70 (1980).
81. T. N. Wade and R. Guedj, *Tetrahedron Lett.*, **20**, 3953 (1979).
82. A. I. Ayi, M. Remli and R. Guedj, *J. Fluorine Chem.*, **18**, 93 (1981).
83. R. Guedj, A. I. Ayi and M. Remli, *Ann. Chim. Fr.*, **9**, 691 (1983).
84. T. N. Wade, *J. Org. Chem.*, **45**, 5328 (1984).
85. T. N. Wade and R. Kheribet, *J. Chem. Res. (S)*, 210 (1980).
86. T. N. Wade and R. Kheribet, *J. Org. Chem.*, **45**, 5333 (1980).
87. A. I. Ayi, M. Remli and R. Guedj, *Tetrahedron Lett.*, **22**, 1505 (1981).
88. A. Ourari, R. Condom and R. Guedj, *Can. J. Chem.*, **60**, 2707 (1982).
89. R. B. Silverman and M. A. Levy, *J. Org. Chem.*, **45**, 815 (1980).
90. M. A. Levy, A. J. Muztar, R. B. Silverman *et al.*, *Biochemistry*, **20**, 1197 (1981).
91. C. -Y. Yuean, C. -N. Chang and I. -F. Yeh, *Yao Hsueh Pao*, **7**, 237 (1959).; *Chem. Abstr.*, **54**, 12096 (1959).
92. S. Backerman, R. L. Martin, A. W. Burgstrahler *et al.*, *Nature* (London) **212**, 849 (1966).
93. A. A. Gottlieb, Y. Fujita, S. Udenfried and B. Witkop, *Biochemistry*, **4**, 2507 (1965).
94. L. V. Aleseeva, B. N. Lundin and N. L. Burde, *Zh. Obshch. Khim.* **38**, 1687 (1968).
95. N. N. Yarovenko and M. A. Raksha, *Zh. Obshch. Khim.*, **29**, 2159 (1959).
96. E. D. Bergmann and L. Chun-Tsu, *Synthesis*, 44 (1973).
97. K. L. Kirk, *J. Org. Chem.*, **45**, 2015 (1980).
98. K. L. Kirk and L. A. Cohen, *J. Am. Chem. Soc.*, **95**, 4619 (1973).

99. K. L. Kirk, W. Nagai and L. A. Cohen, *J. Am. Chem. Soc.*, **95**, 8389 (1973).
100. F. Weygand, W. Steglich, W. Oettmeier *et al.*, *Angew. Chem.*, **78**, 640 (1966).
101. F. Weygand, W. Steglich and W. Oettmeier, *Chem. Ber.*, **103**, 1655 (1970).
102. F. Weygand, W. Steglich and W. Oettmeier, *Chem. Ber.*, **103**, 818 (1970).
103. F. Weygand and W. Steglich, *Chem. Ber.*, **98**, 487 (1965).
104. F. Weygand, W. Steglich and F. Fraunberger, *Angew. Chem.* **79**, 822 (1967).
105. F. Weygand, W. Steglich and F. Fraunberger, *Angew. Chem., Int. Ed. Engl.*, **6**, 808 (1967).
106. U. Andor, N. Agnes and K. Endre, *Magy. Kem. Foly.*, **79**, 333 (1973).
107. V. A. Soloshonok, I. I. Gerus and Yu. L. Yagupol'skii, *Zh. Org. Khim.*, **22**, 1335 (1986).
108. V. A. Soloshonok, I. I. Gerus and Yu. L. Yagupol'skii, *Zh. Org. Khim.*, **23**, 2308 (1987).
109. S. I. Kobzev, V. A. Soloshonok, S. V. Galushko *et al.*, *Zh. Obshch. Khim.*, **59**, 909 (1989).
110. S. N. Osipov, N. D. Chkanikov, A. F. Kolomeits *et al.*, *Izv. Akad. Nauk SSSR, Ser. Khim., 1384 (1986)*.
111. V. A. Soloshonok and V. P. Kukhar', *Zh. Org. Khim.*, **26**, 419 (1990).
112. S. N. Osipov, A. F. Kolomeits and A. V. Fokin, *Izv. Akad. Nauk SSSR, Ser. Khim.*, 132 (1988).
113. I. L. Knunyants, V. V. Shokina, V. V. Tyuleneva *et al.*, *Izv. Akad. Nauk SSSR, Ser. Khim.*, 1831 (1966).
114. I. L. Knunyants, V. V. Shokina, V. V. Tyuleneva *et al.*, *Izv. Akad. Nauk SSSR, Ser. Khim.*, 415 (1968).
115. Yu. V. Zeifman, V. V. Tyuleneva, A. P. Pleshakova *et al.*, *Izv. Akad. Nauk SSSR, Ser. Khim.*, 2732 (1975).
116. Yu. V. Zeifman, L. T. Lantseva and I. L. Knunyants, *Izv. Akad. Nauk SSSR, Ser. Khim.*, 401 (1986).
117. N. P. Gambaryan, E. M. Rokhlin, Yu. V. Zeifman *et al.*, *Dokl. Akad. Nauk SSSR*, **166**, 864 (1966).
118. Yu. V. Zeifman, N. P. Gambaryan, L. A. Simonyan *et al.*, *Zh. Obshch. Khim.*, **37**, 2476 (1967).
119. G. Hoffe and W. Steglich, *Chem. Ber.*, **104**, 1408 (1971).
120. G. Hoffe, W. Steglich and H. Vorbruggen, *Angew. Chem.*, **90**, 602 (1978).
121. K. Burger, D. Huble and P. Gertitschke, *J. Fluorine Chem.*, **27**, 327 (1985).
122. Yu. V. Zeifman and I. L. Knunyants, *Dokl. Akad. Nauk SSSR*, **173**, 354 (1967).
123. I. L. Knunyants and Yu. A. Cheburkov, *Izv. Akad. Nauk SSSR, Ser. Khim.*, 2162 (1960).
124. A. V. Eremeev, I. V. Solodin and F. R. Polyak, *Izv. Akad. Nauk. Latv. SSR, Ser. Khim.*, 345 (1985).
125. I. V. Solodin, A. V. Eremeev, I. I. Chervin *et al.*, *Khim. Geterosikl. Soedin.*, 1359 (1985).
126. A. V. Eremeev, I. V. Solodin and E. E. Liepinsh, *Khim. Geterosikl. Soedin.*, 917 (1984).
127. I. V. Solodin, F. R. Polyak and A. V. Eremeev, *Khim. Geterosikl. Soedin.*, 1335 (1985).
128. I. V. Solodin, A. V. Eremeev and E. E. Liepinsh, *Khim. Geterosikl. Soedin.*, 505 (1984).
129. T. Y. Chen, N. P. Gambaryan and I. L. Knunyants, *Dokl. Akad. Nauk SSSR*, **133**, 1113 (1960).
130. I. Ojlma, *Actual. Chem. Fr.*, 171 (1987).
131. I. Ojima, K. Kato, K. Nakahashi *et al.*, *J. Org. Chem.*, **54**, 4511 (1989).
132. T. Kitszume and K. Murata, *J. Fluorine Chem.*, **36**, 339 (1987).

133. H. Kimoto, S. Fujii and L. A. Cohen, *J. Org. Chem.*, **49**, 1060 (1984).
134. S. Fujii, Y. Maki, H. Kimoto and L. A. Cohen, *J. Fluorine Chem.*, **35**, 437 (1987).
135. M. E. Houston, Jr and J. F. Honek, *J. Chem. Soc., Chem. Commun.*, 761 (1989).
136. V. Soloshonok, V. Kukhar', Y. Pustovit and V. Nazaretian, *Synlett*, 657 (1992).
137. T. Umemoto and O. Miyano, *Toyo Soda Kenkyu Hokoku, Sci. Rep. Toyo Soda Manuf. Co. Ltd.*, **27**, 69 (1983); *Chem Abstr.*, **100**, 67911z (1984).
138. T. Tsushima, S. Ishihara and Y. Fujita, *Tetrahedron Lett.*, **31**, 3017 (1990).
139. D. R. Dohn, J. R. Leininger and L. H. Lash, *J. Pharmacol. Exp. Ther.*, **235**, 51 (1985).
140. M. T. Kolycheva, I. I. Gerus, Yu. L. Yagupol'skii, S. V. Galushko and V. P. Kukhar', *Zh. Org. Khim.*, **27**, 781 (1991).
141. M. T. Kolycheva, I. I. Gerus and V. P. Kukhar', *Amino Acids*, **5**, 99 (1993).

3 Synthesis of Fluorine-containing Amino Acids by Means of Homogeneous Catalysis

IWAO OJIMA and QING DONG
Department of Chemistry, State University of New York at Stony Brook, Stony Brook, NY 11794-3400, USA

3.1 INTRODUCTION

It has been shown that fluorinated analogs of naturally occurring biologically active compounds often exhibit unique physiological activities [1, 2]. For example, fluorinated pyrimidines act as anticancer and antiviral agents, some fluoroaromatic compounds and CF_3-aromatic compounds are being used as non-steroidal antiinflammatory drugs, antifungal agents, human antiparasitic agents, central nervous system agents for psychopharmacology, diuretics and antihypertensive agents and some fluoroamino acids act as 'suicide substrate enzyme inactivators' showing strong antibacterial activities; some of them also act as antihypertensive agents [3]. There has been an increasing interest in the incorporation of fluoroamino acids into peptides [4]. Accordingly, the development of practical synthetic methods which enable us to introduce these fluoro groups effectively and selectively into the desired amino acids from readily available materials is of significant synthetic importance.

In this respect, commercially available fluoroalkenes such as 3,3,3-tri-fluoropropene (TFP), vinyl fluoride (VF) and pentafluorostyrene (PFS) are important starting materials. Hydrocarbonylations of fluoroalkenes provide versatile intermediates for the syntheses of a variety of useful organofluorine compounds, especially fluoroamino acids. Efficient and practical methods have been developed for (i) the syntheses of 2- and 3-trifluoromethylpropanals (2-TFMPA and 3-TFMPA) and 2-pentafluorophenylpropanal (2-PFPPA) based on the extremely regioselective hydroformylation of 3,3,3-trifluoropropene (TFP) and pentafluorostyrene (PFS) and (ii) the synthesis of 2-tri-fluoromethylacrylic acid (TFMAA) through the carboxylation of 2-bromo-3,3,3-trifluoropropene (2-Br-TFP).

The fluoroaldehydes and fluoroacrylic acid thus obtained serve as excellent intermediates for the synthesis of fluoroamino acids. Enzymatic optical resolution and asymmetric hydrogenation catalyzed by chiral rhodium complexes provide enantiomerically pure or enriched fluoroamino acids. A

Fluorine-containing Amino Acids: Synthesis and Properties   Edited by V. P. Kukhar' and V. A. Soloshonok
© 1995 John Wiley & Sons Ltd

palladium-catalyzed carbalkoxylation of perfluoroalkylimidoyl iodides furnishes a convenient route to fluoroamino acids. We review in this chapter efficient syntheses of 4,4,4-trifluorovaline (TFV), 5,5,5-trifluoronorvaline (TFNV), 5,5,5-trifluoroleucine (TFL), 6,6,6-trifluoronorleucine (TFNL), fluoro- and trifluoromethylphenylalanines, 4,5,6,7-tetrafluorotryptophan, α-trifluoromethyl-β-alanine (α-TFM-β-Ala), trifluoroalanine, α-R_f-norleucine, α-CF$_3$-alanine and related compounds.

3.2 SYNTHESES OF FLUORINE-CONTAINING ALDEHYDES AND 2-TRIFLUOROMETHYLACRYLIC ACID THROUGH CARBONYLATIONS

3.2.1 SYNTHESES OF FLUORINE-CONTAINING ALDEHYDES

Hydroformylation of alkenes is an important reaction for the practical synthesis of aldehydes, and a detailed mechanism of this reaction and applications to organic syntheses have been extensively studied [5]. However, little had been known about the reactions of alkenes bearing perfluoroalkyl or perfluoroaryl substituents [6] until Fuchikami and Ojima reported the highly regioselective hydroformylation of these alkenes in 1982 [7b]. Hydroformylation of a variety of fluoroalkenes has shown unusually high regioselectivity and a remarkable dependence of the regioselectivities on the catalyst metal species; this is unique in comparison with the reaction of ordinary alkenes [7, 8].

$$R_f\text{-CH=CH}_2 + H_2 + CO \xrightarrow{\text{catalyst}} R_fCH_2CH_2CHO + R_f(Me)CH\text{-CHO} \qquad (3.1)$$

$$\textbf{(1)} \qquad\qquad\qquad \textbf{(2)}\,(n) \qquad \textbf{(3)}\,(iso)$$

$$R_f = F, CF_3, C_2F_5, C_3F_7, C_8F_{17}, C_6F_5$$

The hydroformylation of TFP has been carried out with $Co_2(CO)_8$, $Ru_3(CO)_{12}$, $Rh_6(CO)_{16}$ and $PtCl_2(DIOP)/SnCl_2$, which are typical hydroformylation catalysts, at 100 °C and 100 atm ($CO/H_2 = 1$) for the cobalt, platinum and ruthenium catalysts and at 80 °C and 110 atm ($CO/H_2 = 1$) for the Rh catalyst. Results are listed in Table 3.1 [7, 9].

As Table 3.1 shows, the $Co_2(CO)_8$-catalyzed reaction of TFP gives (trifluoromethyl)propanals (TFMPA) in 95% yield, where an n-aldehyde, $CF_3CH_2CH_2CHO$ (3-TFMPA), is formed with high regioselectivity (93%). In sharp contrast with $Co_2(CO)_8$, a rhodium carbonyl cluster, $Rh_6(CO)_{16}$, exhibits extremely high catalytic activity and regioselectivity (96%) to give an iso-aldehyde, $CF_3(CH_3)CHCHO$ (2-TFMPA). The platinum catalyst, $PtCl_2(DIOP)/SnCl_2$, favors the formation of the n-aldehyde ($n/iso = 71:29$), while $Ru_3(CO)_{12}$ gives the iso-aldehyde as the main product ($m/iso = 15:85$), and in both cases substantial amounts of hydrogenated product, $CF_3CH_2CH_3$, are formed (25–38%). Addition of triphenylphosphine to the Co, Ru and Rh catalysts considerably decreases their

Table 3.1 Hydroformylation of TFP[a]. (Reprinted with permission from Ojima, *Chem. Rev.*, **88**, 1011. Copyright (1988) American Chemical Society)

Catalyst	TFP/ cat[b]	Pressure (atm)[c] ($CO/H_2 = 1$)	Temperature (°C)	Time (h)	Aldehydes[d] %	iso/n	Alkane[d] (%)
$Co_2(CO)_8$	50	130	100	20	95	7:93	0
$Co_2(CO)_8/PPh_3$	50	130	100	41	3	9:91	1
$PtCl_2(DIOP)/SnCl_2$	100	130	100	4	75	29:71	25
$Ru_3(CO)_{12}$	33	130	100	16	62	85:15	38
$Ru_3(CO)_{12}/PPh_3$	33	130	100	39	25	92:8	1
$Rh_6(CO)_{16}$	1200	110	80	5	98	96:4	2
$Rh_6(CO)_{16}/PPh_3$	1200	110	80	22	93	97:3	7
$Rh_4(CO)_{12}$	1200	110	80	6	97	97:3	3
$Rh–C/P(OPh)_3$	1200	110	80	5	98	96:4	2
$Rh–C$	1200	110	80	5	96	96:4	4
$HRh(CO)(PPh_3)_3$	1200	110	80	5	95	95:5	5
$Rh–C/PPh_3$	1200	110	80	15	90	95:5	10
$RhCl(dppb)$	1200	110	80	22	42	97:3	<4
$RhCl(PPh_3)_3$	1200	110	80	22	30	96:4	<3
$RhCl(CO)(PPh_3)_2$	1200	110	80	22	23	95:5	<2
$RhCl_3.3H_2O/PPh_3$	1200	110	80	22	17	96:4	<2

[a] All experiments were run with 130 mmol of TFL in 20 ml of toluene.
[b] (mol TFP)/(mol metal).
[c] Initial pressure at room temperature.
[d] Determined by GLC.

catalytic activities but somewhat increases the *iso*-aldehyde selectivity. This result contrasts with the case of ordinary alkenes, where the addition of triphenylphosphine increases the *n*-aldehyde selectivity [6]. Since $Rh_6(CO)_{16}$ gave excellent regioselectivity in the formation of 2-TFMPA, several other rhodium catalysts have been employed to examine their catalytic activities and regioselectivities [7, 9]. The results are also listed in Table 3.1.

The results clearly indicate that the rhodium(I) complexes having chlorine as a ligand, such as $RhCl(PPh_3)_3$, are less active than $HRh(CO)$ $(PPh_3)_3$, Rh–C, $Rh_4(CO)_{12}$ and $Rh_6(CO)_{16}$, but the regioselectivity is virtually the same in all cases examined.

Consequently, it is concluded that the nature of the central metal of the catalyst plays a key role in determining the regioselectivity of the reaction. Moreover, it should be noted that in the present reaction the dependence of regioselectivity on the metal species is remarkable compared with that reported for propene [6]. The reported regioselectivities in the formation of butanal using cobalt, platinum, ruthenium and rhodium catalysts are as follows: $Co_2(CO)_8$ (150 atm; $CO/H_2 = 1$; 110 °C) 94%, *iso/n* = 20:80 [10]; $PtCl_2(PPh_3)_2/SnCl_2$ (89 atm; $CO/H_2 = 1$; 66 °C) 90%, *iso/n* = 13:87 [11]; $Ru_3(CO)_{12}$ (150 atm; $CO/H_2 = 1$; 110 °C) 40%, *iso/n* = 26:74 [10]; $Rh_6(CO)_{16}$ (120 atm; $CO/H_2 = 1$; 70 °C) 51%, *iso/n* = 49:51 [12].

In a similar manner, the hydroformylation of PFS has been carried out at 90 °C and 80 atm with the use of cobalt, platinum, ruthenium and rhodium catalysts. The results are shown in Table 3.2 [7, 9]. Rhodium catalysts exhibit high catalytic activity to give the iso-aldehyde, $C_6F_5(CH_3)CHCHO$ (2-PFPPA), with excellent regioselectivity (97–98%) in quantitative yields, whereas $Co_2(CO)_8$ gives the n-aldehyde (3-PFPPA) as the major product, but its regioselectivity is not as high as that observed in the reaction of TFP. The ruthenium catalyst, $Ru_3(CO)_{12}$, shows low catalytic activity, giving the iso-aldehyde as the major isomer and forming a substantial amount of hydrogenated product, $C_6F_5CH_2CH_3$. The platinum catalyst, $PtCl_2(DIOP)/SnCl_2$, shows high catalytic activity, but virtually no regioselectivity is observed and the hydrogenation of PFS occurs as a severe side-reaction. Overall, the metal species dependence of regioselectivity is similar to that for TFP and it is also remarkable compared with that reported for styrene. The reported regioselectivities in the formation of phenylpropanal under typical conditions are as follows: $Co_2(CO)_8$ (80 atm; $CO/H_2 = 1$; 120 °C) 46%, $iso/n = 59:41$ [13]; $PtCl_2(DIOP)/SnCl_2$ (250 atm; $CO/H_2 = 1$; 100 °C) 60%, $iso/n = 57:43$ [14]; $Rh_2Cl_2(CO)_4$ (62 atm; $CO/H_2 = 1$; 130 °C) 93%, $iso/n = 43:57$ [14]; $Rh_2Cl_2(CO)_4/PPh_3$ (62 atm; $CO/H_2 = 1$; 130 °C) 98%, $iso/n = 72:28$ [15].

A kinetic study has been performed on the $Rh_4(CO)_{12}$- and $Co_2(CO)_8$-catalyzed reactions of PFS [7a, 16]. At 100 °C and 82 atm ($CO/H_2 = 1$) with

Table 3.2 Hydroformylation of PFS[a]. (Reprinted with permission from Ojima, *Chem. Rev.*, **88**, 1011. Copyright (1988) American Chemical Society)

Catalyst	PFS/ cat[b]	Pressure (atm)[c] ($CO/H_2 = 1$)	Temperature (°C)	Time (h)	Conversion[d] (%)	Aldehydes[d] %	Aldehydes[d] iso/n	Alkane[d] (%)
$Co_2(CO)_8$	21	80	90	12	67	54	21:79	9
	20	54[e]	120	16	81	59	10:90	22
$PtCl_2$ (DIOP)/ $SnCl_2$	100	80	90	4	100	76	49:51	20
$Ru_3(CO)_{12}$	33	80	90	17	49	22	74:26	25
$Rh_6(CO)_{16}$	5000	80	90	3	100	100	97:3	0
$Ru_4(CO)_{12}$	5000	82[e]	100	2	100	100	98:2	0
$HRh(CO)$ $(PPh_3)_3$	5000	80	90	8	100	100	98:2	0
$RhCl$ $(PPh_3)_3$	333	90	90	20	100	100	97:3	0

[a] Reactions were run with 30–100 mmol of PFS and 15–30 ml of benzene.
[b] (mol PFS)/(mol metal).
[c] Initial pressure at room temperature.
[d] Determined by GLC.
[e] Pressure at the given temperature.

a 1.0×10^{-5} M catalyst concentration, the rhodium-catalyzed reaction is first order in PFS concentration and the apparent rate constant for $Rh_4(CO)_{12}$ is calculated to be 6.2×10^{-4} s^{-1}, i.e. the turnover number is estimated to be 55 800 h^{-1} per rhodium metal. The cobalt-catalyzed reaction with a 1.0×10^{-2} M catalyst concentration at 100 °C and 82 atm (CO/ $H_2 = 1$) is also first order in PFS concentration and the apparent rate constant is calculated to be 1.6×10^{-5} s^{-1}, i.e. the turnover number per cobalt metal is 2.88 h^{-1}. Hence the rhodium catalyst is ca 20 000 times more active than the cobalt catalyst per metal provided that all metal species participate in the catalysis.

Judging from the fact that the addition or the introduction of tertiary phosphines to the catalyst brings about only a slight change in regioselectivity, in sharp contrast with the hydro-formylation of propene or styrene using the same catalysts, both TFP and PFS may well have a large binding constant with catalyst metal species, and thus they should act as important ligands that stabilize the catalysts during the reaction.

In order to examine the scope of this highly regioselective reaction, the hydroformylation of other fluoroalkenes of the type $R_fCH{=}CH_2$ has been carried out, where R_f is C_2F_5 (PFB), C_3F_7 (HPFP) or C_8F_{17} (HPDFP) [7a, 16]. As Table 3.3 shows, the reactions give much lower regioselectivities than that for TFP under standard conditions, i.e. at 80 °C and 100 atm (CO/ $H_2 = 1$); higher selectivities (> 90%) are achieved at 60 °C.

The hydroformylation of vinyl fluoride (VF) promoted by rhodium, ruthenium and cobalt catalysts has also been carried out (equation 3.2). In this reaction, 2-fluoropropanal (2-FPA) is formed exclusively, regardless of the catalyst species used. A mechanism that can accommodate all the results described above in a consistent manner has been proposed [7a, 16].

Table 3.3 Regioselectivities in the hydroformylation of $R_fCH{=}CH_2$ catalyzed by $Rh_4(CO)_{12}{}^a$. (Reprinted with permission from Ojima, *Chem. Rev., **88**, 1011. Copyright (1988) American Chemical Society)

$R_fCH{=}CH_2$	Pressure (atm)[b] (CO/$H_2 = 1$)	Temperature (°C)	Time (h)	Aldehydes[c] (*iso/n*)
$FCH{=}CH_2$	110	80	6	100:0
$CF_3CH{=}CH_2$	110	80	6	97:3
$C_2F_5CH{=}CH_2$	110	80	6	83:17
	110	60	6	95:5
$C_3F_7CH{=}CH_2$	110	80	6	74:26
	110	60	6	91:9
$C_8F_{17}CH{=}CH_2$	110	80	6	73:27
	110	60	6	92:8

[a] Reactions were run with 0.065–0.40 mol% of $Rh_4(CO)_{12}$ in toluene.
[b] Initial pressure at the given temperature.
[c] Determined by GLC.

$$\text{FCH=CH}_2 + \text{CO} + \text{H}_2 \xrightarrow{\text{cat.}} \underset{\text{2-FPA}}{\text{CH}_3\text{CHF-CHO}} \qquad (3.2)$$

$$\text{cat.: Rh}_4(\text{CO})_{12}, \text{HRh}(\text{CO})(\text{PPh}_3)_3, \text{Ru}_3(\text{CO})_{12}, \text{Co}_2(\text{CO})_8$$

In connection with the hydroformylation of R_f-alkenes, the hydrocarboxylation and hydroesterification of TFP and PFS has been studied (equations 3.3 and 3.4) [9]. The hydrocarboxylation of TFP catalyzed by $\text{PdCl}_2(\text{dppf})$–$5\text{SnCl}_2$ [dppf = 1,1′-bis(diphenylphosphino)ferrocene] at 125 °C and 110 atm of carbon monoxide for 70 h gives 3-trifluoromethylpropanoic acid (4) (n-acid) in 93% yield with 99% selectivity. The same yield and selectivity were achieved in the reaction of PFS catalyzed by $\text{PdCl}_2(\text{dppf})$ under the same reaction conditions (48 h). The hydroesterification of TFP with ethanol catalyzed by $\text{PdCl}_2(\text{PPh}_3)_2$ in acetonitrile at 100 °C and 110 atm of carbon monoxide for 40 h gives a 21:79 mixture of n- and iso-esters, 6a and 7a, in 96% yield, whereas the reaction of PFS with methanol in acetone at 100 °C and 120 atm of carbon monoxide for 60 h affords the iso-ester 7b in 80% yield with 95% selectivity. These fluorocarboxylic acids and esters can serve as useful intermediates for fluoroamino acid syntheses.

$$\underset{(1)}{R_f\text{-CH=CH}_2} + \text{CO} + \text{H}_2\text{O} \xrightarrow{\text{catalyst}} \underset{(4)\ (n)}{R_f\text{CH}_2\text{CH}_2\text{COOH}} + \underset{(5)\ (iso)}{R_f(\text{Me})\text{CH-COOH}}$$

$$(3.3)$$

$$\underset{(1)}{R_f\text{-CH=CH}_2} + \text{CO} + \text{ROH} \xrightarrow{\text{catalyst}} \underset{(6)\ (n)}{R_f\text{CH}_2\text{CH}_2\text{COOR}} + \underset{(7)\ (iso)}{R_f(\text{Me})\text{CH-COOR}}$$

$$(3.4)$$

$$R_f = \text{(a) CF}_3; \text{(b) C}_6\text{F}_5$$

3.2.2 SYNTHESIS OF 2-TRIFLUOROMETHYLACRYLIC ACID

The bromination of TFP promoted by photoirradiation followed by dehydrobromination on potassium hydroxide gives 2-bromo-3,3,3-trifluoropropene (2-Br-TFP) in high yield [17]. The carboxylation of 2-Br-TFP catalyzed by a palladium complex, e.g. $\text{PdCl}_2(\text{PPh}_3)_2$ or $\text{PdCl}_2(\text{dppf})$, in the presence of triethylamine in DMF or THF gives 2-trifluoromethylacrylic acid (8) (2-TFMAA) in 65–78% yield (equation 3.5) [18].

$$(3.5)$$

2-Br-TFP   (8) (2-TFMAA)

3.3 SYNTHESIS OF FLUOROAMINO ACIDS THROUGH AMIDOCARBONYLATION OF FLUOROALDEHYDES

3.3.1 4,4,4-TRIFLUOROVALINE (TFV) AND 5,5,5-TRIFLUORONORVALINE (TFNV)

Trifluorovaline (9) (TFV) and trifluoronorvaline (10) (TFNV) have been synthesized via amidocarbonylation of 2-TFMPA and 3-TFMPA, respectively [19]. The amidocarbonylation of 2-TFMPA and 3-TFMPA with acetamide, catalyzed by $Co_2(CO)_8$ at 120 °C and 100 atm of carbon monoxide/hydrogen (1:1), gives N-acetyltrifluorovaline (11) and N-acetyltrifluoronorvaline (12), respectively, in good yields, which are further hydrolyzed to the corresponding free amino acids (equation 3.6 and 3.7). Trifluorovaline (9) thus obtained is a mixture of two diastereomers (*threo*/*erythro* = 38:62).

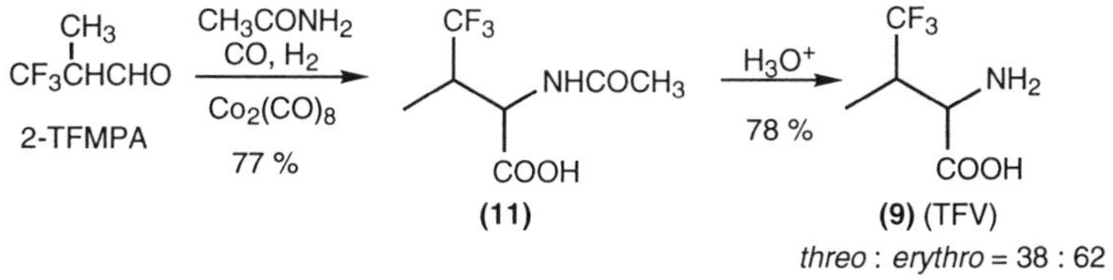

$$\text{(3.6)}$$

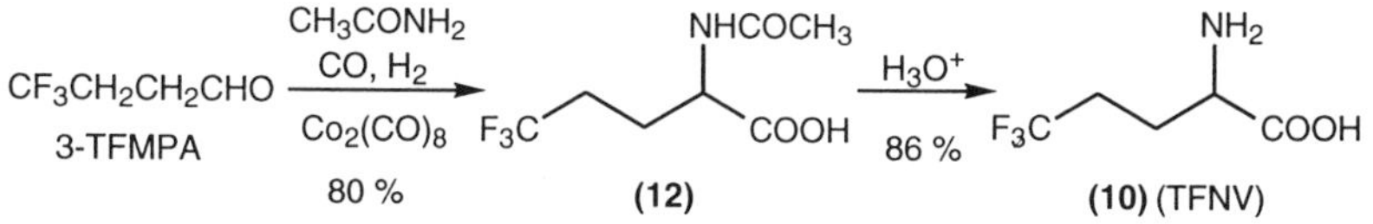

$$\text{(3.7)}$$

The kinetic optical resolution of N-acetyltrifluoronorvaline (12) has been carried out using a porcine kidney acylase I (25 °C, pH 7.0) to give (S)-trifluoronorvaline [10-(S)] and (R)-N-acetyltrifluorovaline [12-(R)] with high enantiomeric purities, which are readily separated (equation 3.8) [19]. The latter is further hydrolyzed to (R)-trifluoronorvaline [10-(R)] with 3 M hydrochloric acid. The enantiomeric purities of 10-(S) and 10-(R) by Mosher's MTPA method [20] (^{1}H and ^{19}F NMR) are >99% e.e. (the other diastereomer is not detected) and 95% e.e., respectively [19].

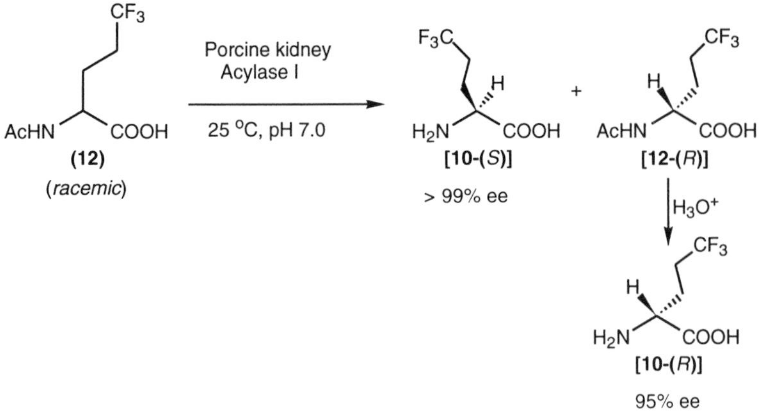

$$
(3.8)
$$

Trifluoronorvaline (10) inhibits the growth of *Escherichia coli* and may be used as a growth regulatory factor in microbiology [21]. Although no significant biological activity of trifluorovaline (9) has been reported to date, it may serve as a modifier of biologically active peptides.

The enzymatic resolution of *N*-acetyltrifluorovaline (11), which contains two chiral centers, is much more complicated. Nevertheless, it has been found that (2*S*,3*S*)- and (2*S*,3*R*)-trifluorovalines [9-(*S*,*S*) and 9-(*S*,*R*)] are yielded exclusively, leaving the 2*R*,3*S*- and 2*R*,3*R*-isomers [11-(*R*,*S*) and 11-(*R*,*R*)] unreacted with the use of an *Aspergillus* acylase I (AA) or a porcine kidney acylase I (PKA) (equation 3.9) [22]. It is noteworthy that 9-(2*S*,3*S*) can be obtained selectively using an *Aspergillus* acylase I with a moderate conversion (Table 3.4).

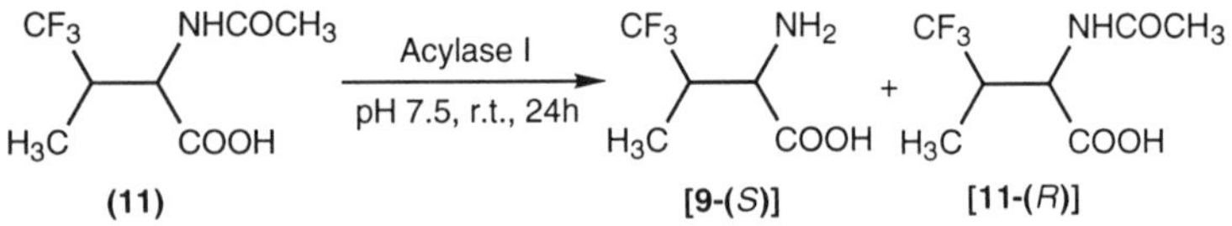

$$
(3.9)
$$

Table 3.4 Enzymatic resolution of *N*-acetyltrifluorovaline (11)

Acylase[a]	Chemical yield %		*erythro/threo*[b]		$[\alpha]_D^{20}(°)$
	9	11	9-(*S*)	11-(*R*)	9-(*S*)
AA	28.0	65.0	1:4.2	2.1:1	+7.36
PKA	39.9	37.3	1.5:1	1.05:1	+6.54

[a] AA = acylase I from *Aspergillus* (Sigma); PKA = acylase I from porcine kidney (Sigma).
[b] The *erythro/threo* ratio was determined by ^{1}H NMR analysis.

3.3.2 5,5,5-TRIFLUOROLEUCINE (TFL) AND 6,6,6-TRIFLUORO-NORLEUCINE (TFNL)

Trifluoroleucine (13) (TFL) and trifluoronorleucine (14) (TFNL) have been synthesized via azlactones starting from 2-TFMPA and 3-TFMPA, respectively (equations 3.10 and 3.11) [19]. For the preparation of the azlactones, the use of lead acetate, N-benzoylglycine and acetic anhydride in tetrahydrofuran (THF) is crucial for good yields [23], since the usual reaction conditions for Erlenmeyer's method using N-acetylglycine and sodium acetate in acetic anhydride do not give the desired azlactones at all. The azlactones thus formed are mixtures of E- and Z-isomers ($E/Z = 13:87$ for 15 and 12:88 for 16).

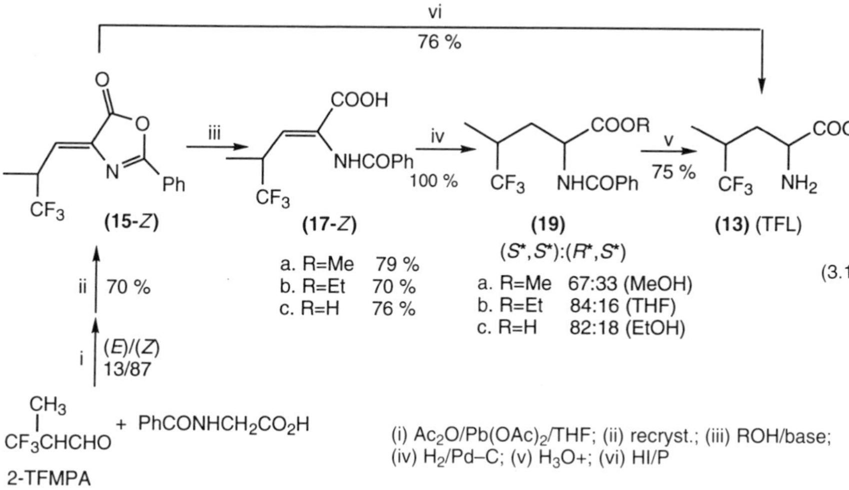

Pure Z-isomers (15-Z and 16-Z) are obtained by recrystallization in 70% and 71% yields, respectively. Pure E-isomers (15-E and 16-E) are isolated through chromatography on silica gel. The azlactones are subjected to alcoholysis to give the corresponding dehydroamino acid esters (17 and 18). A (Z)-dehydroamino acid (17c) is obtained by hydrolysis of 15-Z. Then, the (Z)-dehydroamino acid and esters (17 and 18) are hydrogenated over palladium on carbon followed by hydrolysis to give the corresponding amino acids (13 and

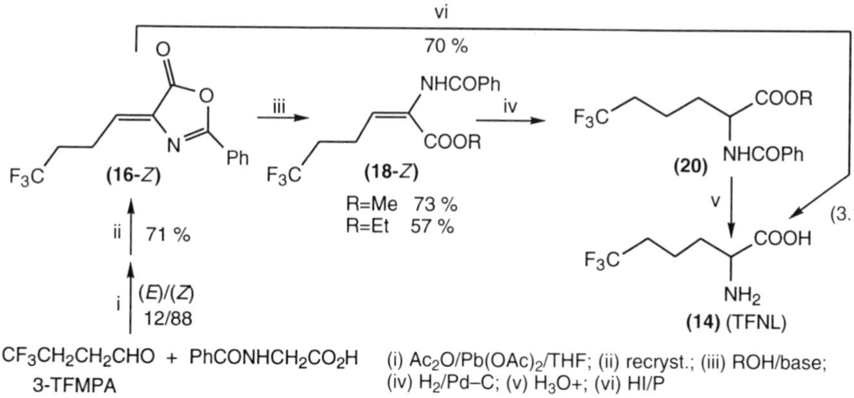

14). The azlactones (**15**-Z and **16**-Z) are also treated with hydriodic acid–red phosphorus to give directly the final amino acids (**13** and **14**).

The stereochemistry of the dehydroamino acid esters **17** and **18** is confirmed on the basis of a large difference in the chemical shifts of C-alkyl protons [24, 25] and of the olefinic protons between E- and Z-isomers, and by NOE experiments. The results also elucidate the stereochemistry of the azlactones **15** and **16**.

(S)- and (R)-trifluoronorleucines (**14**) with >98% e.e. are obtained through enzymatic resolution of N-acetyltrifluoronorleucine (**21**) using a porcine kidney acylase I in a manner similar to that for N-acetyltrifluoronorvaline (**12**) (equation 3.12). The enantiomeric purities are confirmed by the MTPA method.

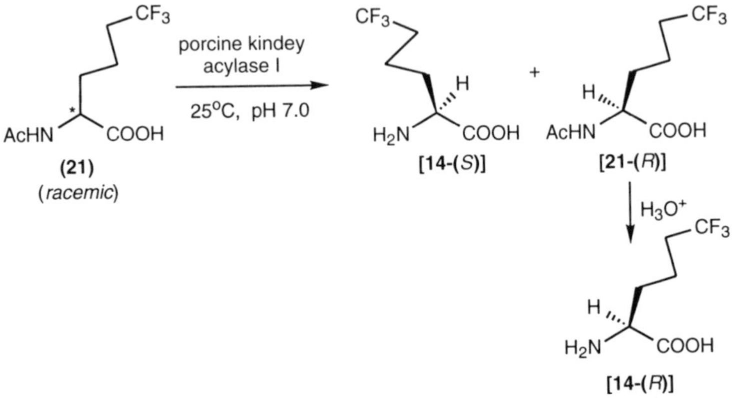

3.3.3 STEREOSELECTIVE HYDROGENATION OF N-BENZOYLDEHYDROTRIFLUOROLEUCINE, ITS ESTERS AND RELATED COMPOUNDS

In the hydrogenation of **17** over palladium on carbon yielding **19**, the trifluoroisopropyl group that has a chiral carbon serves as an effective stereogenic center [19]. Thus, the reductions of **17a**, **17b** and **17c** in THF at ambient temperature and pressure give **19a** (70% d.e.), **19b** (80% d.e.) and **19c** (64% d.e.), respectively. The major diastereomers of **19** thus formed are found to be S*,S*-isomers on the basis of the ^{1}H NMR analysis of trifluoroleucine (**13**) (TFL) obtained by hydrolysis of **19**. Two diastereomers, **19**-(S*,S*) and **19**-(R*,S*), are clearly distinguished in their ^{1}H NMR spectra [26, 27]. It is surprising that the 'chiral isopropyl group' can induce a relatively high degree of stereoselectivity. The solvent effect on the stereoselectivity of the reaction has been examined with **17a** by using methanol, ethanol, propan-2-ol, THF and diethyl ether under the standard conditions. While methanol gives the worst selectivity (38% d.e.), the other solvents give similar selectivities, i.e. EtOH 68% d.e., i-PrOH 72% d.e., THF 70% d.e. and Et$_2$O 70% e.e. [19].

A molecular modeling study has been performed in order to accommodate the observed S^*,S^* selectivity [19]. For the possible ground-state conformers of **17a**, molecular mechanics calculations with SYBYL 5.5 suggest the two low-energy structures, **17a-A** and **17a-B**, and **17a-A** is 4.2 kcal mol^{-1} more stable than **17a-B**. Hence it is reasonable to assume that **17a-A** represents the predominant species on the palladium surface. In this conformation, hydrogens can add across the olefinic bond from either the methyl side or the trifluoromethyl side, and the S^*,S^*-isomer should be formed through the methyl side attack. Judging from the relatively high stereoselectivity observed, it is suggested that the trifluoromethyl group is not only a bulkier substituent than methyl but it also has a unique electronic effect against the palladium surface.

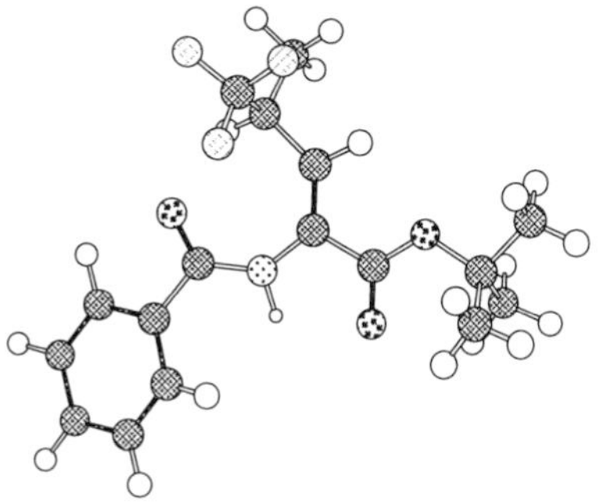

(17 a-A) 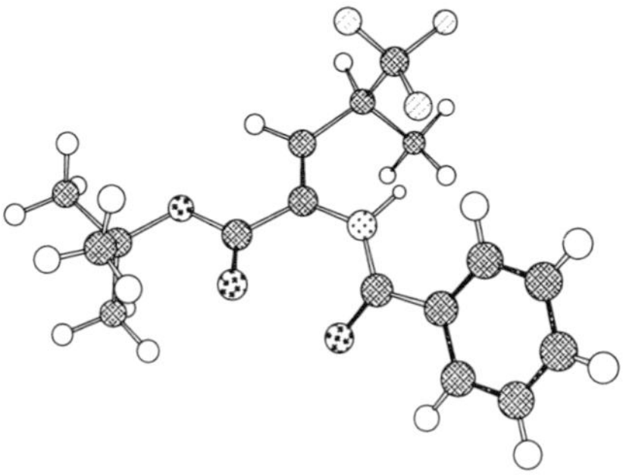   (17 a-B)

In a similar manner, the hydrogenation of (Z)-N-benzoyl-4-penta-fluorophenyldehydronorvaline methyl ester (**22-Z**), which is prepared from 2-PFPPA via an azlactone, is carried out under the standard conditions (equation 3.13) [19]. The reaction gives N-benzoyl-4-penta-fluorophenylnorvaline methyl ester (**23**) with 87% d.e., i.e. (S^*,S^*) $(R^*,R^*) = 93.5:6.5$, which is further hydrolyzed to 4-pentafluorophenyl-norvaline (**24**). Although the stereochemical assignment of two dia-stereomers awaits further elucidation, the ^{1}H NMR patterns of the ester (**23**) and the acid (**24**) suggest that the mode of stereodifferentiation for **23** is the same as that for **17a**.

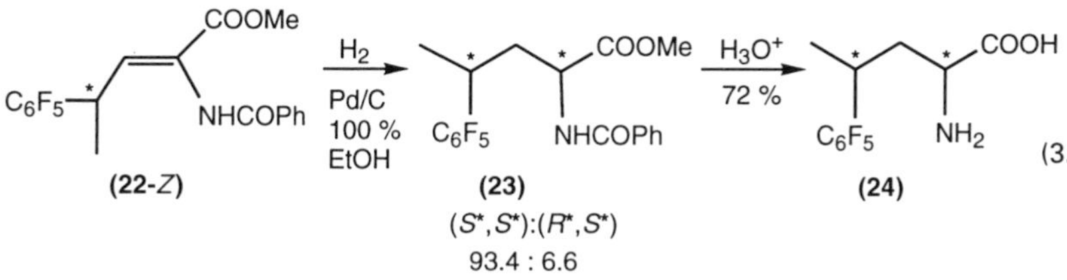

$(22$-$Z)$ (23)   (24)   (3.13)

$(S^*,S^*):(R^*,S^*)$
93.4 : 6.6

3.4 HIGHLY REGIOSELECTIVE SYNTHESIS OF FLUOROAMINO ACIDS THROUGH HYDROFORMYLATION–AMIDOCARBONYLATION OF FLUOROALKENES

A homogeneous hetero-bimetallic catalyst system has been applied for the synthesis of fluoroamino acids, which can sequentially promote catalytic processes including amidocarbonylation as the key reaction. The cobalt-catalyzed amidocarbonylation of aldehydes was discovered in 1971 by Wakamatsu et al. [28] and was developed by Ajinomoto's group, and later reinvestigated in more detail by Pino and co-worker [29]. Further applications of this reaction, e.g. in syntheses of heterocyclic compounds, have been developed by Izawa and co-workers [30].

Since amidocarbonylation requires both H_2 and CO, the reaction conditions are very similar to those of hydroformylation except for the coexistence of an amide, and it is logically possible to combine two reactions. Hydroformylation–amidocarbonylation can provide fluoro-N-acyl-α-amino acids directly from fluoroalkenes. In fact, the possible synthesis of N-acyl-α-amino acids through hydroformylation–amidocarbonylation of alkenes was suggested by Wakamatsu in a review in 1974 [31]. However, it was in 1981 that the proposed process was actually examined by Stern et al. and appeared as a patent [32]. The process is claimed to be useful for the production of C_9–C_{31} straight-chain N-acyl-α-amino acids, although the process gives a mixture of straight-chain and branched isomers. Thus, if excellent regioselectivities are realized in the hydroformylation of alkenes N-acyl-α-amino acids can be synthesized with high regioselectivity. As described above, such high regioselectivities (> 93%) are achieved in the hydroformylation of fluoroalkenes such as TFP and PFS [7, 8]. Based on this finding, Ojima and co-workers have successfully combined highly regioselective hydroformylations of fluoroalkenes with amidocarbonylation for the one-step synthesis of both straight-chain and branched N-acylfluoroamino acids.

3.4.1 HYDROFORMYLATION–AMIDOCARBONYLATION OF TRIFLUOROPROPENE (TFP)

The hydroformylation–amidocarbonylation of TFP gives N-acetyltrifluorovaline (**11**) or N-acetyltrifluoronorvaline (**12**), directly in a highly regioselective manner (equation 3.14) [33].

$$\text{CF}_3\text{CH=CH}_2 \atop +\atop \text{CH}_3\text{CONH}_2 \quad \xrightarrow[120\,^\circ\text{C},\ 10\,\text{h}]{\text{CO, H}_2,\ \text{cat.}} \quad$$

(structures with products **(11)** and **(12)**)

(3.14)

Catalyst	Yield (%)	
	(11)	**(12)**
Co$_2$(CO)$_8$	4	96
Rh$_6$(CO)$_{16}$–Co$_2$(CO)$_8$	94	6

As equation 3.14 shows, Co$_2$(CO)$_8$-catalyzed reaction (initial pressure at 25 °C: CO, 80 atm; H$_2$, 50 atm) at 120 °C for 10 h gives N-acetyltrifluoronorvaline **(12)** with 96% selectivity, whereas the reaction catalyzed by the Rh$_6$(CO)$_{16}$–Co$_2$(CO)$_8$ binary system [Co$_2$(CO)$_8$/Rh$_6$(CO)$_{16}$ = 50] under the same conditions gives N-acetyltrifluorovaline **(11)** with 94% selectivity [33]. The latter result clearly indicates that the rhodium-catalyzed hydroformylation takes place exclusively in the first step to give 2-TFMPA highly selectively, which is then effectively incorporated in the subsequent cobalt-catalyzed amidocarbonylation.

3.4.2 HYDROFORMYLATION–AMIDOCARBONYLATION OF PENTAFLUOROSTYRENE (PFS)

The regioselectivities observed in the hydroformylation–amidocarbonylation of PFS catalyzed by the Co–Rh binary system and also by Co$_2$(CO)$_8$ are different from those obtained for the reactions of TFP [34]. The hydroformylation–amidocarbonylation of PFS catalyzed by Co$_2$(CO)$_8$ at 120 °C and 130 atm of CO and H$_2$ (CO/H$_2$ = 1.6) with acetamide (2.0 equiv.) in dioxane gives N-acetyl-4-pentafluorophenylhomoalanine **(25)** with 90–92% regioselectivity (ca 30% yield) [34]. This selectivity is much higher than that (79%) of the simple hydroformylation in benzene. The reaction catalyzed by a Co–Rh binary system [Co$_2$(CO)$_8$/Rh$_6$(CO)$_{16}$ = 50; CO, 80 atm; H$_2$, 50 atm; 120 °C; acetamide, 2.0 equiv.; dioxane] gives N-acetyl-3-pentafluorophenylhomoalanine **(26)** with only ca 80% regioselectivity (70% yield) [34], which is much lower than the excellent regioselectivity (98%) observed in the simple hydroformylation of PFS in benzene [7, 8].

The detailed study of the reaction has revealed very interesting mechanistic aspects of Co–Rh mixed metal catalyst systems including a novel CoRh(CO)$_7$-catalyzed process [16, 34]. Since the organometallic and cluster chemistry is outside the scope of this chapter, this aspect will not be discussed here; briefly, it has been found that the regioselectivity is dependent on the active catalyst species in the hydroformylation step (equation 3.15) [16, 34].

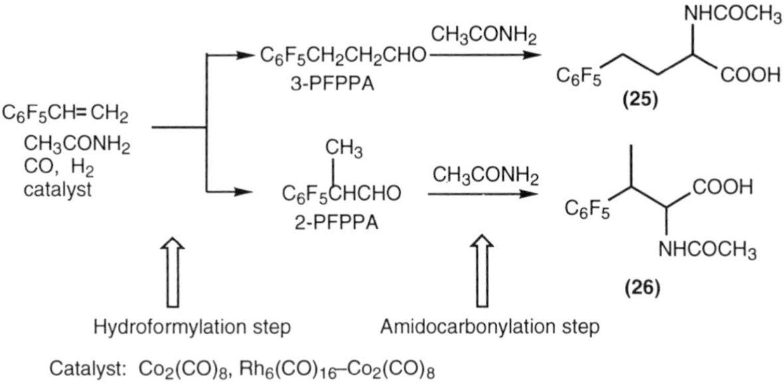

(3.15)

From the synthetic viewpoint, a highly regioselective production of the fluoroamino acids **25** and **26** is the ultimate goal. As for the synthesis of **26**, it can be said that the process is practical. For example, the $Co_3Rh(CO)_{12}$ catalyst (1.0 mol%) gives **26** in 74% yield with 92% regioselectivity at 110 °C and 82 atm $(CO/H_2 = 3:1)$ and the $Co_2(CO)_8$ (5.0 mol%) – $Rh_4(CO)_{12}$ (0.05 mol%) catalyst system gives **26** in 80% yield with 98.2% regioselectivity at 60 °C for 6 h and then 125 °C for 5 h under 75 atm of CO and 48 atm of H_2. With regard to the synthesis of **25**, however, the single $Co_2(CO)_8$-catalyzed process is not very efficient yet; although the regioselectivity of the reaction is high (90–94%), the chemical yield is low (30–35%), mainly owing to the hydrogenation of PFS (30–55%). Accordingly, a search for another mixed metal system which suppresses the hydrogenation without affecting the high straight-chain selectivity of cobalt carbonyl is necessary to overcome this problem.

N-Acetyl-3-pentafluorophenylhomoalanine (**26**) thus obtained can serve as a good precursor for fluoroindoles. The base-promoted cyclization of **26** gives N-acetyl-2-carbohydroxy-3-methyl-2,3-dihydro-4,5,6,7-tetrafluoroindole (**27**) in 92% yield (equation 3.16) [16, 34], which can be converted into a variety of fluoroindoles and fluoroalkaloids.

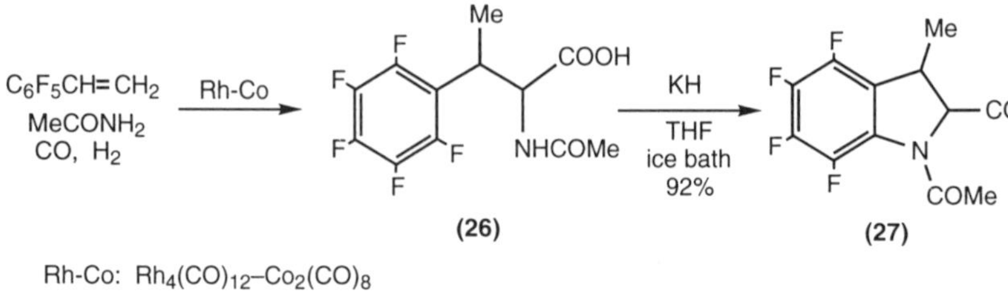

(3.16)

Rh-Co: $Rh_4(CO)_{12}$–$Co_2(CO)_8$

3.5 PREPARATION OF OPTICALLY ACTIVE FLUOROAMINO ACIDS THROUGH ASYMMETRIC HYDROGENATION

3.5.1 6,6,6-TRIFLUORONORLEUCINE

Optically active methyl (S)-N-benzoyltrifluoronorleucinate [20a-(S), 87–89% e.e.] is obtained quantitatively by asymmetric hydrogenation of 18a-Z using a rhodium catalyst with $(1R, 2R)$-1,2-bis[o-anisylphenyl)phosphino]ethane (diPAMP) [35] as the chiral ligand, [(diPAMP)Rh(NBD)]ClO$_4$ (NBD = norbornadiene) (equation 3.17). The optical purity is determined based on specific rotation and confirmed by ^{1}H and ^{19}F NMR analyses of the corresponding MTPA-amino acid methyl ester, i.e. by Mosher's method [20]. Other chiral ligands such as $(3R,4R)$-1-benzyl-3,4-bis(diphenylphosphino)pyrrolidine (Degphos) [36], $(2S,3S)$-2,3-bis(diphenylphosphino)butane (Chiraphos) [37], $(2R,4R)$-1-$tert$-butoxycarbonyl-2-diphenylphosphinomethyl-4-diphenylphosphinopyrrolidine [(+)-BPPM] [38] and (S)-2,2′-bis(diphenylphosphino)-1,1′-binaphthyl [(S)-BINAP] [39] do not give better results. The results are listed in Table 3.5.

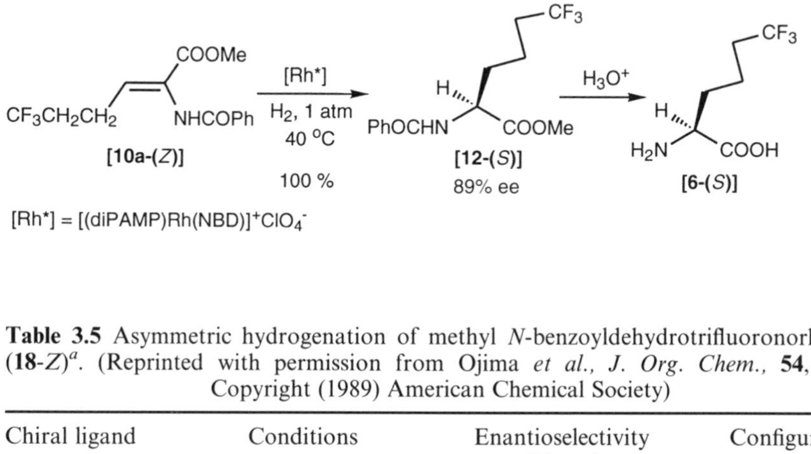

$$(3.17)$$

[Rh*] = [(diPAMP)Rh(NBD)]$^+$ClO$_4^-$

Table 3.5 Asymmetric hydrogenation of methyl N-benzoyldehydrotrifluoronorleucine (18-Z)^a. (Reprinted with permission from Ojima *et al.*, *J. Org. Chem.*, **54**, 4511. Copyright (1989) American Chemical Society)

Chiral ligand	Conditions	Enantioselectivity (% e.e.)	Configuration
DiPAMP	1 atm, 40 °C, 12 h	89	S
	5 atm, 30 °C, 16 h	52	S
Degphos	5 atm, 25 °C, 12 h	67	S
Chiraphos	5 atm, 30 °C, 16 h	54	R
(+)-BPPM	1 atm, 25 °C, 12 h	22	S
(S)-BINAP	5 atm, 25 °C, 48 h	18	R

^{a}All reactions were run with 1.0 mmol of 18-Z and 2.0 mol% of chiral rhodium catalyst in ethanol (10 ml).

3.5.2 FLUORINATED PHENYLALANINES

Optically active *N*-benzoylamino(fluorophenyl)alanines and their methyl esters (**30**) are prepared from the corresponding (*Z*)-*N*-benzoylamino-(fluorophenyl)acrylic acids and their methyl esters (**29**) using rhodium complexes with Propraphos (**32a** or **32b**) and *O,N*-bis(diphenylphosphino)-2-*exo*-hydroxy-3-*endo*-methylaminonorbornane (**32c**) as chiral catalysts; chemical yields are 80–90% and high enantioselectivities of 86–91% e.e. can be obtained except when using the catalyst **32c** (equation 3.18) [40]. Simple recrystallization of the *N*-benzoylamino acid (**30**, R = H) can improve the optical purity up to 99% e.e. Fluorophenylalanines (**31**) are readily obtained by acid hydrolysis of **30**. The (*S*)-catalyst **32a** gives (*R*)-fluorophenylalanines, i.e. the catalyst **32b** affords (*S*)-amino acids. Results are listed in Table 3.6. Although Table 3.6 only shows results using free acid substrates (**30**, R = H), virtually the same results are obtained in the reactions of the corresponding methyl esters (**30**, R = Me) [40].

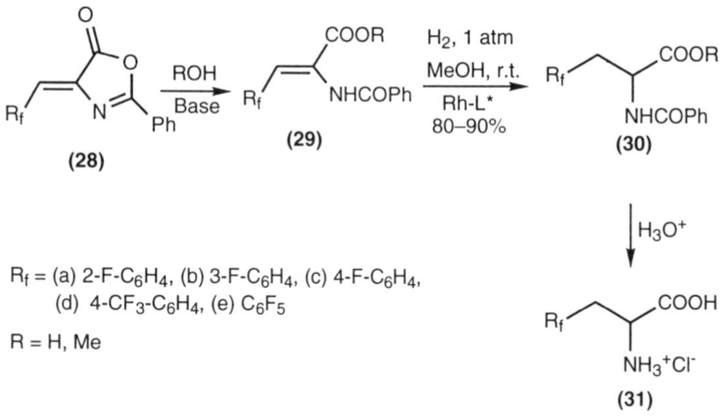

(3.18)

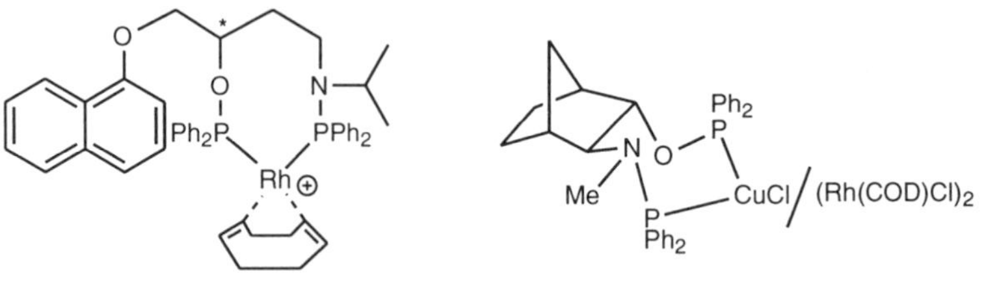

R_f = (a) 2-F-C$_6$H$_4$, (b) 3-F-C$_6$H$_4$, (c) 4-F-C$_6$H$_4$,
         (d) 4-CF$_3$-C$_6$H$_4$, (e) C$_6$F$_5$

R = H, Me

(**32a**): (+)-*S*
(**32b**): (–)-*R*

(**32c**): (2*R*,3*S*)-
          in situ (2:1)

Table 3.6 Asymmetric synthesis of optically active N-benzoyl-amino(fluorophenyl)alanines (**30**, R = H) through enantioselective hydrogenation

Entry	Substrate	Catalyst (configuration)	29/cat. molar ratio	Product (configuration)	Enantioselectivity (% e.e.)^{a,b}
1	**29a**	**32a** (S)	200	**30a** (R)	91 (99)
2	**29a**	**32b** (R)	200	**30a** (S)	91 (99)
3	**29b**	**32a** (S)	200	**30b** (R)	88 (99)
4	**29b**	**32b** (R)	200	**30b** (S)	90
5	**29c**	**32a** (S)	200	**30c** (R)	88 (99)
6	**29c**	**32b** (R)	200	**30c** (S)	88 (89)
7	**29b**	**32b** (R)	1000	**30b** (S)	89
8	**29c**	**32a** (S)	3000	**30c** (R)	86
9	**29c**	**32a** (S)	2000	**30c** (R)	90
10	**29d**	**32a** (S)	200	**30d** (R)	90 (92)
11	**29e**	**32b** (R)	200	**30e** (S)	86
12	**29a**	**32c** (S)	200	**30a** (R)	75
13	**29b**	**32c** (S)	200	**30b** (R)	71
14	**29c**	**32c** (S)	200	**30c** (R)	75

^{a} The enantiomeric purity was determined by GLC.
^{b} The values in parentheses are the enantiomeric purity of **30** after recrystallization.

3.6 SYNTHESES OF α-AMINOPERFLUOROALKANOIC ACIDS THROUGH PALLADIUM-CATALYZED CARBALKOXYLATION OF PERFLUOROALKYLIMIDOYL IODIDES

The palladium-catalyzed carbalkoxylation of perfluoroalkylimidoyl iodides (**34**) has been successfully applied to the syntheses of α-iminoperfluoro-alkanoates (**35**), which are converted into α-aminoperfluoroalkanoic acids (**36** or **37**) [41].

Trifluoroacetimidoyl chloride (**33**, R_f = CF_3) is prepared in good yield by reaction of a mixture of trifluoroacetic acid (TFA), 4-methoxyaniline, triphenylphosphine and triethylamine in refluxing tetrachloromethane. The iodide **34** is quantitatively obtained by treating **33** with sodium iodide in acetone (equation 3.19).

$$R_fCOOH + H_2N\text{-}PMP \xrightarrow[\substack{CCl_4 \uparrow\downarrow \\ 80\text{-}90\%}]{Ph_3P,\ Et_3N} \underset{(33)}{R_f} \overset{Cl}{\underset{N}{\diagup}} PMP \xrightarrow[100\%]{NaI} \underset{(34)}{R_f} \overset{I}{\underset{N}{\diagup}} PMP \quad (3.19)$$

PMP = 4-methoxyphenyl

(**33**) a. R_f = CF_3
b. R_f = C_2F_5
c. R_f = C_7F_{15}

Carbalkoxylation of **34** catalyzed by $Pd_2(dba)_3$ (dba = dibenzylidene-acetone) at room temperature and atmospheric pressure of carbon monoxide gives the corresponding α-imino ester **35** in excellent yield (equation 3.20). The α-iminoperfluoroalkanoates **35** serve as useful intermediates for the syntheses of fluoro-α-amino acids. For example, **35a** is converted into 2-amino-3,3,3-trifluoropropanoic acid (trifluoroalanine), (**36a**) through reduction followed by removal of PMP (4-methoxylphenyl) with CAN [cerium(IV) ammonium nitrate] and benzyl ester by hydrogenolysis over palladium on carbon; **35a** is also transformed into 2-alkyl-2-amino-3,3,3-trifluoropropanoates (**37a–d**) through alkylation followed by deprotection (equation 3.20) [41].

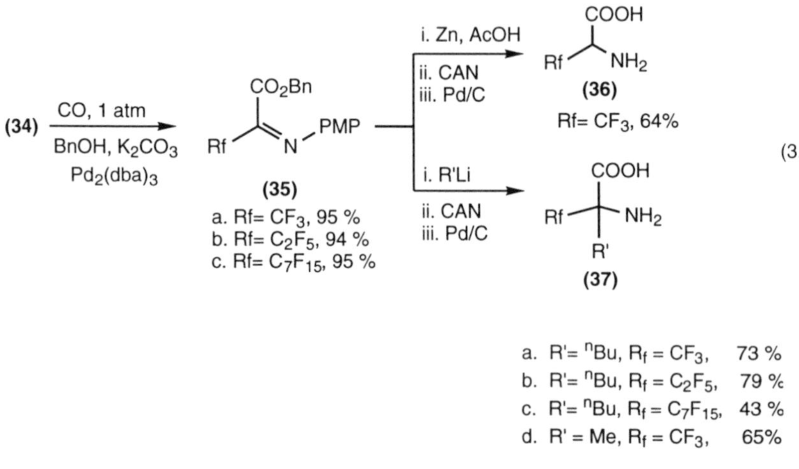

$$
(34) \xrightarrow[\substack{BnOH, K_2CO_3 \\ Pd_2(dba)_3}]{CO, 1\ atm}
$$

Rf—⟨CO₂Bn⟩=N—PMP

(**35**)

a. Rf= CF_3, 95 %
b. Rf= C_2F_5, 94 %
c. Rf= C_7F_{15}, 95 %

i. Zn, AcOH →
ii. CAN
iii. Pd/C

COOH
Rf—CH(NH₂)
(**36**)
Rf= CF_3, 64%

i. R'Li
ii. CAN
iii. Pd/C

COOH
Rf—C(R')(NH₂)
(**37**)

a. R'= ^nBu, $R_f = CF_3$,   73 %
b. R'= ^nBu, $R_f = C_2F_5$,   79 %
c. R'= ^nBu, $R_f = C_7F_{15}$, 43 %
d. R' = Me, $R_f = CF_3$,   65%

(3.20)

3.7 4,5,6,7-TETRAFLUOROTRYPTOPHAN AND RELATED COMPOUNDS

Because of the importance of biologically active compounds containing the indole ring such as tryptophan, tryptamine, indoleacetic acid and alkaloids, the syntheses of tetrafluoro analogs of indoles are of particular interest. A useful synthetic intermediate, 3-methyl-4,5,6,7-tetrafluoroindole (**41**), is synthesized from 2-PFPPA through reaction with allylamine followed by cyclization using lithium diisopropylamide (LDA) as the base and deprotection of the indole nitrogen (equation 3.21) [19, 42].

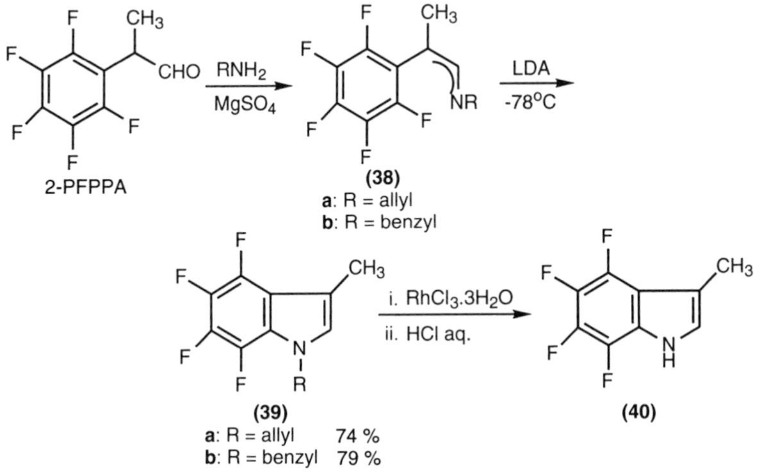

2-PFPPA
RNH2
MgSO4
NR
LDA
-78°C
(38)
a: R = allyl
b: R = benzyl
i. RhCl3.3H2O
ii. HCl aq.
(39)
a: R = allyl 74 %
b: R = benzyl 79 %
(40)

(3.21)

3-Formyl-4,5,6,7-tetrafluoroindoles (**42**) and 3-acetoxymethyl-4,5,6,7-tetrafluoroindoles (**43**) are key intermediates in the syntheses of tetrafluoro analogs of tryptophan, tryptamine and indoleacetic acid. The synthesis of **42** is performed through selenium dioxide oxidation of 1-acyl-3-methyltetrafluoroindoles (**41**) [19, 42]. Direct oxidation of **40** with selenium dioxide results in the decomposition of the indole skeleton. Therefore, the 1-position must be protected by an acetyl, benzoyl or tosyl group.

Selenium dioxide oxidation of **41a** and **41b** give 3-formyl-4,5,6,7-tetrafluoroindole (**42a**) in 86% and 74% yields, respectively, i.e. the acetyl and benzoyl protecting groups are removed during the reaction (equation 3.22). On the other hand, the 1-tosyl group is tolerant of the reaction conditions, thus the reaction of **41c** gives 1-tosyl-3-formyl-4,5,6,7-tetrafluoroindole (**42b**) in 81% yield (equation 3.22).

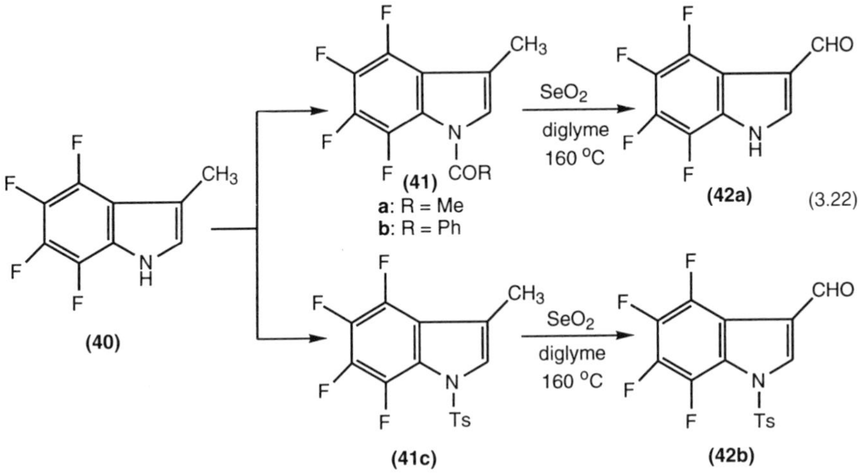

(40)
(41)
a: R = Me
b: R = Ph
SeO2
diglyme
160 °C
(42a)
(41c)
(42b)

(3.22)

The oxidation of **41** in the presence of acetic anhydride gives 1-acetyl- or 1-benzoyl-3-acetoxymethyl-4,5,6,7-tetrafluoroindole (**43a** or **43b**) in 50–60% yield after chromatographic separation to remove a small amount of **42a** that is formed as a side-product (equation 3.23) [19, 42].

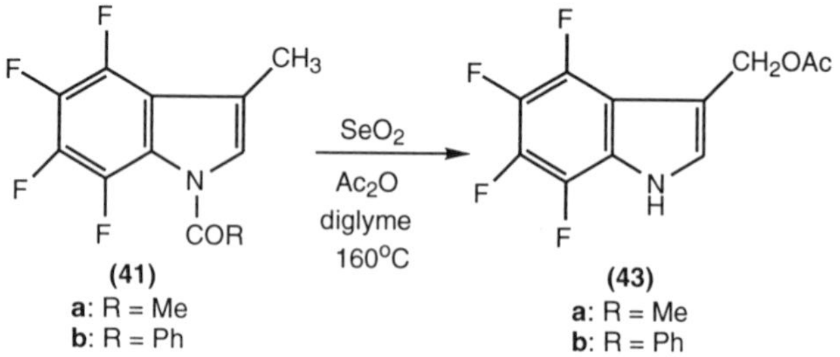

(3.23)

The usefulness of **42** and **43** is demonstrated by the following examples [19, 42]. (i) The reaction of **43a** with piperidine (a large excess) at room temperature for 20 h gives 3-piperidinomethyl-4,5,6,7-tetrafluoroindole (**44**) [43] in 97% yield, which is a known key intermediate for the synthesis of 4,5,6,7-tetrafluorotryptophan (**45**) (equation 3.24). (ii) The reaction of N-methyl-N-(4,5,6,7-tetrafluoroindol-3-ylmethyl)piperidinium sulfate (**46**) [44] with potassium cyanide (4 equiv.) in aqueous dimethylformamide (DMF) under reflux for 2 h gives 3-cyanomethyl-4,5,6,7-tetrafluoroindole (**47**) in 96% yield, which is readily converted into 4,5,6,7-tetrafluoroindole acetic acid (**48**) [43a] by hydrolysis with aqueous potassium hydroxide; the N-methylpiperidinium salt **46** is obtained by treating **44** with dimethyl sulfate (equation 3.25). (iii) 4,5,6,7-Tetrafluorotryptamine (**50**) [43a] was obtained from **42a** through condensation with nitromethane followed by LiAlH$_4$ reduction in 83% overall yield (equation 3.26). (iv) 4,5,6,7-Tetrafluorotryptophan (**45**) was obtained from **42a** through Erlenmeyer's azlactone method [45] in four steps in 51% overall yield (equation 3.27).

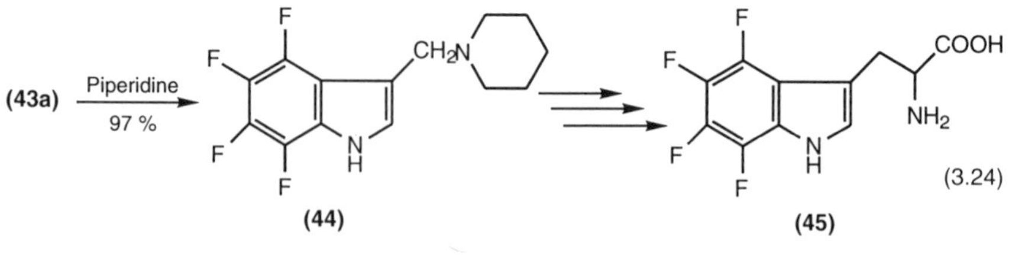

(3.24)

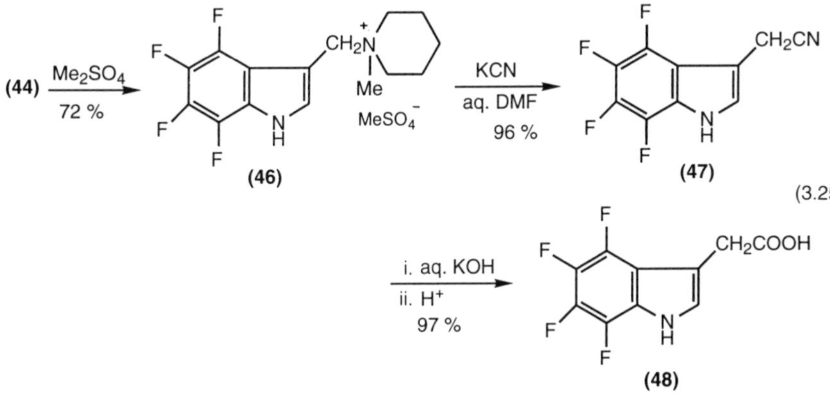

(3.25)

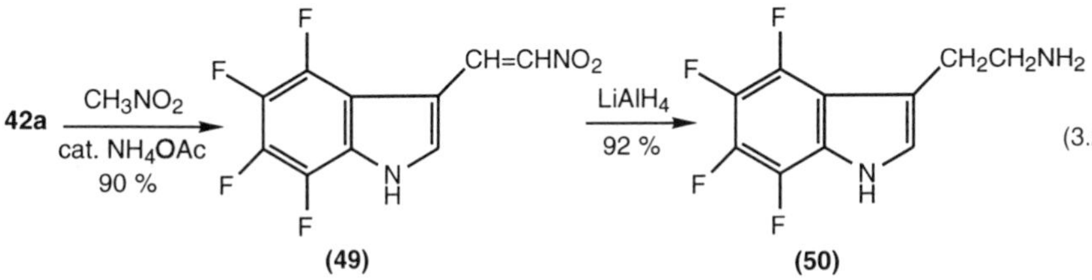

(3.26)

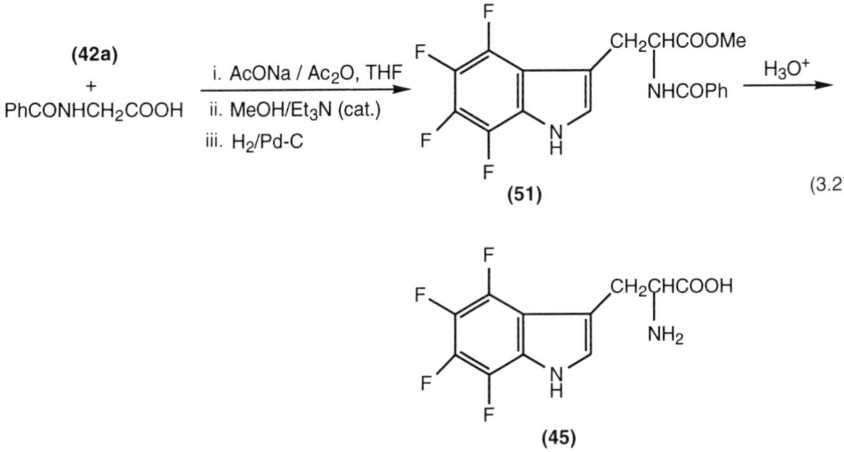

(3.27)

Since 4,5,6,7-tetrafluorotryptophan (45) strongly inhibits both the trypto-phanyl hydroxamate and aminoacyl t-RNA formation [46, 47] the tetrafluoro analogs of tryptamine, indoleacetic acid and other indole derivatives are expected to have interesting physiological activities.

3.8 α-TRIFLUOROMETHYL-β-ALANINE (α-TFM-β-Ala)

Addition of gaseous ammonia to 2-trifluoromethylacrylic acid (TFMAA) at 0–5 °C in dichloromethane gives a new β-amino acid, α-trifluoromethyl-β-alanine (α-TFM-β-Ala), in excellent yield [19]. The reaction sometimes gives double and triple Michael addition products, i.e. $HN[CH_2CH(CF_3)COOH]_2$ and $N[CH_2(CF_3)COOH]_3$, depending on the reaction conditions. The use of hexamethyldisilazane (HMDS) in an attempt to protect the C-terminus of TFMAA with a trimethylsilyl (TMS) group resulted in the addition of monotrimethylsilylamine, probably generated *in situ*, to O-TMS-TFMAA giving N-O-bis-TMS-α-TFM-β-Ala (52) in quantitative yield. No trace of N,N,O-tris-TMS-α-TFM-β-Ala has been detected. Michael addition of HMDS to the methyl and benzyl esters of TFMAA does not proceed at all, which is reasonable since HMDS is well known as a bulky non-nucleophilic amine. Accordingly, this silylation–Michael addition process is noteworthy. N,O-bis-TMS-α-TFM-β-Ala (52) is treated with methanol to give α-TFM-β-Ala in nearly quantitative yield (equation 3.28) [19].

$$\text{α-TFMAA} \xrightarrow[\text{CH}_2\text{Cl}_2]{(\text{Me}_3\text{Si})_2\text{NH}} \textbf{(52)} \xrightarrow{\text{MeOH}} \text{α-TFM-β-Ala}$$

$$\text{α-TFMAA} \xrightarrow[\text{CH}_2\text{Cl}_2]{\text{NH}_3\,(g)} \text{α-TFM-β-Ala} \tag{3.28}$$

α-TFM-β-Ala does not have any antibacterial activity. However, an enkephalin analog bearing an α-TFM-β-Ala fragment, Tyr-D-Ala-(α-TFM-β-Ala)-Phe-Met, has shown fairly strong analgesic effects [48]. This suggests that the new fluoro-β-amino acid can serve as a modifier for a variety of peptide hormones and other physiologically active peptides.

3.9 CONCLUSION

Efficient methods have been developed for the syntheses of 4,4,4-trifluorovaline (TFV), 5,5,5-trifluoronorvaline (TFNV), 5,5,5-trifluoroleucine (TFL), 6,6,6-trifluoronorleucine (TFNL), 4,5,6,7-tetrafluorotryptophan and α-trifluoromethyl-β-alanine (α-TFM-β-Ala) through homogeneous carbonylations catalyzed by transition metal complexes. Enzymatic optical resolution has been successfully applied to the syntheses of (S)- and (R)-TFNV and (S)-and (R)-TFNL with virtually 100% e.e. Diastereoselective hydrogenation has been studied and an interesting effect of the trifluoromethyl group on diastereoselectivity is observed. Enantioselective hydrogenation of dehydrofluoroamino acids and their esters catalyzed by chiral rhodium complexes has proved to be useful for the asymmetric synthesis of optically active fluoroamino acids.

3.10 ACKNOWLEDGMENTS

This research was supported in part by grants from the National Institutes of Health (NIGMS), National Science Foundation and Center for Biotechnology, which is sponsored by New York State Science and Technology Foundation. Generous support from Japan Halon Co. Ltd, Ajinomoto Co. Inc. and Fuji Chemical Industries Ltd is also gratefully acknowledged. The authors thank Dr K. Kato, Dr K. Nakahashi, Dr F. A. Jameison, Dr B. Peté, Mr J. D. Conway, Dr Z. Zhang and Ms C. Hanson for their excellent contributions to this research. One of the authors (I.O.) also thanks Dr T. Fuchikami and Dr M. Fujita of Sagami Chemical Research Center for their productive collaboration. The authors are grateful to Dr W. S. Knowles (Monsanto) for the generous gift of diPAMP and also to Dr G. Prescher (Degussa) for generously providing Degphos.

3.11 REFERENCES

1. J. T. Welch, *Tetrahedron*, **43**, 3123–3197 (1987).
2. (a) R. Filler, *J. Fluorine Chem.*, **33**, 361–375 (1986); (b) R. Filler (Ed.), *Biochemistry Involving Carbon–Fluorine Bonds*, ACS Symposium Series, No. 28, American Chemical Society, Washington, DC, 1976; (c) F. A. Smith, *CHEMTECH*, 422-429 (1973); (d) R. Filler, *CHEMTECH*, 752–757 (1973).
3. R. Filler and Y. Kobayashi (Eds), *Biomedicinal Aspects of Fluorine Chemistry*, Elsevier Biomedical, Amsterdam (1982).
4. As a review, B. Imperialli, in A. Mizrahi (Ed.), *Advances in Biotechnological Processes, Vol. 10, Synthetic Peptides in Biotechnology*, Alan R. Liss, New York, 1988, pp. 97–131. As articles, e.g. (a) D. Rasnick, *Anal. Biochem.*, **149**, 461–465 (1985); (b) B. Imperialli and R. H. Abeles, *Tetrahedron Lett.*, **27**, 135–138 (1986); (c) M. Kolb, J. Barth and B. Neises, *Tetrahedron Lett.*, **27**, 1579–1582 (1986); (d) P. Rauber, H. Angliker, B. Walker and E. Shaw, *Biochem. J.*, **239**, 633-640 (1986); (e) B. Imperialli and R. H. Abeles, *Biochemistry*, **26**, 3760–3767 (1986); (f) L. Lamden and P. A. Bartlett, *Biochem. Biophys. Res. Commun.*, **112**, 1085–1090 (1983); (g) M. H. Gelb, J. P. Svaren and R. H. Abeles, *Biochemistry*, **24**, 1813–1817 (1985); (h) S. Thaisrivongs, D. T. Pals, W. M. Kati, S. R. Turner, L. M. Thomasco and M. Watt, *J. Med. Chem.*, **29**, 2080–2087 (1986); (i) S. J. Hocart and D. H. Coy, in C. M. Deber, V. J. Hruby, V. J. and K. D. Kopple (Eds), *Peptides: Structure and Function, Proceedings of the Ninth Peptide Symposium*, Pierce, Rockford, IL, 1985, pp. 461–464; (j) H. Tanaka, F. Osakada, S. Ohashi, M. Shiraki and E. Munekata, *Chem. Lett.*, 391–394 (1986); (k) G. Feuerstein, D. Lozovsky, L. A. Cohen, V. M. Labroo, K. Kirk, I. J. Kopkin and A. I. Faden, *Neuropeptides*, **4**, 303–310 (1984).
5. (a) P. Pino, F. Piacenti and M. Bianchi, in I. Wender and P. Pino (Eds), *Organic Syntheses via Metal Carbonyls*, Vol. 2, Wiley–Interscience, New York, 1977,

136 FLUORINE-CONTAINING AMINO ACIDS

pp. 43–231; (b) B. Cornils, in J. Falbe (Ed.), *New Syntheses with Carbon Monoxide,* Springer, Berlin, 1980, pp. 1–225.

6. Hydroformylation of hexafluoropropene was reported to give a mixture of hexa-fluoropropane (50%), alcohols (40%), and aldehydes (5–8%); see D. M. Rudkovskii, N. S. Imayanitov and V. Yu. Gankin, *Tr. Vses. Nauchno.-Issled. Inst. Neftekhim. Protsessov,* 121–124 (1961); *Chem. Abstr.,* **57**, 10989 (1962). A patent claimed the reaction of heptadecafluorodecene, $CF_3(CF_2)_7CH=CH_2$, catalyzed by $Co_2(CO)_8$, which gave the corresponding alcohols or aldehydes; see F. Roehrscheid (Hoechst), *Ger. Offen.,* 2 163 752 *(1973); Chem. Abstr.,* **79**, 78110m (1973).

7. (a) I. Ojima, K. Kato, M. Okabe and T. Fuchikami, *J. Am. Chem. Soc.,* **109**, 7714–7720 (1987); (b) T. Fuchikami and I. Ojima, *J. Am. Chem. Soc.,* **104**, 3527–3529 (1982).

8. I. Ojima and T. Fuchikami, *US Pat.,* 4 370 504 (1983).

9. T. Fuchikami, K. Ohishi and I. Ojima, *J. Org. Chem.,* **48**, 3803–3807 (1983).

10. P. Pino, F. Piacenti, M. Bianchi and R. Lazzaroni, *Chim. Ind. (Milan),* **50**, 106–118 (1968).

11. I. Shwager and J. F. Knifton (Texaco Development), *Ger. Offen.,* 2 322 751 (1973); *Chem. Abstr.,* **80**, 70327m (1974).

12. B. L. Booth, M. J. Else, R. Fields and R. N. Haszeldine, *J. Organomet. Chem.,* **27**, 119–131 (1971).

13. C. Botteghi, G. Consiglio and P. Pino, *Chimia,* **26**, 141–143 (1972).

14. Y. Kawabata, T. Suzuki and I. Ogata, *Chem. Lett.,* 361–362 (1978).

15. I. Ogata, Y. Ikeda and T. Asakawa, *Kogyo Kagaku Zasshi,* **74**, 1839–1841 (1971).

16. I. Ojima, *Chem. Rev.,* **88**, 1011–1030 (1988).

17. A. L. Henne and M. Nager, *J. Am. Chem. Soc.,* **73**, 1042–1043 (1951).

18. T. Fuchikami, A. Yamanouchi and I. Ojima, *Synthesis,* 766–768 (1984).

19. I. Ojima, K. Kato, K. Nakahashi, T. Fuchikami and M. Fujita, *J. Org. Chem.,* **54**, 4511–4522 (1989).

20. J. A. Dale and H. S. Mosher, *J. Org. Chem.,* **34**, 2543–2549 (1970); (b) S. Yamaguchi, J. A. Dale and H. S. Mosher, *J. Org. Chem.,* **37**, 3174–3176 (1972); (c) S. Yamaguchi and H. S. Mosher, *J. Org. Chem.,* **38**, 1870–1877 (1973).

21. H. M. Walborsky, M. Baum and D. F. Loncrini, *J. Am. Chem. Soc.,* **77**, 3637–3640 (1955).

22. I. Ojima and Q. Dong, unpublished data.

23. E. Baltazzi and R. Robinson, *Chem. Ind. (London),* 191 (1954).

24. C. Shin, M. Hayakawa, T. Suzuki, A. Ohtsuka and J. Yoshimura, *Bull. Chem. Soc. Jpn.,* **51**, 550–554 (1978).

25. J. W. Scott, D. Keith, G. Nix, Jr, D. R. Parrish, S. Remington, G. P. Roth, J. M. Townsend, D. Valentine, Jr, and R. Yang, *J. Org. Chem.,* **46**, 5086–5093 (1981).

26. K. Weinges and E. Kromm, *Liebigs Ann. Chem.,* 90–102 (1985).

27. T. Taguchi, A. Kawara, S. Watanabe, Y. Oki, H. Fukushima, Y. Kobayashi, M. Okada, K. Ohta and Y. Iitaka, *Tetrahedron Lett.,* **27**, 5117–5120 (1986).

28. H. Wakamatsu, J. Uda and N. J. Yamakami, *Chem. Soc., Chem. Commun.,* 1540 (1971).

29. J. J. Parnaud, G. Camperi and P. Pino, *J. Mol. Catal.,* **6**, 341–350 (1979).

30. K. Izawa, *J. Mol. Catal.,* **41**, 135–146 (1987), and references cited therein.

31. H. Wakamatsu, *Sekiyu Gakkai Shi,* **17**, 105–110 (1974).

32. R. Stern, A. Hirschauer, D. Commereuc and Y. Chauvin, *US Pat.,* 4 264 515 (1981).

33. I. Ojima, K. Hirai, M. Fujita and T. Fuchikami, *J. Organomet. Chem.,* **279**, 203–214 (1985).

34. I. Ojima, M. Okabe, K. Kato, H. B. Kwon annd I. T. Horvath, *J. Am. Chem. Soc.,* **110**, 150–157 (1988).

35. W. S. Knowles, M. J. Sabacky, B. D. Vineyard and D. J. Weinkauff, *J. Am. Chem. Soc.*, **97**, 2567–2568 (1975).
36. (a) U. Nagel, *Angew. Chem., Int. Ed. Engl.*, **23**, 435–436 (1984); (b) U. Nagel, E. Kinzel, J. Andrade and G. Prescher, *Chem. Ber.*, **119**, 3326–3343 (1986).
37. M. D. Fryzuk and B. Bosnich, *J. Am. Chem. Soc.*, **99**, 6262–6267 (1977).
38. (a) I. Ojima, T. Kogure, N. Yoda, T. Suzuki, M. Yatabe and T. Tanaka, *J. Org. Chem.*, **47**, 1329–1334 (1982); (b) G. Baker, S. J. Fritschel, J. R. Stille and J. K. Stille, *J. Org. Chem.*, **46**, 2954–2960 (1981).
39. (a) A. Miyashita, A. Yasuda, H. Takaya, K. Toriumi, T. Ito, T. Souchi and R. Noyori, *J. Am. Chem. Soc.*, **102**, 7932–7934 (1980); (b) A. Miyashita, H. Takaya, T. Souchi and R. Noyori, *Tetrhedron*, **40**, 1245–1253 (1984).
40. H. -W. Krause, H. -J. Kreuzfeld, C. Dobler and S. Taudien, *Tetrahedron: Asymmetry*, **3**, 555–566 (1992).
41. H. Watanabe, Y. Hashizume and K. Uneyama, *Tetrahedron Lett.*, **33**, 4333–4336 (1992).
42. M. Fujita and I. Ojima, *Tetrahedron lett.*, **24**, 4573–4576 (1983).
43. (a) T. D. Petrova, T. I. Savchenko, T. F. Ardyukova and G. G. Yakobson, *Khim. Geterotsikl. Soedin.*, **7**, 213–214 (1971); (b) H. M. Rajh, J. H. Uitzetter, L. W. Westerhuis, C. L. Dries and G. I. Tesser, *Int. J. Pept. Protein Res.*, **14**, 68–79 (1979).
44. T. D. Petrova, T. I. Savchenko, L. N. Shchegoleva, T. F. Ardyukova and G. G. Yakobson, *Khim. Geterotsikl. Soedin.*, **10**, 1344–1347 (1970).
45. E.g. R. M. Herbst and D. Shemin, *Org. Synth., Coll. Vol. 2*, 1–5 (1943).
46. D. G. Knorre, O. I. Lavrik, T. D. Petrova, T. I. Savchenko and G. G. Yakobson, *FEBS Lett.*, **12**, 204–206 (1971).
47. G. A. Nevinsky, O. O. Favorova, O. I. Lavrik, T. D. Petrova, L. L. Kochkina and T. I. Savachenko, *FEBS Lett.*, **43**, 135–140 (1974).
48. I. Ojima and K. Nakahashi, *Jpn. Pat., Kokai Tokkyo Koho*, S61-286353 (1986).

4 Synthesis of β-Fluorine-containing Amino Acids

NORBERT SEWALD and KLAUS BURGER
Universität Leipzig, Institut für Organische Chemie, Leipzig, Germany

4.1 INTRODUCTION

4.1.1 STRATEGIES FOR THE INTRODUCTION OF FLUORINE, PARTIALLY FLUORINATED OR PERFLUORINATED SUBSTITUENTS INTO α-AMINO ACIDS

There are two fundamentally different strategies by which fluorine, partially fluorinated or perfluorinated substituents can be introduced into amino acids.

Direct introduction is achieved via direct replacement of hydrogen by fluorine, partially fluorinated or perfluorinated substituents in a late step of the reaction sequence or via functional group transformation (e.g. fluorodehydroxylation) in a late step of the reaction sequence. Although this approach seems to be more straightforward (provided that suitable fluorinating agents are available), control of regio- and stereochemistry often is difficult to achieve. Functional groups additionally present in the molecule may be transformed in an undesired way. Therefore, they have to be appropriately protected, which requires additional synthetic steps. Further, many of the reagents currently used for the direct introduction of fluorine or fluorinated moieties are expensive, toxic, corrosive and sometimes explosive.

Consequently, *the building block strategy* represents an attractive alternative concept. This method often allows the regio- and stereoselective introduction of fluorine or fluorinated substituents.

Valuable reviews covering the references for both types of methodology up to 1990 are available [1–7]. Since the chemistry of fluorinated amino acids is a rapidly developing area of interdisciplinary interest, this chapter presents a comprehensive review, focusing on β-fluoro containing α-amino acids.

4.1.2 PHYSICAL AND CHEMICAL FEATURES OF FLUORINE SUBSTITUENTS

Fluorine or fluorinated substituents strategically positioned in biologically active compounds may change their chemical properties, biological activity

Fluorine-containing Amino Acids: Synthesis and Properties   Edited by V. P. Kukhar' and V. A. Soloshonok
© 1995 John Wiley & Sons Ltd

and selectivity considerably [3,8,9]. The exchange of hydrogen by fluorine is commonly regarded as being an isosteric substitution, because the van der Waals radii are similar (H, 1.20 Å; F, 1.35 Å). Thus, fluorine is small in size, but it is still significantly larger than hydrogen [10]. The important differences in chemical reactivity of fluorinated compounds are mainly based on the difference in the carbon–fluorine (456–486 kJ mol^{-1}) and the carbon–hydrogen bond energies (356–435 kJ mol^{-1}) and on the different electronegativities of fluorine and hydrogen (Pauling scale: H, 2.1; F, 4.0). Fluorine substituents exert considerable polarization effects on adjacent functional groups; their reaction behaviour may be changed or even inverted. An impressive example is the difference in the pK_a values of natural amino acids and α-trifluoromethyl-substituted amino acids (TFMXaa): Ala-CO$_2$H, 2.34; TFMAla-CO$_2$H, 1.98; Ala-NH$_2$, 9.87; TFMAla-NH$_2$, 5.91 [11].

Introduction of fluorine into strategic positions of biologically active compounds may also block metabolic degradation.

Fluorine substituents are capable of participating in hydrogen bonding or of acting as coordinative sites in metal complexes because of the availability of the three non-bonding electron pairs.

4.1.3 UNIQUE PROPERTIES OF THE TRIFLUOROMETHYL (TFM) GROUP

The often postulated quasi-isosterism between CH$_3$ and CF$_3$ groups is still a controversial issue [12]. The van der Waals radii are quite similar (2.0 and 2.7 Å), whereas the van der Waals volumes differ significantly (16.8 and 42.6 Å^3) [13]. The steric bulk of a trifluoromethyl group seems to be close to that of an isopropyl group.

With increasing degree of fluorination, the C—C bond length shortens and, consequently, the bond strength increases. Owing to this unique effect, the introduction of fluorine or perfluoroalkyl groups stabilizes molecules.

Other properties of a trifluoromethyl group are an electronegativity similar to that of oxygen and high lipophilicity (lipophilicity scale F < CF$_3$ < OCF$_3$ < SCF$_3$) [14] enhancing *in vivo* absorption and transport rates of biologically active compounds and improving their permeability through certain body barriers. However, the lipophilic effect of a trifluoromethyl group seems to depend very much on its position in a molecule. According to the octanol/water partition coefficients of several aliphatic trifluoromethyl-substituted alcohols, only the α-trifluoromethyl substituted derivatives are more lipophilic than their non-fluorinated pendants. The polarization effects of the trifluoromethyl group decrease the basicity of the hydroxyl group and increase the lipophilicity of the molecule [15]. Similar effects are to be expected in the case of α-trifluoromethylamino acids.

Because of the high electron density, the trifluoromethyl group is capable of participating in strong hydrogen bonding. Further, it can act as a ligand for metals. Another attractive feature of the trifluoromethyl group is the relatively low toxicity and high stability compared with mono- and difluoromethyl groups [16].

Although the trifluoromethyl group is often considered to be chemically inert, it is known to undergo a variety of reactions. The stability of a CF_3 group depends very much on its position in a molecule. Trifluoromethyl groups attached to carbon atoms bearing acidic protons are readily susceptible to hydrolytic degradation in basic media to give a carboxylic group [17]. Thus, 3,3,3-trifluoroalanine (TFMGly) is unstable above pH 6 even at room temperature and undergoes hydrolysis via stepwise deprotonation and fluoride elimination [18, 19].

4.1.4 BIOLOGICALLY ACTIVE β-FLUORO-α-AMINO ACIDS

Several β-fluorine-containing α-amino acids are biologically active, exhibiting antibacterial, antihypertensive, cancerostatic and cytotoxic effects. 2-(Difluoromethyl)ornithine (DFMOrn) seems to be a superior drug for the treatment of 'sleeping sickness'. β-Fluoro-substituted amino acids in general are potent irreversible inhibitors of pyridoxal phosphate-dependent enzymes, e.g. decarboxylases, transaminases and racemases. The inhibitory mechanism has been studied in detail [20].

Regulation of enzymatic decarboxylation reactions of amino acids by using highly specific inhibitors is of fundamental therapeutic interest. Some typical physiologically important amines are dopamine, serotonine, histamine and tyramine. The catecholamines take part in controlling blood pressure [21]. Elevated histamine levels are observed in allergies, hypersensitivities, gastric ulcers, inflammations, etc. [22]. High putrescine levels are characteristic of rapid cell development, including tumor growth [23].

Since an increasing number of enzymes have been characterized in terms of their three-dimensional structure, and since mechanisms for the reactions at the active sites have been elucidated, a rational design of mechanism-based fluorinated drugs should be possible.

In this context, β-fluorine-containing amino acids are currently one of the most promising types of low molecular mass bioregulators [24].

4.1.5 ANALYTICAL APPLICATIONS OF FLUORINATED AMINO ACIDS

The natural isotope ^{19}F is an NMR-active nucleus with spin $\frac{1}{2}$, high relative receptivity and a chemical shift range of ca 300 ppm, representing a sensitive and powerful probe for NMR studies. ^{19}F NMR may provide a very efficient method for conformational studies of fluorine-containing peptides and proteins or the observation of metabolic processes.

The artificial isotope ^{18}F decays with emission of positrons ($\tau_{1/2} = 110$ min). Positron emission tomography (PET) is a useful non-invasive technique for the survey of living tissue, allowing real-time analysis of metabolic processes [25]. One major application of ^{18}F-PET is brain imaging of Parkinsonian patients.

4.2 AMINO ACIDS WITH FLUORINE SUBSTITUENTS IN THE SIDE-CHAIN

4.2.1 MONOFLUORO DERIVATIVES

4.2.1.1 Fluorodehydrogenation

Introduction of fluorine into amino acids can be achieved by using molecular fluorine or suitable fluorinating agents (e.g. perchloryl fluoride [26–28] or trifluoromethyl hypofluorite [29]). Direct fluorination of organic compounds mostly suffers from low selectivity. The most efficient method for the direct introduction of fluorine into the β-position of amino acids proceeds via radical photofluorination with trifluoromethyl hypofluorite in liquid hydrogen fluoride or trifluoroacetic acid to give, e.g., L-3-fluoro-alanine (54%), D-3-fluoroalanine (57%), L-2-deuterio-3-fluoroalanine (59%) [29], β-fluoroaspartic acid [30] or 3-fluoroazetidine-2-carboxylic acid [31]. Fully protected cysteine derivatives (e.g. methyl S-benzyl-N-tri-fluoroacetylcysteinate) react with xenon difluoride to give a mixture of methyl S-benzyl N-trifluoroacetyl-3-fluorocysteinate, which is labile towards elimination of hydrogen fluoride, and methyl S-(α-fluorobenzyl)-N-tri-fluoroacetylcysteinate [32].

4.2.1.2 Fluorodehydroxylation

Displacement of hydroxyl groups in the side-chain of amino acids by fluorine is achieved on reaction with sulfur tetrafluoride in liquid hydrogen fluoride at low temperature. The reactivity of SF_4 towards a variety of hydroxyl groups (but not towards carbonyl or carboxyl groups) is increased considerably when liquid hydrogen fluoride is used as the solvent. Simultaneously, the amino group is protected by protonation. Monofluoroalanine (R = H [33–35]), 3-fluoroaspartic acid (R = CO_2H [34, 36]), L-threo-/L-erythro-3-fluoro [1,2-^{14}C]aspartic acid [36] and 2-amino-3-fluorobutyric acid [34, 37] have been obtained among others via this route.

(i) SF$_4$, liq. HF, $-78\,^{\circ}$C

R	Product Configuration	Yield (i) (%)	d.e. (%)	Ref.
L-Ser	H	2R	51		33, 34
[2-^{2}H]Ser	H		80		35
L-Thr	CH$_3$	2R,3S	48	84	34
D-Thr	CH$_3$	2S,3R			34
L-allo-Thr	CH$_3$	2R,3R	57	56	34
D-allo-Thr	CH$_3$	2S,3S	60		34
threo-DL-(3-OH)Phe	C$_6$H$_5$		65		34
threo-DL-(3-OH)Asn	CONH$_2$	threo	27	a	36

^{a} Only one isomer according to NMR and amino acid analysis

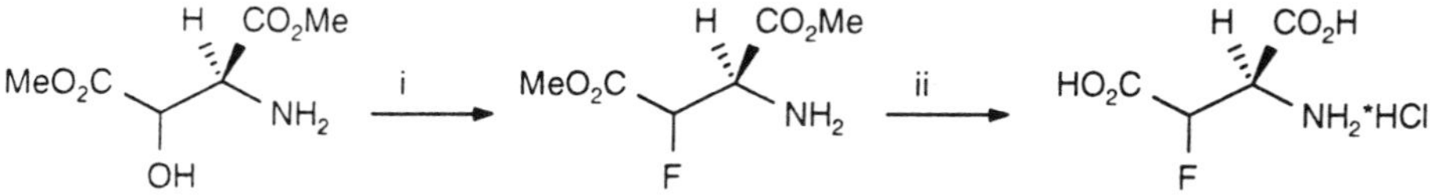

(i) SF$_4$, liq. HF, $-78\,^{\circ}$C $\rightarrow$ r.t.; (ii) 4M HCl, 80 $^{\circ}$C

Amino acid	Product Configuration	Yield (i, ii) (%)	d.e. (%)	Ref.
threo-DL-(3-OH)Asp^a		25^b	20	36
erythro-DL-(3-OH)Asp^a	erythro	42^c	d	34
threo-DL-(3-OH)Asp^a	threo	71^c	d	34

^{a} Fluorodehydroxylation occurs only with the diester.
^{b} As free acid.
^{c} As hydrochloride.
^{d} Only one isomer according to NMR and amino acid analysis.

Incomplete conversion and lower yields due to the intermediate formation of serine fluorosulfites can be overcome by higher dilution [33] (see p. 144).

Fluorodehydroxylation of 3-hydroxyglutamic acid with SF$_4$/liquid hydrogen fluoride gives 3-fluoropyroglutamic acid [38], which cannot be transformed into 3-fluoroglutamic acid without elimination of hydrogen fluoride. DL-erythro-N-Acetyl-3-hydroxyglutamic acid is converted into N-acetyl-3-fluoroglutamic acid on treatment with sulfur tetrafluoride in liquid hydrogen fluoride. Resolution of the diastereoisomers via anion exchange chromatography and subsequent enzymatic resolution of the enantiomers via deacetylation with acylase I yields the L-threo and L-erythro isomers of 3-fluoroglutamic acid, respectively. The fluorodehydroxylation presumably proceeds via an S_N2 mechanism, as a high percentage of inversion is observed (88% d.e. starting from the *threo* isomer and 74% d.e. from the *erythro* isomer [38]).

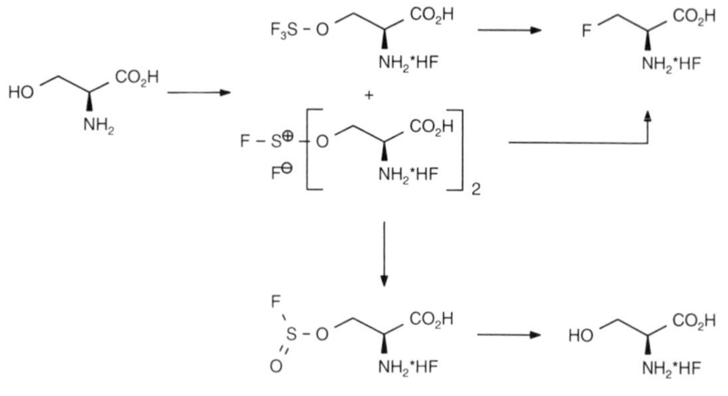

Application of Schöllkopf's strategy provides access to a variety of β-fluoro-substituted α-amino acids with varying optical yields via hydroxyalkylation of the bislactim ether and subsequent fluorodehydroxylation with diethylaminosulfur trifluoride (DAST) [39]. Fluoroalanine, however, cannot be obtained via this route. A major drawback of this methodology is the facile elimination of hydrogen fluoride to give the alkylidene bislactim ether or pyrazine derivatives.

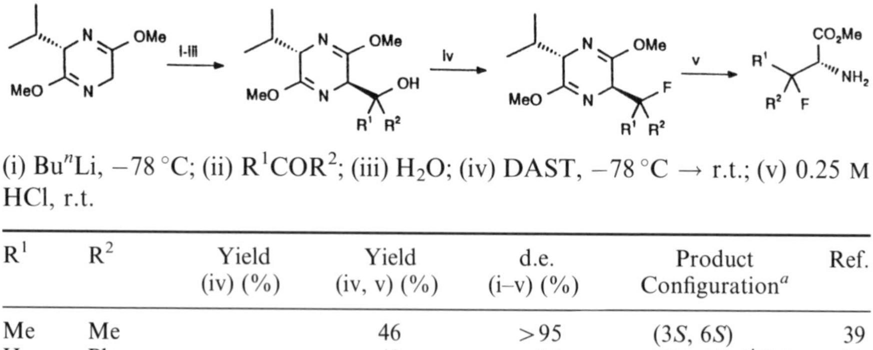

(i) Bu^{n}Li, $-78\,°C$; (ii) R^{1}COR^2; (iii) H$_2$O; (iv) DAST, $-78\,°C \rightarrow$ r.t.; (v) 0.25 M HCl, r.t.

R^1	R^2	Yield (iv) (%)	Yield (iv, v) (%)	d.e. (i–v) (%)	Product Configuration[a]	Ref.
Me	Me		46	>95	(3S, 6S)	39
H	Ph		40	71	(3S, 6S, 3′SR)	39
Me	Ph	96		>95	(3S, 6S, 3′SR)	39

[a] Bislactim ether; The CIP nomenclature in ref. 39 is incorrect.

The transformation of β-hydroxy-α-amino acids into β-fluoro-α-amino acids by fluorodehydroxylation with diethylaminosulfur trifluoride proved to be successful only in those cases where the nucleophilicity of the amino group is decreased significantly, for instance by protection as 3*H*-oxazol-2-one. Thus, β-fluoro-α-amino acids can be synthesized from natural β-hydroxy-α-amino acids under mild conditions and in good yields [40].

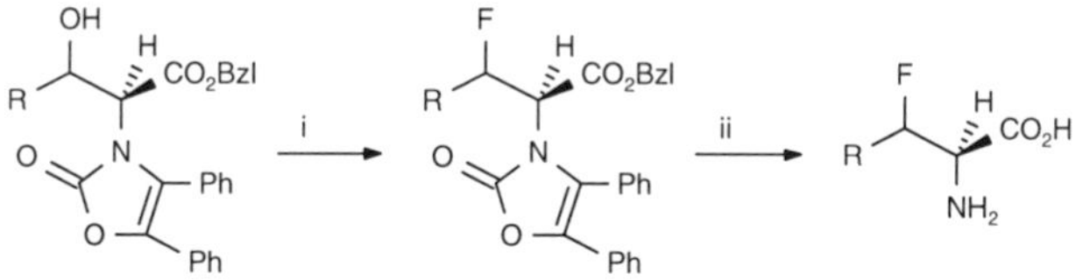

(i) DAST, CH_2Cl_2, $-78\,°C \to$ r.t.; (ii) H_2, Pd/C, EtOH

R	Yield (i) (%)	Yield (ii) (%)	Ref.
Me	45	99	40
Et	48	88	40
^{n}Pr	65	89	40
^{i}Bu	64	91	40

In several other cases, treatment of β-hydroxy-α-amino acids with DAST results in the formation of α-fluoro-β-amino acids via an intermediate aziridinium ion [41]; the initial stereochemistry (*threo* or *erythro*) is retained.

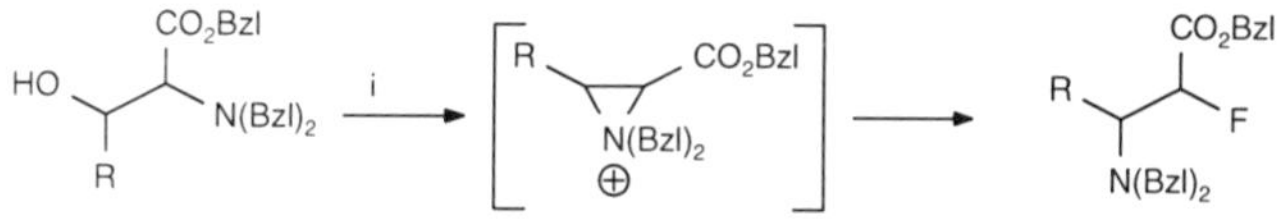

(i) DAST, THF, r.t.

Several homochiral stereoisomers of 3-fluorodiaminopimelic acid (2,6-diamino-3-fluoroheptanedioic acid) are prepared from corresponding heterocyclic 3-hydroxy acid derivatives by treatment with DAST [42].

Treatment of *threo*-β-phenylserine with DAST in liquid hydrogen fluoride does not proceed stereospecifically, but results in the formation of a diastereoisomeric mixture of β-fluorophenylalanine [43].

Fluorodehydroxylation is also achieved in a two-step process via O-tosylation and nucleophilic displacement of tosylate by fluoride [44].

β-Hydroxy-α-amino acids having a tertiary amino group undergo fluorodehydroxylation with Yarovenko's reagent [(2-chloro-1,1,2-trifluoroethyl)diethylamine] [45]. If the amino group is not protected suitably, the reaction results in formation of the corresponding 2-(chlorofluoromethyl)-2-oxazoline. Application of N-phthaloyl-protected α-amino-β-hydroxy acid esters gives N-protected β-fluoroamino acids, but deprotection of the amino group cannot be achieved without loss of the fluorine substituent [46].

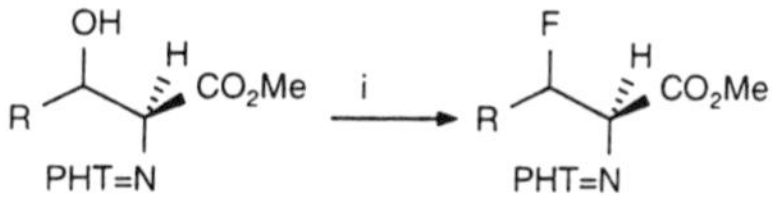

(i) $ClFCHCF_2$-NEt_2, CH_2Cl_2, $0\,°C$ (80%)

4.2.1.3 Fluorodesulfuration

Photofluorination of D-penicillamine in liquid hydrogen fluoride at $-78\,°C$ with fluoroxytrifluoromethane does not afford the expected γ-fluoro derivatives. Instead, fluorodesulfuration takes place to give β-fluorovaline in 94% yield. The reaction proceeds equally well without irradiation or with chlorine or N-chlorosuccinimide in liquid hydrogen fluoride (β-fluoro-phenylalanine from β-mercaptophenylalanine in 34% yield) instead of fluoro-xytrifluoromethane. The reaction cannot be applied to cysteine, however, because of oxidation of the mercapto group.

(i) CF_3OF, liq. HF, $0\,°C$; (ii) conc. HCl (94% for i, ii)

Instead, treatment of cysteine with a fluorine/helium mixture in liquid hydrogen fluoride at $0\,°C$ gives 3-fluoroalanine (33%) and 3,3-difluoroalanine (3%) [47].

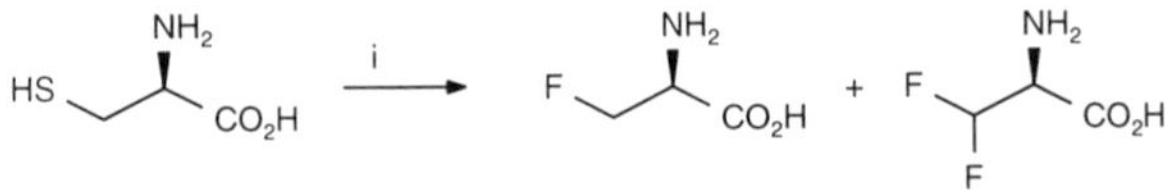

(i) F_2, He, liq. HF, BF_3, $0\,°C$, then ion-exchange chromatography

4.2.1.4 Fluorodehalogenation

3-Fluoroalanine was first obtained via nucleophilic displacement of chloride in 4-chloromethylene-2-phenyloxazolin-5-one with potassium fluoride, subsequent hydrolysis and reduction of the dehydro amino acid [48].

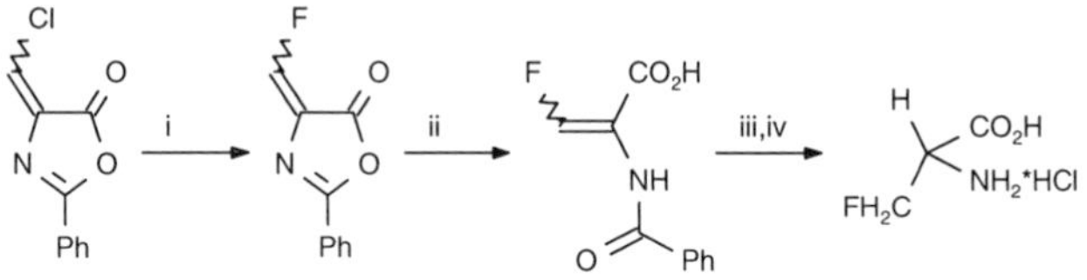

(i) KF, acetylacetone, hν, $70\,°C$; (ii) $NaHCO_3/H_2O$, reflux (86% for i, ii); (iii) H_2, PtO_2, MeOH, HCl, $40\,°C$; (iv) HCl, HCO_2H

N-Acetyl-β-fluoroglutamic acid is formed on fusion of diethyl N-acetyl-β-chloroglutamate with antimony trifluoride [49].

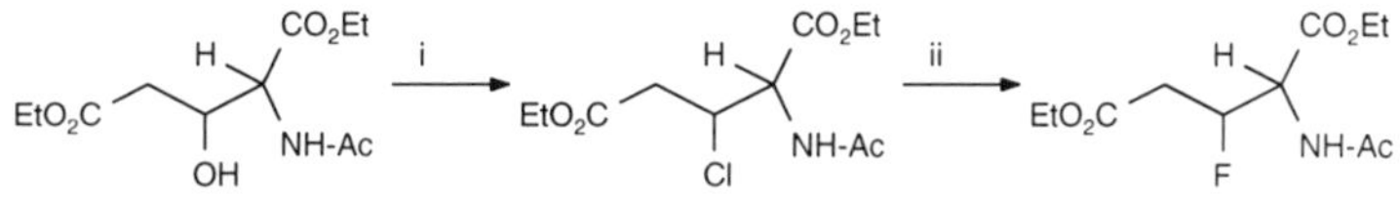

(i) $SOCl_2$, r.t. (84%), or PCl_5, $CHCl_3$, 6 °C; (ii) SbF_3, 100 °C (43%)

4.2.1.5 Reductive amination of β-fluoro-substituted α-keto acids

Several β-fluoro-substituted amino acids are obtained via a two-step reductive amination of β-fluoro-substituted α-keto acids [50].

An E/Z mixture of β-fluoro-α,β-dehydroaspartates results on elimination of hydrogen fluoride from 3,3-difluoroaspartate with 1,5-diazabicyclo[4.3.0]non-5-ene (DBN) (Z/E ratio 20:1) or by enamine formation from mono-fluorooxaloacetate (Z/E ratio 1:1). Reduction of the Z/E mixture with $NaBH_3CN$ yields predominantly the *erythro* isomer. Protonation of the amino group gives a racemic mixture of the iminium salt; the stereoselectivity is explained by Coulomb repulsion between the fluorine atom and the incoming nucleophile (see below). The diastereoisomers can be separated by crystallization or chromatography. β-Esterification of *erythro*-β-fluoroaspartic acid and subsequent ammonolysis gives *erythro*-β-fluoroasparagine [51].

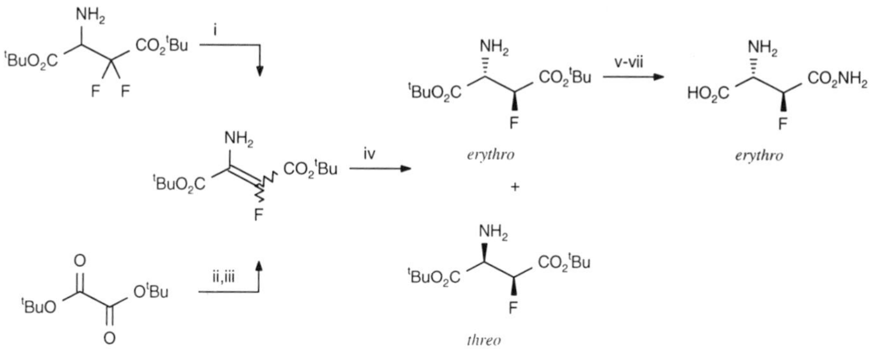

(i) DBN, THF, reflux (96%, $Z:E = 20:1$); (ii) Bu^{t}OK, Bu^{t}OH, $FCH_2CO_2Bu^t$, reflux (83%); (iii) NH_4OAc, MeOH, r.t. (85%, $Z:E = 1:1$); (iv) $NaBH_3CN$, MeOH, AcOH, r.t. (55% *erythro*, 1% *threo*); (v) CF_3CO_2H, reflux (75%); (vi) MeOH, $SOCl_2$, -18 °C → r.t. (90%); (vii) MeOH, NH_3, 4 °C (83%).

erythro-3-Fluorophenylalanine and its derivatives can be synthesized similarly [52, 53].

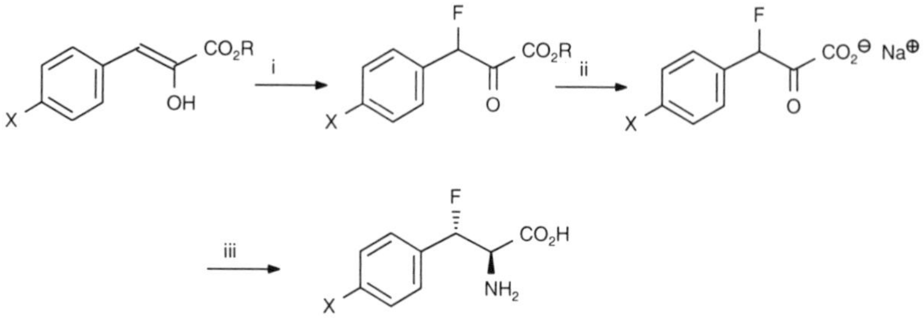

(i) F_2, N_2, CH_3CN, $-5\,°C$; (ii) $NaHCO_3$, Pr^iOH/H_2O, $40\,°C$ (75% for i, ii); (iii) aq. NH_3, $NaBH_4$, $10 \rightarrow 30\,°C$ or NH_4Br, $NaBH_3CN$, MeOH, r.t., then ion-exchange chromatography

X	Reducing agent	Yield (iii) (%)	d.e. (%)	Ref.
H	$NaBH_4$	52	>90	52
H	$NaBH_3CN$	18	>90	52
Cl	$NaBH_4$	60	>90	52
NO_2	$NaBH_4$	26	>90	52

This method is only limited by the availability of the corresponding β-fluoro-α-keto acids, which can be prepared by direct fluorination of α-keto acids. Another problem is the base lability of the β-fluoro-α-amino acids, which precludes alkaline hydrolysis of the carboxylic acid protective group in the final step. The method of choice is ester hydrolysis at the pyruvate stage with sodium hydrogencarbonate in aqueous propan-2-ol and reductive amination of the sodium salt with $NaBH_4$, reported to proceed with 90% diastereoselectivity favouring the *erythro* isomer. The selectivity observed is explained either by Coulomb repulsion between the fluorine substituent and the attacking hydride ion (see above) [51] or by stabilization effects between the fluorine substituent and the iminium group. A fluorine substituent is assumed to act as an electron-pair donor in hydrogen bridges. Thus, transition state A accounting for the formation of the *erythro* isomer is stabilized compared with B. The observed selectivity seems to be entirely different from that for other α-halo ketone reactions [54, 55].

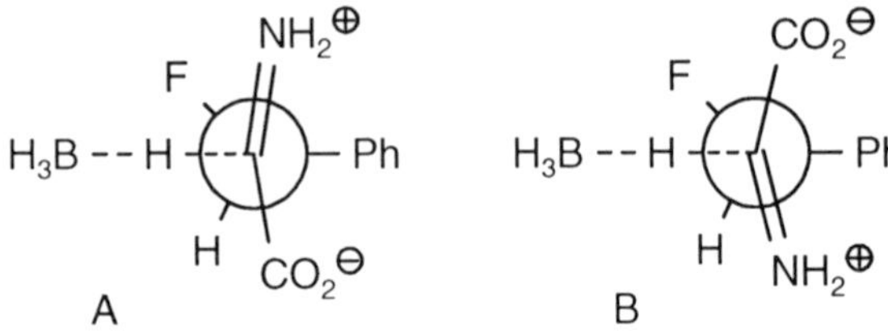

Similar stabilization effects have been proposed to control the rotamer equilibria of *erythro*- and *threo*-3-fluorophenylalanine according to NMR spectroscopic conformational analysis [52]. The asymmetric synthesis of 3-fluoroalanine can be achieved via reduction of chiral imines derived from 3-fluoropyruvic acid [56].

2-Deuterio-3-fluoroalanine (fludalanine, yield 70%) can be synthesized analogously using $NaBD_4$ as reducing agent. Similarly, [^{15}N]-DL-3-fluoroalanine is obtained in 38% yield on treatment of the corresponding ketone with liquid $^{15}NH_3$ and subsequent reduction [57]. Resolution is achieved via crystallization of diastereoisomeric quinine salts [58].

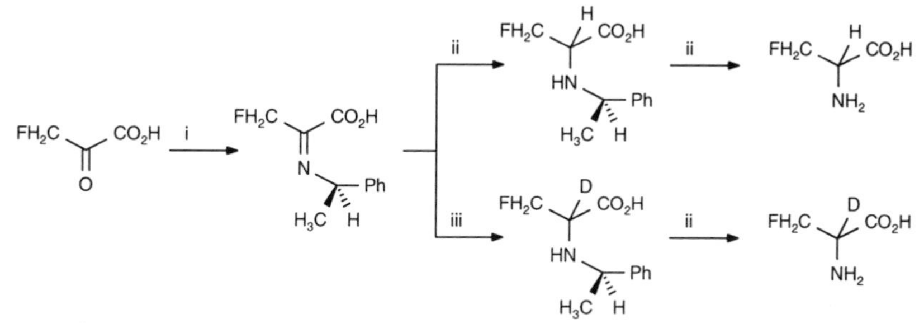

(i) (R)-PhCH(Me)NH$_2$; (ii) H$_2$, Pd/C; (iii) NaBD$_4$

An enzymatic approach towards 3-fluoroglutamic acid utilizes reductive amination of racemic 3-fluoro-2-ketoglutarate by glutamate dehydrogenase to give a diastereoisomeric mixture of $(2R,3S)$- and $(2R,3R)$-fluoroglutamic acid (96% yield) that can be resolved by anion-exchange chromatography [59].

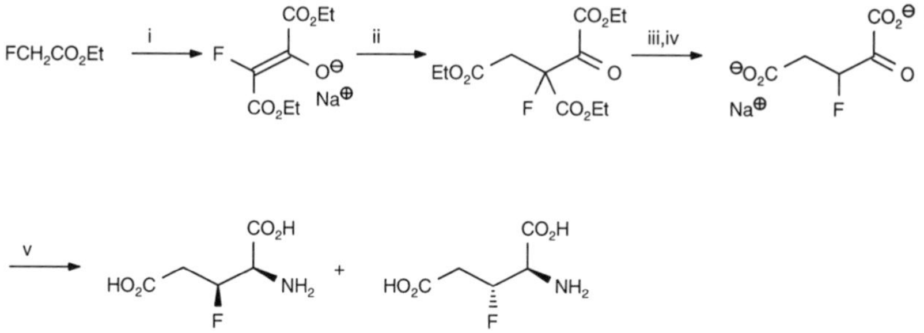

(i) (EtO$_2$C)$_2$, NaOEt, EtOH (82%); (ii) BrCH$_2$CO$_2$Et, DMF, 0 °C → r.t. (52%); (iii) AcOH/HCl; (iv) 50 °C, then NaHCO$_3$ (66% for iii, iv); (v) glutamate dehydrogenase, yeast alcohol dehydrogenase, NAD$^+$, cofactors, pH 7, 30 °C [96%, exclusively $(2R,3R)$ and $(2R,3S)$ isomers]

4.2.1.6 Amination of β-fluoro-containing α,β-unsaturated carboxylic acids

The enzyme 3-methylaspartate ammonia lyase (EC 4.3.1.2) catalyzes the reversible α,β-elimination of ammonia from $(2S,3S)$-3-methylaspartic acid. The retro reaction can be utilized for the synthesis of 3-substituted aspartic acid derivatives. However, only prolonged incubation of 2-fluorofumarate with a large excess of enzyme is reported to give traces of the desired product [60].

A synthetic procedure for 3-fluoroaspartic acid starts from dibenzyl difluoromaleate prepared from difluoromaleic anhydride. The reaction with dibenzylamine proceeds via an addition/elimination process to give the fluorinated enamine; reduction and deprotection yield predominantly *threo*-3-fluoroaspartic acid [61].

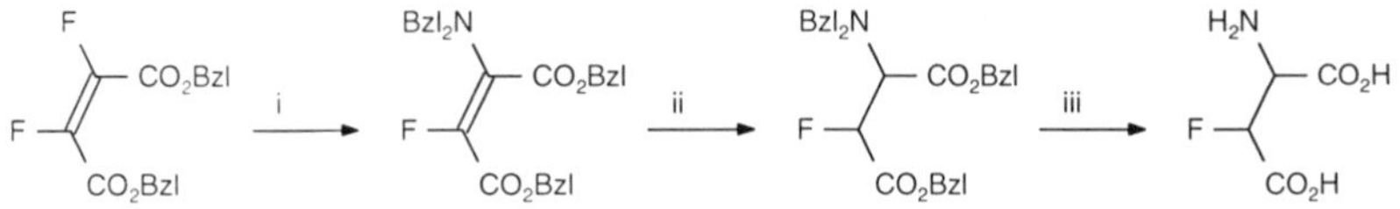

(i) $(Bzl)_2NH$, NEt_3, CCl_4, r.t. (90%); (ii) $NaBH_3CN$, HCl, THF, EtOH, r.t. (89%, 70% d.e.); (iii) H_2, Pd/C, EtOH, r.t. (33%)

4.2.1.7 Nucleophilic amination of β-fluoro-containing α-halo acid or α-hydroxy acid derivatives

Fluoroalkyl groups (especially perfluoroalkyl groups) usually prevent nucleophilic substitution at the neighboring carbon atom. Therefore, nucleophilic displacement reactions in the α-position are only feasible to a limited extent in the case of β-monofluoro-α-halo acids. This type of compound can be synthesized by fluorobromination of α,β-unsaturated carboxylic acids with *N*-bromoacetamide (NBA) in liquid hydrogen fluoride. Conversion of the resulting α-bromo-β-fluorocarboxylic acids into the corresponding amino acids is achieved by treatment with ammonia [62].

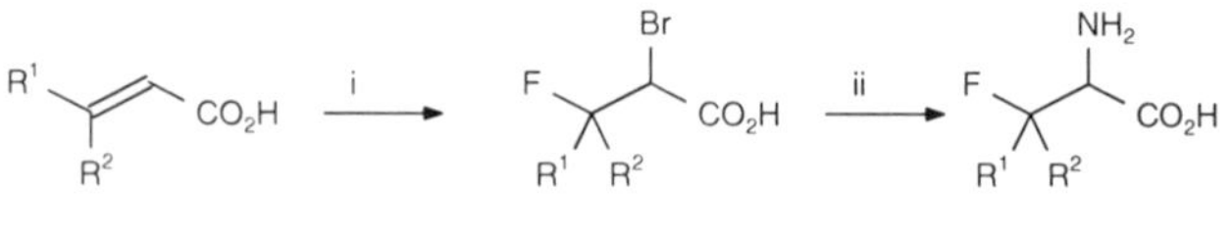

(i) NBA, liq. HF, $-30\,^\circ$C → r.t.; (ii) NH_3, autoclave, r.t.

R^1	R^2	Yield (i) (%)	Yield (ii) (%)	Ref.
H	H	44	30	62
Me	H	70	36	62
Me	Me	63	62	62
Et	H	19	16	62
Pr^n	H	6	36	62
Bu^n	H	30	10	62

Similarly, 3-fluoro-2-bromopropionic acid reacts in liquid ammonia to give 3-fluoroalanine in 66% yield [63]. 3-Fluoroalanine also can be synthesized selectively starting from 2,3-dibromopropionate via subsequent displacement of the bromine substituent in the 3-position by fluorine (with BrF₃) and, under phase-transfer conditions, in the 2-position by azide. Catalytic hydrogenation gives racemic methyl 3-fluoroalaninate; the N-Boc derivative is resolved enzymatically by papain to give N-Boc-3-fluoro-L-alanine [64].

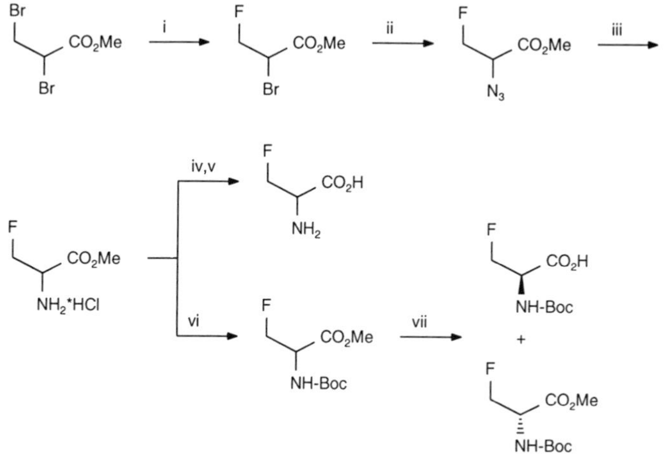

(i) BrF_3, $SnCl_4$, CCl_2FCClF_2 (70%); (ii) NaN_3, $Bu_4N^+Br^-$, CH_2Cl_2/H_2O (60%); (iii) H_2, Pd/C, MeOH, HCO_2H, HCl (85%); (iv) 12M HCl, reflux; (v) Pr^iOH, propylene oxide (60% for iv, v); (vi) $(Boc)_2O$, DMF, NEt_3 (87%); (vii) papain (91% D; 95% L)

Oxiranecarbonitriles, obtained from ketones and chloroacetonitrile via glycidonitrile synthesis, react with hydrogen fluoride/pyridine to give β-fluorocyanohydrins, which can be converted into β-fluoroamino acids by treatment with ammonia and subsequent acid hydrolysis [65, 66].

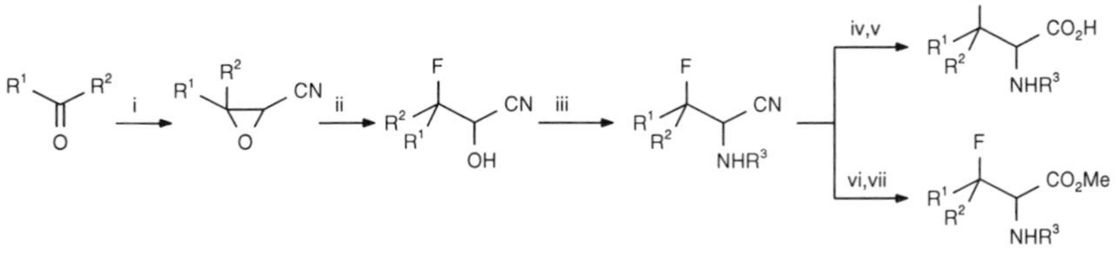

(i) $ClCH_2CN$, Bu^tOK, Bu^tOH; (ii) HF/pyridine, CH_2Cl_2; (iii) R^3NH_2, MeOH, $MgSO_4$, r.t.; (iv) 12M HCl, reflux; (v) pyridine; (vi) HCl, MeOH, reflux; (vii) pyridine

R^1	R^2	R^3	Yield (ii) (%)	Yield (iii) (%)	d.e. (%)	Yield (iv, v) (%)	Yield (vi, vii) (%)	Ref.
Me	Me	H	92	85		65	80	65
Et^a	Me	H		85	50	60	72	65
Et^b	Me	H		82	60			65
Et	Et	H		85		65	85	65
—(CH$_2$)$_4$—		H	85	85		50	65	65
—(CH$_2$)$_5$—		H	88	85		55	65	65
Ph	Me	H	91	80		50		65
Me	Me	Me		95		60		65
Me	Me	PhCH(Me)		90				65
Me	Me	(S)-PhCH(Me)		41	70	21		65
Me	Me	(R)-PhCH(Me)		38	80	17		65

^{a} *threo* isomer.
^{b} *erythro* isomer.

4.2.1.8 Fluoroalkylation of nucleophilic glycine synthons

β-Fluorophenylalanine is synthesized via alkylation of a glycine-derived Schiff base with α-bromo-α-fluorotoluene [67].

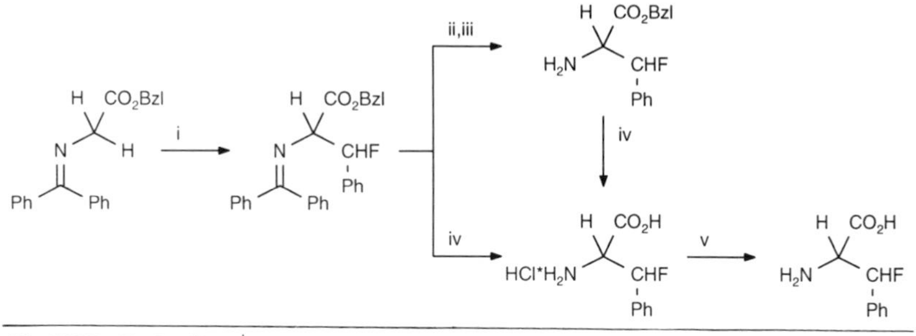

(i) PhCHBrF, Bu$^n_4$N$^+$HSO$_4^-$, CH$_2$Cl$_2$, NaOH, 5 °C (71%); chromatography, separation of diastereoisomers; (ii) 1M HCl, Et$_2$O, r.t.; (iii) NaHCO$_3$, H$_2$O; (iv) Me$_3$SiI, CHCl$_3$, 60 °C, HCl (70%); (v) propylene oxide, Pr^{i}OH, r.t. (59%)

Alkylation of the Schiff base derived from aminomalonate affords derivatives of 3-chloro-3-fluoroalanine. However, acidolytic deprotection is not possible; under the reaction conditions applied, elimination and hydrolysis to monofluoropyruvic acid are observed [68].

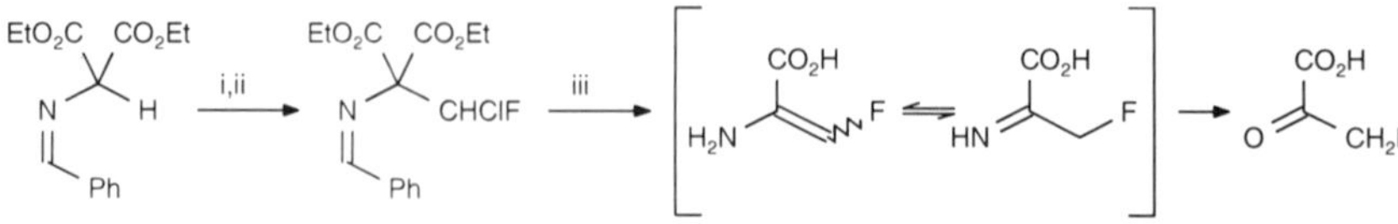

(i) NaN(SiMe$_3$)$_2$, THF, r.t.; (ii) CHCl$_2$F, THF, 10 °C; (iii) 3M HCl, 80 °C (31% for i–iii)

The bislactim ether derivative of monofluoroalanine is obtained in 80% yield (45% d.e.) on reaction of Schöllkopf's reagent with fluoroiodomethane [39]. Fully protected β-fluoro-α,β-dehydroalanine is formed on elimination of hydrogen fluoride from the corresponding 3,3-difluoroalanine derivative [68].

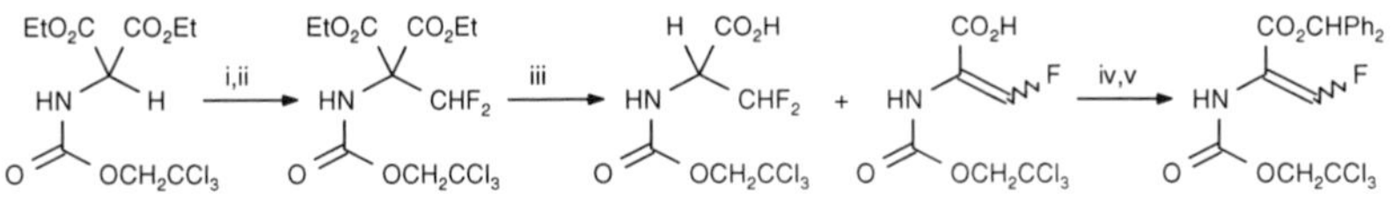

EtO₂C CO₂Et → EtO₂C CO₂Et → H CO₂H → CO₂H → CO₂CHPh₂

(i) NaN(SiMe₃)₂, THF, r.t.; (ii) CHClF₂, THF, −30 °C (41% for i, ii); (iii) Me₃SiI, reflux; (iv) (Ph)₂CN₂; CH₂Cl₂; (v) NEt₃, CH₂Cl₂ (58% for iii–v)

4.2.1.9 Fluoroalkylation of electrophilic glycine synthons

Trifluorovinylation of *in situ*-prepared methyl *N*-carbamoyliminoglyoxylate with trifluorovinylmagnesium bromide and subsequent deprotection of the amino and carboxylic groups results in the formation of 2-amino-3,4,4-tri-fluorobut-3-enoic acid, which readily decarboxylates to give 2,3,3-tri-fluoroprop-2-enylamine. Catalytic hydrogenation of the trifluorovinylamino acid hydrochloride yields a diastereoisomeric mixture of 2-amino-3-fluoro-butyric acid hydrochloride [69, 70].

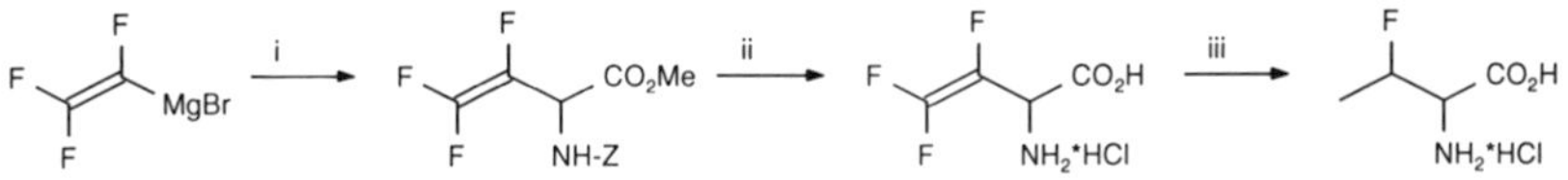

(i) BzlO₂CNHCHClCO₂Me, THF, −70 °C (85%); (ii) 6M HCl, reflux; (iii) H₂, Pd/C, H₂O (80% d.e.)

4.2.1.10 Ring opening of aziridines

Regioselective ring opening of aziridinecarboxylates [71], aziridine-car-boxamides or aziridinecarbonitriles [72] with hydrogen fluoride/pyridine results in the formation of β-fluoro-α-amino acid derivatives (predominantly *threo* isomers, even on application of *trans*-aziridines), providing a versatile methodology for the synthesis of this class of compound [71–77].

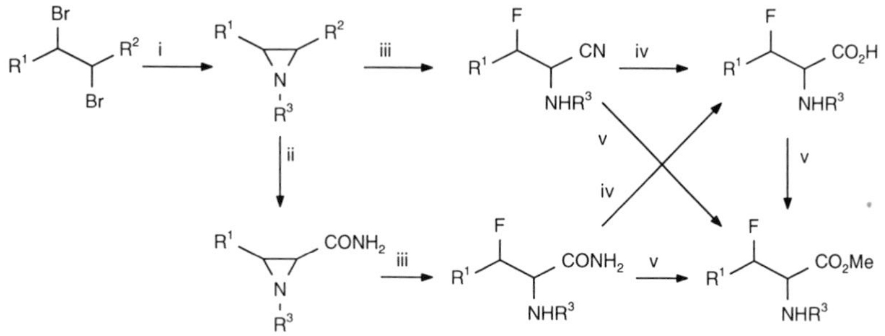

(i) R^3NH_2, DMSO, r.t. (70–80%); (ii) KOH, $H_2O/EtOH$, r.t. ($R^2 = CN \rightarrow$ $R^2 = CONH_2$); (iii) HF/pyridine, CH_2Cl_2; (iv) 12M HCl, reflux; (v) MeOH, HCl, reflux, then pyridine, r.t.

R^1	R^2	R^3	Yield (i) (%)	Yield (i, ii) (%)	Yield (iii) (%)	d.e. (%)	Yield (iv) (%)	Yield (v) (%)	Ref.
H	CN	Bu^t			45		40		72
H	CN	Bzl			50				72
Me	CN	Me			2		55		72
Ph	CN	H	70		80	14	60		72
Ph	CN	Me	75		75	13	57		72
p-ClC$_6$H$_4$	CN	H	75		80	14	55		72
p-ClC$_6$H$_4$	CN	Me	70		78	13	55		72
Ph	CONH$_2$	H		55	80	>98		70	72
Ph	CONH$_2$	Me		65	76	>98			72
p-ClC$_6$H$_4$	CONH$_2$	H		60	85	>98	60	80	72
p-ClC$_6$H$_4$	CONH$_2$	Me		58	80	>98			72

4.2.1.11 Miscellaneous

Diazotization of diaminosuccinic acid in liquid hydrogen fluoride at low temperature gives β-fluoroaspartic acid in 25% yield [78].

(i) NaNO$_2$, liq. HF, $-30 \rightarrow -10\,^\circ$C (25%)

4.2.2 DIFLUORO DERIVATIVES

4.2.2.1 Fluorodehydrogenation

3,3-Difluoroalanine is prepared by photofluorination of alanine with trifluoromethyl hypofluorite [30].

4.2.2.2 Reductive amination of β,β-difluoro-substituted α-keto acids

Both enantiomeric forms of β,β-difluoroaspartic acid are obtained via a reaction sequence consisting of electrophilic fluorination of oxaloacetate with perchloryl fluoride, reductive amination and resolution of the diastereo-isomeric brucine salts [79].

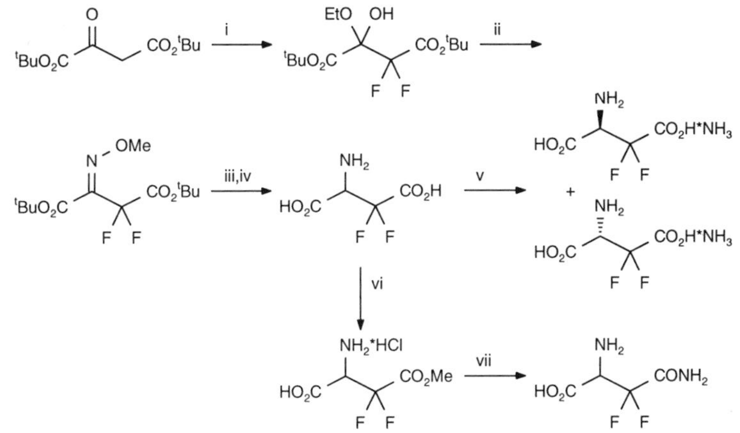

(i) EtONa, EtOH, $FClO_3$ (78%); (ii) $MeONH_2*HCl$, EtOH, r.t. (65%); (iii) Al/ Hg (45%); (iv) CF_3CO_2H, reflux, (95%); (v) (−)-brucine, H_2O, acetone, then conc. aq. NH_3; (vi) MeOH, $SOCl_2$, −15 °C → r.t. (72%); (vii) NH_3, MeOH, 0 °C → r.t. (85%)

4.2.2.3 Fluoroalkylation of nucleophilic glycine synthons

Alkylation of the metallated aminomalonate-derived Schiff base with chloro-difluoromethane affords 3,3-difluoroalanine [68, 80].

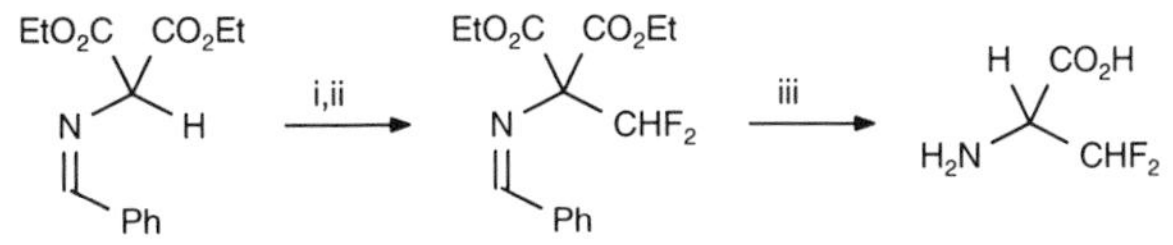

(i) $NaN(SiMe_3)_2$, THF; (ii) $CHClF_2$, −30 °C (50–65% for i, ii); (iii) 3M HCl, 80 °C (36% for i–iii)

4.2.2.4 Fluoroalkylation of electrophilic glycine synthons

Fluoroalkylation of in situ-prepared methyl glyoxylate N-carbamoylimine with aromatic 1,1-difluoro-2-silaenol ethers and subsequent deprotection give 4-aryl-3,3-difluoro-4-oxo-α-amino acids as an equilibrium mixture of the keto form and the cyclic semiacylal according to ^1H and ^{19}F NMR spectroscopy. In the presence of water, the geminal diol is also observed [81].

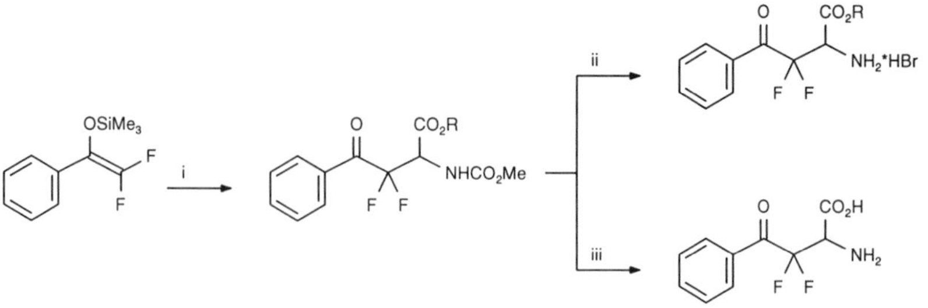

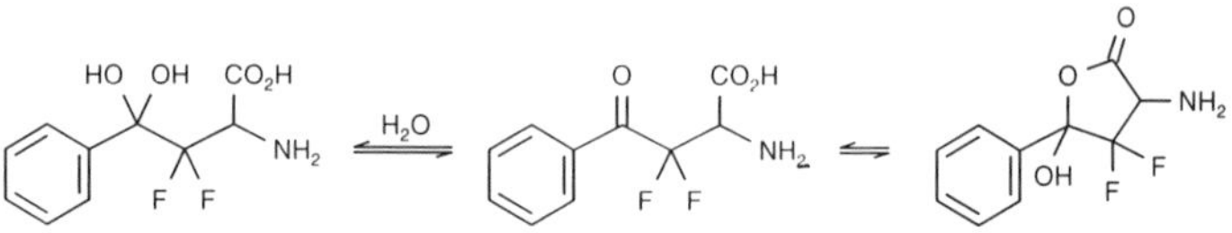

(i) MeO$_2$CNHCHClCO$_2$R, AlCl$_3$, CH$_2$Cl$_2$, $-5\,^{\circ}$C $\rightarrow$ r.t. (R = Me, 49%; R = Bzl, 40%); (ii) R = Me, HBr/HOAc; (iii) R = Bzl, Me$_3$SiI, CHCl$_3$, 40 $^{\circ}$C (68%)

Adaptation of a strategy originally developed by Weygand and co-workers for the synthesis of 3,3,3-trifluoroalanine (see below) gives 2,2-difluoroalanine hydrochloride in 10% overall yield via amidoalkylation of vinylmagnesium bromide with *in situ*-prepared N-benzyloxycarbonyldifluoroacetaldimine and subsequent oxidative cleavage of the double bond [82].

β,β-Difluorohomoserine and β,β-difluorohomocysteine can be prepared via a multi-step procedure starting from N-benzyl (S)-glyceraldimine acetonide and ethyl bromodifluoroacetate as a fluorine containing building block [83].

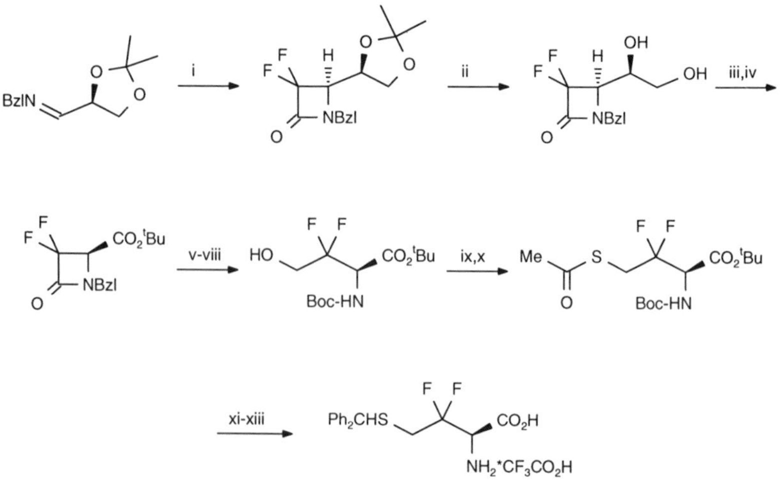

(i)　　BrF_2CCO_2Et, Zn, THF, reflux;
(ii)　　p-TsOH, MeOH, 55 °C;
(iii)　$CrCO_3$, H_5IO_6, acetone/H_2O, 0 °C;
(iv)　$Cl_3CC(=NH)OBu'$, BF_3*OEt_2, CH_2Cl_2/c-C_6H_{12}, 0 °C;
(v)　　DIBAL-H, Et_2O, −78 °C;
(vi)　NaBH$_4$, EtOH, 0 °C;
(vii)　H_2, Pd/C, EtOH;
(viii) $(Boc)_2O$, $KHCO_3$, dioxane/H_2O;
(ix)　$(CF_3SO_2)_2O$, pyridine, CH_2Cl_2, 0 °C;
(x)　　$MeCOS^-K^+$, DMF, 0 °C;
(xi)　0.2 M NaOH, MeOH;
(xii)　6 M HCl, dioxane;
(xiii) Ph_2CHOH, CF_3CO_2H

4.2.2.5 Ring opening of 2H-azirines

The reaction of 2H-azirinecarboxylates with hydrogen fluoride/pyridine provides an elegant route to β,β-difluoro-substituted amino acids [84, 85]. The reaction probably proceeds via aziridinium ions. 2H-Azirines are more reactive than aziridines towards addition of hydrogen fluoride.

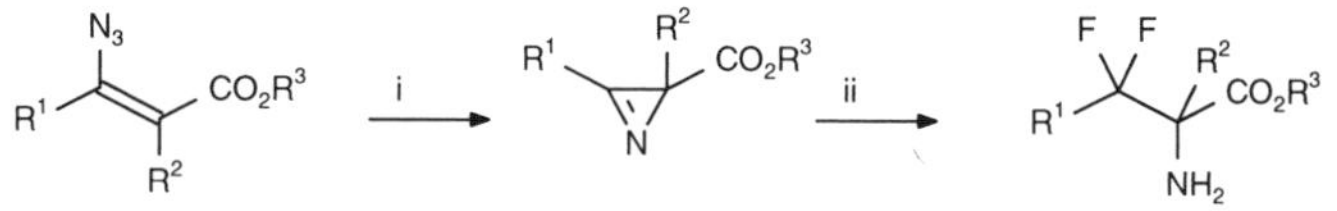

(i) heptane or hexane/benzene, reflux; (ii) HF/pyridine, benzene, 5 °C → r.t.

R^1	R^2	R^3	Yield (i, ii) (%)	Ref.
Ph	H	Me	32	85
Ph	D	Me	40	85
Me	H	Et	43	85

4.2.2.6 Functional group transformation

2-Hydroxyprolinol can be utilized for the synthesis of 3,3-difluoroglutamic acid. A bicyclic 2-hydroxyprolinol derivative is oxidized and then converted into the corresponding 3,3-difluoro compound via treatment with DAST and then oxidized to give N-protected 3,3-difluoropyroglutamate, which is subsequently hydrolyzed to give 3,3-difluoroglutamic acid [86].

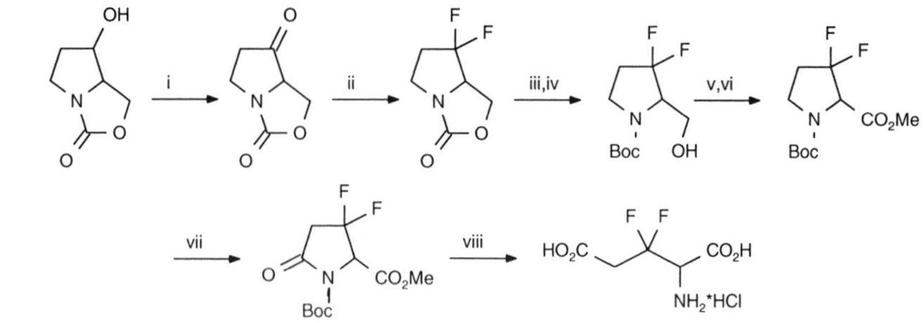

(i) $(CF_3CO)_2O$, DMSO, NEt_3, CH_2Cl_2 (73%);
(ii)   DAST, CH_2Cl_2, $-78\,^{\circ}C \rightarrow$ r.t. (64%);
(iii) $6M$ HCl, reflux;
(iv) $(Boc)_2O$, $NaHCO_3$, $CHCl_3$, H_2O (92% for iii, iv);
(v) RuO_2*xH_2O, 10% $NaIO_4$, EtOAc;
(vi) CH_2N_2, Et_2O (93% for v, vi);
(vii) RuO_2*xH_2O, 10% $NaIO_4$, EtOAc;
(viii) conc. HCl, reflux (40% for vii, viii)

4.2.3 TRIFLUORO DERIVATIVES (3,3,3-TRIFLUOROALANINE, TFMGly)

4.2.3.1 Fluorodehydrogenation

Photofluorination of alanine with trifluoromethyl hypofluorite in liquid hydrogen fluoride gives 3,3,3-trifluoroalanine [30].

4.2.3.2 Amidoalkylation of carboxylic group synthons with trifluoroacetaldimines

The first syntheses of 3,3,3-trifluoroalanine were described by Weygand and Steglich. *N*-Acyl trifluoroacetaldimines obtained from the reaction of tri-fluoroacetaldehyde with carboxamides and subsequent elimination of water [18] or via oxidation of *N*-acyl-2,2,2-trifluoro-1-(ethylmercapto)ethylamines and elimination of ethylsulfinic acid [87] are highly electrophilic species. They can either be isolated or be prepared *in situ* and readily add various carbon nucleophiles. The adducts formed with anhydrous hydrogen cyanide [88] or copper(I) cyanide (CuCN, DMF, 100 °C, 60%) [89] can be saponified to give 3,3,3-trifluoroalanine hydrochloride.

The amidoalkylation products of vinylmagnesium bromide with *N*-acyl-3,3,3-trifluoroacetaldimines afford, on subsequent oxidative cleavage of the CC-double bond, *N*-acyl 3,3,3-trifluoroalanine derivatives [90, 91]. Dipeptides with 3,3,3-trifluoroalanine in the C-terminal position are obtained via a similar route (i, v–vii) using acylimines derived from homochiral *N*-protected amino acid amides (R = ZHNCHR'—) [92].

(i)   X = Cl,Br,SO$_2$Et, NEt$_3$, THF, r.t;
(ii)   HCN, THF, −50 °C → r.t. (R = Ph, 81%; R = BzlO, 98% for i, ii);
(iii) 6M HCl, HOAc, reflux;
(iv)   NEt$_3$, CHCl$_3$, then HOAc (R = BzlO, 81% for iii, iv);
(v)   CH$_2$CHMgBr, THF, 0 °C → r.t.;
(vi)   H$_3$O$^+$ (R = Ph, 73%; R = BzlO, 76% for v, vi);
(vii) KMnO$_4$, acetone/H$_2$O (1:3, v/v), H$_3$O$^+$ (R = Ph, 92%; R = BzlO, 75%);
(viii) R = Ph, conc. HCl/HOAc, reflux or sealed tube, 95 °C; R = BzlO, 6M HCl or CF$_3$CO$_2$H, reflux;
(ix)   NEt$_3$, CHCl$_3$, then HOAc (75% for viii, ix)

Derivatives with a longer perfluorinated side-chain are not obtained via the corresponding aldimines, because of the facile elimination of hydrogen fluoride even in the absence of base. Instead, vinylation of protected α-amino sulfones and permanganate oxidation of the double bond yield the corresponding amino acids [91].

(i) CH$_2$CHMgBr, THF, 0 °C → r.t.; (ii) H$_3$O$^+$, (R$_F$ = C$_2$F$_5$, 76%; R$_F$ = C$_3$F$_7$, 62% for i, ii); (iii) KMnO$_4$, 1.5M H$_2$SO$_4$, acetone, H$_2$O, 0 °C → r.t. (R$_F$ = C$_2$F$_5$, 82%; R$_F$ = C$_3$F$_7$, 82%); (iv) conc. HCl, HOAc, reflux; (v) NEt$_3$, CHCl$_3$, then HOAc (R$_F$ = C$_2$F$_5$, 62%, R$_F$ = C$_3$F$_7$, 54%)

Amides of 3,3,3-trifluoroalanine are obtained via [4 + 1] cycloaddition of isocyanides to the acylimines and cleavage of the resulting 5-imino-2-oxazoline ring with dilute acid. When isonitriles synthesized from N-formylamino acid esters are used as dienophiles, the [4 + 1] cycloadducts yield 3,3,3-trifluoro-alanyl peptides on hydrolysis [90, 93].

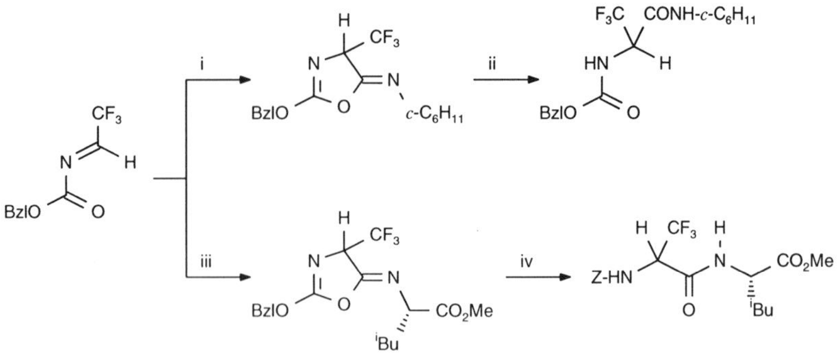

(i) c-$C_6H_{11}NC$; (ii) H_3O^+ (80% for i, ii); (iii) $CNCHR(Bu^i)CO_2Me$; (iv) H_3O^+ (64%, 44% d.e.)

The N-acyl group can usually be removed acidolytically. The trifluoromethyl group of 3,3,3-trifluoroalanine derivatives is base-labile: a stepwise elimination of hydrogen fluoride occurs at pH > 6 to give glycine via aminomalonate [19,92].

4.2.3.3 Transformation of trifluoroacetic acid

A mechanistically interesting approach to 3,3,3-trifluoroalanine hydrochloride starting from trifluoroacetic acid involves the intermediacy of trifluoromethyl-substituted 2H-oxazol-5-ones and oxazoles [94, 95].

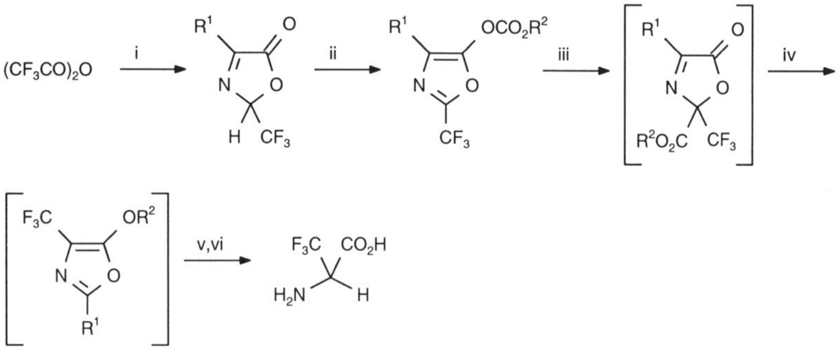

(i) $H_2NCHR^1CO_2H$, THF; (ii) R^2OCOCl, NEt_3, THF; (iii) cat. DMAP, THF (96%); (iv) 230 °C; (v) conc. HCl, HOAc, reflux; (vi) NEt_3, $CHCl_3$, then HOAc ($R^1 = Bu^i$, $R^2 = CO_2Bu^i$, 44% for ii–vi)

4.2.3.4 Amination of 2-bromo-3,3,3-trifluoropropionates

2-Bromo-3,3,3-trifluoropropionic acid already possesses the complete fluorinated carbon atom backbone of 3,3,3-trifluoroalanine. Esterification and nucleophilic amination provide derivatives of 3,3,3-trifluoroalanine. However, the cleavage of the N-protective group seems to be problematic [96].

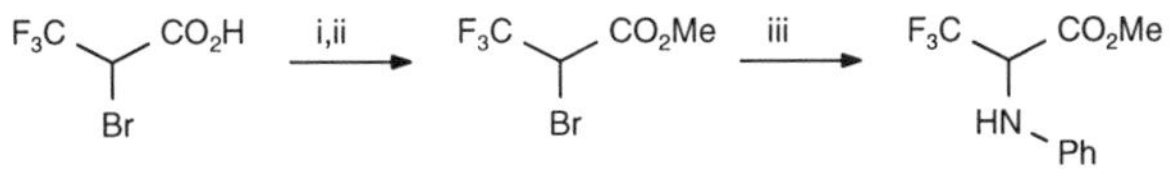

(i) SOCl₂; (ii) MeOH; (iii) PhNH₂

4.2.3.5 2H-Perfluoroisobutyric acid as precursor of 3,3,3-trifluoroalanine: transformation of a trifluoromethyl group into an amino group

The introduction of orthogonal protective groups during amino acid synthesis permits selective deprotection of the amino or the carboxylic group. The combination of methyl ester and urethane as protective groups for 3,3,3-trifluoroalanine is achieved via a several-step synthesis starting from 2H-perfluoroisobutyric acid. One trifluoromethyl group is partially hydrolyzed into a carbonyl fluoride, which undergoes Curtius rearrangement on treatment with azide [97].

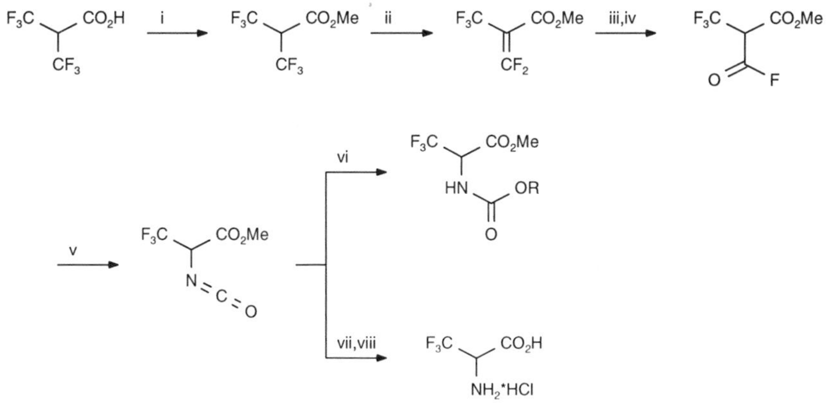

(i) MeOH, H₂SO₄; (ii) BF₃*NEt₃; (iii) AcOH; (iv) ΔT; (v) Me₃SiN₃, Et₂O, 4 → 60 °C (88%) or NaN₃, benzene, 20 °C → reflux (70%); (vi) ROH, benzene, 60°C (R = Et, 97%; R = Prⁱ, 93%, R = Bzl, 81%; R = Ph, 87%); (vii) conc. HCl, r.t. (74%); (viii) 6M HCl, 90 °C (72%)

4.2.3.6 Hexafluoroacetone as precursor of 3,3,3-trifluoroalanine: reductive amination and transformation of a trifluoromethyl group into a carboxy group

The complete carbon atom skeleton of 3,3,3-trifluoroalanine is already present in hexafluoroacetone. The N-acylimines of hexafluoroacetone [98, 99] can be transformed into 5-fluoro-4-trifluoromethyloxazoles in a one-pot procedure with tin(II) chloride [100–102] or zinc on sonication [103].

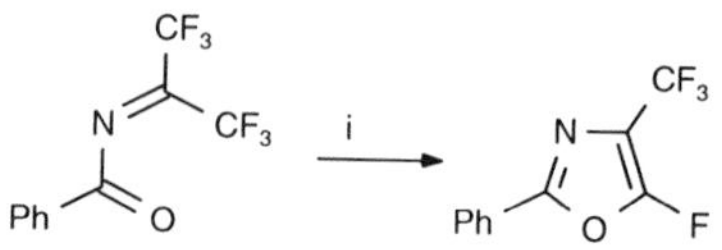

(i) SnCl$_2$, xylene/THF (3:1, v/v), reflux (68%) or Zn, THF, sonication, r.t. (87%)

The reaction sequence consists of a [4 + 1] cycloaddition of the heterodiene with tin(II) chloride to give a five-membered metallacycle, which on heating undergoes heterolytic ring cleavage, fluoride elimination, electrocyclization and finally aromatization via elimination of hydrogen fluoride [104].

5-Fluoro-2-phenyl-4-trifluoromethyloxazole represents a cyclic derivative of N-benzoyl-3,3,3-trifluoroalanyl fluoride. However, the yields of the direct acidolysis are low. The transformation into 3,3,3-trifluoroalanine is achieved via controlled ring cleavage: nucleophilic displacement of fluoride in the 5-position by alcohols, cleavage of the oxazole ring with HBr/HOAc or iodotrimethylsilane and finally acidolytic deprotection of the amino function on heating with concentrated hydrochloric acid gives 3,3,3-trifluoroalanine hydrochloride [6, 105].

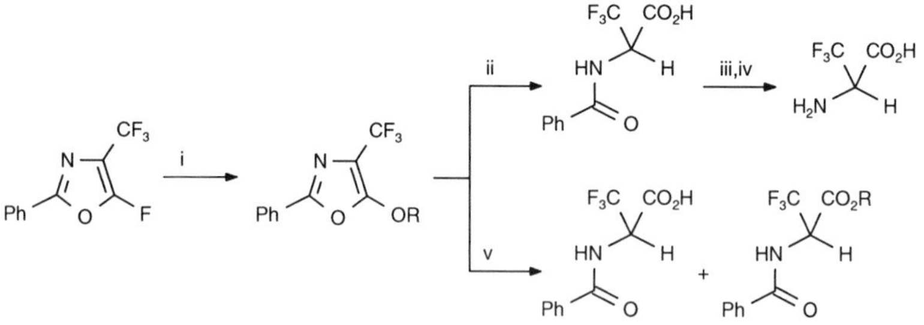

(i) ROH, KOH, (R = Et: 89%); (ii) (Me$_3$Si)$_2$, I$_2$, CH$_2$Cl$_2$, reflux (46%); (iii) conc. HCl, reflux; (iv) NEt$_3$, CHCl$_3$ then HOAc (72% for iii, iv); (v) HBr/HOAc, reflux (30% + 41% ester)

Yields are improved considerably when 5-tert-butoxy-2-phenyl-4-trifluoromethyloxazole is treated with aqueous trifluoroacetic acid at room temperature. N-Benzoyl-3,3,3-trifluoroalanine is obtained in nearly quantitative yield [6].

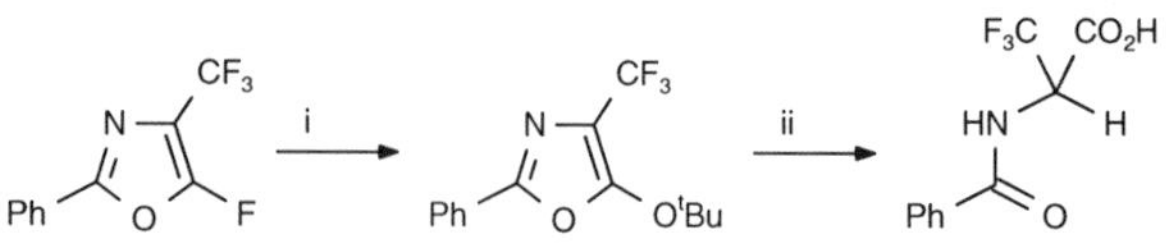

(i) KOBu^t, dioxane, r.t. (97%); (ii) CF$_3$CO$_2$H, H$_2$O, r.t. (98%)

Various derivatives of N-benzoyl-3,3,3-trifluoroalanine (e.g. esters, amides and thioamides) are also readily available via the 'oxazole route' [6].

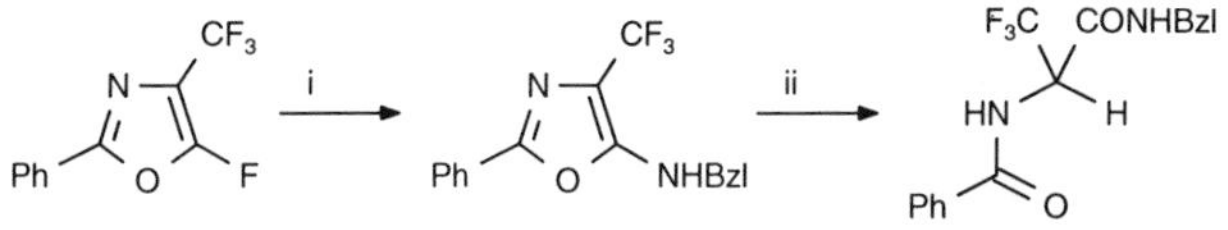

(i) BzlNH$_2$, dioxane, r.t. (87%); (ii) CHCl$_3$, HCl, 80 °C (76%)

5-Azido-4-trifluoromethyloxazoles represent a synthetic equivalent of N-protected 3,3,3-trifluoroalanyl azide [106]. The transformation into the nitrile derivative of 3,3,3-trifluoroalanine is achieved on catalytic hydrogenation. The nitrile is converted into the thioamide on treatment with hydrogen sulfide and base or into 3,3,3-trifluoroalanine hydrochloride on acidolysis [6].

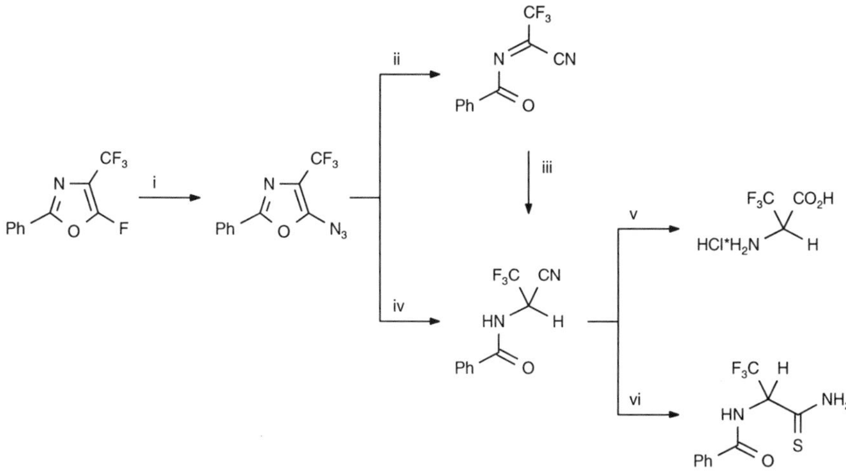

(i) Me$_3$SiN$_3$, THF, 10°C (63%); (ii) ΔT; (iii) H$_2$, Pd/C, Et$_2$O (76% for ii, iii); (iv) H$_2$, Pd/C, MeOH, −40 °C (48%); (v) conc. HCl, reflux; (vi) H$_2$S, pyridine, NEt$_3$ (52%)

4.2.3.7 Reductive amination of 3,3,3-trifluoropyruvates

The most direct synthetic approach towards 3,3,3-trifluoroalanine derivatives starts from 3,3,3-trifluoropyruvates. Only one functional group transformation is necessary.

3,3,3-Trifluoropyruvates are readily available from hexafluoropropene oxide [107–112] or from oxalates [113, 114].

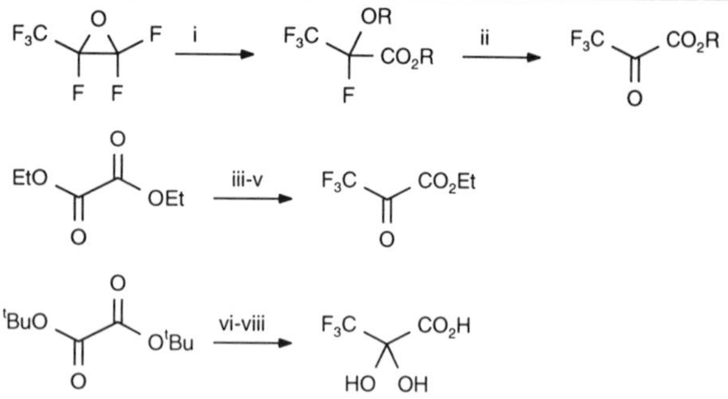

(i)   ROH, r.t. (83–96%);
(ii)   conc. H_2SO_4, SiO_2, 100–120 °C (83%);
(iii) CF_3Br, Zn, pyridine, 20 °C;
(iv) H_3O^+;
(v)   conc. H_2SO_4 (38% for iii–v);
(vi) Me_3SiCF_3, cat. TBAF, THF (78%);
(vii) HCl, CH_2Cl_2, r.t.;
(viii) H_2O (100% for vii, viii)

Most imines of 3,3,3-trifluoropyruvate are not obtained directly on treatment with ammonia or amines. The stable hemiaminals can be dehydrated only in the case of α-phenylethylamine or aniline [115].

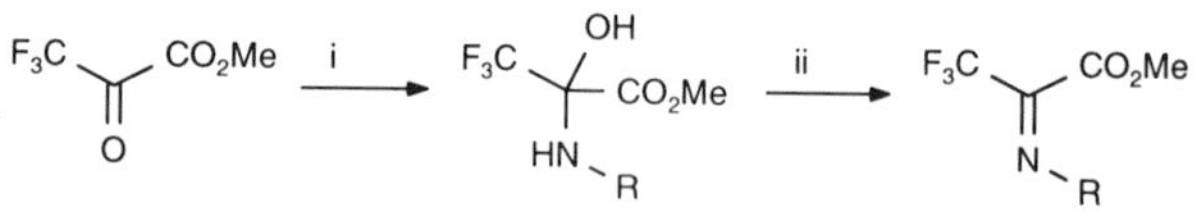

(i) RNH_2, toluene, −15 °C; (ii) cat. p-TsOH, azeotropic distillation

R	Yield (i) (%)	Yield (i, ii) (%)	Ref.
Bzl	91		115
PhCHMe	95	70	115
Ph		35	116
2,4,6-$Me_3C_6H_2$		93	116

A preparatively efficient method for the synthesis of imines derived from 3,3,3-trifluoropyruvate is the Staudinger reaction [115, 117, 118].

(i) $Ph_3P{=}NR^2$, benzene or CCl_4

R^1	R^2	T (°C)	Yield (%)	Ref.
Me	MeO_2C	60	95	117
Me	Ph	4 → 50	85	118
Me	Bu^s	4 → 50	79	118
Me	Bu^n	−15 → r.t.	70	115

Palladium-catalyzed carbonylation of N-aryltrifluoroacetimidoyliodides in the presence of different alcohols gives the corresponding 2-imino-3,3,3-trifluoropropionates [119].

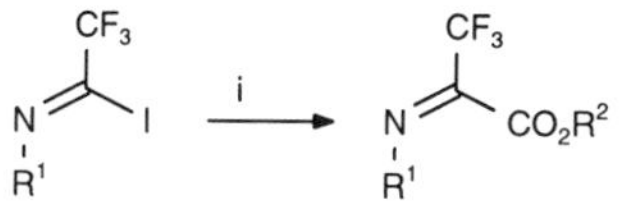

(i) R^2OH, CO, cat. $Pd_2(dba)_3{*}CHCl_3$, K_2CO_3, toluene, r.t.

R^1	R^2	Yield (%)	Ref.
p-$MeOC_6H_4$	Et	98	119
p-ClC_6H_4	Et	82	119
p-$NO_2C_6H_4$	Et	27	119
o,o-$Me_2C_6H_3$	Et	48	119
o-MeC_6H_4	Et	96	119
m-MeC_6H_4	Et	89	119
p-MeC_6H_4	Et	89	119
p-MeC_6H_4	Bu^n	98	119
p-MeC_6H_4	Pr^i	57	119
p-MeC_6H_4	Bzl	85	119
p-$MeOC_6H_4$	Bzl	95	119
p-MeC_6H_4	CH_2CHCH_2	44	119
p-MeC_6H_4	$ClCH_2CH_2$	43	119

The most general method for the synthesis of N-acylimines or N-sulfonylimines of 3,3,3-trifluoropyruvates is the addition of carboxamides, urethanes [6, 120–124] or sulfonamides [125, 126] to give stable 1:1 adducts. They can be dehydrated on treatment with thionyl chloride, phosphorus oxytrichloride or trifluoroacetic anhydride in the presence of base. Via this route, urethane protective groups (Boc-, Z-) are introduced, permitting orthogonal protective group strategy.

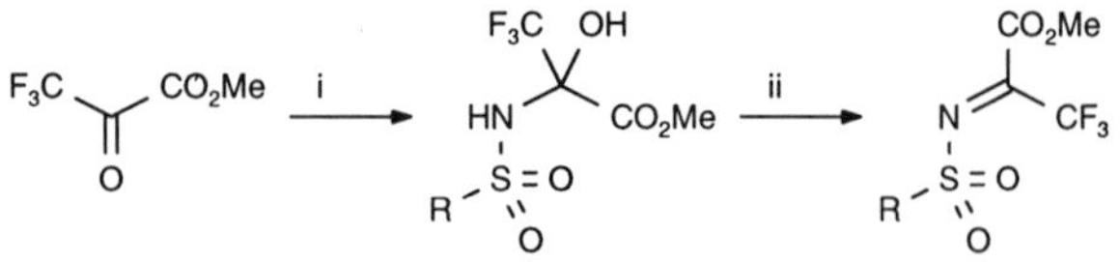

(i) RSO_2NH_2, 70 °C (R = Ph, 95%; R = Me, 98%); (ii) $SOCl_2$, cat. pyridine, 80 °C (R = Ph, 78%; R = Me, 72%)

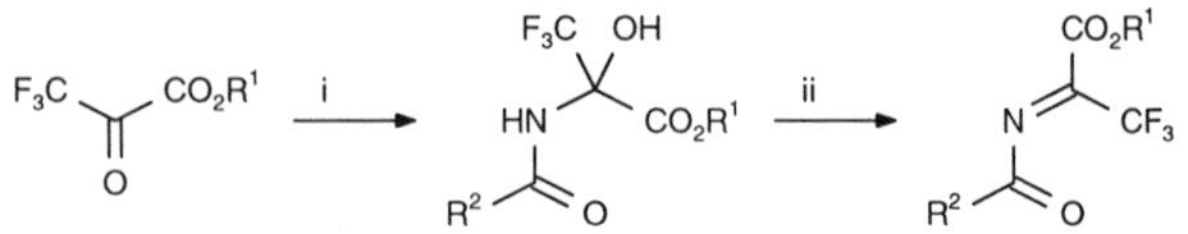

(i) R^2CONH_2, CH_2Cl_2, r.t.; (ii) $(CF_3CO)_2O$, pyridine or quinoline, 0 °C, or $SOCl_2$, base, r.t., or $POCl_3$, base, r.t.

R^1	R^2	Yield (i) (%)	Yield (ii) (%)	Ref.
Me	Me	81	40	6
Et	Ph	99	58	6
Et	Bzl	98	80	6
Me	Bu^tO	98	84	6
Et	Bu^tO	98	70	6
Et	Cl_3CCMe_2O	99	75	6
Me	BzlO	88	68	6
Et	BzlO	98	74	6
Me	CF_3	92	75	124

The N-formylimine obtained on elimination of water from the formamide-trifluoropyruvate hemiamidal spontaneously rearranges to give the corresponding isocyanate [127].

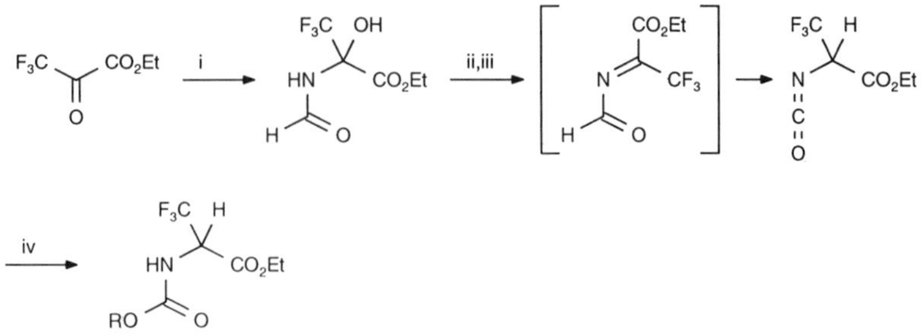

(i) $HCONH_2$, 80 °C (70%); (ii) $SOCl_2$, reflux (85%); (iii) NEt_3, Et_2O, r.t. (36%); (iv) ROH, Et_2O, r.t. (R = Bzl, 76%)

Some perfluoroalkyl-substituted imines are known to undergo base-catalyzed azomethinimine → azomethinimine isomerization, which can be utilized for the synthesis of 3,3,3-trifluoroalanine [128].

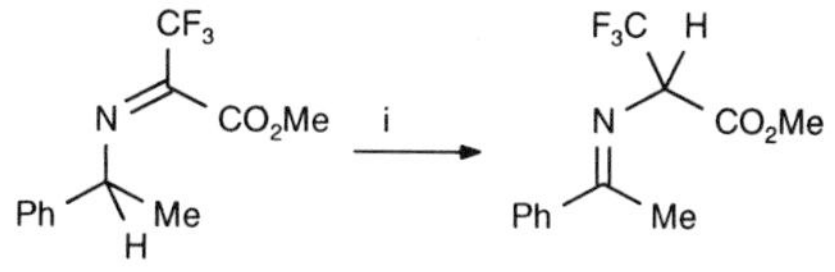

(i) $CF_3CH_2NH_2$, 60–80 °C (95%)

Selective hydrogenation of the CN double bond of imines is accomplished by a number of alternative routes [6]. Catalytic hydrogenation seems to be the method of choice, because the yield is nearly quantitative and workup is unproblematic even in large-scale runs. N-Protected 3,3,3-trifluoroalaninates are formed via this route when appropriate N-protective groups are used or via reduction with sodium borohydride [6]. Application of sodium borodeuteride provides access to N-protected 2-deuterio-3,3,3-trifluoroalaninates in high yield [6].

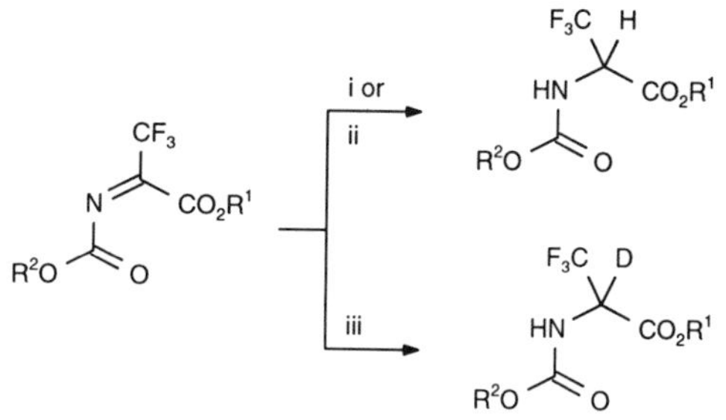

(i) H_2, Pd/C, hexanes, r.t.; (ii) $NaBH_4$, Et_2O, r.t.; (iii) $NaBD_4$, Et_2O, r.t.

R^1	R^2		Yield (%)	Ref.
Me	Me	(i)	65	6
Me	Cl_3CCMe_2O	(ii)	67	6
Me	BzlO	(ii)	88	6
Me	Bu^tO	(iii)	73	6
Et	Bu^tO	(i)	99	6
Et	Bu^tO	(ii)	80	6

N-Alkyl- and *N*-aryl-substituted 3,3,3-trifluoroalanine derivatives can be synthesized via the same strategy. In addition to catalytic hydrogenation and reduction with sodium borohydride (see above), reduction of the CN double bond with zinc/acetic acid or zinc/formic acid is highly efficient [115]. Deblocking of the amino group is accomplished in the case of $R^1 = p$-MeOC$_6$H$_4$ via ammonium cerium(IV) nitrate oxidation; the benzyl ester is cleaved hydrogenolytically [119].

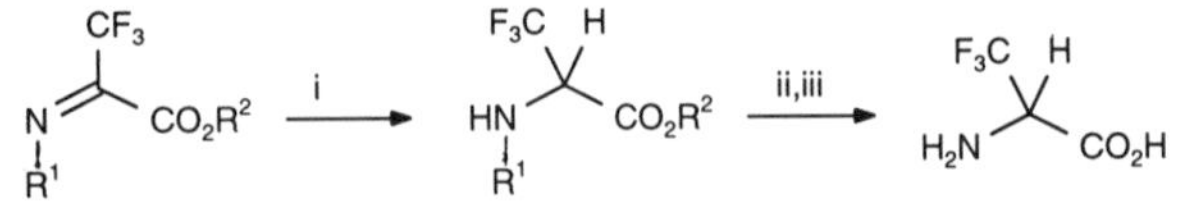

(i) Zn, HOAc, reflux; (ii) $R^1 = p$-MeOC$_6$H$_4$, $R^2 = $ Bzl, ammonium cerium(IV) nitrate, H$_2$O, CH$_3$CN; (iii) H$_2$, Pd/C, MeOH

R^1	R^2	Yield (i) (%)	Yield (i–iii) (%)	Ref.
MeO$_2$C	Me	65		115
Ph	Me	75		115
Bu^s	Me	80		115
PhCHMe	Me	83		115
4-MeOC$_6$H$_4$	Bzl		65	119

The reduction of *in situ*-prepared 2-acylimino-3,3,3-trifluoropropionates derived from homochiral *N*-protected amino acid amides or peptidyl-amides proceeds diastereoselectively to give di- or tripeptides with 3,3,3-trifluoroalanine in the *C*-terminal position (Z-Xaa-Yaa-TFMGly-OMe) [129].

N-Acylimines of 3,3,3-trifluoropyruvates represent a class of highly reactive heterodienes. They are readily susceptible to cycloaddition reactions of the [4 + 2] and [4 + 1] type. Trivalent phosphorus species add to give phosphoranes. This [4 + 1] cycloaddition reaction represents a redox process: the phosphorus species is oxidized ($P^{III} \rightarrow P^V$), while the heterodiene substructure undergoes reduction. Consequently, on hydrolysis the phosphoranes are transformed into *N*-protected 3,3,3-trifluoroalaninates. Hydrolysis with D$_2$O provides a preparatively simple access to 2-deuterio-3,3,3-trifluoroalaninates [6].

The phosphoranes decompose in a [3 + 2] cycloreversion process on heating to give nitrile ylides, which spontaneously undergo electrocyclization [6, 130–133] to 5-alkoxy-4-trifluoromethyloxazoles [134]. They are transformed into 3,3,3-trifluoroalanine hydrochloride on heating with concentrated hydrochloric acid [6].

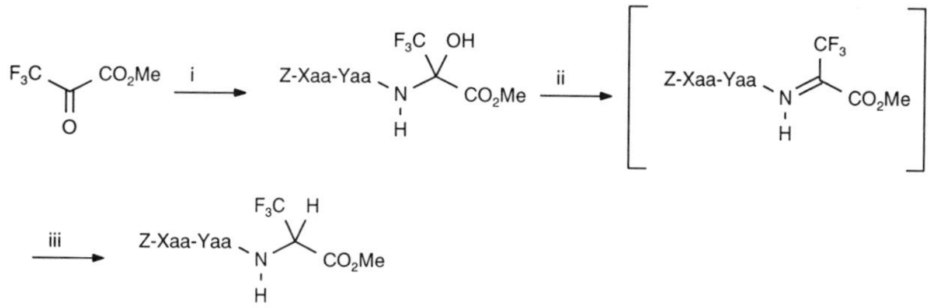

(i) Z-Xaa-Yaa-NH$_2$, Et$_2$O or CH$_2$Cl$_2$; (ii) (CF$_3$CO)$_2$O, quinoline, Et$_2$O, $-78\,^{\circ}$C; (iii) NaBH$_4$, $-78\,^{\circ}$C $\rightarrow$ r.t.

Xaa	Yaa	Yield (i) (%)	Yield (ii, iii) (%)	d.e. (%)	Ref.
—	Ala	70	34	10	129
—	Val	89	55	52	129
—	Phe	99	38	29	129
—	Pro	83	26	47	129
Leu	Ala	98	56	59	129
Gly	Phe	99	32	57	129

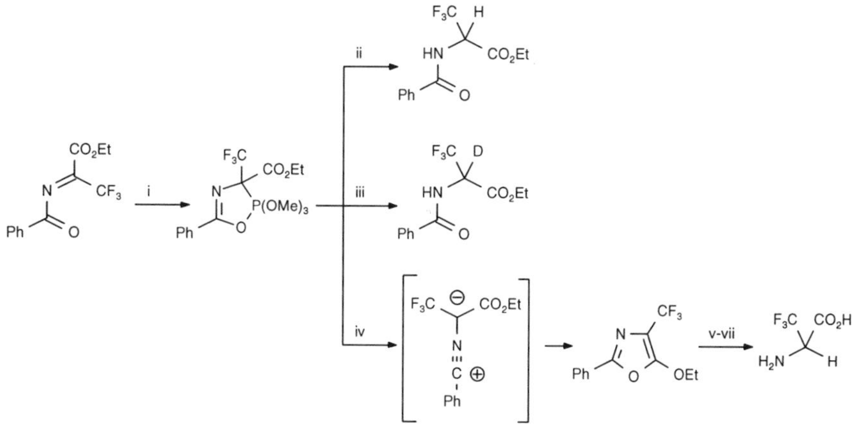

(i) (MeO)$_3$P, hexanes, $0\,^{\circ}$C (75%); (ii) THF, H$_2$O, r.t. (50%); (iii) THF, D$_2$O, r.t. (44%); (iv) $50\,^{\circ}$C (60%); (v) (Me$_3$Si)$_2$, I$_2$, CHCl$_3$, $70\,^{\circ}$C (46%); (vi) conc. HCl, reflux; (vii) NEt$_3$, CHCl$_3$, r.t., then AcOH (72% for vi, vii)

A reaction sequence of a very similar mechanistic type is involved in the [4 + 1] cycloaddition of the N-benzoylimines of 3,3,3-trifluoropyruvates with tin(II) chloride and subsequent hydrolysis of the stannacycles [6, 104].

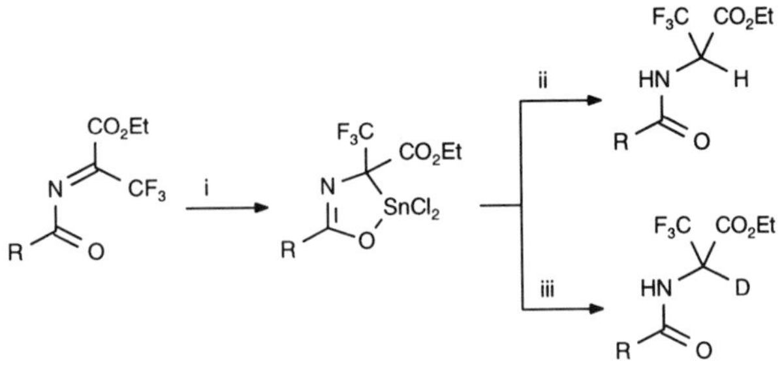

(i) $SnCl_2$, xylene/THF (4:1, v/v), r.t. (75%); (ii) H_2O, xylene/THF, r.t.; (iii) D_2O, xylene/THF, r.t.

R		Yield (%)	Ref.
Bzl	(ii)	68	6
Bu^{t}O	(ii)	45	6
BzlO	(ii)	50	6
BzlO	(iii)	40	6

Reaction of N-acylimines of 3,3,3-trifluoropyruvates with trimethylsilyl-carbonitrile and subsequent desilylation on filtration through a silica gel layer or direct addition of anhydrous hydrogen cyanide provides N-protected 2-cyano-3,3,3-trifluoroalaninates. They are hydrolyzed and decarboxylated on heating with concentrated hydrochloric acid to give 3,3,3-trifluoroalanine hydrochloride. Treatment of the nitrile with dihydrogen sulfide in the presence of triethylamine/pyridine forms N-protected 3,3,3-trifluoroalaninates via the unstable thioamide intermediate [6].

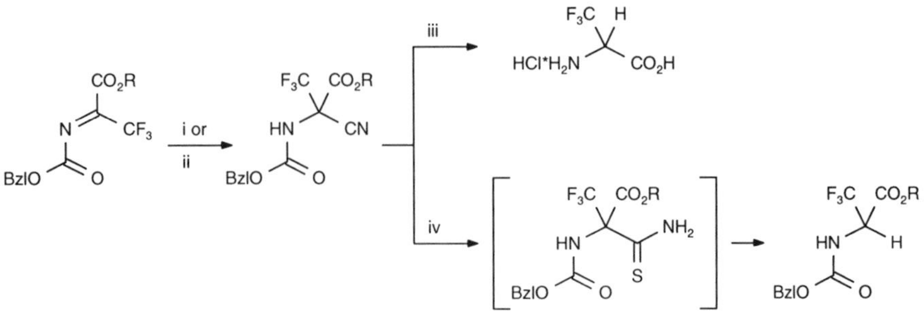

(i) Me_3SiCN, 80 °C, (R = Me, 75%; R = Et, 78%); (ii) HCN, hexanes, cat. NEt_3, 0 °C → r.t. (R = Me, 78%); (iii) conc. HCl, reflux (65%); (iv) H_2S, pyridine, NEt_3, r.t. (55%)

N-heteroaryl-substituted trifluoroalanine derivatives are obtained from 5-azido-4-trifluoromethylthiazoles via Staudinger reaction and borohydride reduction [6].

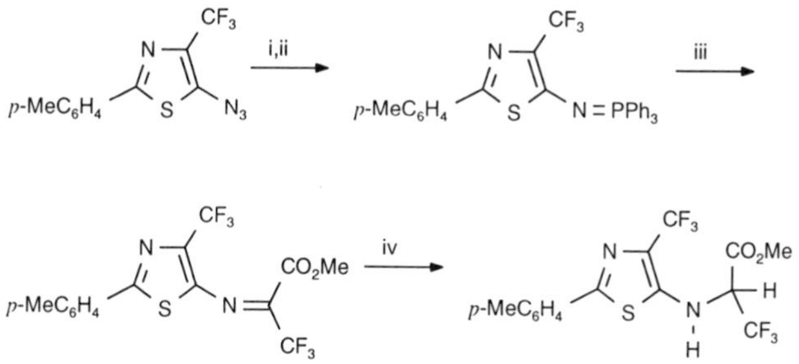

(i) Ph$_3$P, Et$_2$O, r.t. (81%); (ii) THF, r.t. (80%); (iii) CF$_3$COCO$_2$Me, THF, 60 °C (65%); (iv) NaBH$_4$, Et$_2$O, −78 °C → r.t. (46%)

The reduction of N-acyliminopropionates or N-carbamoyliminopropionates with LiAlH$_4$ provides a preparatively simple, high-yielding route to 2-alkylamino-3,3,3-trifluoropropanols or 2-carbamoylamino-3,3,3-trifluoro-propanols [6].

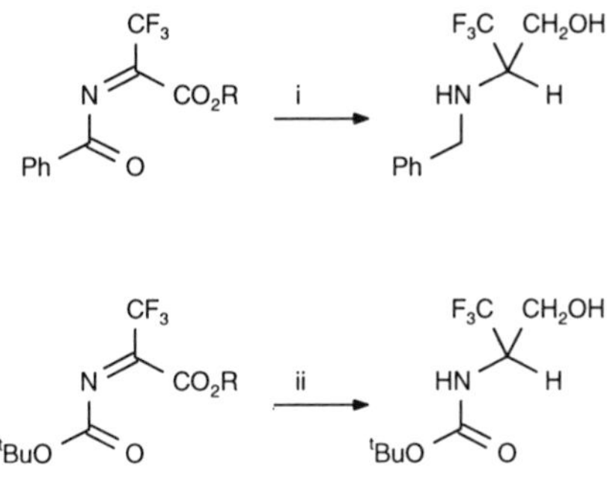

(i) LiAlH$_4$, Et$_2$O, −78 °C → r.t. (84%); (ii) LiAlH$_4$, Et$_2$O, −78 °C → r.t. (59%)

4.3 AMINO ACIDS WITH FLUOROALKYL SUBSTITUENTS IN α-POSITION (α-BRANCHED AMINO ACIDS)

α-Halomethyl-α-amino acids represent a class of compounds which are generally not readily available via the Bucherer or Strecker reactions usually employed for the preparation of α-amino acid analogs.

α,α-Dialkylamino acids frequently induce conformational restrictions on the peptide chain and promote helical secondary structures in peptides [135]. Incorporation of α-fluoroalkyl-substituted amino acids (e.g. α-trifluoromethyl amino acids) into biologically active peptides should doubtlessly result in severe conformational restrictions, increased lipophilicity and higher metabolic stability. Hence this approach could be used to design proteolytically stable drugs with well defined secondary structures.

4.3.1 α-MONOFLUOROMETHYL DERIVATIVES (MFM AMINO ACIDS)

4.3.1.1 Fluorodehydrogenation

α-(Fluoromethyl) glutamic acid can be prepared from α-methylglutamic acid by photofluorination with trifluoromethyl hypofluorite [136].

4.3.1.2 Fluorodehydroxylation of α-hydroxymethyl-α-amino acids

α-Hydroxymethyl-α-amino acids can be prepared easily from the parent α-amino acids [137]. However, halodehydroxylation is achieved only in a few cases. The fluorodehydroxylation proceeds in good yields using the highly toxic sulfur tetrafluoride in liquid hydrogen fluoride [138] and several aliphatic, aromatic and heterocyclic α-fluoromethyl-substituted amino acids are obtained [139, 140].

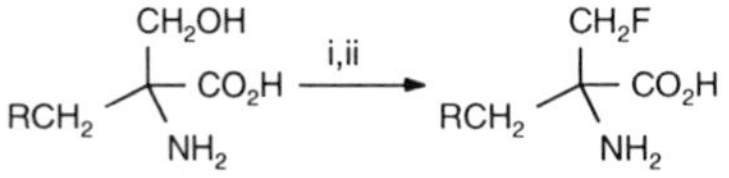

(i) SF$_4$, liq. HF, $-80\,^\circ$C → r.t.; (ii) BF$_3$, R = 3,4-(HO)$_2$C$_6$H$_3$, 4-HOC$_6$H$_4$, Ph, indol-3-yl, 5-hydroxyindol-3-yl, imidazol-4-yl

α-(Fluoromethyl)histidine is resolved via crystallization as a salt with D-α-bromocamphorsulfonic acid [141].

An enantiospecific synthesis of α-(S)-(fluoromethyl)tryptophan, α-(S)-(fluoromethyl)-5-hydroxytryptophan and their antipodes proceeds via face-selective hydroxyalkylation and subsequent fluorodehydroxylation of cyclic tryptophan derivatives [142].

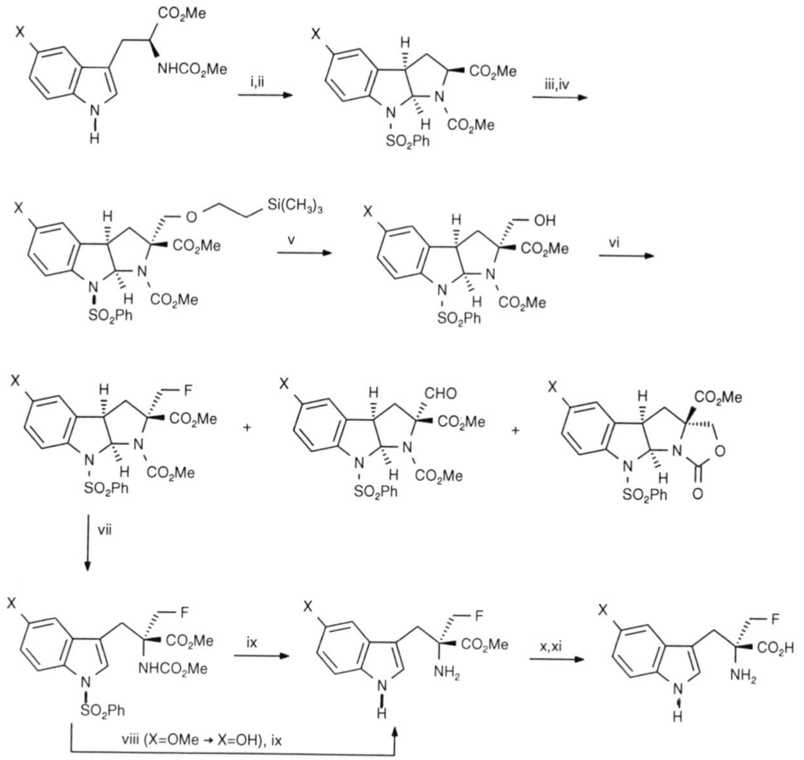

(i) 85% H_3PO_4; (ii) $PhSO_2Cl$; (iii) LDA, THF, $-78\,°C$; (iv) SEMCl, $-78\,°C \rightarrow$ r.t.; (v) BF_3*OEt_2, CH_2Cl_2, $0\,°C$; (vi) DAST, CH_2Cl_2, r.t.; (vii) CF_3CO_2H, r.t.; (viii) BBr_3, CH_2Cl_2, $-78\,°C \rightarrow$ r.t.; (ix) Na, liq. NH_3; (x) Me_3SiI, CH_3CN, reflux; (xi) NaOH, H_2O, r.t., then ion-exchange chromatography

X	Starting isomer	Yield (iii, iv) (%)	Yield (v) (%)	Yield (vi) (%)	Yield (vii) (%)	Yield (viii, ix) (%)	Yield (ix) (%)	Yield (x, xi) (%)	Ref.
H	(+)	96	85	59	95		90	56	142
H	(−)	84	56	57	97		82	52	142
OMe	(+)	53	69	50		46^a			142
OMe	(−)	97	62	32	94	43^a			142

^{a} X = OH

Fluorodehydroxylation of vicinal N-acylaminohydroxy compounds frequently gives 2-oxazoline derivatives [143].

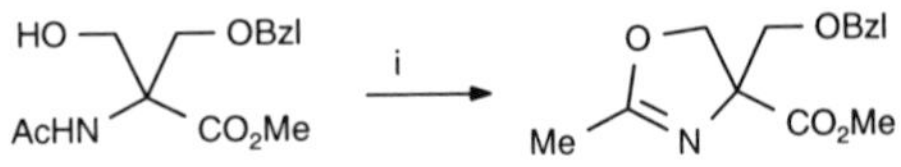

(i) DMST (dimethylaminosulfur trifluoride), $-45\,^{\circ}C$ (61%)

Application of the fluorodehydroxylation reaction with DAST to a 6α-hydroxymethylpenicillin-derived Schiff base gives the corresponding 6α-(fluoromethyl)penam [144].

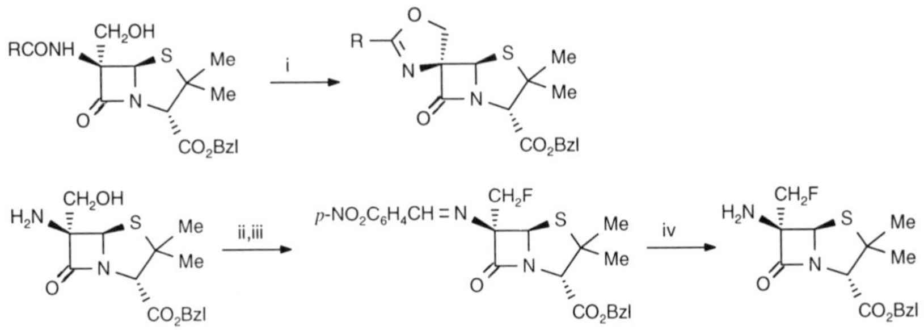

(i) DAST, $CaCO_3$, CH_2Cl_2, $-70 \rightarrow -20\,^{\circ}C$ (55%); (ii) p-$NO_2C_6H_4CHO$; (iii) DAST, $CaCO_3$, CH_2Cl_2, $-20\,^{\circ}C$; (iv) p-TsOH, Girard T reagent, CH_2Cl_2, $0\,^{\circ}C$ (19%)

4.3.1.3 Fluoroalkylation of nucleophilic amino acid synthons

α-Fluoromethyl-α-amino acids have been obtained in good yields via regio-selective alkylation of N-benzylidene-protected amino acid enolates with chlorofluoromethane [145–147]. Similarly, α-(chlorofluoromethyl) derivatives of amino acids are obtained on reaction of amino acid Schiff base anions [148] or of suitably protected malonates {α-(chlorofluoromethyl)-ornithine, -m-tyrosine and -glutamic acid [149]} with the readily available dichloro-fluoromethane (Freon 21) (p. 175).

The diastereoisomers can be separated by MPLC; enantiomeric mixtures are resolved successfully by HPLC with a chiral eluent [149]. Side-chain transformation and deprotection are achieved according to standard procedures (derivatives of ornithine and m-tyrosine [149]). Reduction of α-chlorofluoromethylamino acids with tri-n-butylstannane in the presence of azobisisobutyronitrile (AIBN) provides α-fluoromethyl-α-amino acids in good yields. This procedure avoids the use of the less common chloro-fluoromethane (p. 175).

A stereoselective route to α-(fluoromethyl)histidine proceeds via asymmetric fluoromethylation of Seebach's amino acid-derived 1,3-imidazolin-5-one [150] (p. 176).

$$
\begin{array}{c}
\overset{R^1}{\underset{\underset{Ph}{\parallel}}{N}}\!\!-\!CO_2R^2,\ H
\ \xrightarrow{i,ii}\
\overset{R^1}{\underset{\underset{Ph}{\parallel}}{N}}\!\!-\!CO_2R^2,\ CHXF
\ \xrightarrow{iii}\
\underset{H_2N}{\overset{R^1}{}}\!\!-\!CO_2R^2,\ CHXF
\ \xrightarrow{iv}\
\underset{H_2N}{\overset{R^1}{}}\!\!-\!CO_2H,\ CHXF
\end{array}
$$

(down arrow v from the H_2N / CO_2R^2 / CHXF intermediate)

$$
\underset{Boc}{\overset{R^1}{HN}}\!\!-\!CO_2R^2,\ CHXF
\ \xrightarrow{vi}\
\underset{Boc}{\overset{R^1}{HN}}\!\!-\!CO_2R^2,\ CH_2F
$$

(i) Na or K or NaH, THF or HMPA; (ii) CHClXF; THF or THF/HMPA, r.t.; (iii) 2M HCl, Et$_2$O, r.t.; (iv) HBr or HCl, reflux, then EtOH, propylene oxide; (v) (Boc)$_2$O, base; (vi) X = Cl: Bu$^n_3$SnH, AIBN, benzene or toluene, reflux

R^1	R^2	X	Yield (i, ii) (%)	Yield (i–iii) (%)	Yield (iv) (%)	Yield (v, vi) (%)	Ref.
Me	Et	Cl		32	60[b]	78	148
CH$_2$=CH(CH$_2$)$_2$	Me	Cl		20			148
Bzl	Me	Cl		25	90[b]	57	148
Bzl	Me	H			77		146
PhCH=N(CH$_2$)$_3$	Me	H	87[a]		85[c]		146
Me	Me	H	72[a]		72		146

[a] Crude yield.
[b] As hydrochloride.
[c] R^1 = H$_2$N(CH$_2$)$_3$—, as hydrochloride.

$$
\underset{CO_2^tBu}{\overset{CO_2Et}{R}}\!\!-\!H
\ \xrightarrow{i,ii}\
\underset{CO_2^tBu}{\overset{CO_2Et}{R}}\!\!-\!CHClF
$$

(branch iii–vii) $\longrightarrow \underset{NH_2\cdot HCl}{\overset{CO_2H}{R}}\!\!-\!CHClF$

(branch viii) $\longrightarrow \underset{CO_2^tBu}{\overset{CO_2Et}{R}}\!\!-\!CH_2F$

(i) NaH, THF, r.t.; (ii) CHCl$_2$F, −20 °C; (iii) CF$_3$CO$_2$H, r.t.; (iv) SOCl$_2$, reflux; (v) NaN$_3$, H$_2$O/acetone; (vi) benzene, reflux; (vii) H$_3$O$^+$, reflux; (viii) Bu$^n_3$SnH, AIBN, benzene, reflux

R	Yield (i, ii) (%)	Yield (iii–vii) (%)	Yield (viii) (%)	Ref.
CH$_2$=CHCH(CH$_3$)	68		83	148
CH$_2$=C(CH$_3$)	36		80	148
m-CH$_3$OC$_6$H$_4$CH$_2$	48	38		148,149
PHT=N(CH$_2$)$_3$	30	50[a]		148,149
MeO$_2$CCH$_2$	47		90	148

[a] R = H$_2$N(CH$_2$)$_3$, as hydrochloride.

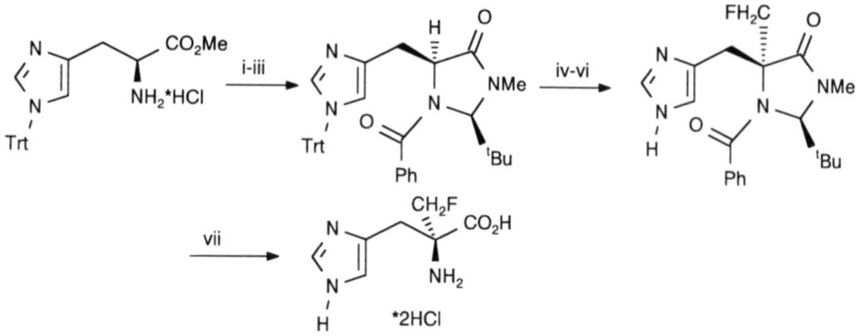

(i) MeNH$_2$, EtOH, r.t. (92%); (ii) K$_2$CO$_3$, NaHCO$_3$, mol. sieve, Bu^{t}CHO, CH$_2$Cl$_2$ (94%); (iii) (PhCO)$_2$O (61%); (iv) LDA, THF/HMPA, mol. sieve, $-70 \rightarrow -10\,°C$; (v) TMEDA, CH$_2$ClF, $-10\,°C \rightarrow$ r.t. (94% for iv, v); (vi) 6M HCl/toluene, reflux (81%); (vii) 12M HCl, sealed tube, 150–160 °C (69%)

4.3.1.4 Strecker synthesis

In contrast to non-fluorinated ketones, the Strecker reaction of highly electron-deficient ketones frequently stops at the stage of the corresponding cyanohydrin.

Fluoromethyl ketones are obtained on treatment of monofluoroacetonitrile with Grignard compounds and subsequent acid hydrolysis of the intermediate imines. Strecker synthesis with these ketones affords the racemic α-fluoro-methylamino acids. Several α-fluoromethyl analogs of phenylalanine [151], tyrosine [152], (dihydroxyphenyl)alanine (DOPA) [153] and (E)-2,5-diamino-pentenoic acid [(E)-2-(fluoromethyl)dehydroornithine] [154, 155], among others [156], can be synthesized via this route.

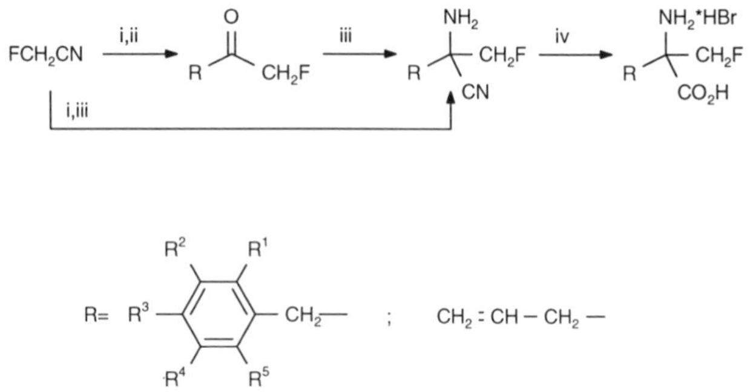

R^1 = OH, H, Cl, OMe, F, OBzl; R^2 = Me, OH, H, OMe; R^3 = H, Me, OH, OMe, Et, F, Cl, OBzl; R^4 = H, OH, OMe, OBzl, Cl; R^5 = H, Me, OH, Et, Bu^t (i) RMgX, $-30 \rightarrow -40\,°C$; (ii) 12M HCl; (iii) NaCN, NH$_3$, NH$_4$Cl, H$_2$O, r.t.; (iv) HBr, 100 °C

α-(Fluoromethyl)-2,6-diaminopimelic acid is prepared via derivatives of α-(fluoromethyl)-substituted 2-amino-5-hydroxypentanenitrile and 2-amino-5-iodopentanenitrile. The α-fluoromethyl-α-amino acid skeleton is formed via Strecker synthesis with a monofluoromethylketimine formed *in situ* [157].

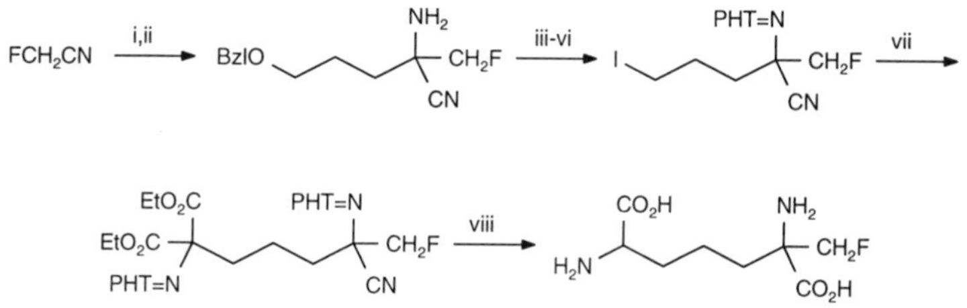

(i) BzlO(CH$_2$)$_3$MgCl, Et$_2$O; (ii) NH$_4$Cl, NaCN, H$_2$O (57% for i, ii); (iii) phthaloyl dichloride, NEt$_3$, CH$_2$Cl$_2$, r.t. (51%); (iv) Me$_3$SiI, CH$_2$Cl$_2$ (99%); (v) MeSO$_2$Cl, NEt$_3$, CH$_2$Cl$_2$, $-10 \rightarrow 0\,^{\circ}$C (93%); (vi) NaI, acetone, reflux (95%); (vii) diethyl phthalimidomalonate, Bu^{t}OK, DMF, 60 $^{\circ}$C (71%); (viii) 12M HCl, reflux, then propylene oxide (79%)

Monofluoroacetone also can be used in Strecker-type syntheses of α-fluoromethylamino acids {e.g. α-(fluoromethyl)alanine, 50% [158], 64% [159]}. Similarly, application of the Strecker protocol to 1,3-difluoroacetone gives α-amino-β,β'-difluoroisobutyric acid in 9% overall yield [158].

Hydantoins are obtained via Bucherer–Bergs reaction with (fluoromethyl)ketones [160].

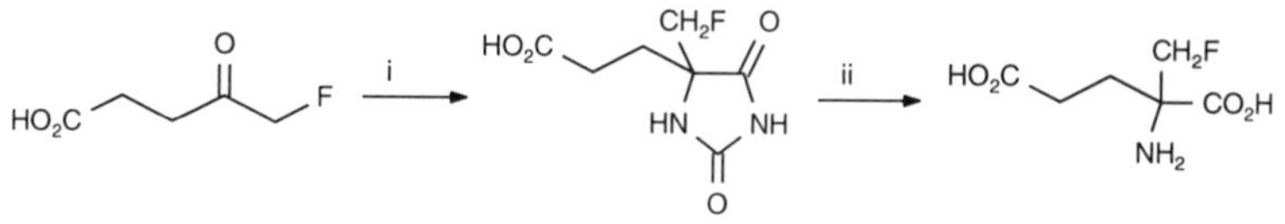

(i) KCN, (NH$_4$)$_2$CO$_3$, H$_2$O, 60 $^{\circ}$C (75%); (ii) 12M HCl, reflux, then ion-exchange chromatography (30%)

α-(Monofluoromethyl)serine is obtained in a multi-step procedure starting from *O*-benzylglycidol via epoxide ring opening with KHF$_2$, Swern oxidation and Strecker synthesis [161]. This procedure avoids the use of the highly toxic and not commercially available monofluoroacetonitrile.

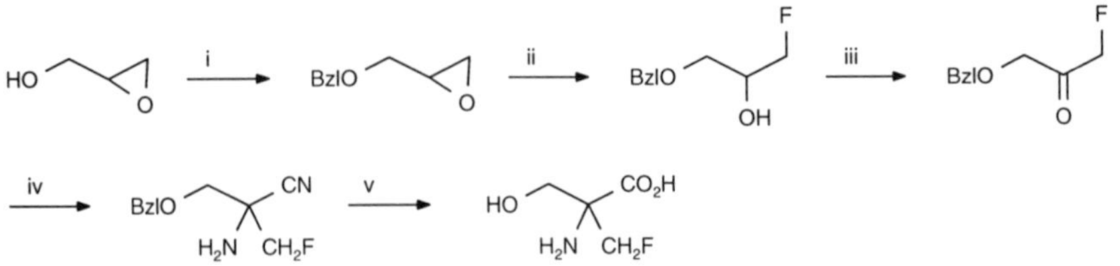

(i) BzlBr, KOBu^t, THF, 0 °C (82%); (ii) KHF$_2$, triethylene glycol, 130 °C, (80%); (iii) (COCl)$_2$, DMSO, CH$_2$Cl$_2$, −60 °C, then NEt$_3$ (92%); (iv) NaCN, NH$_4$Cl, H$_2$O (80%); (v) HCl, H$_2$O, reflux, then propylene oxide, Pr^{i}OH (79%)

4.3.1.5 Functional group transformation

α-(Monofluoromethyl)serine is a highly valuable building block for the synthesis of α-fluoromethyl-substituted amino acids with further func-tionalities in the side-chain [e.g. (E)-dehydro-α-(monofluoromethyl)ornithine] [161].

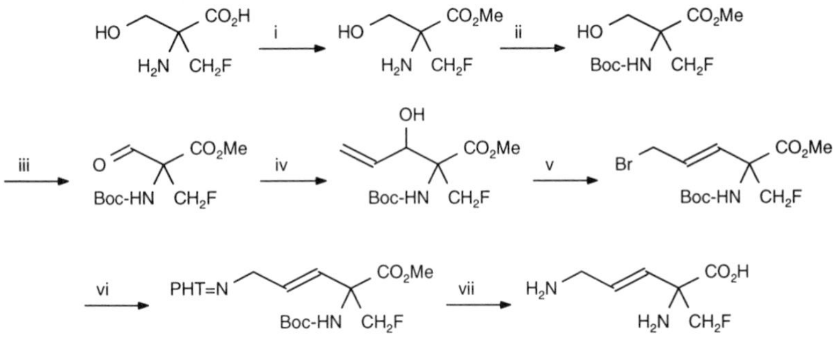

(i) MeOH, CH(OMe)$_3$, HCl, reflux; (ii) (Boc)$_2$O, NEt$_3$, THF (84% for i, ii); (iii) pyridinium chlorochromate, molecular sieves, CH$_2$Cl$_2$, (80%); (iv); CH$_2$=CHMgBr, THF, −78 °C, then HOAc (70%); (v) Me$_2$C=C(Br)NMe$_2$, CH$_2$Cl$_2$, (70%); (vi) potassium phthalimide, DMF, 80 °C (84%); (vii) conc. HCl, reflux, then EtOH, propylene oxide (82%)

α-(Chlorofluoromethyl)glutamic acid and α-(chlorofluoromethyl)ornithine are obtained from α-(chlorofluoromethyl)homoallylglycine [148,149]:

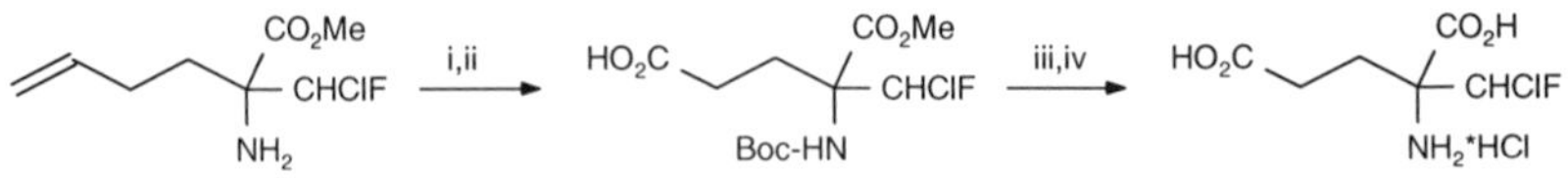

(i) (Boc)$_2$O, THF, reflux; (ii) KMnO$_4$, AcOH; (iii) LiOH, H$_2$O/DME; (iv) HCl, Et$_2$O (27% for i–iv)

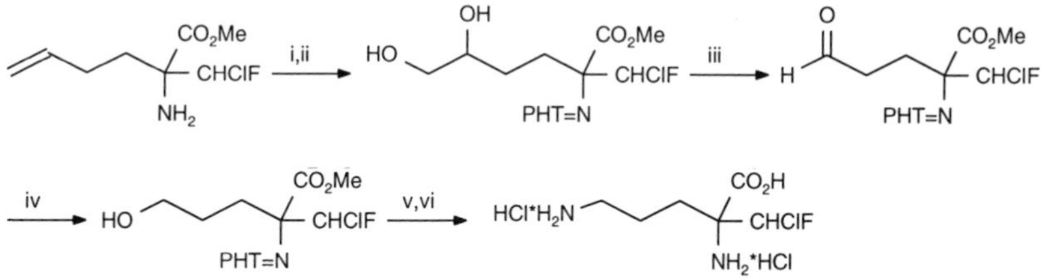

(i) phthaloyl dichloride, NEt$_3$, CH$_2$Cl$_2$; (ii) OsO$_4$, N-methylmorpholine N-oxide (NMMO), Bu^{t}OH; (iii) NaIO$_4$, THF/H$_2$O; (iv) NaBH$_3$CN, MeOH/H$_2$O; (v) DEAD, PPh$_3$, phthalimide; (vi) AcOH, 12M HCl, reflux (30% for i–vi)

2-(Fluoromethyl)-2-phthalimidopent-3-enenitrile is transformed into α-(fluoromethyl)-(E)-3,4-dehydroornithine via allylic bromination and nucleophilic substitution of bromide by phthalimide [154].

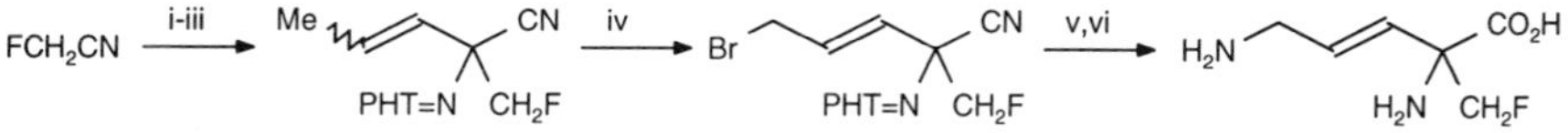

(i) CH$_3$CH=CHMgBr, THF, $-40\,^\circ$C; (ii) NaCN, NH$_4$Cl, H$_2$O, $0\,^\circ$C $\rightarrow$ r.t. (64% for i, ii); (iii) phthaloyl dichloride, NEt$_3$, CH$_2$Cl$_2$, r.t. $\rightarrow$ reflux (80%); (iv) NBS, cat. (PhCO$_2$)$_2$, hv, CCl$_4$, reflux (56%); (v) potassium phthalimide, DMF, $80\,^\circ$C (87%); (vi) conc. HCl, reflux, then EtOH, propylene oxide (58%)

α-(Fluoromethyl)glutamic acid is transformed into α-(fluoromethyl)tryptophan or α-(fluoromethyl)-5-hydroxytryptophan via Fischer indole synthesis [162].

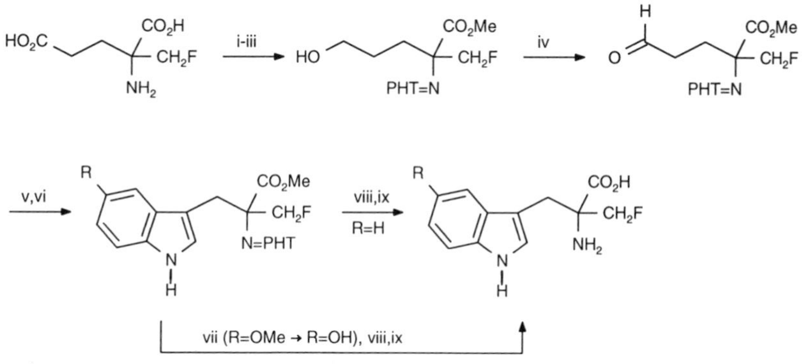

(i) MeOH, SOCl$_2$, $0\,^\circ$C $\rightarrow$ reflux (61%); (ii) phthaloyl dichloride, NEt$_3$, CH$_2$Cl$_2$, $0\,^\circ$C $\rightarrow$ reflux (79%); (iii) DIBAL-H, THF, $-78\,^\circ$C $\rightarrow$ r.t. (93%); (iv) (COCl)$_2$, DMSO, NEt$_3$, CH$_2$Cl$_2$, $-60\,^\circ$C $\rightarrow$ r.t.; (v) p-RC$_6$H$_4$NHNH$_2$, THF/H$_2$O, 1M AcOH (pH 5), r.t.; (vi) H$_2$SO$_4$, MeOH, reflux (R = H, 46%, R = OMe, 38% for iv–vi); (vii) BBr$_3$, CH$_2$Cl$_2$, $-78\,^\circ$C $\rightarrow$ r.t.; (viii) H$_2$NNH$_2$ (1M in EtOH), reflux (R = H, 42%; R = OH, 71%); (ix) 0.1M NaOH/THF, r.t., then 0.1 M HCl

4.3.2 α-DIFLUOROMETHYL DERIVATIVES (DFM AMINO ACIDS)

4.3.2.1 Difluoroalkylation of nucleophilic amino acid synthons

α-Difluoroalkylamino acids can be prepared in close analogy to α-fluoroalkyl amino acids on reaction of amino acid Schiff base anions with chlorodifluoromethane [145, 146, 163]. Deprotonation is usually achieved with lithium diisopropylamide (LDA), sodium hexamethyldisilazanide, sodium hydride, potassium *tert*-butoxide, or *n*-butyllithium in THF, sometimes with HMPA as an additive.

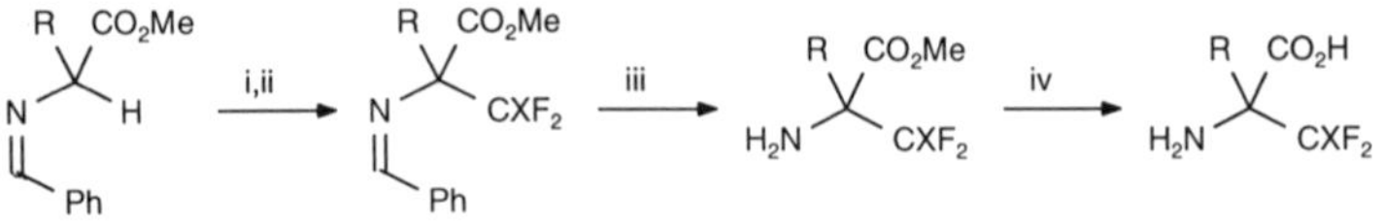

(i) base; (ii) CHClF$_2$ (X = H) or CBr$_2$F$_2$ (X = Br), 40 °C or r.t.; (iii) 2M HCl, Et$_2$O, r.t.; (iv) 6M HCl, reflux, then EtOH, propylene oxide

R	X	Yield (i, ii) (%)	Yield (iii) (%)	Yield (iv) (%)	Ref.
CH$_3$	H	83		91	146
CH$_2$=CH(CH$_2$)$_2$	H	69	70		162
Bzl	H	72		62	146
p-HOC$_6$H$_4$CH$_2$	H	78			146
m,p-(CH$_3$O)$_2$C$_6$H$_3$CH$_2$	H	86		45	146
C$_6$H$_5$CH=N(CH$_2$)$_3$	H	97[a]		70[b]	146,163
C$_6$H$_5$CH=N(CH$_2$)$_4$	H	80[a]	85[b]	65[c]	146
MeSCH$_2$CH$_2$	H	72[a]		75	146
(*N*-Trt-5-imidazolyl)CH$_2$	H	90		10[d]	146
Me	Br				146

[a] Crude yield
[b] R = H$_2$N(CH$_2$)$_3$, as hydrochloride.
[c] R = H$_2$N(CH$_2$)$_4$, as hydrochloride.
[d] R = (5-imidazolyl)CH$_2$, as hydrochloride.

The ornithine derivatives can be resolved by crystallization of the corresponding six-membered lactams with homochiral binaphthylphosphoric acid; cleavage of the lactam is achieved by refluxing in 6M hydrochloric acid (40–60%) [146]

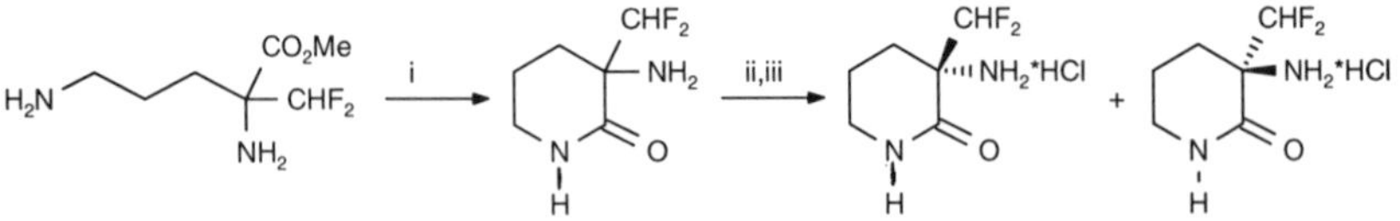

(i) NaOMe, MeOH, r.t. (73%); (ii) (−)-binaphthylphosphoric acid, EtOH, crystallization; (iii) 3M HCl, r.t.

Presumably, the difluoromethylation with chlorodifluoromethane proceeds via a difluorocarbene based chain mechanism. However, the mechanism of the halomethylation reaction seems to depend on the nature of the alkylating agent; for reactions with bromochloromethane and chlorofluoromethane, an S_N2 process is postulated [145, 146].

Difluoromethylation of N-benzylidene-protected dimethylglutamate gives the corresponding α-difluoromethyl derivative. In its unprotected form, it cyclizes to give α-(difluoromethyl)pyroglutamic acid. The latter is cleaved on treatment with trifluoromethanesulfonic acid/thionyl chloride in methanol and dimethyl α-(difluoromethyl)glutamate is obtained as trifluoromethanesulfonate [80].

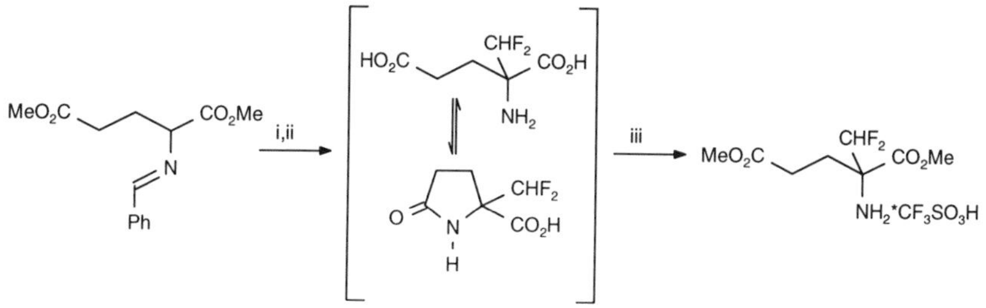

(i) $CHClF_2/NaF$, THF; (ii) 12M HCl, reflux; (iii) CF_3SO_3H, $SOCl_2$, MeOH (38% for i–iii)

Difluoroalkylation of diaminoheptanedioate gives a mixture of all four possible stereoisomeric forms of 2-(difluoromethyl)-2,6-diaminopimelate [164]; application of the difluoroalkylation procedure to 2,5-diaminopentanoic acid derivatives gives α-(difluoromethyl)-2,5-diaminopentanoic acid [α-(difluoromethyl)ornithine] [165].

A stereoselective synthesis of α-(difluoromethyl)alanine and -phenylalanine proceeds via difluoromethylation of homochiral amidines derived from (S)-(−)1-di(methoxy)methyl-2-methoxymethylpyrrolidine (SDMP) [166, 167].

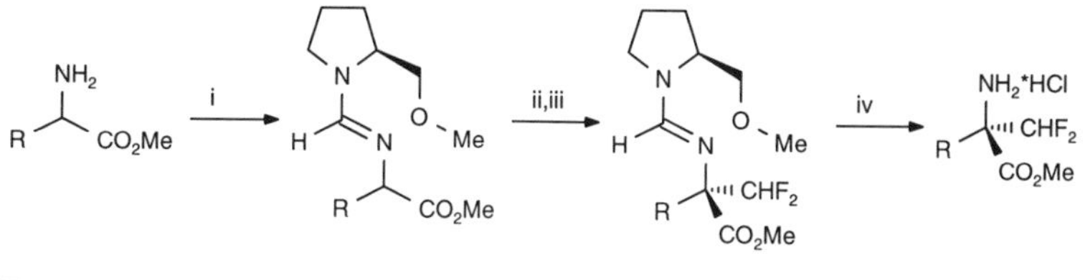

(i) SDMP, 110 °C (R = Me, 66%; R = Bzl, 74%); (ii) LDA, THF, −78 → 50 °C; (iii) $CHClF_2$, 50 °C (R = Me, 36%; R = Bzl, 54% for ii, iii); (iv) 3M HCl, MeOH/H_2O, reflux (R = Me, 59%; R = Bzl, 58%, no e.e. values given)

Application of the difluoromethylation protocol to monoalkylated ethyl tert-butyl malonates gives the corresponding difluoromethyl derivatives; hydrolysis of the tert-butyl ester with trifluoroacetic acid and transformation

of the carboxy group into an amino function via Curtius rearrangement gives
α-difluoromethyl-substituted amino acids [168].

(i) NaH, THF, r.t.; (ii) CHClF$_2$, r.t.; (iii) CF$_3$CO$_2$H, r.t.; (iv) SOCl$_2$, reflux; (v) NaN$_3$, H$_2$O/acetone, 0 °C; (vi) benzene, reflux; (vii) HCl, reflux

R	Yield (i, ii) (%)	Yield (iii–vii) (%)	Ref.
Bzl	89	76	168
m-(CF$_3$)C$_6$H$_4$CH$_2$	85		168
m,p-Cl$_2$C$_6$H$_3$CH$_2$	70		168
m-MeOC$_6$H$_4$CH$_2$	70		168
m,p-(MeO)$_2$C$_6$H$_3$CH$_2$	58		168
PHT=N(CH$_2$)$_3$	54	85[a]	168
CH$_2$=CHCH$_2$	75		154

[a] The PHT group is cleaved off during acidolysis.

4.3.2.2 Strecker synthesis

Difluoromethyl ketones, obtained from reaction of difluoroacetates and α-metallated arylacetic acid, can be used in Strecker-type syntheses of difluoromethyl-substituted hydantoins, which are cleaved by concentrated hydrochloric acid to give derivatives of 2-(difluoromethyl)-3-arylalanine [169].

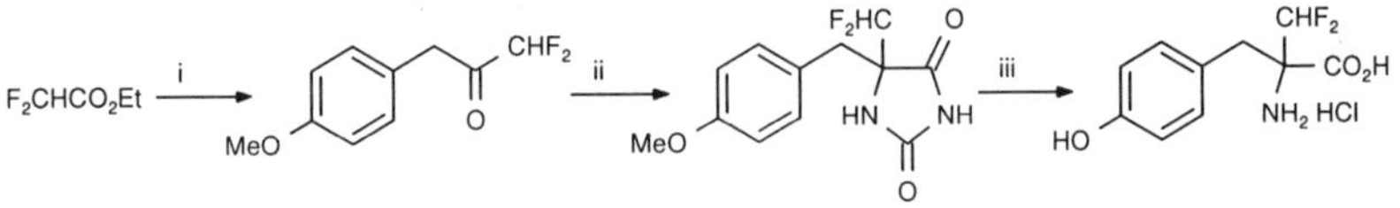

(i) LDA, THF, p-MeOC$_6$H$_4$CH$_2$CO$_2$H, −78 °C; (ii) (NH$_4$)$_2$CO$_3$, NaCN, H$_2$O; (iii) 12M HCl, 130 °C

4.3.2.3 Functional group transformation

Allyldifluoromethyl malonates are brominated under radical conditions. Nucleophilic substitution of bromide by phthalimide (30–35% yield) and the reaction sequence for the conversion of a carboxylic group into an amino group (see Section 4.3.2.1; ester hydrolysis, formation of the acyl azide via the acid chloride, subsequent Curtius rearrangement and cleavage of the phthalimido and the ester group) gives (E)-2-(difluoromethyl)ornithine (12%) [154].

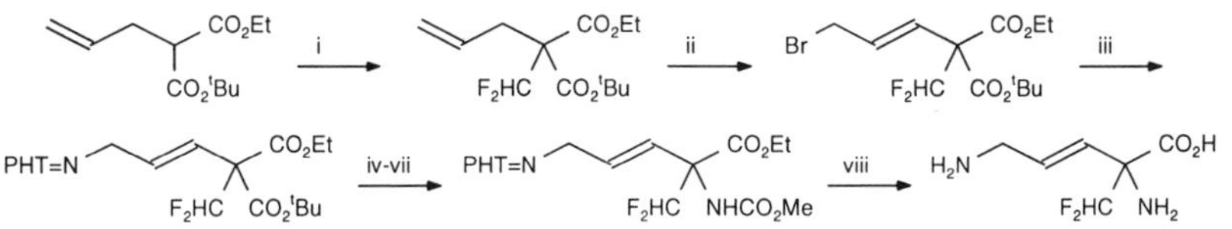

(i) NaH, THF, $CHClF_2$, 45 °C (75%); (ii) NBS, cat. $(PhCO_2)_2$, benzene, reflux (30%); (iii) potassium phthalimide, DMF, 80 °C (50%); (iv) CF_3CO_2H, r.t.; (v) $SOCl_2$, reflux; (vi) NaN_3, acetone, H_2O; (vii) MeOH, reflux (60% for iv–vii); (viii) HOAc, conc. HCl, reflux, then EtOH, propylene oxide (40%)

DL-2-Amino-2-(difluoromethyl)piperidone is prepared from 4-cyano-2-difluoromethyl-2-methoxycarbonylbutanoate [170]. $(-)$-2-(Difluoromethyl) ornithine is resolved enzymatically starting from DL-2-amino-2-(difluoromethyl) piperidone [171].

α-(Difluoromethyl)homoallylglycine is transformed into α-(difluoro-methyl)tryptophan or α-(difluoromethyl)-5-hydroxytryptophan via Fischer indole synthesis [162].

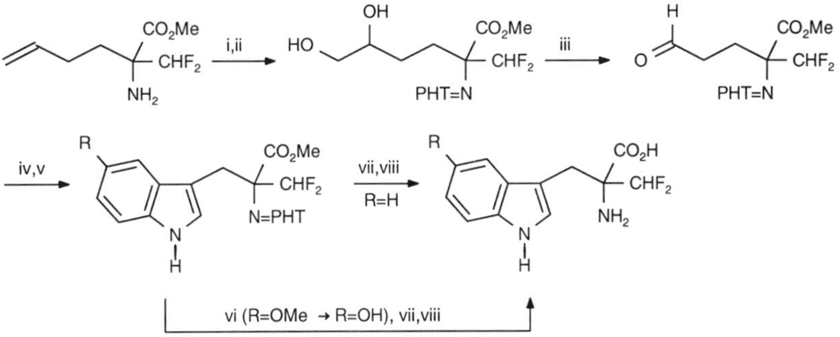

(i) phthaloyl dichloride, NEt_3, CH_2Cl_2, 0 °C $\rightarrow$ reflux (75%); (ii) OsO_4, N-methylmorpholine N-oxide (NMMO), Bu^tOH, H_2O/acetone, reflux (83%); (iii) $NaIO_4$, THF/H_2O, r.t. (93%); (iv) p-$RC_6H_4NHNH_2$, THF/H_2O, 1M AcOH (pH 5), r.t.; (v) H_2SO_4, MeOH, reflux, (R = H: 29%; R = OMe, 60% for iv, v); (vi) BBr_3, CH_2Cl_2, -78 °C $\rightarrow$ r.t. (20%); (vii) H_2NNH_2 (1M in EtOH), reflux (R = H, 63%; R = OH, 74%); (viii) 0.1M NaOH/THF, r.t. (R = H,

α-(Difluoromethyl)ornithine and ethylthiouronium bromide react to give α-(difluoromethyl)arginine [146].

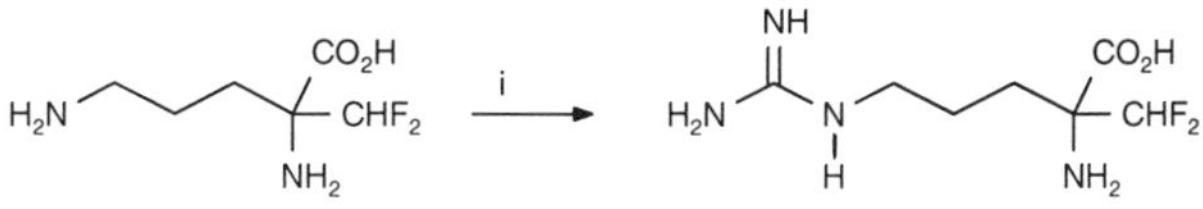

(i) $EtSC(=NH)NH_3^+$ Br^-, NaOH, pH 10.5, then ion-exchange chromatography (50%)

α-(Difluoromethyl)methionine is converted into S-benzyl-α-(difluoro-methyl)homocysteine [146].

(i) BzlCl, 12M HCl, reflux (39%)

Conjugates with nucleotides (e.g. adenosine derivatives) are obtained via reductive debenzylation and S-alkylation of α-(difluoromethyl)homocysteine and $5'$-O-tosyladenosine [146, 172].

4.3.3 α-TRIFLUOROMETHYL DERIVATIVES (TFM AMINO ACIDS)

4.3.3.1 Trifluoromethylation of nucleophilic α-amino acid synthons

2-Trifluoromethyl-substituted α-amino acids are reported to be accessible from α-metallated N-benzylidene amino acid esters on trifluoromethylation with bromotrifluoromethane [145].

4.3.3.2 Strecker synthesis

Low yields are characteristic $\{(\alpha$-trifluoromethyl)alanine, 20% [159]; (α-tri-fluoromethyl)butanoic acid, 21% [173]$\}$ when Strecker synthesis is performed with trifluoromethyl ketones because of the considerable stability of the intermediate trifluoromethyl-substituted cyanohydrins [1, 174]. However, the Bucherer–Bergs modification via hydantoins represents a preparatively useful method for the synthesis of 2-trifluoromethyl-α-amino acids (R = CH$_2$-aryl, CHR$'$-aryl [175]; R = Me [176]).

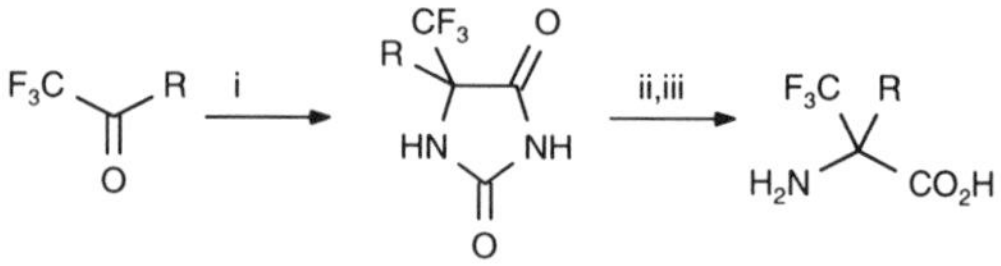

(i) KCN, (NH$_4$)$_2$CO$_3$, EtOH, H$_2$O, r.t. $\rightarrow$ 60 °C (R = Me, 23%); (ii) 48% HBr, reflux, aryl methyl ethers are cleaved, or 60% H$_2$SO$_4$, reflux; (iii) ethylene oxide, acetone, r.t. (R = Me, 58% for ii, iii)

Hog kidney aminoacylase catalyzes the hydrolysis of N-acyl-2-trifluoro-methylalanine (TFMAla) with high enantioselectivity, providing a method for the kinetic resolution of the racemate [177].

4.3.3.3 Oxazole route: Cope rearrangement *versus* radical 1,3-shift

5-Fluoro-4-trifluoromethyloxazoles possess the complete skeleton atom pattern of *N*-protected 3,3,3-trifluoroalanine (see Section 4.2.3.6). The fluoro substituent at C-5 can be readily displaced by various types of nucleophiles [178]. When 1,5-diene substructures are formed via this route, they spontaneously undergo a low-temperature Claisen rearrangement to form 4-allyl-, 4-allenyl- or 4-heteroaryl-4-trifluoromethyloxazol-5(4H)-ones which are hydrolyzed *in situ* to yield *N*-protected 2-allyl-, 2-allenyl- or 2-heteroaryl-3,3,3-trifluoroalanine derivatives [179, 180].

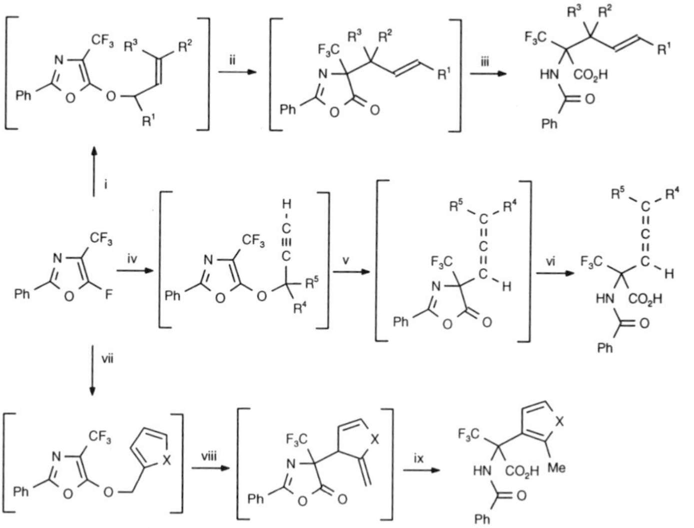

(i) $R^2R^3C{=}CHCHR^1OH$, KOH, dioxane, r.t.; (ii) r.t.; (iii) H_2O, r.t.; (iv) $HC{\equiv}CCR^4R^5OH$, KOH, dioxane, r.t.; (v) r.t.; (vi) H_2O, r.t. ($R^4 = $ Et, $R^5 = $ H, 64%; $R^4 = $ Me, $R^5 = $ Et, 62% for iv–vi); (vii) (c-$C_4H_3X)CH_2OH$, KOH, dioxane, r.t.; (viii) r.t.; (ix) H_2O, r.t. (X = O, 47% for vii–ix; X = S, 30% for vii–ix)

R^1	R^2	R^3	Yield (i–iii) (%)	d.e. (%)	Ref.
H	H	H	78		179
Me	H	H	54		179
Et	H	H	49		179
Pr^n	H	H	49		179
H	Pr^n	H	61	50	179
H	H	Pr^n	36	100	181
H	Me	H	69		179
H	H	Et	42	100	181
H	Ph	H	40	84	181
H	Me	Me	46		179
H	Me$_3$Si	H	87^a	32	181
CH$_2{=}$CH	H	H	89		182

^{a} Yield of oxazolone.

The Cope rearrangement of *cis*-configurated allyl vinyl ethers proceeds with high diastereoselectivity [181].

Deprotection is achieved on boiling with concentrated hydrochloric acid. Derivatives of furan or thiophene undergo decarboxylation under these reaction conditions and give 1-heteroaryl-2,2,2-trifluoroethylamine hydrochlorides [180].

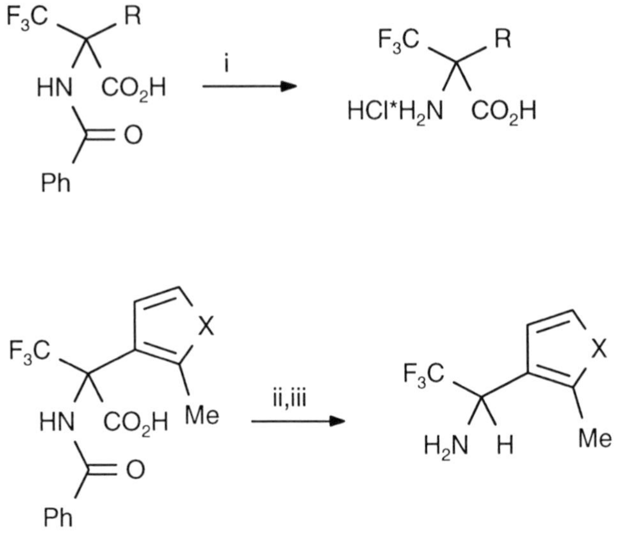

(i)   conc. HCl, reflux (R = CH$_2$CH=CH$_2$, 38%);
(ii) conc. HCl, dioxane, reflux;
(iii) 1M NaOH, r.t. (X = S, 63% for ii, iii)

4-Benzyl-4-trifluoromethyl-5(4*H*)-oxazolones are available on reaction of benzyl alcohols (the allyl substructure being a part of the aromatic system) with 5-fluoro-4-trifluoromethyloxazoles. In contrast to five-membered heteroarylmethyl alcohols, the substitution products initially formed in the key step of the reaction sequence undergo a 1,3-benzyl shift from oxygen to carbon. The isolation of mixed products in crossing experiments supports the suggestion that this reaction is a non-concerted process which proceeds via a radical pair mechanism [183].

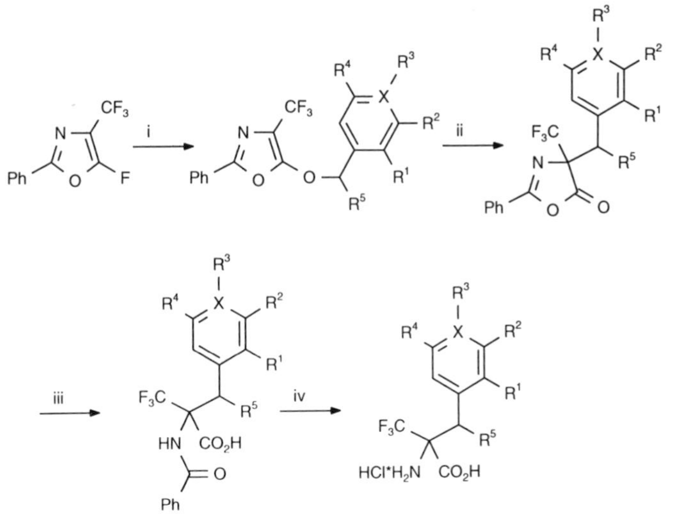

(i) Benzyl alcohol derivative, KOH, dioxane, r.t.; (ii) CHCl$_3$, toluene or xylene, ΔT; (iii) 1M HCl, 50 °C; (iv) conc. HCl, reflux

R^1	R^2	R^3	R^4	R^5	X	Yield (i) (%)	Yield (ii) (%)	Yield (iii) (%)	Yield (iv) (%)	Ref.
H	H	H	H	H	C	69	81	87	42	180
I	H	H	H	H	C	77	40			180
H	MeO	MeO	H	H	C	86	66	73		180
H	MeO	MeO	MeO	H	C	98	99	99		180
H	—OCH$_2$O—		H	H	C	85	73	99		180
H	H	MeO	H	H	C	54	99			180
H	H	F	H	H	C	78	42	99		180
H	H	Cl	H	H	C	87	46	99		180
H	H	Br	H	H	C	84	54			180
H	H	NO$_2$	H	H	C	58	43	99		180
H	H	Ph	H	H	C	97	8			180
H	H	^{t}Bu	H	H	C	85	7	24		181
H	H	H	H	Me	C	73	41			180
H	H	H	H	CO$_2$H	C	82	61			180
H	H	—	H	H	N	82	53			180

Ferrocenylmethanol behaves like benzyl alcohols, giving derivatives of 3-ferrocenyl-2-trifluoromethylalanine [181].

4-Trifluoromethyl-5(4H)-oxazolones represent N-protected, carboxylic group-activated derivatives of 2-trifluoromethyl-substituted α-amino acids. 4-Benzyl-4-trifluoromethyl-5(4H)-oxazolones are suitable precursors for the synthesis of 2-(trifluoromethyl)phenylalanine derivatives (e.g. amides and di- or tripeptides) with variable substitution patterns in the aromatic ring [184].

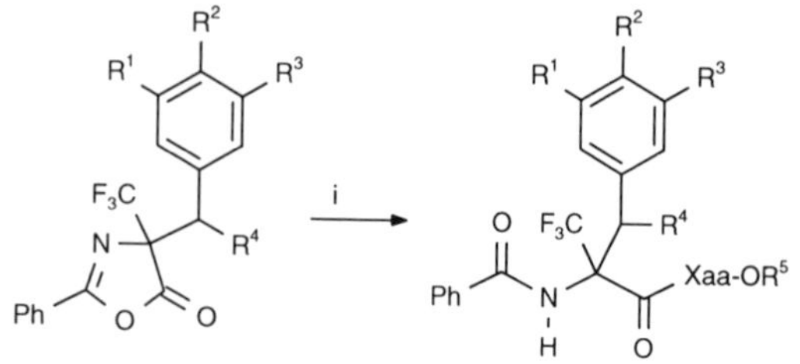

(i) H-Xaa-OR^5, Et$_2$O, r.t., or CHCl$_3$, 60°C

R^1	R^2	R^3	R^4	Xaa-OR^5	Yield (i) (%)	Ref.
H	MeO	H	H	Ala-OMe	42	184
MeO	MeO	MeO	H	Leu-OMe	45	184
MeO	MeO	H	H	Gly-Gly-OEt	67	184
H	H	H	CO$_2$H	Phe-OEt	31	184

4.3.3.4 Trifluoropyruvate route: alkylation and arylation of 2-imino-3,3,3-trifluoropropionates

Alkylation and arylation with organometallic reagents

N-Alkyl-, N-aryl- [118], N-acyl- [6], N-carbamoyl- [6, 118], and N-sulfonylimines [125] derived from 3,3,3-trifluoropyruvates (see Section 4.2.3.7) are readily available, highly electrophilic synthons for α-trifluoromethyl amino acids (TFMXaa) [185]. An enormous number of structurally different types of side-chains can be introduced on reaction with organometallic reagents. Magnesium [118, 122, 186], lithium [119, 181], sodium [186], zinc [118] and cadmium organyls [118] are applied preferentially. The nucleophilic addition prodeeds regiospecifically and results in alkylation or arylation of the CN double bond.

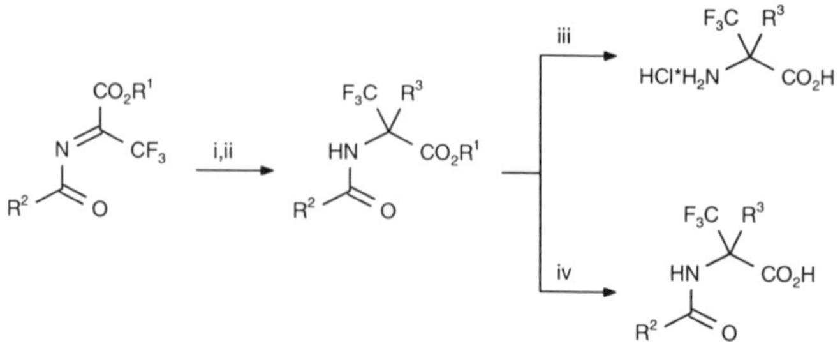

(i) R^3M, Et_2O; (ii) H_3O^+; (iii) conc. HCl, HCO_2H (1:1), reflux ($R^2 = Me$, 71%); (iv) 1M KOH, MeOH, r.t. (57–93%)

M	T (°C)	R^1	R^2	R^3	Yield (i) (%)	Ref.
CdMe	4 → reflux	Me	Ph	Me	69	118
CdMe	4 → reflux	Me	MeO	Me	64	118
CdPh	4 → reflux	Me	MeO	Ph	60	118
MgI	−198 → r.t.	Me	MeO	Me	80	117
MgCl	−70 → r.t.	Me	BzlO	Me	95	181
MgBr	−70 → r.t.	Me	Bu^tO	Me	72	187
MgBr	−198 → r.t.	Me	MeO	Pr^n	90	118
MgBr	−198 → r.t.	Me	MeO	Bu^n	92	118
MgI	−198 → r.t.	Me	MeO	Bu^i	62	118
MgBr	−70 → r.t.	Et	BzlO	Bu^i	97	181
MgBr	−70 → r.t.	Me	Bu^tO	Bu^i	98	187
MgCl	−198 → r.t.	Me	MeO	Bzl	79	118
MgCl	−70 → r.t.	Et	BzlO	Bzl	90	181
MgCl	−70 → r.t.	Me	Bu^tO	Bzl	80	187
MgBr	−198 → r.t.	Me	MeO	Ph	53	117
MgCl	−70 → r.t.	Me	BzlO	CH_2SiMe_3	97	188
Li	−70 → r.t.	Me	BzlO	$2\text{-}(c\text{-}C_4H_3S)$	54	181
MgBr	−70 → r.t.	Me	BzlO	$CH_2{=}CH$	98	122
MgBr	−70 → r.t.	Me	BzlO	$CH_2{=}CHCH_2$	92	122
MgBr	−70 → r.t.	Me	Bu^tO	$CH_2{=}CHCH_2$	51	181
MgBr	−70 → r.t.	Me	BzlO	$CH_2{=}CH(CH_2)_2$	73	122
MgBr	−70 → r.t.	Me	Bu^tO	$CH_2{=}CH(CH_2)_2$	48	181
MgBr	−70 → r.t.	Me	BzlO	$CH_2{=}CH(CH_2)_3$	95	122
MgBr	−70 → r.t.	Me	BzlO	$CH_2{=}CH(CH_2)_4$	62	122
Na	−78 → r.t.	Me	BzlO	$C{\equiv}CH$	80	186
MgCl	−78 → r.t.	Me	BzlO	$C{\equiv}CPh$	75	186
MgCl	−78 → r.t.	Et	BzlO	$C{\equiv}CPh$	71	186
MgCl	−78 → r.t.	Me	BzlO	$C{\equiv}C(c\text{-}C_6H_{11})$	70	186
MgCl	−78 → r.t.	Me	BzlO	$C{\equiv}C(c\text{-}C_6H_9)$	78	186
MgCl	−78 → r.t.	Me	BzlO	$C{\equiv}CBu^t$	97	186
MgCl	−78 → r.t.	Et	BzlO	$C{\equiv}CSiMe_3$	80	186
ZnI	reflux	Me	MeO	CH_2CO_2Et	54	118

The products of a Grignard reduction are observed on addition of sterically demanding Grignard reagents bearing a proton in the α-position. For instance, on reaction of urethane-protected imines derived from 3,3,3-trifluoropyruvate with isopropylmagnesium bromide, a 3:2 mixture of α-(trifluoromethyl)valine (TFMVal) and 3,3,3-trifluoroalanine (TFMGly) is obtained [11, 181]. Reduction is suppressed when the less reactive diisopropylcadmium is used [11].

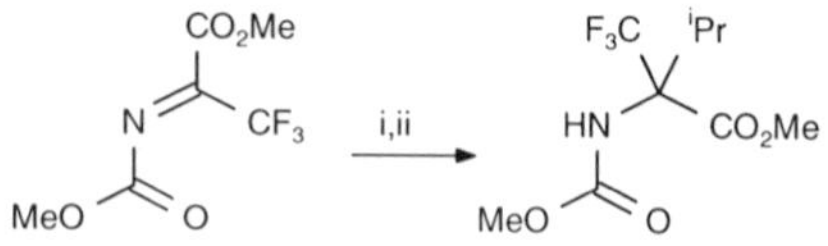

(i) Pr^i_2Cd, 4°C r.t.; (ii) H_3O^+ (50% for i, ii)

Many functional groups that are tolerated by the organometallic reagent applied can be introduced simultaneously. Especially CC double bonds [122, 189] and CC triple bonds [186] and also trimethylsilyl [188] and carboxylate groups [118] are very suitable for further functional group transformations of the side-chain to give multifunctional α-trifluoromethylamino acids.

The hemiamidals obtained from reaction of amides or urethanes with 3,3,3-trifluoropyruvates can also be used after O-acetylation for the in situ generation of the corresponding acylimine. However, two equivalents of the organometallic reagent are necessary; one equivalent acts as a base to generate the acylimine and the second equivalent reacts with the acylimine [118].

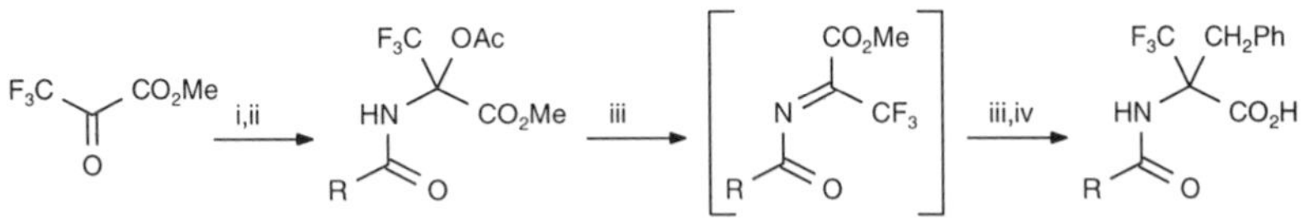

(i) $RCONH_2$, Et_2O, r.t. (R = MeO, 88%; R = Me, 72%); (ii) Ac_2O, dioxane, reflux (R = Me, 70%); (iii) BzlMgCl, Et_2O/THF, r.t.; (iv) H_3O^+ (59% for iii, iv)

Orthogonal protective groups can be introduced during the synthesis of α-trifluoromethylamino acids via the pyruvate route. The amino acid esters obtained via this route exhibit considerable stability towards acidolysis, which can only be achieved on prolonged (40 h) refluxing in concentrated hydrochloric acid/formic acid. Therefore, alkaline hydrolysis is the method of choice for deblocking of the carboxylic group [117, 118]. Deprotection of the α-amino group is performed in good yields according to standard procedures when urethane protective groups (Boc, Z) are employed [187].

Experiments to alkylate N-alkylimines of 3,3,3-trifluoropyruvates with organometallic reagents have so far failed. No N-alkyl-substituted α-trifluoromethylamino acids have been obtained. However, N-arylimines can be alkylated successfully with dimethylcadmium [118] or organolithium compounds [119]. The 4-methoxyphenyl group is removed oxidatively and the benzyl ester by hydrogenation.

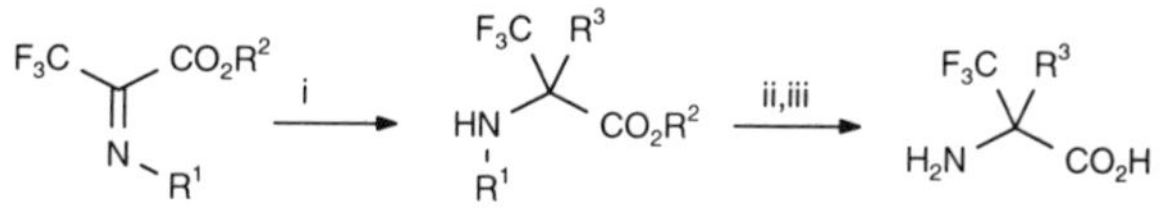

(i) R^3M, Et$_2$O or THF; (ii) $R^1 = p\text{-MeOC}_6\text{H}_4$, $R^2 = $ Bzl: ammoniumcerium(IV) nitrate, H$_2$O, CH$_3$CN; (iii) H$_2$, Pd/C, MeOH

M	$T(°)$	R^1	R^2	R^3	Yield (i) (%)	Yield (i–iii) (%)	Ref.
CdMe	$4 \rightarrow 50$	Ph	Me	Me	70		118
Li		$p\text{-MeOC}_6\text{H}_4$	Bzl	Me		65	119
Li		$p\text{-MeOC}_6\text{H}_4$	Bzl	Bu^n		73	119

Allyl transfer from a homochiral titanium compound [190] to a 3,3,3-trifluoropyruvate-derived acylimine results only in moderate enantioselectivity [122].

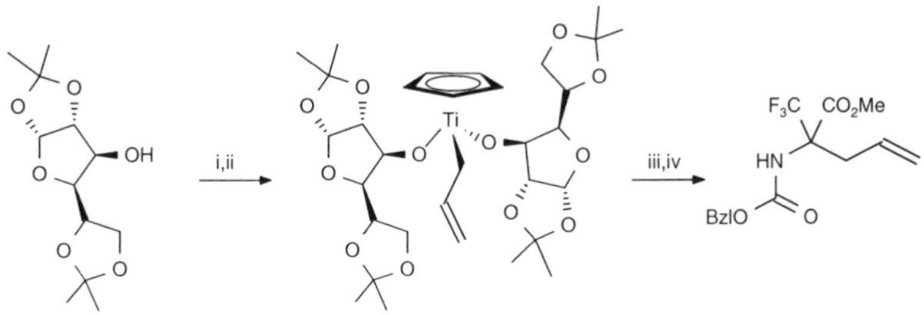

(i) CpTiCl$_3$, NEt$_3$, Et$_2$O, r.t.; (ii) CH$_2$=CHCH$_2$MgBr, 0 °C; (iii) BzlO$_2$CN=C(CF$_3$)CO$_2$Me, -78 °C $\rightarrow$ r.t.; (iv) H$_3$O$^+$ (40%, 22% e.e. for iii, iv)

A promising strategy for the diastereoselective synthesis of α-trifluoromethylamino acids proceeds via amidoalkylation of carbon nucleophiles with *in situ*-formed cyclic homochiral acylimines. The corresponding dioxopiperazines are formed with good stereoselectivity [191].

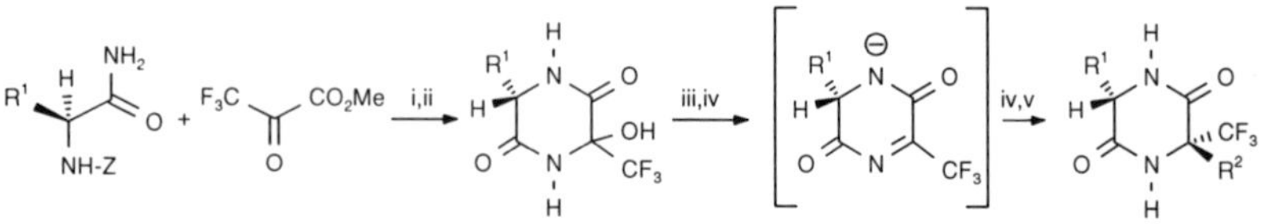

(i) r.t.; (ii) $H_2/Pd/C$; (iii) $(CF_3CO)_2O$, THF, $0\,^\circ C$; (iv) R^2MgX, $-100 \rightarrow -35\,^\circ C$; (v) phosphate buffer, pH 7

R^1	R^2	d.e. (crude) (iii–v) (%)	Yield (iii–v) (%)	d.e.[a] (%)	Ref.
Bu^i	Me	56	51	>99	191
Bzl	Me	64	62	71	191
Bu^i	Bzl	70	58	80	191
Bzl	Bzl	67	40	94	191

[a] d.e. after workup to quoted yield.

Alkylation with CH acidic compounds

C-Alkylation of acylimines derived from 3,3,3-trifluoropyruvates with CH acidic compounds represents a general access to α-trifluoromethylamino acids. This method is especially useful for the synthesis of α-trifluoromethylamino acids with multifunctional or heterocyclic side-chains. Regiospecific C-alkylation is achieved on reaction with 2-methylpyridine [192] or 2-methylbenzothiazole [187].

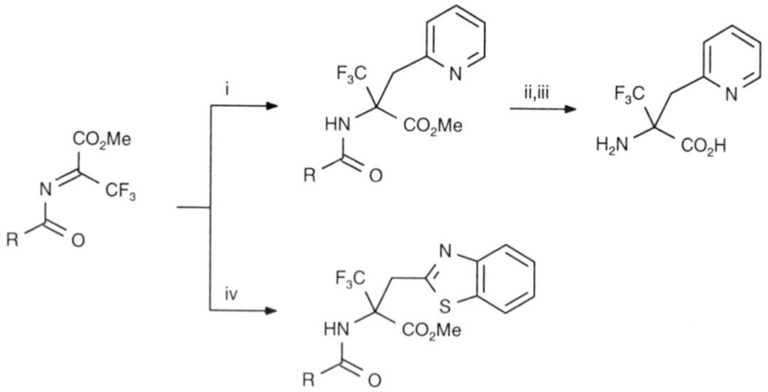

(i) 2-methylpyridine, $F_2ClCCFCl_2$, reflux (80%); (ii) NaOH; (iii) HBr (100% for ii, iii); (iv) 2-methylbenzothiazole, Bu^nLi, hexanes/THF, $-78\,^\circ C \rightarrow$ r.t. (38%)

Moderate yields of 1:1 adducts are obtained when diethyl malonate at room temperature or acetylacetone at $90\,^\circ C$ is treated with 2-(benzenesulfonyl-imino)-3,3,3-trifluoropropionate. The reaction of diethyl malonate with the 2-trifluoroacetylimino derivative occurs only above $90\,^\circ C$ in the presence of acid [193]. Higher yields are obtained even with the less activated 2-(benzoylimino)- or 2-(carbamoylimino)-3,3,3-trifluoropropionates in the presence of sodium hydride. (Z)-α-(trifluoromethyl)aspartic acid is obtained on ester hydrolysis and decarboxylation of the 1:1 adduct with concentrated hydrochloric acid [187]. Open-chain or cyclic adducts are formed with aceto-acetates (i or iv), depending on the reaction temperature.

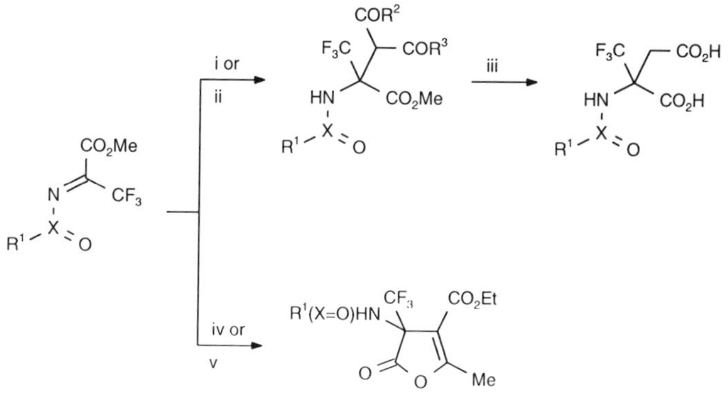

(i) $CH_2(COR^2)(COR^3)$;
(ii) $NaCH(COR^2)(COR^3)$, THF, 0 °C → r.t.;
(iii) 5M NaOH, MeOH, r.t., then 6M HCl (X = C, R^1 = BzlO, 41%);
(iv) $CH_2(CO_2Et)(COMe)$, 100 °C (X = S=O, R^1 = Ph, 31%);
(v) $NaCH(CO_2Et)(COMe)$, 0 °C → r.t.

X	R^1	R^2	R^3	T (°C)	Yield (i, ii) (%)		Ref.
S=O	Ph	EtO	EtO	r.t.	(i)	22	193
S=O	Ph	Me	Me	90	(i)	30	193
C	CF_3	EtO	EtO	90	(i)^a	30	193
C	CF_3	EtO	Me	r.t.	(i)	78	193
C	CF_3	Me	Me	r.t.	(i)	92	193
C	BzlO	MeO	MeO		(ii)	72	187
C	Bu^{t}O	MeO	MeO		(ii)	54	187
C	Ph	MeO	MeO		(ii)	58	187

^{a} In the presence of p-TsOH.

The Z-protected imine derived from 3,3,3-trifluoropyruvate reacts with acetyl chloride/triethylamine to give methyl 2-benzyloxy-4,5-dihydro-6-oxo-4-trifluoromethyl-1,3-oxazine-4-carboxylate, a β-activated equivalent of α-(trifluoromethyl)aspartic acid. This reagent provides a versatile route to β-derivatized α-(trifluoromethyl)aspartic acid via nucleophilic ring cleavage. α-(Trifluoromethyl)isoaspartyl peptides are obtained with amino acid esters as nucleophiles [194].

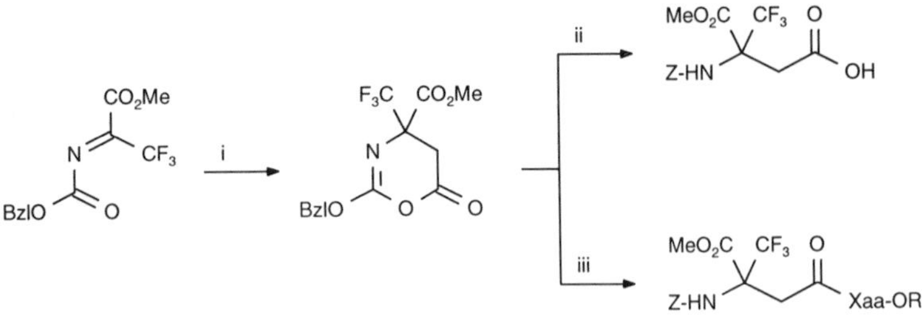

(i)   AcCl, NEt$_3$, Et$_2$O, 0 °C (72%);
(ii) 1M HCl, dioxane, r.t. (62%);
(iii) H-Xaa-OR, Et$_2$O or CH$_2$Cl$_2$, 0 °C → r.t. (Xaa-OR = Ala-OBu^t, 46%; Phe-OMe, 64%; Pro-OBzl, 40%)

Alkylation via ene reaction

The ene reaction represents a powerful concept for CC bond formation [195, 196]. The electron-deficient CN double bond of 2-(trifluoroacetylimino)-3,3,3-trifluoropropionates acts as an excellent enophile. The reaction with isobutene provides *N*-protected 2-trifluoromethyl-4,5-dehydroleucine [124].

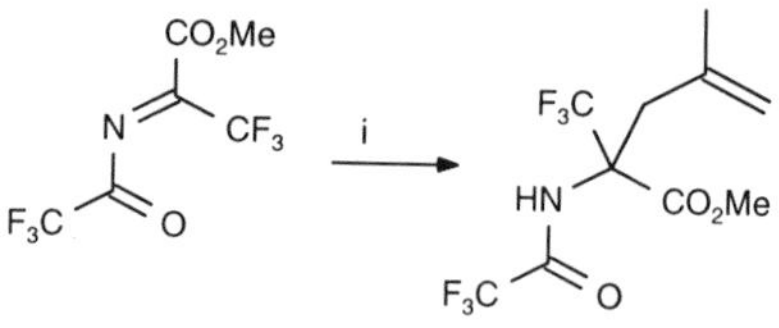

(i) CH$_2$=C(CH$_3$)$_2$, sealed tube, r.t.

Arylation via electrophilic aromatic substitution

2-Acylimino-3,3,3-trifluoropropionates as highly reactive electrophilic species are very suitable for electrophilic aromatic substitution reactions. Activated aromatic (e.g. *N,N*-dialkylanilines) or heteroaromatic (e.g. pyrrole, furan, thiophene, indole) compounds undergo regioselective amidoalkylation to give the corresponding 2-aryl- or 2-heteroaryl-substituted α-trifluoromethylamino acids [125,126]. Ester hydrolysis is achieved in good yield (60–70%) on heating with 5% aqueous NaOH at 70 °C. However, in some cases the free acids readily decarboxylate (e.g. on sublimation) to give the corresponding amines (61–72%) [197].

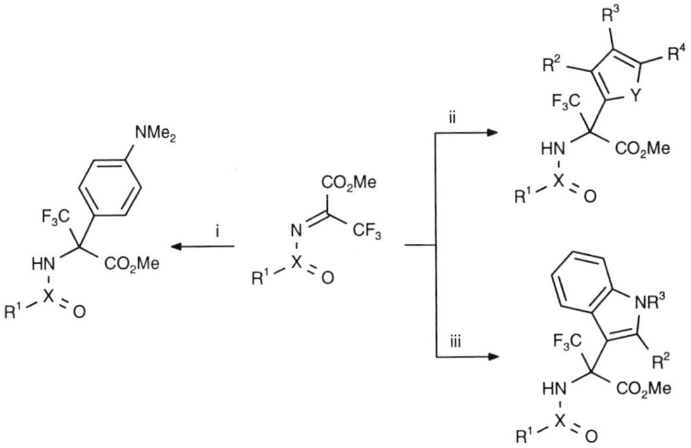

(i) *N,N*-dimethylaniline, CHCl$_3$ or F$_2$ClCCFCl$_2$, $-50\,^\circ$C $\rightarrow$ r.t.; (ii) thiophene derivative, CCl$_4$, reflux or furan/pyrrole derivative, CHCl$_3$, $-50\,^\circ$C $\rightarrow$ r.t.; (iii) indole derivative, CHCl$_3$, $-50\,^\circ$C $\rightarrow$ r.t.

Aniline derivatives:

X	R^1	Yield (i) (%)	Ref.
C	CF$_3$	60	126
S=O	Me	53	126
S=O	Ph	91	125

Furan/thiophene/pyrrole derivatives:

X	Y	R^1	R^2	R^3	R^4	Yield (ii) (%)	Ref.
C	S	CF$_3$	H	*p*-MeC$_6$H$_4$	H	62	197
C	S	CF$_3$	Me	H	Ph	87	197
C	S	CF$_3$	Me	Ph	H	84	197
C	S	CF$_3$	Me	Ph	Ph	91	197
C	S	CF$_3$	Ph	H	Me	83	197
S=O	S	Ph	H	H	H	40	125
C	O	CF$_3$	H	H	H	71	197
C	NMe	CF$_3$	H	H	H	68	197

Indole derivatives:

X	R^1	R^2	R^3	Yield (iii) (%)	Ref.
C	CF$_3$	H	H	65	197
C	CF$_3$	Me	H	81	197
S=O	Ph	H	H	72	125
S=O	Ph	Me	H	85	125
S=O	Ph	Ph	Me	87	125
S=O	Ph	Ph	H	72	125

Addition of electron-rich CC multiple bonds

The addition of electron-rich CC double bonds of non-aromatic heterocyclic compounds (e.g. pyrazolones, pyrimidones) provides α-trifluoromethyl amino acids with a heterocyclic substituent in the 2-position in a regioselective manner [197, 198].

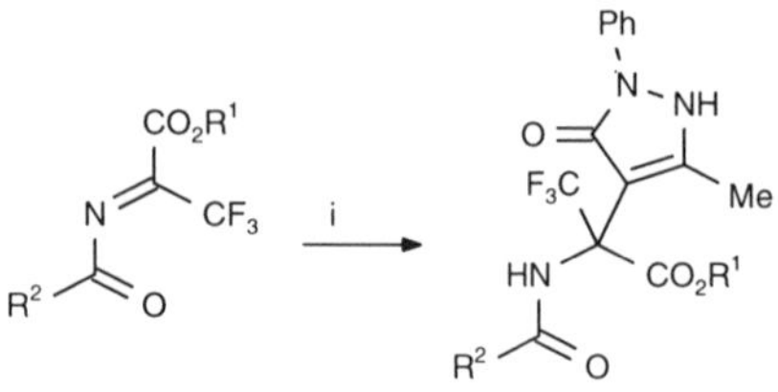

(i) 5-methyl-2-phenyl-1*H*-pyrazol-3-one, CHCl₃, r.t. (90%)

Addition of heteronucleophiles

Many α-heteroatom substituted α-amino acids have been isolated from natural products [199]. They show a broad variety of biological activity [200]. The presence of a trifluoromethyl group at C-2 renders them more stable. Alcohols, thiols [201], *N*-alkylamines, *N*-arylamines [126] and PH compounds [193] add to the imino function of 2-(acylimino)-3,3,3-trifluoropropionates to give stable adducts.

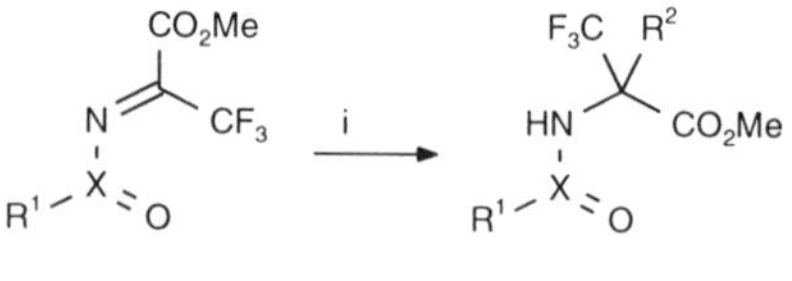

(i) R²H, CH₂Cl₂, r.t., or reflux (P compounds) or F₂ClCCFCl₂, −20 °C → r.t. (amines)

X	R^1	R^2	Yield (i) (%)	Ref.
C	MeO	OH	89	117
C	MeO	OPr^i	76	117
C	BzlO	OPr^i	95	201
C	BzlO	OBzl	96	201
C	BzlO	SBu^n	96	201
C	CF_3	NHPh	65	126
C	CF_3	NMePh	81	126
C	BzlO	NMe(OMe)	90	201
S=O	Me	NHPh	68	126
S=O	Ph	NHPh	70	126
C	CF_3	$P(=O)(O-n-C_5H_{11})_2$	51	193
C	CF_3	$P(=O)(OPh)_2$	50	193
S=O	Ph	$P(=O)(OMe)_2$	85	193
S=O	Ph	$P(=O)(OEt)_2$	80	193

In the case of N-alkyl- and N-arylanilines with sterically demanding substituents at the nitrogen atom, the addition process is slowed considerably. Consequently, electrophilic aromatic substitution competes efficiently. 3H-2-Oxoindoles are formed on reaction with N-isopropylaniline or N,N-diphenylamine [126].

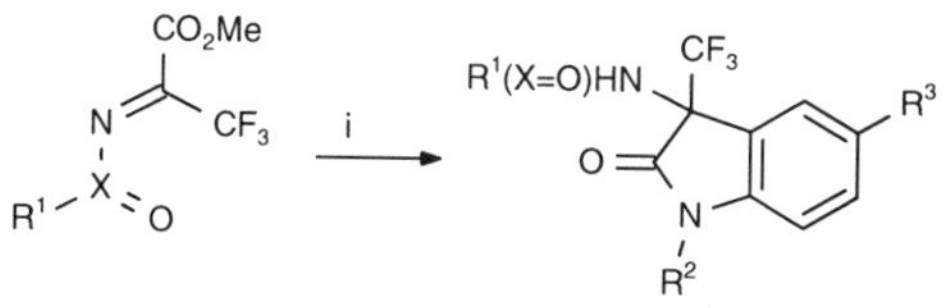

(i) p-$R^3C_6H_4NHR^2$, $F_2ClCCFCl_2$ or CCl_4, reflux

X	R^1	R^2	R^3	Yield (i) (%)	Ref.
C	CF_3	Pr^i	H	20	126
S=O	Ph	Pr^i	Me	51	126
C	CF_3	Pr^i	Me	60	126
C	CF_3	Pr^i	MeO	60	126
C	CF_3	Ph	H	30	126
S=O	Ph	Ph	H	15	126
S=O	Me	Ph	H	50	126

Amidoalkylation of alcohols, thiols or amines is also observed with 2-chloro-2-formylamino-3,3,3-trifluoropropionates in the presence of base [127].

Alkylation via cycloaddition reactions

[4 + 2] Cycloaddition reactions

2-Acylimino-3,3,3-trifluoropropionates represent a class of highly reactive 1,3-hetero-1,3-dienes. Their polarity pattern causes a reaction behaviour similar to that of 1,4-dipoles. Because of the low-lying LUMO, cycloaddition reactions preferentially occur with electron rich multiple bond systems [202].

On ring opening, the [4 + 2] cycloadducts with enol ethers [124, 198] give cyclic acetals of 4-oxo-2-trifluoromethylbutyric acid, which can be transformed into α-(trifluoromethyl)aspartic acid on methanolysis and subsequent oxidation with $KMnO_4$ [203].

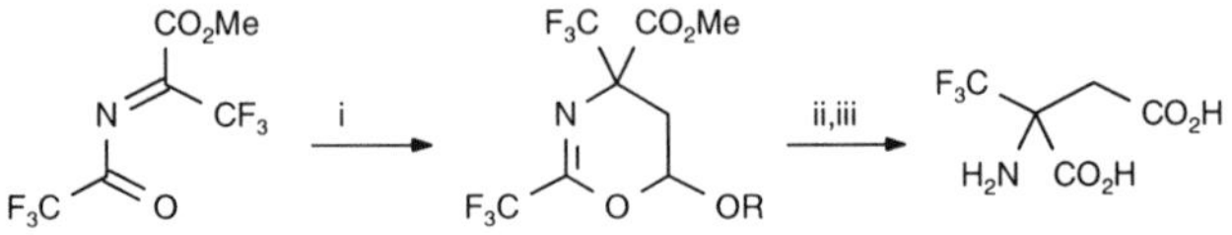

(i) $ROCH=CH_2$, $F_2ClCCFCl_2$, (R = Ac, $0\,°C \to$ r.t., 75%; R = Bu^n, $-30\,°C \to$ r.t., 51%);
(ii) HCl, MeOH (R = Ac, 60%);
(iii) H_2SO_4, $KMnO_4$ (52%)

Alkenes add regioselectively to 2-(trifluoroacetylimino)-3,3,3-trifluoropropionates to give 5,6-dihydro-4*H*-1,3-oxazines [124], which on treatment with acid in methanol undergo a rearrangement to yield (α-trifluoromethyl)homoserine lactone [204].

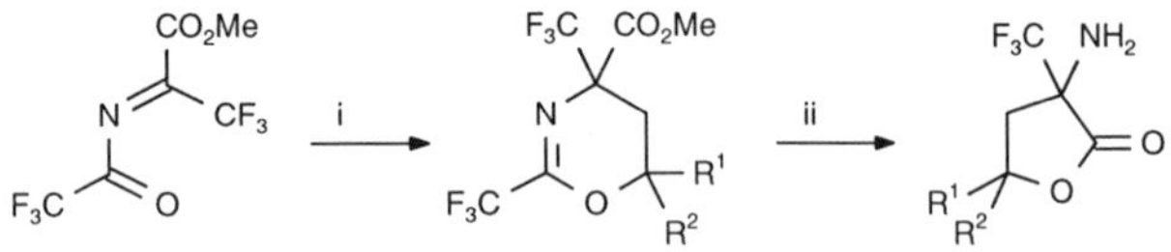

(i) $CH_2=CR^1R^2$, sealed tube, r.t., R^1 = H, Me, CH_2Cl, R^2 = Me, CH_2I, CH_2OMe, CH_2Cl;
(ii) MeOH, cat. HCl/H_2O, r.t

Bicyclic systems result when cyclic alkenes are used as dienophiles [124]. However, the CN double bond of the acylimines can also act as a dienophile: an azanorbornene (*endo/exo* ratio 1:1) is formed on reaction with cyclopentadiene [124].

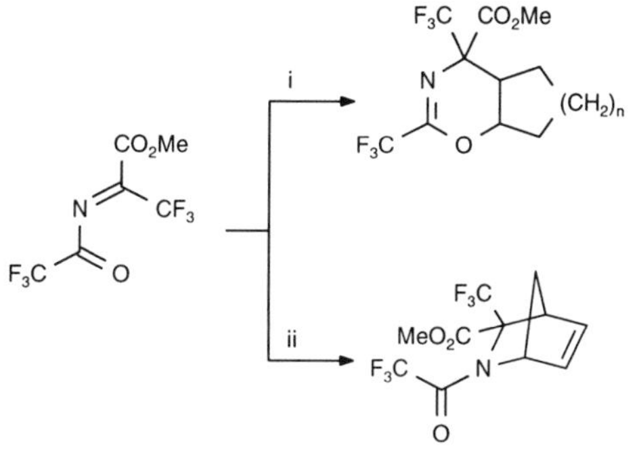

(i) $c\text{-}C_5H_8$ ($n = 1$) or $c\text{-}C_6H_{10}$ ($n = 2$), 150 °C, sealed tube ($n = 1$, 50%; $n = 2$, 70%); (ii) $c\text{-}C_5H_6$, cat. hydroquinone, Et_2O, $-50 \rightarrow 50$ °C (60%)

[4 + 1] Cycloaddition reactions
Isocyanides react with 2-(acylimino)-3,3,3-trifluoropropionates to give [4 + 1] cycloadducts, which are transformed into derivatives of 2-trifluoromethyl-2-aminomalonic acid on acid hydrolysis. Dipeptides are obtained when isonitriles prepared from *N*-formylamino acid esters are employed.

Hydrolytic cleavage of [4 + 1] cycloadducts from bis(trifluoromethyl)-substituted heterodienes [205] and isonitriles [206] or orthoformates [207] gives *N*-protected 2-amino-3,3,3,3',3',3'-hexafluoroisobutyric acid or dipeptides when amino acid-derived isonitriles are used as dienophiles [93].

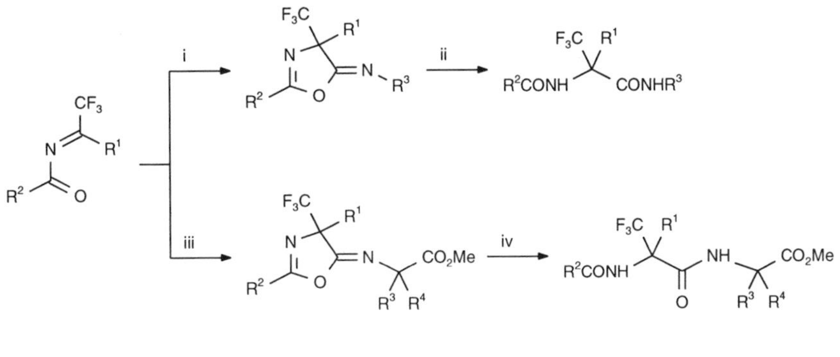

(i) CNR^3, 0 °C; (ii) conc. H_2SO_4, r.t., then H_2O, 0 °C; (iii) $CNCR^3R^4CO_2Me$, 60 °C, (iv) MeOH or Bu^tOH, cat. HCl/H_2O, r.t. (for R^n, see p. 200)

R^1	R^2	R^3	R^4	Solvent	Yield (i)/(iii) (%)	Yield (ii)/(iv) (%)	Ref.
CF_3	Ph	$c\text{-}C_6H_{11}$	—	heptane	83	66	206
CF_2NO_2	Ph	$c\text{-}C_6H_{11}$	—		90	56	206
CF_3	EtO	$c\text{-}C_6H_{11}$	—	Et_2O	83		206
CF_3	Ph	CF_3	Bu^i	toluene	68	65	93
CF_3	Ph	CF_3	Ph	toluene	65	52	93
CO_2Me	BzlO	H	H	toluene	83	48	93
CO_2Me	BzlO	CF_3	Me	toluene	49		93
CO_2Me	BzlO	CF_3	Ph	toluene	79		93

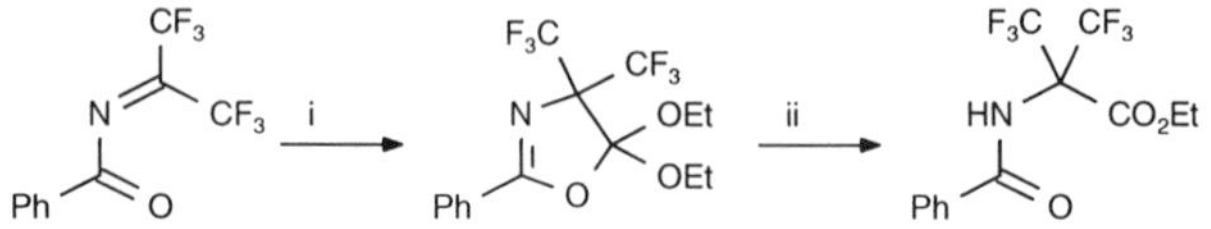

(i) $(EtO)_3CH$, 150 °C (92%); (ii) conc. HCl, reflux (74%)

[3 + 2] Cycloaddition reactions

The CN double bond of 3,3,3-trifluoropyruvate-derived N-acylimines can also act as a highly reactive dipolarophile in [3 + 2] cycloaddition reactions. [3 + 2] Cycloadducts representing aza analogues of α-(trifluoromethyl)proline are formed regioselectively with diazoalkanes. The adduct from 2-(benzoylimino)-3,3,3-trifluoropropionate and diazomethane is unstable at room temperature: aziridines are the products of the [3 + 2] cycloreversion process. The stability of the 1,2,3-triazolines depends considerably on the substitution pattern. For instance, the [3 + 2] cycloadducts from 2-(benzyloxycarbonylimino)-3,3,3-trifluoropropionate and diazomethane or diazoacetate are stable at room temperature. It is noteworthy that the latter [3 + 2] adduct shows reversed regiochemistry. Both cycloadducts undergo [3 + 2] cycloreversion on heating to give aziridines [124,181]. With arylnitrile oxides, 4-substituted methyl 3-aryl-5-trifluoromethyl-1,2,4-oxadiazol-2-ine-5-carboxylates are obtained [198].

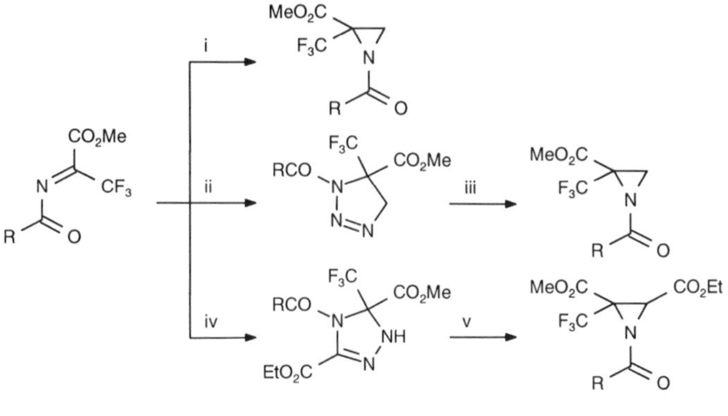

(i) CH_2N_2, Et_2O, $-30\,°C$ ($R = CF_3$, 60%);
(ii) CH_2N_2, Et_2O, $-20\,°C \rightarrow$ r.t. ($R = BzlO$, 72%);
(iii) ΔT ($R = BzlO$, 99%; $R = Ph$, 86%);
(iv) EtO_2CCHN_2, $-30\,°C \rightarrow$ r.t., Et_2O;
(v) $150°C$ ($R = CF_3$: 70% for iv, v)

4.3.3.5 Ring opening of epoxides

Nucleophilic ring-opening reactions of trifluoromethyl-substituted three-membered ring systems can be utilized to synthesize trifluoromethyl-substituted amino acids.

The reaction of perfluoroisobutene oxide with ammonia affords 2-amino-3,3,3',3',3'-hexafluoroisobutyramide [208]. Alkaline hydrolysis of the amide results in formation of 2-amino-3,3,3',3',3'-hexafluoroisobutyrate, which

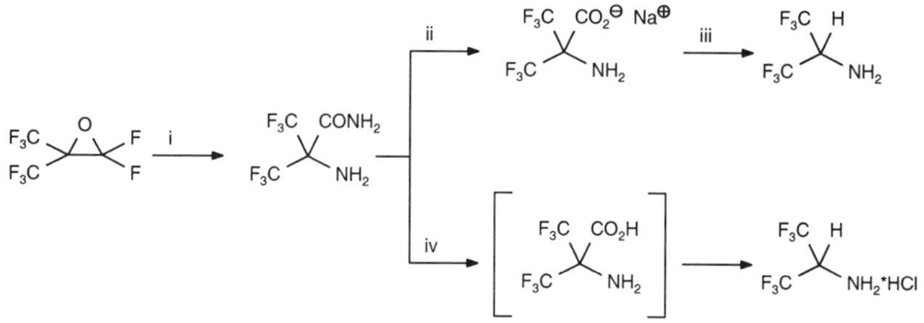

(i) NH_3, Et_2O, $-10\,°C$ (64%);
(ii) $4M$ NaOH, reflux (60%);
(iii) $4M$ HCl, r.t., then NaOH (46%); (iv) conc. HCl, r.t. (60%)

decarboxylates on acidification to give hexafluoroisopropylamine hydrochloride [209].

It is noteworthy that derivatives of 2-aminohexafluoroisobutyric acid do not form stable salts even with sulfuric acid, which is explained by the strong electron accepting effect of the two trifluoromethyl groups adjacent to the amino group.

4.3.3.6 Ring opening of 2H-azirines

Esters and N,N-dialkylamides of 2-amino-3,3,3',3',3'-hexafluoroisobutyric acid are obtained on hydrolysis of suitably substituted 2H-azirines, which are available on reaction of the corresponding alkoxy- or dialkylaminoperfluoroisobutene derivatives with sodium azide [210, 211].

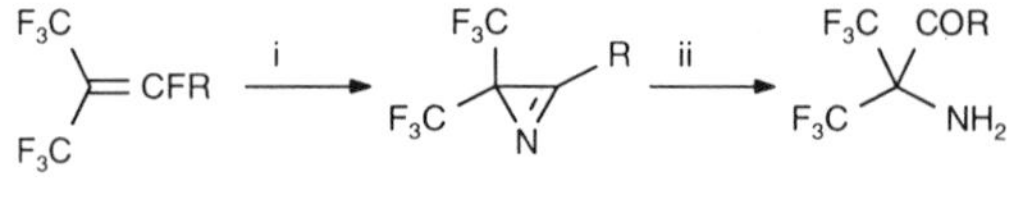

(i) NaN$_3$, tetraglyme or DMF, r.t.; (ii) conc. H$_2$SO$_4$, T, then H$_2$O, 0 °C

R	Yield (i) (%)	T (°C)	Yield (ii) (%)	Ref.
EtO	78	r.t	82	211
Me$_2$N	56	100	86	210
c-C$_5$H$_{10}$N	67	100	77	210

4.3.3.7 Functional group transformation

Side Chain Transformation

Functional groups in the side-chain of α-trifluoromethylamino acids can be derivatized further. Functionalization of the CC double bond in 2-allyl-3,3,3-trifluoroalanine and other compounds with an unsaturated side-chain provides a versatile methodology for the synthesis of multifunctional α-trifluoromethyl-substituted α-amino acids. For instance, oxidation with KMnO$_4$ gives α-trifluoromethyl-α-amino-α,ω-dicarboxylic acids, like α-(trifluoromethyl)aspartic acid and its 3-alkyl or 3,3-dialkyl derivatives. Deprotected α-(trifluoromethyl) aspartic acid is a stable compound, whereas α-(trifluoromethyl)glutamic acid and α-(trifluoromethyl)-α-aminoadipic acid undergo lactam formation even at room temperature. Therefore, they have to be stored in their N-protected form [189,122].

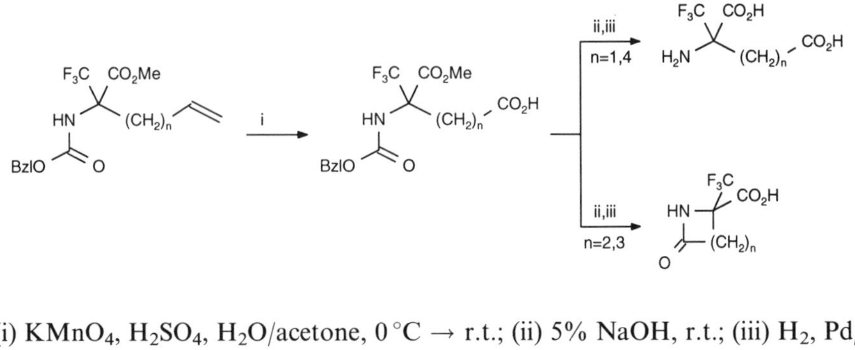

(i) KMnO$_4$, H$_2$SO$_4$, H$_2$O/acetone, 0 °C → r.t.; (ii) 5% NaOH, r.t.; (iii) H$_2$, Pd/ C, MeOH, H$_2$O, 30 °C

n	Yield (i) (%)	Yield (ii) (%)	Yield (iii), $n = 1,4$ (%)	Yield (iii), $n = 2,3$ (%)	Ref.
1	95	92	75	—	122
2	75	99	—	70	122
3	96	74	—	86	122
4	41	42	78	—	122

Many other types of functionalizations of CC double bonds have been studied, such as the reaction with NBS [93], addition of dihalogens [181], PH compounds [212] and epoxidation reactions [93].

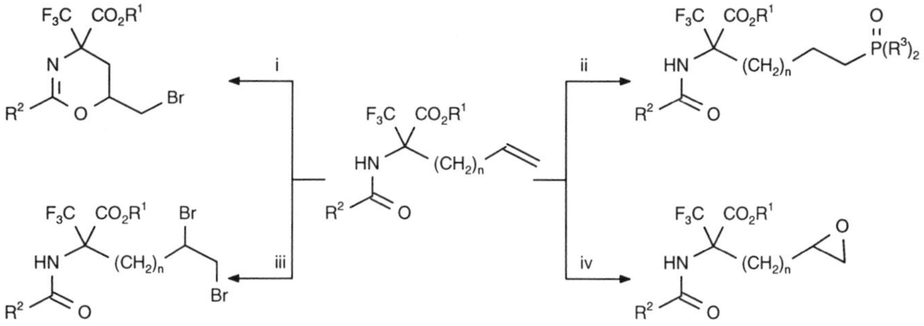

(i) NBS, dioxane/H_2O, r.t., ($n = 1$, 65%, 60% d.e.); (ii) $(R^3)_2(P{=}O)H$, cat. AIBN, benzene, 70 °C; (iii) Br_2, $CHCl_3$, 0 °C, ($R^1 = Me$, $R^2 = BzlO$, $n = 2$, 95%); (iv) mCPBA, $CHCl_3$, 0 °C → r.t. ($R^1 = Me$, $R^2 = BzlO$, $n = 2$, 82%; $n = 3$, 90%)

Phosphonates					
n	R^1	R^2	R^3	Yield (ii) (%)	Ref.
0	Me	BzlO	Ph	76	212
0	Me	BzlO	Me	42	212
1	Me	BzlO	Ph	37	212
1	Me	BzlO	Me	48	212
1	Me	Ph	Ph	58	212

In situ acid-catalyzed ring opening of some epoxides ($n = 1$) gives oxazine derivatives; only in the case of sterically hindered derivatives are α-(trifluoromethyl)hydroxyproline derivatives obtained [93].

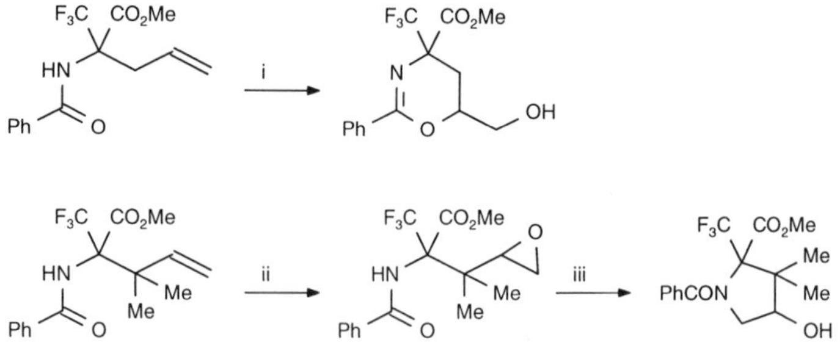

(i) *m*CPBA, CHCl₃, 0 °C → r.t. (72%, 50% d.e.); (ii) *m*CPBA, CHCl₃, 0 °C → r.t. (61%); (iii) HClO₄, H₂O, Et₂O, r.t. (60%)

The reaction of 3-hydroxypenta-1,4-dienes with 5-fluoro-4-(trifluoromethyl)oxazoles provides access to 3,3,3-trifluoroalanine derivatives with a diene substructure. *Inter alia*, the large synthetic repertoire of [4 + 2] and [4 + 1] cycloaddition reactions [213] can be applied in order to introduce carbocyclic and heterocyclic ring systems into the side-chain of 2-trifluoromethylamino acids. The alkyne cycloadduct (a ≡ b: dimethylacetylene dicarboxylate) and the naphthoquinone cycloadduct (a = b: naphthoquinone) aromatize on oxidation [181, 182].

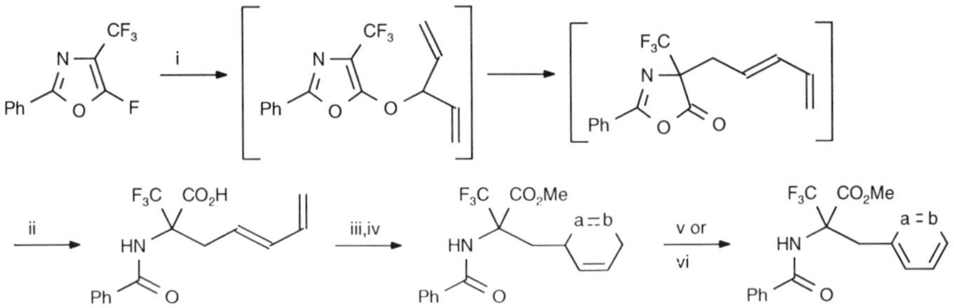

(i) (CH₂=CH)₂CHOH, KOH, dioxane, r.t.; (ii) H₂O, r.t. (89% for i, ii); (iii) CH₂N₂, Et₂O, r.t. (92%); (iv) a = b, see table; or a ≡ b, dimethylacetylene dicarboxylate, toluene, 100 °C; (v) a=b, naphthoquinone, O₂, r.t. (53% for iv, v); (vi) a≡b, dimethylacetylene dicarboxylate, DDQ, benzene, 80 °C (69% for iv, vi)

a=b	Solvent	T (°C)	Yield (iv) (%)	d.e. (%)	Ref.
Maleic anhydride	toluene	100	56	40	182
Tetracyanoethylene	toluene	r.t.	87	20	182
Azodicarboximide	CH₂Cl₂	r.t.	72	40	182
Diethyl azodicarboxylate	CH₂Cl₂	r.t.	65	50	182
Phthalazine-1,4-dione	CH₂Cl₂	0	65	90	182
Benzo[*g*]phthalazin-1,4-dione	CH₂Cl₂	0	60	90	182
Nitrosobenzene	CH₂Cl₂	r.t.	60	20	182

α-Trifluoromethylamino acids with a triple bond in the side-chain react with 1,3-dipoles (nitrile oxides, diazoalkanes, phenyl azide) in a regioselective manner to give derivatives of α-trifluoromethylamino acids with heterocyclic side-chains [214].

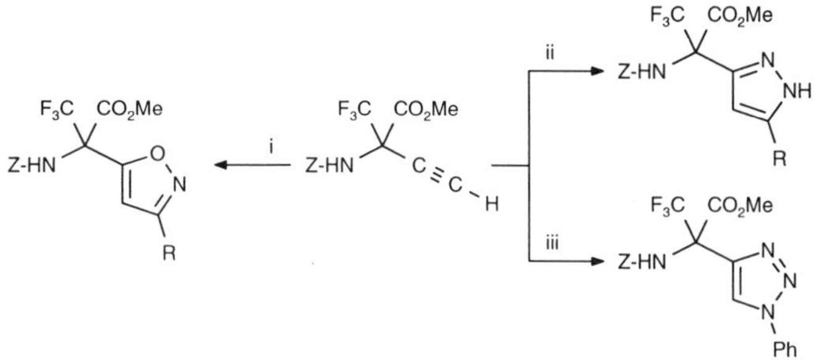

(i) RCX=NOH, NEt$_3$, Et$_2$O or NaHCO$_3$, EtOAc (R = Ph, X = Cl, 48%; R = p-ClC$_6$H$_4$, X = Cl, 39%; R = p-FC$_6$H$_4$, X = Cl, 56%; R = X = Br, 41%); (ii) RCHN$_2$, (R = H, Et$_2$O, r.t., 50%; R = CO$_2$Et, CHCl$_3$, 50 °C, 89%); (iii) PhN$_3$, THF, cat. DMF, reflux (79%)

Hydrostannation of the alkyne moiety gives (E)-vinylstannanes, which undergo palladium-catalyzed cross-coupling with acid chlorides according to the Stille protocol [188, 215]. Iododestannation with I$_2$ gives (E)-vinyl iodides.

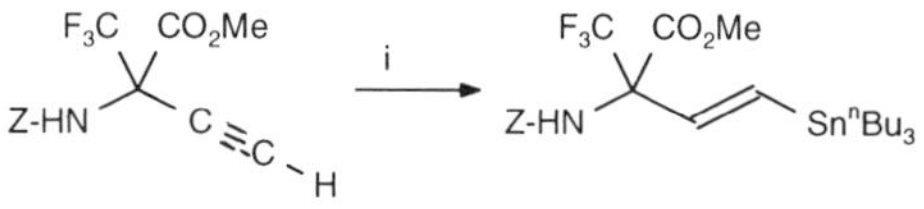

(i) Bu$^n_3$SnH, cat. AIBN, benzene, 60 °C (47%)

The alkyne group reacts with dicobalt octacarbonyl to give the stable alkyne–dicobalt complex. Complexes of this type are well known as preparatively valuable building blocks for the regio- and stereoselective synthesis of cyclopentenones. Application of the Pauson–Khand protocol [216] to the cobalt complex of the racemic α-trifluoromethylamino acid gives exclusively a diastereoisomeric mixture of a cyclopentenone-substituted α-trifluoro-methylamino acid derivative [188].

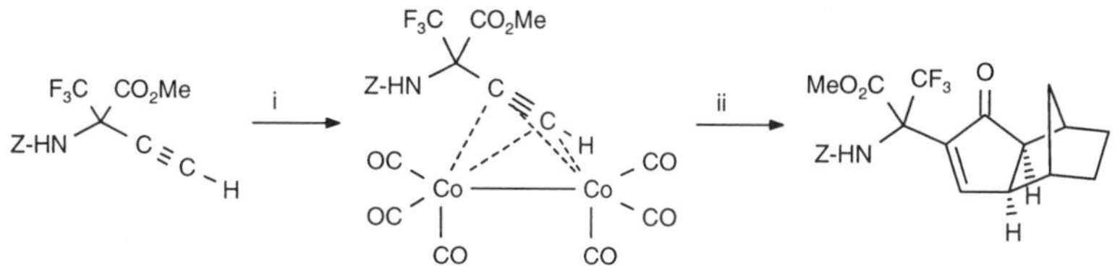

(i) Co$_2$(CO)$_8$, hexanes, r.t.; (ii) norbornene, CO, toluene, 100 °C (54% for i, ii)

Methyl 6-chloromethyl-5,6-dihydro-2,4-bis(trifluoromethyl)-1,3-oxazin-4-carboxylate, readily obtained via [4 + 2] cycloaddition of allyl chloride to methyl 2-(trifluoroacetylimino)-3,3,3-trifluoropropionate, undergoes ring contraction on methanolysis in the presence of acid. The overall reaction represents an elegant route to α-trifluoromethyl-4-hydroxyproline [217].

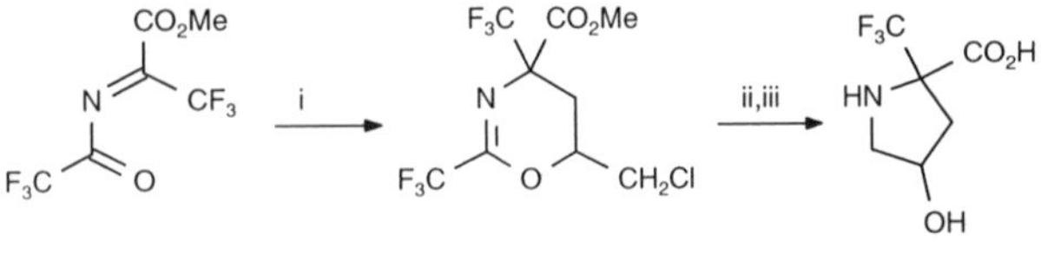

(i) CH_2=$CHCH_2Cl$, sealed tube, 90 °C; (ii) MeOH, cat. HCl/H_2O, r.t.; (iii) NaOH, H_2O (68% for ii, iii)

N-Alkylation of α-trifluoromethylamino acids

A preparatively efficient route to *N*-alkylated α-trifluoromethylamino acids is available via *N*-alkylation of the corresponding *N*-carboxy anhydrides (NCA, Leuchs anhydrides). They are obtained on treatment of a Z-protected α-trifluoromethylamino acid with thionyl chloride and represent *N*-protected, carboxylic group-activated derivatives of α-trifluoromethylamino acids. Deprotonation with sodium hydride and *N*-alkylation with alkyl halides offer a preparatively simple route to *N*-alkyl-α-trifluoromethylamino acids. The carboxylic group can be derivatized subsequently on reaction with nucleophiles [218].

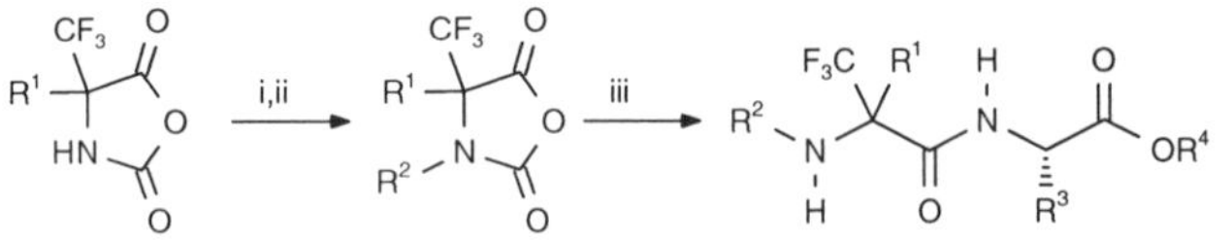

(i) NaH, DMF, r.t.; (ii) R^2I, DMF, r.t.; (iii) H-Xaa-OR^4, r.t.

R^1	R^2	R^3	R^4	Yield (i, ii) (%)	Yield (iii) (%)	Yield (i–iii) (%)	Ref.
Bzl	Me	Me	Bu^t			63	218
Bzl	Me	Bzl	Bu^t	56	71		218
Ph	Me	H	Me			49	218
Ph	Me	Me	Bu^t	42	87		218
Ph	CH_2CO_2Et	H	Bzl	49	77		218
Bu^i	Me	Bzl	Bu^t	47	95		218

4.3.4 PEPTIDE SYNTHESIS WITH α-TRIFLUOROMETHYLAMINO ACIDS

4.3.4.1 Protective group strategy

The first peptides containing α-trifluoromethylamino acids (e.g. with 3,3,3-trifluoroalanine, TFMGly) were described by Weygand *et al.* [90]. α-Trifluoromethylamino acids with orthogonal protective groups (Boc/Z and OMe) are obtained using N-(carbamoylimino)-3,3,3-trifluoropropionate as an electrophilic synthon (see Section 4.3.3.4). Alkaline hydrolysis with 1M KOH/methanol (1:1, v/v) gives the free carboxylic acids Boc-TFMXaa-OH or Z-TFMXaa-OH. Hydrogenolytic or acidolytic cleavage of the Z or the Boc group yields H-TFMXaa-OMe. However, the presence of the electron-withdrawing CF_3 substituent in the α-position exerts considerable electronic (pK_a) and steric effects on the reactivity of both the carboxylic and the amino groups. The low basicity of the amino group prevents the formation of dioxopiperazines; esters of α-trifluoromethylamino acids can even be distilled without oligomerization or decomposition [219].

Table 4.1. pK_a values of α-trifluoromethylamino acids compared with the natural amino acids [220]

	TFMXaa pK_a (CO_2H)	TFMXaa pK_a (NH_2)	Xaa pK_a (CO_2H)	Xaa pK_a (NH_2)
TFMGly	1.22	5.37	2.35	9.78
TFMAla	1.98	5.91	2.34	9.87
TFMVal	1.90	5.76	2.32	9.62
TFMPhe	1.90	5.25	2.58	9.24
TFMAsp	1.77	6.65	2.09	9.92

4.3.4.2 Amino group activation

Therefore, the activation of the amino group of α-trifluoromethylamino acids is very difficult to achieve. So far, satisfactory results have only been obtained with methyl α-(trifluoromethyl)alaninate (H-TFMAla-OMe, with mixed anhydrides or Fmoc-Yaa-Cl). For the bulkier α-trifluoromethylamino acids, all classical methods fail or result in substantial epimerization of the non-fluorinated amino acid. Peptide bond formation under drastic reaction conditions is useful only with substrates where epimerization is not possible [185].

PHT=Gly-Cl + H-TFMPhe-Gly-OMe $\xrightarrow{\text{i}}$ PHT=Gly-TFMPhe-Gly-OMe

(i) CH_2Cl_2, NEt_3, r.t (49%)

In some cases, dipeptides with C-terminal α-trifluoromethylamino acids can be obtained on reaction of PHT=Yaa—OH with isocyanates derived from α-trifluoromethylamino acids [218, 219].

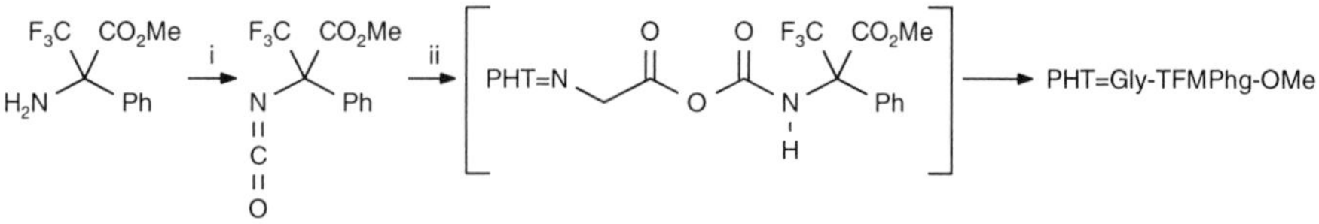

(i) CCl$_3$OCOCl, dioxane, 70 °C; (ii) PHT=Yaa—OH, toluene, pyridine, reflux (33%)

Isocyanates derived from amino acids [219] are valuable components for the synthesis of azapeptides, which are obtained in good yields on reaction of the isocyanate with amino acid hydrazides. Azatripeptides with α-trifluoromethylamino acids in N-terminal (H-TFMXaa-Agly-Yaa-OR), C-terminal (Z-Xaa-Agly-TFMYaa-OMe) or in both positions (H-TFMXaa-Agly-TFMYaa-OMe) can be synthesized [221].

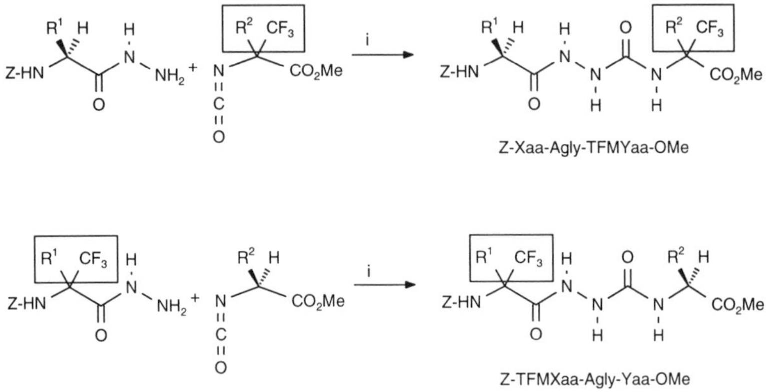

Z-Xaa-Agly-TFMYaa-OMe

Z-TFMXaa-Agly-Yaa-OMe

(i) CHCl$_3$, or Et$_2$O, 0 °C → r.t.

Similarly, α-trifluoromethylamino acids can subsequently be N-formylated and dehydrated to give the corresponding isonitriles [222], which can be used for the synthesis of tripeptides with C-terminal α-trifluoromethylamino acids via the Ugi reaction [93].

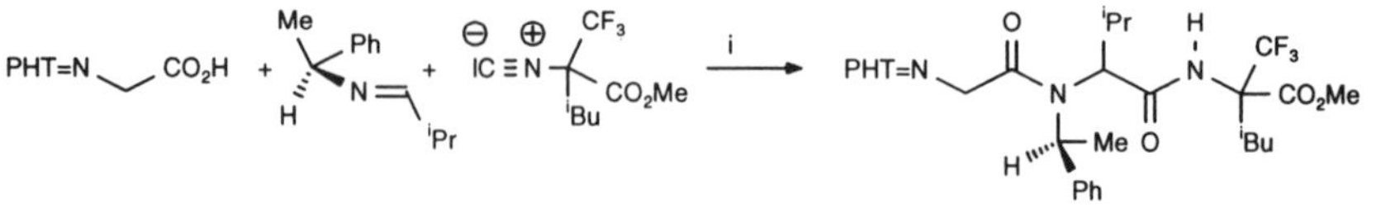

(i) MeOH, r.t.

4.3.4.3 Carboxylic group activation

Carboxylic group activation is achieved via the formation of mixed anhydrides with alkyl chloroformates [19] or of Leuchs anhydrides (N-carboxy anhydrides, NCA) [223]. The mixed anhydrides formed primarily cyclize spontaneously to the surprisingly stable oxazolones, which are also formed on treatment of Z-TFMXaa-OH with DCCI, even in the presence of HOBt. Epi- merization at the stage of the oxazolone, however, is not a problem with α,α- disubstituted amino acids, as there is no α-proton. Formation of dipeptides Z- TFMXaa-Yaa-OR occurs on ring opening of the oxazolones with amino acid esters H-Yaa-OR [19].

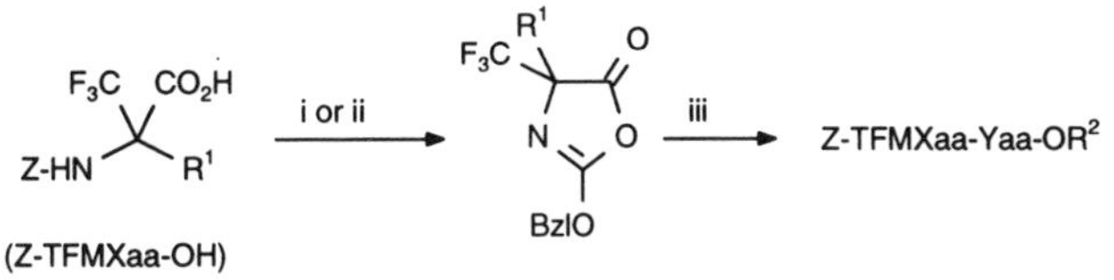

(i) EtO_2CCl, base; CH_2Cl_2, r.t.; (ii) DCCI, HOBt, CH_2Cl_2, r.t.; (iii) H-Yaa- OR^2, CH_2Cl_2, r.t.

The NCA derivatives of α-trifluoromethylamino acids are obtained in very good yields on heating Z-TFMXaa-OH with PCl_5, diphosgene or thionyl chloride [223].

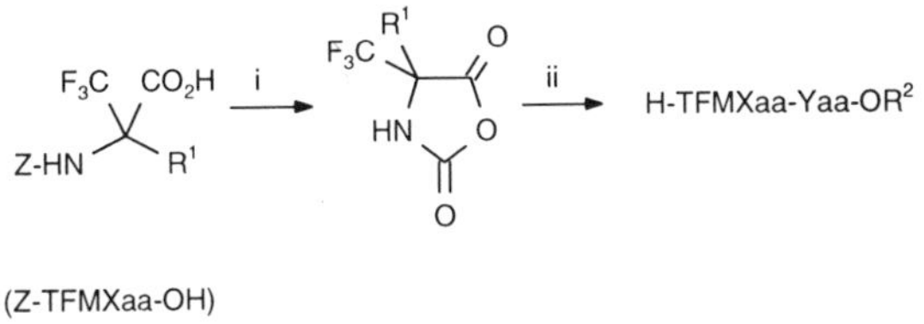

(i) $SOCl_2$, reflux; (ii) H-Yaa-OR^2, CH_2Cl_2, r.t.

The major disadvantage of the NCA method in classical peptide chemistry is the high tendency towards oligomerization, because the amino group of the peptide formed during the reaction can compete with the amino acid ester component. This problem does not arise on ring opening of NCAs derived from α-trifluoromethylamino acids owing to the low pK_a of the newly formed amino function [11].

4.3.4.4 Hydrolytic and proteolytic stablility of Z-TFMXaa-OMe - protease-catalyzed peptide synthesis with α-trifluoromethylamino acids

Z-TFMGly-OMe (methyl N-benzyloxycarbonyl-3,3,3-trifluoroalaninate) is very unstable at pH > 6. The presence of an α-proton severely destabilizes the CF_3 group and leads to sequential base-catalyzed elimination of hydrogen fluoride. All other α-trifluoromethylamino acids are lacking an α-proton and are therefore stable towards base. Their rate of alkaline ester hydrolysis (pH 9) is decreased considerably. After 20 min at pH 9, only 4% of Z-TFMPhe-OMe is hydrolyzed to the acid, whereas Z-Phe-OMe is hydrolyzed completely after 5 min under the same conditions. This shows that chemical hydrolysis is slowed by a factor of ca 12 on introduction of a CF_3 group in the α-position. Proteases such as subtilisin, α-chymotrypsin or papain accept α-trifluoromethylamino acids only to a very limited extent. Both the hydrolysis rate and the turnover decrease in the order Z-TFMGly-OMe > Z-TFMAla-OMe > Z-TFMPhe-OMe. The last amino acid is not turned over at all. These data exclude the application of enzyme catalyzed peptide synthesis to α-trifluoromethylamino acids. However, some dipeptide esters with N-terminal α-trifluoromethylamino acids are accepted as substrates by proteolytic enzymes. H-TFMPhg-Phe-OMe is converted by α-chymotrypsin or subtilisin within 20 min into the tripeptide H-TFMPhg-Phe-Leu-NH$_2$ in the presence of H-Leu-NH$_2$ [19].

4.3.5 MISCELLANEOUS

Ring expansion of perfluorinated α-lactams with carbonyl compounds or nitriles gives polyfluorinated imidazolinones or imidazolones, respectively. Acid-catalyzed ring opening yields derivatives of 2-amino-3,3,3′,3′,3′-hexafluoroisobutyrate [224].

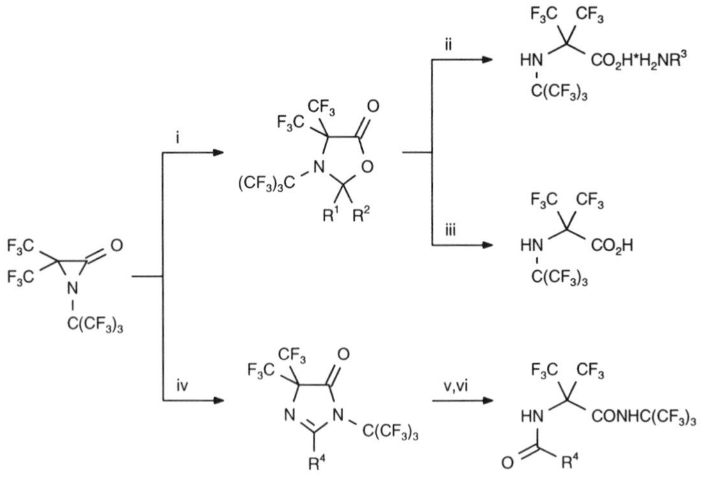

(i) R^1COR^2, 100 °C, sealed tube (R^1 = Ph, R^2 = H, 86%; R^1 = p-MeOC$_6$H$_4$, R^2 = H, 70%; R^1 = Me, R^2 = Me, 63%); (ii) R^3NH_2, 100 °C, sealed tube (R^1 = Ph, R^2 = H, R^3 = Ph, 59%; R^1 = p-MeOC$_6$H$_4$, R^2 = H, R^3 = Ph, 42%); (iii) NaOH, MeOH, 120 °C, sealed tube (53%); (iv) R^4CN, 100 °C, sealed tube (R^4 = Me, 96%; R^4 = Ph, 64%); (v) conc. H$_2$SO$_4$, r.t.; (vi) H$_2$O (R^4 = Me, 55% for v, vi)

A derivative of 2-amino-2H-perfluorooctanoic acid is described as a product of the boron trifluoride-catalyzed perfluorohexylation of a glyoxaldimine with perfluorohexyllithium, generated *in situ* from perfluorohexyl iodide and methyllithium/lithium bromide [225].

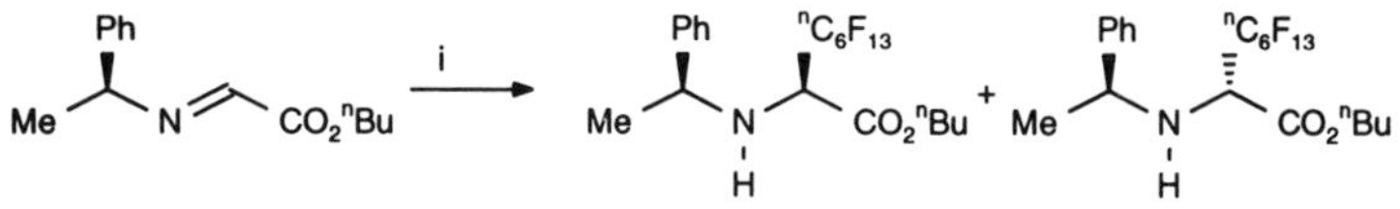

(i) $n\text{-}C_6F_{13}I$, $BF_3 * OEt_2$, MeLi, LiBr, Et_2O, $-78\,^\circ C$ (53%, 54% d.e.)

Benzyl 2-aryliminoperfluoroalkanoates are formed on palladium-catalyzed carbonylation of N-arylperfluoroimidoyl iodides in benzyl alcohol. Subsequent reductive alkylation with organolithium compounds provides a general access to 2-perfluoroalkylamino acids. Deblocking of the amino group is accomplished via ammonium cerium (IV) nitrate oxidation; the carboxylic group can be deprotected hydrogenolytically [119].

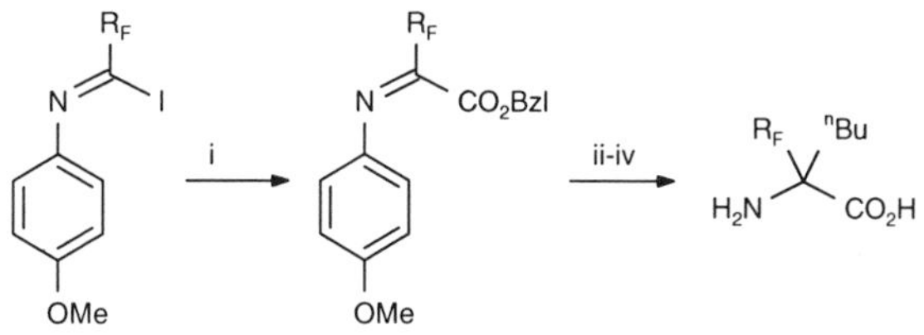

(i) $PhCH_2OH$, CO, cat. $Pd_2(dba)_3 * CHCl_3$, K_2CO_2, toluene, ($R_F = C_2F_5$, 94%, $R_F = C_7F_{15}$, 95%);
(ii) Bu^nLi, THF;
(iii) ammonium cerium(IV) nitrate, H_2O, CH_3CN;
(iv) H_2, Pd/C ($R_F = C_2F_5$, 79%; $R_F = C_7F_{15}$, 43% for ii, iii)

α-Fluoroalkyl derivatives of histidine are obtained via chlorine–fluorine exchange with silver fluoride [226].

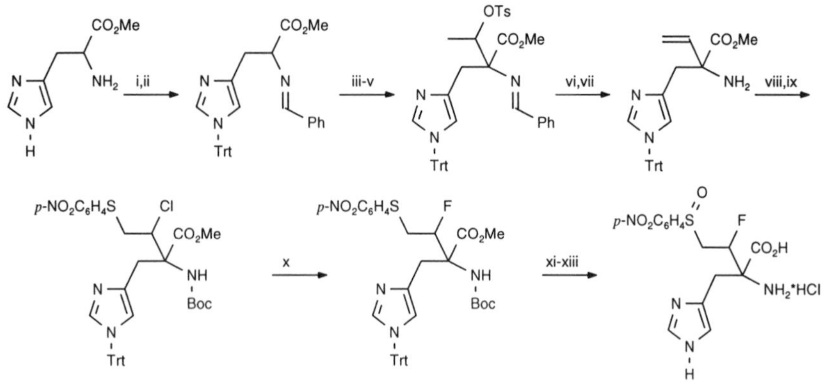

(i) PhCHO; (ii) tritylation; (iii) PhLi; (iv) CH_3CHO; (v) pTsCl, base; (vi) DBN; (vii) pTsOH, $NaHCO_3$; (viii) $(Boc)_2O$; (ix) p-$NO_2C_6H_4SCl$; (x) AgF; (xi) mCPBA; (xii) saponification; (xiii) deprotection

4.4 PERFLUORINATED AMINO ACIDS

Perfluoroglycyl fluoride, perfluoroalanyl fluoride, perfluoroaminoisobutyryl fluoride or their derivatives are obtained from terminal perfluoroalkenes [227–230] or bis(trifluoromethyl)ketene [231] with N-fluorinated hydrazine or hydroxylamine derivatives.

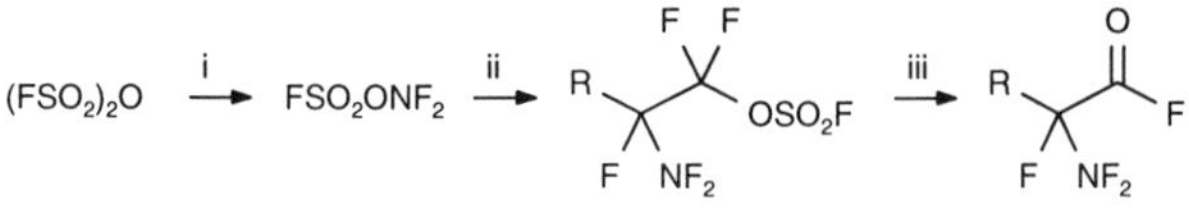

(i) N_2F_2; (ii) $RCF{=}CF_2$ (R = F, CF_3); (iii) KF

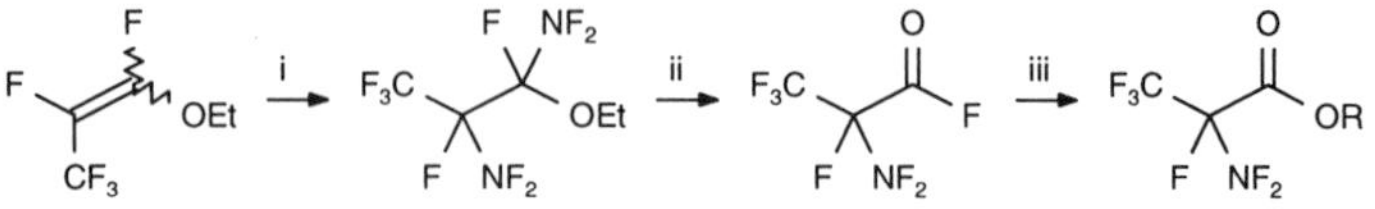

(i) N_2F_4 (80%); (ii) SbF_5 (72%); (iii) ROH (100%)

Electrochemical fluorination of N,N-disubstituted alanine derivatives affords among other products the corresponding perfluorinated amino acid fluorides [232]. In contrast to the electrofluorination of N,N-dimethyl derivatives [233], no cyclized byproducts are observed. Aminolysis of the acid fluorides with (−)-phenylethylamine allows the resolution of the diastereoisomers and optically active perfluoro derivatives of alanine are obtained on cleavage of the amide with sodium hydroxide [234].

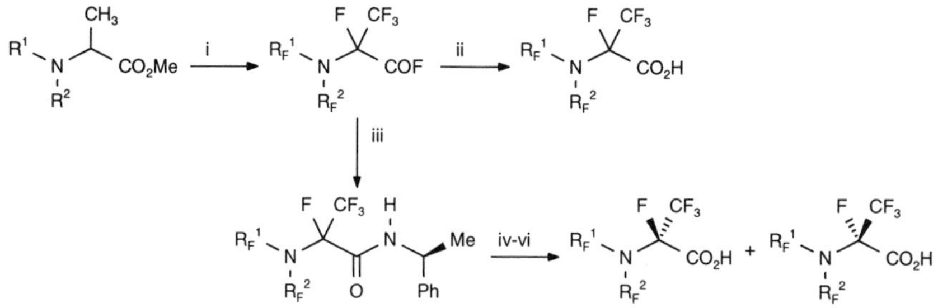

(i) electrofluorination; (ii) H_2O; (iii) $(-)$-PhCH(Me)NH$_2$, NEt$_3$, MeCN, r.t.; (iv) chromatographic resolution; (v) H_2SO_4, r.t.; (vi) 6M NaOH, reflux

R_F^1	R_F^2	Yield (i) (%)	Ref.
$-(CF_2)_4-$		20	232
$-(CF_2)_2O(CF_2)_2-$		14	232
$-(CF_2)_5-$		14	232
$-(CF_2)_6-$		21	232
$-(CF_2)_2(NCF_3)(CF_2)_2-$		3	232

4.5 REFERENCES

1. F. Weygand and W. Oettmeier, *Usp. Khim.*, **39**, 622 (1970); *Russ. Chem. Rev.*, **339**, 290 (1970), and references cited therein.
2. D. F. Loncrini and R. Filler, *Adv. Fluorine Chem.*, **6**, 43 (1970), and references cited therein.
3. J. Kollonitsch, in R. Filler and Y. Kobayashi (Eds), *Biomedicinal Aspects of Fluorine Chemistry,* Kodansha, Tokyo, and Elsevier Biomedical Press, Amsterdam, 1982, pp. 93*ff*, and references cited therein.
4. J. T. Welch, *Tetrahedron*, **43**, 3123 (1987), and references cited therein.
5. V. P. Kukhar', Y. L. Yagupol'skii and V. A. Soloshonok, *Usp. Khim.*, **59**, 149 (1990); *Russ. Chem. Rev.*, **59**, 89 (1990), and references cited therein.
6. K. Burger, E. Höss, K. Gaa, N. Sewald and C. Schierlinger, *Z. Naturforsch. Teil B,* **46**, 361, (1991).
7. J. T. Welch and S. Eswarakrishnan, *Fluorine in Bioorganic Chemistry,* Wiley, New York, 1991, pp. 7*ff*, and references cited therein.
8. N. Ishikawa (Ed.), *Synthesis and Reactivity of Fluorocompounds,* CMC, Tokyo, 1987, and references cited therein.
9. J. T. Welch and S. Eswarakrishnan, *Fluorine in Bioorganic Chemistry*, Wiley, 1991, and references cited therein.
10. D. A. Dixon and B. E. Smart, in J. T. Welch (Ed.), *Selective Fluorination in Organic and Bioorganic Chemistry,* ACS Symposium Series, No. 456, American Chemical Society, Washington, DC, 1991, pp. 18*ff* and references cited therein.
11. S. P. Kobzev, V. A. Soloshonok, S. V. Galushko, Y. L. Yagupol'skii and V. P. Kukhar', *Zh. Obshch. Khim.*, **59**, 909 (1989); *J. Gen. Chem. USSR,* **59**, 801 (1989).
12. T. Nagai, G. Nishioka, M. Koyama, A. Ando, T. Miki and I. Kumadaki, *J. Fluorine Chem.,* **57**, 229 (1992).

13. D. Seebach, *Angew. Chem.*, **102**, 1363 (1990); *Angew. Chem., Int. Ed. Engl.*, **29**, 1320 (1990), and references cited therein.
14. T. Fujita, *Prog. Phys. Org. Chem.*, **14**, 75 (1983).
15. N. Muller, *J. Pharm. Sci.*, **75**, 987 (1986).
16. J. T. Welch, in J. T. Welch (Ed.), *Selective Fluorination in Organic and Bioorganic Chemistry*, ACS Symposium Series, No. 456, American Chemical Society, Washington, DC, 1991, and references cited therein.
17. T. Kitazume and T. Ohnogi, *Synthesis* 614 (1988).
18. F. Weygand, W. Steglich, I. Lengyel, F. Fraunberger, A. Maierhofer and W. Oettmeier, *Chem. Ber.*, **99**, 1944 (1966).
19. K. Burger, K. Mütze, W. Hollweck, B. Koksch, P. Kuhl, H.-D. Jakubke, J. Riede and A. Schier, *J. Prakt. Chem.*, **335**, 321, 1993.
20. J. T. Welch and S. Eswarakrishnan, *Fluorine in Bioorganic Chemistry*, Wiley, New York, 1991, pp. 54*ff*, and references cited therein.
21. D. H .G. Versteeg, M. Palkouts, J. Van der Gugten, H. J. L. M. Wijnen, G. W. M. Smeets and W. DeJong, *Prog. Brain Res.*, **47**, 111 (1977).
22. W. W. Douglass, in L. S. Goodman and A. Gilman (Eds), *The Pharmacological Basis of Therapeutics*, MacMillan, New York, 1975, pp. 590*ff*.
23. D. H. Russell, in D. H. Russell (Ed.), *Polyamines in Normal and Neoplastic Growth*, Raven, New York, 1973, pp. 1*ff*.
24. R. H. Abeles and A. L. Maycock, *Acc. Chem. Res.*, **9**, 313 (1976).
25. R. Dagani, *Chem. Eng. News*, **66** (33), 26 (1988).
26. V. Tolman and K. Veres, *Tetrahedron Lett.*, 3909 (1966).
27. V. Tolman and K. Veres, *Collect. Czech. Chem. Commun.*, **32**, 4460 (1967).
28. L. V. Alekseeva, B. N. Lundin and N. L. Burde, *Zh. Obshch. Khim.*, **37**, 1754 (1967); *J. Gen. Chem. USSR*, **37**, 1671 (1967).
29. J. Kollonitsch and L. Barash, *J. Am. Chem. Soc.*, **98**, 5591 (1976).
30. J. Kollonitsch, *US Pat.*, 4 030 994 (1977); *Chem. Abstr.*, **87**, 83904 (1977).
31. J. Kollonitsch, L. Barash and G. A. Doldouras, *J. Am. Chem. Soc.*, **92**, 7494 (1970).
32. X. Huang, B. J. Blackburn, S. C. F. Au-Yeung and A. F. Janzen, *Can. J. Chem.*, **68**, 477 (1990).
33. A. W. Douglas and P. J. Reider, *Tetrahedron Lett.*, **25**, 2851 (1984).
34. J. Kollonitsch, S. Marburg and L. M. Perkins, *J. Org. Chem.*, **44**, 771 (1979).
35. P. J. Reider, R. S. Conn, P. Davis, V. J. Grenda, A. J. Zambito and E. J. Grabowski, *J. Org. Chem.*, **52**, 3326 (1987).
36. A. M. Stern, B. M. Foxman, A. H. Tashjian, Jr, and R. H. Abeles, *J. Med. Chem.*, **25**, 544 (1982).
37. R. S. Loy and M. Hudlicky, *J. Fluorine Chem.*, **7**, 421 (1976).
38. A. Vidal-Cros, M. Daudry and A. Marquet, *J. Org. Chem.*, **50**, 3163 (1985).
39. U. Groth and U. Schöllkopf, *Synthesis* 673 (1983).
40. S. V. Pansare and J. C. Vederas, *J. Org. Chem.*, **52**, 4804 (1987).
41. L. Somekh and A. Shanzer, *J. Am. Chem. Soc.*, **104**, 5836 (1982).
42. M. H. Gelb, Y. Lin, M. A. Pickard, Y. Song and J. C. Vederas, *J. Am. Chem. Soc.*, **112**, 4932 (1992).
43. T. Tsushima, T. Sato and T. Tsuji, *Tetrahedron Lett.*, **21**, 3591 (1980).
44. A. A. Gottlieb, Y. Fujita, S. Udenfriend and B. Witkop, *Biochemistry*, **4**, 2507 (1965).
45. A. Cohen and E. D. Bergman, *Tetrahedron*, **22**, 3545 (1966).
46. E. D. Bergman and A. Cohen, *Isr. J. Chem.*, **8**, 925 (1970).
47. J. Kollonitsch, S. Marburg and L. M. Perkins, *J. Org. Chem.*, **41**, 3107 (1976).
48. C.-Y. Yuan, C.-N. Chang and I.-F. Yeh, *Yao Hsueh Pao*, **7**, 237 (1959); *Chem. Abstr.*, **54**, 12096 (1960).

49. L. V. Alekseeva, N. L. Burde and B. N. Lundin, *Zh. Obshch. Khim.*, **38**, 1687 (1968); *J. Gen. Chem. USSR*, **38**, 1645 (1968).
50. *Jpn. Pat.*, 81 122 337 (1981); *Chem. Abstr.*, **96**, 85987 (1981).
51. M. J. Wanner, J. J. M. Hageman, G.-J. Koomen and U. K. Pandit, *J. Med. Chem.*, **23**, 85 (1980).
52. T. Tsushima, K. Kawada, J. Nishikawa, T. Sato, K. Tori, T. Tsuji and S. Misaki, *J. Org. Chem.*, **49**, 1163 (1984).
53. T. Tsushima, J. Nishikawa, T. Sato, H. Tanida, K. Tori, T. Tsuji, S. Misaki and M. Suefuji, *Tetrahedron Lett.*, **21**, 3593 (1980).
54. J. W. Cornforth, R. H. Cornforth and K. K. Mathew, *J. Chem. Soc.*, 112 (1959).
55. M. Cherest and H. Felkin, *J. Chem. Soc.*, 2205 (1968).
56. D. F. Reinhold, *Fr. Pat.*, 2 170 180 (1974); *Chem. Abstr.*, **80**, 83648 (1974).
57. U. H. Dolling, A. W. Douglas, E. J. Grabowski, E. F. Schoenewaldt, P. Sohar and M. Sletzinger, *J. Org. Chem.*, **43**, 1634 (1978).
58. G. Gal, J. M. Chemerda, D. F. Reinhold and R. M. Purick, *J. Org. Chem.*, **42**, 142 (1977).
59. A. Vidal-Cros, M. Gaudry and A. Marquet, *J. Org. Chem.*, **54**, 498 (1989).
60. M. Akhtar, N. P. Botting, M. A. Cohen and D. Gani, *Tetrahedron*, **43**, 5899 (1987).
61. M. Hudlicky, *J. Fluorine Chem.*, **40**, 99 (1988).
62. H. Gershon, M. W. McNeil and E. D. Bergmann, *J. Med. Chem.*, **16**, 1407 (1973).
63. H. Lettre and U. Wolcke, *Liebigs Ann. Chem.*, **708**, 75 (1967).
64. I. I. Gerus, Y. L. Yagupol'skii, V. P. Kukhar', L. S. Boguslavskaya, N. N. Chuvatkin, A. V. Kartashov and Y. V. Mitin, *Zh. Org. Khim.*, **27**, 537 (1991); *J. Org. Chem. USSR*, **27**, 465 (1991).
65. A. I. Ayi and R. Guedj, *J. Fluorine Chem.*, **24**, 137 (1984).
66. A. I. Ayi, M. Remli and R. Guedj, *Tetrahedron Lett.*, **22**, 1505 (1981).
67. M. J. O'Donnell, C. L. Barney and J. R. McCarthy, *Tetrahedron Lett.*, **26**, 3067 (1985).
68. T. Tsushima and K. Kawada, *Tetrahedron Lett.*, **26**, 2445 (1985).
69. A. L. Castelhano, S. Horne, R. Billedeau and A. Krantz, *Tetrahedron Lett.*, **27**, 2435 (1986).
70. A. L. Castelhano, S. Horne, G. J. Taylor, R. Billedeau and A. Krantz, *Tetrahedron*, **44**, 5451 (1988).
71. T. N. Wade and R. Khéribet, *J. Chem. Res. (S)*, 210 (1980).
72. A. I. Ayi and R. Guedj, *J. Chem. Soc., Perkin Trans. 1*, 2045 (1983).
73. A. I. Ayi, M. Remli and R. Guedj, *J. Fluorine Chem.*, **18**, 93 (1981).
74. T. N. Wade, *J. Org. Chem.*, **45**, 5328 (1980).
75. T. N. Wade, F. Gaymard and R. Guedj, *Tetrahedron Lett.*, 2681 (1979).
76. A. Barama, R. Condom and R. Guedj, *J. Fluorine Chem.*, **16**, 183 (1980).
77. M. L. M. Alcaniz, N. Patino, R. Condom, A. I. Ayi and R. Guedj, *J. Fluorine Chem.*, **35**, 70 (1980).
78. K. Matsumoto, Y. Osaki, T. Iwasaki, H. Horikawa and M. Miyoshi, *Experientia*, **35**, 850 (1979).
79. J. M. Hageman, M. J. Wanner, G.-J. Koomen and U. K. Pandit, *J. Med. Chem.*, **20**, 1677 (1977).
80. T. Tsushima, K. Kawada, S. Ichihara, N. Uchida, O. Shiratori, J. Higaki and M. Hirata, *Tetrahedron*, **44**, 5375 (1988).
81. J. P. Whitten, C. L. Barney, E. W. Huber, P. Bey and J. R. McCarthy, *Tetrahedron Lett.*, **30**, 3649 (1989).
82. H. d'Orchymont, *Synthesis*, 961 (1993).
83. J. E. Baldwin, G. P. Lynch and C. J. Schofield, *J. Chem. Soc., Chem. Commun.*, 736 (1991).

84. T. N. Wade and R. Guedj, *Tetrahedron Lett.*, 3953 (1979).
85. T. N. Wade and R. Khéribet, *J. Org. Chem.*, **45**, 5333 (1980).
86. B. P. Hart and J. C. Coward, *Tetrahedron Lett.*, **34**, 4917 (1993).
87. F. Weygand and W. Steglich, *Chem. Ber.*, **98**, 487 (1965).
88. F. Weygand, W. Steglich and F. Fraunberger, *Angew. Chem.*, **79**, 822 (1967); *Angew. Chem., Int. Ed. Engl.*, **6**, 807 (1967).
89. A. Uskert, A. Neder and E. Kasztreiner, *Magy. Kem. Fol.*, **79**, 333 (1973); *Chem. Abstr.*, **79**, 79147 (1973).
90. F. Weygand, W. Steglich, W. Oettmeier, A. Maierhofer and R. S. Loy, *Angew. Chem.*, **78**, 640 (1966); *Angew. Chem., Int. Ed. Engl.*, **5**, 600 (1966).
91. F. Weygand, W. Steglich and W. Oettmeier, *Chem. Ber.*, **103**, 818 (1970).
92. F. Weygand, W. Steglich and W. Oettmeier, *Chem. Ber.*, **103**, 1655 (1970).
93. K. Mütze, *PhD Thesis*, Technische Universität München (1993).
94. G. Höfle and W. Steglich, *Chem. Ber.*, **104**, 1408 (1971).
95. W. Steglich and G. Höfle, *Tetrahedron Lett.*, 4727 (1970).
96. C. A. Hendrick and B. A. Gracia, *Ger. Offen.*, 2 812 169 (1979); *Chem. Abstr.*, **90**, 122072 (1979).
97. Y. L. Yagupol'skii, V. A. Soloshonok and V. P. Kukhar', *Zh. Org. Khim.*, **22**, 517 (1986); *J. Org. Chem. USSR*, **22**, 459 (1986).
98. W. Steglich, K. Burger, M. Dürr and E. Burgis, *Chem. Ber.*, **107**, 1488 (1974).
99. N. P. Gambaryan, E. M. Rokhlin, Y. V. Zeifman, C. Ching-Yun and I. L. Knunyants, *Angew. Chem.*, **78**, 1008 (1966); *Angew. Chem., Int. Ed. Engl.*, **5**, 947 (1966).
100. K. Burger, K. Geith and D. Hübl, *Synthesis*, 189 (1988).
101. K. Burger, K. Geith and D. Hübl, *Synthesis*, 194 (1988).
102. K. Burger, K. Geith and D. Hübl, *Synthesis*, 199 (1988).
103. K. Burger and B. Helmreich, *Chem.-Ztg.*, **115**, 253 (1991).
104. K. Burger, K. Geith and N. Sewald, *J. Fluorine Chem.*, **46**, 105 (1990).
105. K. Burger, D. Hübl and P. Gertitschke, *J. Fluorine Chem.*, **27**, 327 (1985).
106. K. Burger, K. Geith and E. Höss, *Synthesis*, 352 (1990).
107. H. Millauer, W. Schwertfeger and G. Siegemund, *Angew. Chem.*, **97**, 164 (1985); *Angew. Chem., Int. Ed. Engl.*, **24**, 161 (1985).
108. I. L. Knunyants, V. V. Shokina and V. V. Tyuleneva, *Dokl. Akad. Nauk SSSR, Ser. Khim.*, **169**, 594 (1966); *Proc. Acad. Sci. USSR, Chem. Sect.*, **169**, 722 (1966).
109. A. Pasetti and D. Sianesi, *Gazz. Chim. Ital.*, **98**, 265 (1968).
110. A. Pasetti, F. Tarli and D. Sianesi, *Gazz. Chim. Ital.*, **98**, 277 (1968).
111. D. Sianesi, A. Pasetti and F. Tarli, *J. Org. Chem.*, **31**, 2312 (1966).
112. V. I. Saloutin, K. I. Pashkevich and I. Y. Postovskii, *Usp. Khim.*, **51**, 1287 (1982); *Russ. Chem. Rev.*, **51**, 736 (1982).
113. C. Francese, M. Tordeux and C. Wakselman, *Tetrahedron Lett.*, **29**, 1029 (1988).
114. V. Broicher and D. Geffken, *Tetrahedron Lett.*, **30**, 5243 (1989).
115. V. A. Soloshonok, Y. L. Yagupol'skii and V. P. Kukhar', *Zh. Org. Khim.*, **24**, 1638 (1988); *J. Org. Chem. USSR*, **24**, 1478 (1988).
116. A. E. Zelenin, N. D. Chkanikov, A. F. Kolomiets and A. V. Fokin, *Izv. Akad. Nauk SSSR, Ser. Khim.*, 231 (1987); *Bull. Acad. Sci. USSR, Chem. Sci.*, 209 (1987).
117. V. A. Soloshonok, I. I. Gerus and Y. L. Yagupol'skii, *Zhur. Org. Khim.*, **22**, 1335 (1986); *J. Org. Chem. USSR*, **22**, 1204 (1986).
118. V. A. Soloshonok, I. I. Gerus, Y. L. Yagupol'skii and V. P. Kukhar', *Zh. Org. Khim.*, **23**, 2308 (1987); *J. Org. Chem. USSR*, **23**, 2034 (1987).
119. H. Watanabe, Y. Hashizume and K. Uneyama, *Tetrahedron Lett.*, **33**, 4333 (1992).
120. K. Burger and N. Sewald, *Ger. Offen.*, 3 917 836 (1990); *Chem. Abstr.*, **114**, 247793 (1991).

121. K. Burger, E. Höss and K. Gaa, *Chem.-Ztg.*, **113**, 243 (1989).
122. K. Burger and K. Gaa, *Chem.-Ztg.*, **114**, 101 (1990).
123. I. Malassa and D. Matthies, *Chem.-Ztg.*, **111**, 253 (1987).
124. S. N. Osipov, A. F. Kolomiets and A. V. Fokin, *Izv. Akad. Nauk SSSR, Ser. Khim.*, 132 (1986); *Bull. Acad. Sci. USSR, Chem. Sect.*, 122 (1986).
125. S. N. Osipov, N. D. Chkanikov, A. F. Kolomiets and A. V. Fokin, *Izv. Akad. Nauk SSSR, Ser. Khim.*, 1384 (1986); *Bull. Acad. Sci. USSR, Chem. Sect.*, 1256 (1986).
126. S. N. Osipov, N. D. Chkanikov, A. F. Kolomiets and A. V. Fokin, *Izv. Akad. Nauk SSSR, Ser. Khim.*, 1648 (1989); *Bull. Acad. Sci. USSR, Chem. Sect.*, 1512 (1989).
127. D. Matthies and S. Siewers, *Liebigs Ann. Chem.*, 159 (1992).
128. A. B. Khotkevich, V. A. Soloshonok and Y. L. Yagupol'skii, *Zh. Obshch. Khim.*, **60**, 1005 (1990); *J. Gen. Chem. USSR*, **60**, 885 (1990).
129. E. Höss, M. Rudolph, L. C. Seymour, C. Schierlinger and K. Burger, *J. Fluorine Chem.*, **61**, 163 (1993).
130. K. Burger, in J. I. G. Cadogan (Ed.), *Organophosphorus Reagents in Organic Synthesis*, Academic Press, London, 1979, pp. 467ff.
131. K. Burger, H. Goth, K. Einhellig and A. Gieren, *Z. Naturforsch., Teil B*, **36**, 345 (1981).
132. R. Huisgen, *Angew. Chem.*, **92**, 979 (1980); *Angew. Chem., Int. Ed. Engl.*, **19**, 947 (1980).
133. E. C. Taylor and I. J. Turchi, *Chem. Rev.*, **79**, 181 (1979).
134. E. Höss, *PhD Thesis*, Technische Universität München (1990).
135. G. R. Marshall, J. D.Clarc, J. B. Dunbar, Jr, G. D. Smith, J. Zabrocki, A. S. Redlinski and M. T. Leplawy, *Int. J. Pept. Protein. Res.*, **32**, 544 (1988).
136. J. Kollonitsch, *Dutch Pat.*, 75 14 240 (1976); *Chem. Abstr.*, **86**, 44236 (1977).
137. Z. K. Kaminski, M. T. Leplawy and J. Zabrocki, *Synthesis*, 792 (1973).
138. J. Kollonitsch, A. A. Patchett, S. Marburg, A. L. Maycock, L. M. Perkins, G. A. Doldouras, D. E. Duggan and S. D. Aster, *Nature (London)*, **274**, 906 (1978).
139. J. Kollonitsch, A. A. Patchett and S. Marburg, *S. Afr. Pat.*, 78 3 121 (1979); *Chem. Abstr.*, **93**, 8515 (1980).
140. J. Kollonitsch and S. Marburg, *US Pat.*, 4 215 221 (1980); *Chem. Abstr.*, **94**, 16081 (1981).
141. J. Kollonitsch, L. Perkins, G. A. Doldouras and S. Marburg, *US Pat.*, 4 347 347 (1982); *Chem. Abstr.*, **98**, 17049 (1983).
142. D. E. Zembower, J. A. Gilbert and M. M. Ames, *J. Med. Chem.*, **36**, 305 (1993).
143. N. Devi, M. D. Threadgill and S. J. B. Tendler, *Synth. Commun.*, **18**, 1545 (1988).
144. A. W. Guest, P. H. Milner and R. Southgate, *Tetrahedron Lett.*, **30**, 5791 (1989).
145. P. Bey and J. P. Vevert, *Tetrahedron Lett.*, 1215 (1978).
146. P. Bey, J. P. Vevert, V. Van Dorsselaer and M. Kolb, *J. Org. Chem.*, **44**, 2732 (1979).
147. P. Bey, M. Jung, *US Pat.*, 4 743 691 (1988); *Chem. Abstr.*, **110**, 76067 (1989).
148. P. Bey, J. B. Ducep and D. Schirlin, *Tetrahedron Lett.*, **25**, 5657 (1984).
149. D. Schirlin, J. B. Ducep, S. Batzer, P. Bey, F. Piriou, J. Wagner, J. M. Hornsperger, J. G. Heydt, M. J. Jung, C. Danzin, R. Weiss, J. Fischer, A. Mitschler and A. DeCian, *J. Chem. Soc., Perkin Trans. I*, 1053 (1992).
150. K. G. Grozinger, R. W. Kriwacki, S. F. Leonard and T. P. Pitner, *J. Org. Chem.*, **58**, 709 (1993).
151. P. Bey and M. Jung, *Ger. Offen.*, 3 012 581 (1980); *Chem. Abstr.*, **95**, 25625 (1981).
152. P. Bey and M. Jung, *Eur. Pat.*, 40 150 (1981); *Chem. Abstr.*, **96**, 143335 (1982).
153. P. Bey and M. Jung, *Ger. Offen.*, 3 012 602 (1980); *Chem. Abstr.*, **94**, 140164 (1981).

154. P. Bey, F. Gerhart, V. Van Dorsselaer and C. Danzin, *J. Med. Chem.*, **26**, 1551 (1983).
155. F. Gerhart, *Eur. Pat.*, 46 710 (1982); *Chem. Abstr.*, **97**, 34498 (1982).
156. I. Van Assche, A. Haemers and M. Hooper, *Eur. J. Med. Chem.*, **26**, 363 (1991).
157. F. Gerhart, W. Higgins, C. Tardif and J.-B. Ducep, *J. Med. Chem.*, **33**, 2157 (1990).
158. E. D. Bergman and A. Shani, *J. Chem. Soc.*, 3462 (1963).
159. H. N. Christensen and D. L. Oxender, *Biochim. Biophys. Acta*, **74**, 386 (1963).
160. D. Kuo and R. R. Rando, *Biochemistry*, **20**, 506 (1981).
161. L. Van Hijfte, V. Heydt and M. Kolb, *Tetrahedron Lett.*, **34**, 4793 (1993).
162. D. Schirlin, F. Gerhart, J. M. Hornsperger, M. Hamon, J. Wagner and M. J. Jung, *J. Med. Chem.*, **31**, 30 (1988).
163. B. W. Metcalf, P. Bey, C. Danzin, M. J. Jung, P. Casara and J. P. Vevert, *J. Am. Chem. Soc.*, **100**, 2551 (1978).
164. J. G. Kelland, L. D. Arnold, M. M. Palcic, M. A. Pickard and J. C. Vederas, *J. Biol. Chem.*, **261**, 13216 (1986).
165. P. Bey and M. Jung, *US Pat.*, 4 413 141 (1977); *Chem. Abstr.*, **100**, 103892 (1984).
166. M. Kolb and J. Barth, *Tetrahedron Lett.*, 2999 (1979).
167. M. Kolb and J. Barth, *Liebigs Ann. Chem.*, 1668 (1983).
168. P. Bey and D. Schirlin, *Tetrahedron Lett.*, 5225 (1978).
169. J. Kollonitsch and A. A. Patchett, *Eur. Pat.*, 7 600 (1980); *Chem. Abstr.*, **93**, 11497 (1980).
170. H. Mettler and E. Greth, *Swiss Pat.*, 672 124 (1989); *Chem. Abstr.*, **112**, 178093 (1990).
171. A. T. Au and N. L. Boardway, *Eur. Pat.*, 357 029 (1990); *Chem. Abstr.*, **113**, 38911 (1990).
172. P. Bey and M. Jung, *Belg. Pat.*, 868 882 (1978); *Chem. Abstr.*, **90**, 187335 (1979).
173. J. W. Keller and K. O. Dick, *J. Chromatogr.*, **367**, 187 (1986).
174. J. G. Dingwall, *Eur. Pat.*, 298 029 (1989); *Chem. Abstr.*, **111**, 115172 (1989).
175. M. Sletzinger, N. Plainfield and W. A. Gaines, *US Pat.*, 3 046 300 (1962); *Chem. Abstr.*, **57**, 16740 (1960).
176. J. F. Lontz and M. S. Raasch, *US Pat.*, 2 662 915 (1953); *Chem. Abstr.*, **48**, 12795 (1954).
177. J. W. Keller and B. J. Hamilton, *Tetrahedron Lett.*, **27**, 1249 (1986).
178. K. Burger, *Actual. Chim.*, 168 (1987).
179. K. Burger, K. Geith and K. Gaa, *Angew. Chem.*, **100**, 860 (1988); *Angew. Chem., Int. Ed. Engl.*, **27**, 848 (1988).
180. K. Burger, K. Gaa, K. Geith and C. Schierlinger, *Synthesis*, 850 (1989).
181. K. Gaa, *PhD Thesis*, Technische Universität München (1990).
182. K. Burger, K. Gaa and K. Mütze, *Chem.-Ztg.*, **115**, 292 (1991).
183. K. Burger, C. Schierlinger, K. Gaa, K. Geith, N. Sewald and G. Müller, *Chem.-Ztg.*, **113**, 277 (1989).
184. K. Burger, K. Gaa and E. Höss, *J. Fluorine Chem.*, **47**, 89 (1990).
185. N. Sewald, W. Hollweck, K. Mütze, C. Schierlinger, L. C. Seymour, K. Gaa, K. Burger, B. Koksch and H.-D. Jakubke, *Amino Acids*, in press.
186. K. Burger and N. Sewald, *Synthesis*, 115 (1990).
187. C. Schierlinger, *PhD Thesis*, Technische Universität München (1991).
188. N. Sewald, K. Gaa and K. Burger, *Heteroatom Chem.*, **4**, 253 (1993).
189. K. Burger, K. Geith and K. Gaa, *J. Fluorine Chem.*, **41**, 429 (1988).
190. M. Riediker and R. O. Duthaler, *Angew. Chem.*, **101**, 488 (1989); *Angew. Chem., Int. Ed. Engl.*, **28**, 494 (1989).
191. N. Sewald, L. C. Seymour, K. Burger, S. N. Osipov, A. V. Kolomiets and A. F. Fokin, *Tetrahedron: Asymmetry*, **5**, 1051 (1994).

4. SYNTHESIS OF β-FLUORINE-CONTAINING AMINO ACIDS 219

192. S. N. Osipov, N. D. Chkanikov, A. F. Kolomiets and A. V. Fokin, *Izv. Akad. Nauk SSSR, Ser. Khim.*, 213 (1989); *Bull. Acad. Sci. USSR, Chem. Sect.*, 201 (1989).
193. S. N. Osipov, V. B. Sokolov, A. F. Kolomiets, I. V. Martynov and A. V. Fokin, *Izv. Akad. Nauk SSSR, Ser. Khim.*, 1185 (1987); *Bull. Acad. Sci. USSR, Chem. Sect.*, 1098 (1987).
194. N. Sewald, J. Riede, P. Bissinger and K. Burger, *J. Chem. Soc., Perkin Trans. I*, 267 (1992).
195. W. Oppolzer and V. Snieckus, *Angew. Chem.*, **90**, 506 (1978); *Angew. Chem., Int. Ed. Engl.*, **17**, 486 (1978).
196. H. M .R. Hoffmann, *Angew. Chem.*, **81**, 597 (1969); *Angew. Chem., Int. Ed. Engl.*, **8**, 566 (1969).
197. S. N. Osipov, N. D. Chkanikov, Y. V. Shklyaev, A. F. Kolomiets and A. V. Fokin, *Izv. Akad. Nauk SSSR, Ser. Khim.*, 2131 (1989); *Bull. Acad. Sci. USSR, Chem. Sect.*, 1962 (1989).
198. V. A. Soloshonok and V. P. Kukhar', *Zh. Org. Khim.*, **26**, 419 (1990); *Chem. Abstr.*, **113**, 59045 (1990).
199. R. Nagarajan, L. D. Boeck, M. Gorman, R.L. Hamill, C. E. Higgens, M. M. Hoehn, W. M. Stark and J. G. Whitney, *J. Am. Chem. Soc.*, **93**, 2308 (1971).
200. P. G. Sammes, *Fortschr. Chem. Org. Naturst.*, **32**, 51 (1975).
201. N. Sewald, *PhD Thesis*, Technische Universität München (1991).
202. J. Sauer and R. Sustmann, *Angew. Chem.*, **92**, 773 (1980); *Angew. Chem., Int. Ed. Engl.*, **19**, 779 (1980).
203. S. N. Osipov, A. F. Kolomiets and A. V. Fokin, *Izv. Akad. Nauk SSSR, Ser. Khim.*, 746 (1989); *Bull. Acad. Sci. USSR, Chem. Sect.*, 673 (1989).
204. S. N. Osipov, A. F. Kolomiets and A. V. Fokin, *Izv. Akad. Nauk SSSR, Ser. Khim.*, 1130 (1991); *Chem. Abstr.*, **115**, 92186 (1991).
205. S. N. Osipov, A. F. Kolomiets and A. V. Fokin, *Usp. Khim.*, **61**, 1457 (1992); *Russ. Chem. Rev.*, **61**, 798 (1992).
206. N. P. Gambaryan, E. M. Rokhlin, Y. V. Zeifman, L. A. Simonyan and I. L. Knunyants, *Dokl. Akad. Nauk SSSR*, **166**, 864 (1966); *Proc. Acad. Sci. USSR*, **166**, 161 (1966).
207. Y. V. Zeifman, N. P. Gambaryan, L. A. Simonyan, R. B. Minasyan and I. L. Knunyants, *Zh. Obshch. Khim.*, **37**, 2476 (1967); *J. Gen. Chem. USSR*, **37**, 2355 (1967).
208. I. L. Knunyants, V. V. Shokina, V. V. Tyuleneva, Y. A. Cheburkov and Y. E. Aronov, *Izv. Akad. Nauk SSSR, Ser. Khim.*, 1831 (1966); *Bull. Acad. Sci. USSR, Chem. Sect.*, 1764 (1966).
209. I. L. Knunyants, V. V. Shokina and V. V. Tyuleneva, *Izv. Akad. Nauk SSSR, Ser. Khim.*, 415 (1968); *Bull. Acad. Sci. USSR, Chem. Sect.*, 405 (1968).
210. Y. V. Zeifman, V. V. Tyuleneva, A. P. Pleshkova, R. G. Kostyanovskii and I .L. Knunyants, *Izv. Akad. Nauk SSSR, Ser. Khim.*, 2732 (1975); *Bull. Acad. Sci. USSR, Chem. Sect.*, 2618 (1975).
211. Y. V. Zeifman, L. T. Lantseva and I. L. Knunyants, *Izv. Akad. Nauk SSSR, Ser. Khim.*, 401 (1986); *Bull. Acad. Sci. USSR, Chem. Sect.*, 371 (1986).
212. K. Burger, K. Gaa and K. Mütze, *Chem.-Ztg.*, **115**, 328 (1991).
213. J. Hamer (Ed.), *1,4-Cycloaddition Reactions*, Academic Press, New York, 1967.
214. N. Sewald and K. Burger, *Liebigs Ann. Chem.*, 947 (1992).
215. J. K. Stille, *Angew. Chem.*, **98**, 504 (1986); *Angew. Chem., Int. Ed. Engl.*, **25**, 508 (1986).
216. P. L. Pauson, *Tetrahedron*, **41**, 5855 (1985).
217. S. N. Osipov, A. F. Kolomiets and A. V. Fokin, *Izv. Akad. Nauk SSSR, Ser. Khim.*, 2456 (1990); *Bull. Acad. Sci. USSR, Chem. Sect.*, 2228 (1990).

218. W. Hollweck, *PhD Thesis*, Technische Universität München (1994).
219. K. Burger, C. Schierlinger, W. Hollweck and K. Mütze, *Liebigs Ann. Chem.*, 399 (1994).
220. V. P. Kukhar', *Izv. Akad. Nauk SSSR, Ser. Khim.*, 2290 (1990); *Bull. Acad. Sci. USSR, Chem. Sect.*, 2083 (1990).
221. K. Burger, C. Schierlinger, K. Mütze, W. Hollweck and B. Koksch, *Liebigs Ann. Chem.*, 407 (1994).
222. K. Burger, C. Schierlinger and K. Mütze, *J. Fluorine Chem.*, **65**, 149 (1993).
223. C. Schierlinger and K. Burger, *Tetrahedron Lett.*, **33**, 193 (1992).
224. D. P. Del'tsova, N. P. Gambaryan and E. I. Mysov, *Izv. Akad. Nauk SSSR, Ser. Khim.*, 2343 (1981); *Bull. Acad. Sci. USSR, Chem. Sect.*, 1927 (1981).
225. H. Uno, S. Okada, T. Ono, Y. Shiraishi and H. Suzuki, *J. Org. Chem.*, **57**, 1504 (1992).
226. R. A. Firestone, *US Pat.*, 4 332 813 (1982); *Chem. Abstr.*, **97**, 115331 (1982).
227. M. Lustig and J. K. Ruff, *Inorg. Chem.*, **4**, 1441 (1965).
228. I. L. Knunyants, B. L. Dyatkin and R. A. Bekker, *Izv. Akad. Nauk SSSR, Ser. Khim.*, 1124 (1966); *Bull. Acad. Sci. USSR, Chem. Sect.*, 1086 (1966).
229. R. A. Bekker, B. L. Dyatkin and I. L. Knunyants, *Izv. Akad. Nauk SSSR, Ser. Khim.*, 1060 (1967); *Bull. Acad. Sci. USSR, Chem. Sect.*, 1022 (1967).
230. I. L. Knunyants, B. L. Dyatkin and R. A. Bekker, *Dokl. Akad. Nauk SSSR*, **170**, 337 (1966); *Proc. Acad. Sci. USSR*, **170**, 874 (1966).
231. D. T. Meshri and J. M. Shreeve, *J. Am. Chem. Soc.*, **90**, 1711 (1968).
232 T. Abe, E. Hayashi, H. Fukaya and H. Baba, *J. Fluorine Chem.*, **50**, 173 (1990).
233. J. A. Young and R. D. Dresdner, *J. Am. Chem. Soc.*, **80**, 1889 (1958).
234. E. Hayashi, H. Fukaya, T. Abe and K. Omori, *Chem. Lett.*, 737 (1990).

5 Asymmetric Synthesis of Fluorine-containing Amino Acids

VALERY P. KUKHAR'
Institute of Bioorganic Chemistry and Petrochemistry, National Academy of Sciences of Ukraine, Murmanskaya 1, 253660 Kiev-94, Ukraine

GIUSEPPE RESNATI
CNR, Centro Studio Sostanze Organiche Naturali, Dipartimento Chimica, Politecnico, Via Mancinelli 7, I-20131 Milan, Italy

VADIM A. SOLOSHONOK
Institute of Bioorganic Chemistry and Petrochemistry, National Academy of Sciences of Ukraine, Murmanskaya 1, 253660 Kiev-94, Ukraine

5.1 INTRODUCTION

The great interest in the synthesis and development of bioactive compounds as single enantiomers is due to the final acknowledgement of the relevance of chirality to biological activity [1-3]. Examples in which the property of a compound is strongly related to a given absolute configuration can be drawn from several different classes of products, such as drugs, pheromones, food additives, perfumes and crop protection agents. For instance, (S)-limonene has a lemon odour and the R-antipode has an orange odour; (R)-asparagine tastes sweet whereas the S-enantiomer has a bitter taste; (S,S)-ethambutol is a tuberculostatic agent and the R,R-antipode causes blindness; (S)-propranolol is a β-blocker while the R-isomer is a contraceptive. In general, the differences in the biological activities of enantiomers are due to the fact that such activities originate from a recognition process between the small molecule (drug) and a large, usually proteinaceous, counterpart (receptor site). This counterpart is a highly asymmetric environment and enantioselective interactions with other molecules are a rule, not an exception. It becomes clear why in FDA guidelines enantiomers may be considered as impurities and data on safety and efficacy have to be produced for each stereoisomer [4, 5]. Similarly, for agrochemicals the non-useful stereoisomer can be considered as an environmental pollutant because it can interact with the whole ecological system [6–8].

Considering that most fluorine-containing amino acids have been designed and synthesized as compounds with potential biological activity, the development of efficient methods for preparing pure enantiomers becomes a requirement for their biological evaluation.

Fluorine-containing Amino Acids: Synthesis and Properties Edited by V. P. Kukhar' and V. A. Soloshonok
© 1995 John Wiley & Sons Ltd

This chapter is devoted to the asymmetric synthesis of fluorinated amino acids as one of the different routes for producing optically active compounds. The material included in this chapter has been selected by considering asymmetric synthesis as a process which allows optically active compounds to be prepared from a prochiral starting material under the influence of a chiral auxiliary which is temporarily connected with the prochiral compound or is a part of a reagent or a catalyst [9]. Neither the enzymatic resolution of racemic fluorinated amino acids nor the direct introduction of fluorine, or fluorine-containing groups, into optically active amino acids, or their derivatives, will be discussed.

Mainly for the asymmetric synthesis of polyfluoroalkyl-substituted amino acids there are some specific methods which have no analogy in the asymmetric synthesis of fluorine-free amino acids. However, most of the asymmetric syntheses of fluorinated amino acids reported hitherto in the literature consist in the exploitation, on fluorinated substrates or reagents, of general methods for the asymmetric synthesis of natural amino acids. The possibility of applying these methodologies in the asymmetric synthesis of fluorinated amino acids is dependent on the nature and the extent of the effect of fluorine atom(s), or fluorine containing group(s), on the reaction centres of the molecule and also on the stability of these groups towards the reaction conditions. Fluorine possesses a strong electron withdrawing effect and in several cases this property proved to be critical for reactions proceeding through the formation of carbocations [10]. As a consequence of this effect, reactions can be disfavoured, facilitated and follow different chemical and/or stereochemical pathways. However, in most cases, the application of fluorinated substrates has given the expected results.

All the material presented in this chapter has been organized according to the nature of the bond formed during the creation of the new chiral centre. Special attention will be paid to the stereoselectivity of the process and to the influence of fluorine. This subject has not been discussed previously in the literature [10–12]. Greater emphasis will be given to general, modern methods.

5.2 DIASTEROSELECTIVE C—C BOND-FORMING REACTIONS

An interesting area of asymmetric synthesis of fluorinated amino acids is the alkylation of chiral glycine derivatives. Using a variety of reaction types, a great diversity of fluorinated amino acids have been obtained in high optical purity through this approach.

Recently, the imine **1** of *tert*-butyl glycinate and (*R*)-camphor has been used in an efficient asymmetric synthesis [13] of fluoro-substituted phenylalanines **2** (equation 5.1). The imine **1** presumably exists in the *E*-configuration and easily affords the corresponding chelated enolate on treatment with lithium diiso-propylamide (LDA). The addition of benzyl bromides leads to the formation of products **2** through the preferential entrance of the alkylating agent from the *Re* face of the intermediate enolate (opposite side from camphor C-8). The reaction occurs under kinetic control with good diastereoselectivity (d.e. > 79%).

These stereochemical results can be explained through a π–π interaction

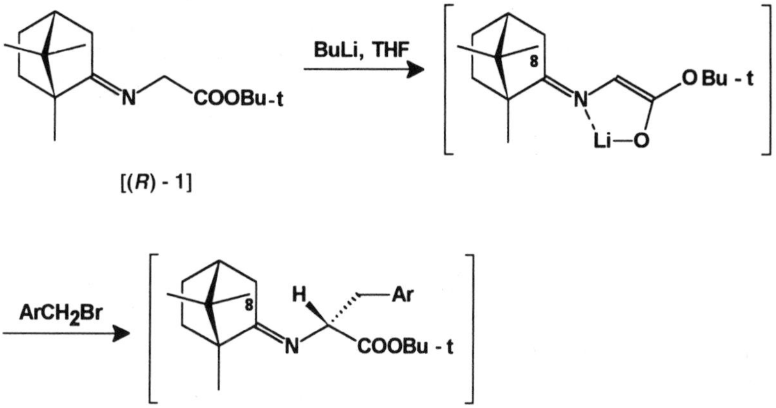

$$[(R) \text{-} 1]$$

(5.1)

(2)

Ar = 4-FC$_6$H4; 4-CF$_3$C$_6$H$_4$

developing in the transition state between with the electron-rich π system of the enolate and the π system of benzyl bromides, which possesses a positive character induced by the partial breaking of the C—Br bond. This leads to a reacting conformation in which the oppositely charged π systems are coplanar and eclipsed. Another explanation involves complexation of the π system of the electrophile with the lithium atom of the enolate. The fact that electron-rich π systems in the alkylating agent favour high stereoselectivities suggests an association with a positively charged centre. Whether one views the stereo-selectivity as arising from π–π or π–Li association, the conclusion that electron-rich systems should increase association remains constant. This approach seems particularly useful for the asymmetric synthesis of ring-polyfluoro-substituted phenylalanines.

Usually, conformationally restricted, cyclic chiral synthones provide higher d.e.s than open-chain analogues. Examples of this general trend in the fluor-oalkylation reaction of glycine enolate equivalents have been reported by various groups [14–19].

Both enantiomers of 1-benzoyl- and 1-(*tert*-butoxycarbonyl)-2-(*tert*-butyl)-3-methyl-4-imidazolidinones (**3a** and **b**) are commercially available (Aldrich). The alkylation of the lithium enolate of **3a** with fluorobenzyl bromides provides alkylated products with excellent diastereoselectivity (>95%) (equation 5.2). Final unprotected phenylalanines (**4**) have been obtained through acid hydrolyses [14, 15]. The very high *trans/cis* ratio in the alkylation reaction has been explained by the steric requirements of the *tert*-butyl group, which hinders the approach to the *cis* face. Attachment of electrophiles with relative topicity *lk*-1,3 is favoured so that L-fluorophenylalanines have been prepared starting from the (*S*)-imidazolidin-4-one **3a**. The method has also been employed for the preparation of α-disubstituted amino acids [16].

The aldol condensation of the lithium enolate of (*S*)-imidazolidininone (**3b**) with trifluoroacetaldehyde affords (2*S*,3*S*)-4,4,4-trifluorothreonine (**6**) (equation 5.2). In this case also the aldehyde enters the enolate from the side opposite to the *tert*-butyl group (*lk*-1,3 induction). The addition is assumed to

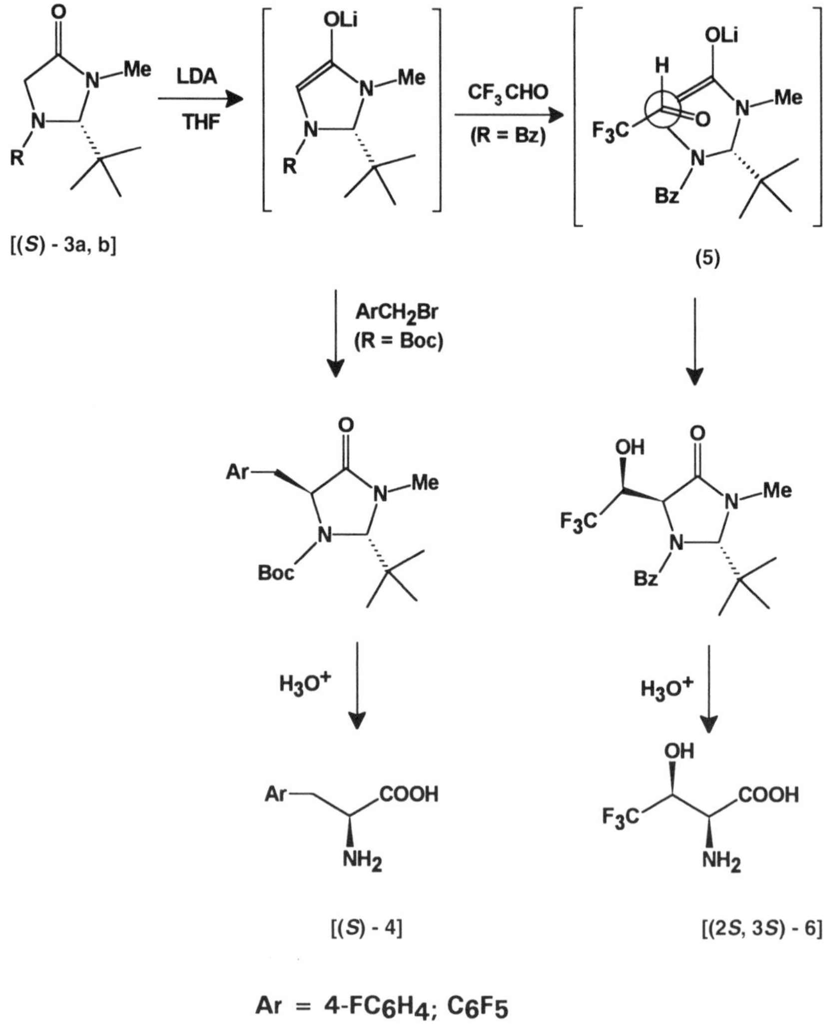

(5.2)

Ar = 4-FC$_6$H$_4$; C$_6$F$_5$

take place via the transition-state orientation **5** in which three criteria are met. First, the double bonds of the donor and of the acceptor are synclinal and not antiperiplanar, preventing the separation of developing opposite charges and still allowing for O—Li····O chelation; second, H-C$_5$ of the enolate is antiperiplanar to the acceptor carbonyl bond and not to the electronegative N-benzoyl group; third, there is a maximum overlap of the donor and acceptor π systems [17]. Exchange of lithium for boron, magnesium or titanium while leaving unaltered the stereochemical outcome of the reaction lowered the yields in the desired aldol products.

The two enantiomers of 3-carbobenzyloxy-2-*tert*-butyloxazolidin-5-one (**7**) can be obtained, in pure form and in preparative quantities, through chromatographic resolution of the racemate on a chiral stationary phase. On treatment with lithium hexamethyldisilazide (LiHMDS) these oxazolidones (**7**) afford the corresponding enolates, which react with fluoro-substituted benzaldehydes to give the condensation products with essentially complete *threo* selectivity (equation 5.3) [18]. Free amino acids (**8**) are isolated in high overall yields (>85%) after removal of the carbobenzyloxy group and successive hydrolyses. Starting from oxazolidinones **7** having *R* and *S* absolute configurations, the *threo* ring-fluorinated phenylalanines **8**, having 2*S*,3*R* and 2*R*,3*S* configurations, respectively, are obtained. On the assumption that the aldol condensation reaction occurs under kinetic control, the staggered approach (chair-like) reported in equation 5.3 accounts for the stereochemical results. The preference of the aryl group of

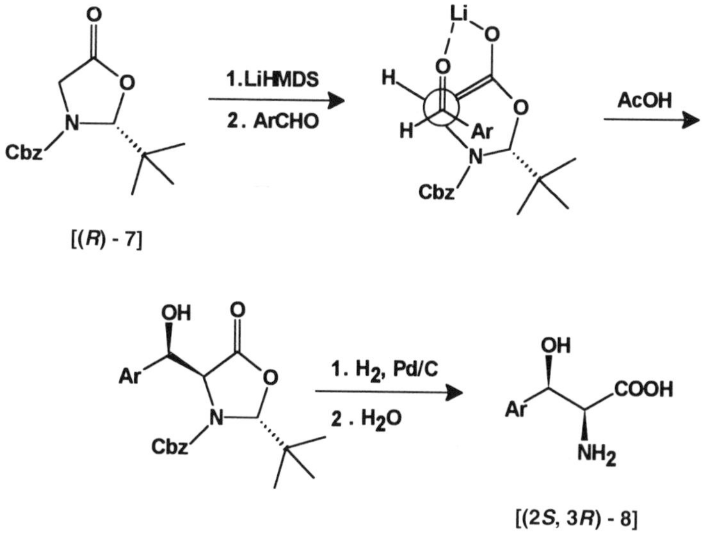

[(*R*) - **7**]

(5.3)

[(2*S*, 3*R*) - **8**]

Ar = 2-FC$_6$H$_4$; 4-CF$_3$C$_6$H$_4$; 2-Cl-6-FC$_6$H$_3$

Cbz = PhCH$_2$OCO

the aldehyde for a pseudo-axial rather than a pseudo-equatorial disposition in the chair may result from an unfavourable non-bonding interaction between the N-carbobenzyloxy group and the aryl group in the equatorial position.

Ni(II) complexes of Schiff bases derived from (S)-2-[N-(N'-ben-zylprolyl)amino]benzophenone (BPB) and glycine or alanine (9a and 9b, respectively) are interesting chiral glycine and alanine anion equivalents [19]. These complexes are stable and neutral, with the charge of the central metal ion neutralized by the two negative charges of ionized carboxyl and amide groups. All such complexes are red, crystalline compounds, soluble in organic solvents and diamagnetic so that they can be easy studied through NMR spectroscopy. The acidity of the CH_2 glycine moiety lies in pH range 17–19 (in DMSO) and only these protons are acidic whereas the proton of the proline moiety is usually inert towards bases. The coordination is essentially square planar, but a pronounced puckering of the ligand chelate ring is induced by the presence of the asymmetric centre in the proline residue. This puckering is responsible for the significant asymmetric induction observed in base-catalysed reactions of the complexes 9 with electrophiles.

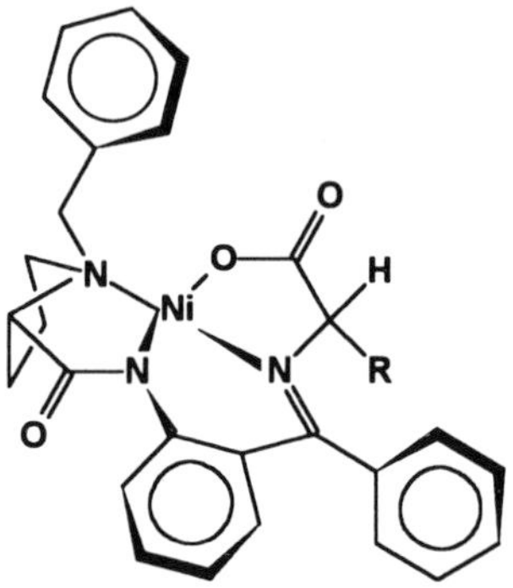

(9a) R = H

(9b) R = Me

Specifically, (S)-phenylalanines 11a and their α-methyl analogues 11b bearing fluorine atom(s) or fluorine-containing substituent(s) in different positions on the phenyl ring have been synthesized in optically active form by alkylation of complexes (S)-9a and (S)-9b (complex 9b has been used as a mixture of two dia-stereoisomers in a 9:1 ratio) with benzyl halides in DMF in the presence of powdered NaOH or KOH (equation 5.4) [20, 21]. Under these conditions epi-merization of the diastereoisomeric alkylation products (S,2R)-10a and (S,2S)-

10a takes place and the latter epimer greatly prevails in the thermodynamic mixture (d.e. 80–95%). Free amino acids **11**, formed through mild acid hydrolysis, can be obtained in optically pure form if the mixture of diastereo-isomeric complexes $(S,2R)$-**10** and $(S,2S)$-**10** is resolved into pure compounds through crystallization or chromatography. The initial chiral auxiliary (S)-BPB is recovered in 80–98% chemical yield without loss of optical activity.

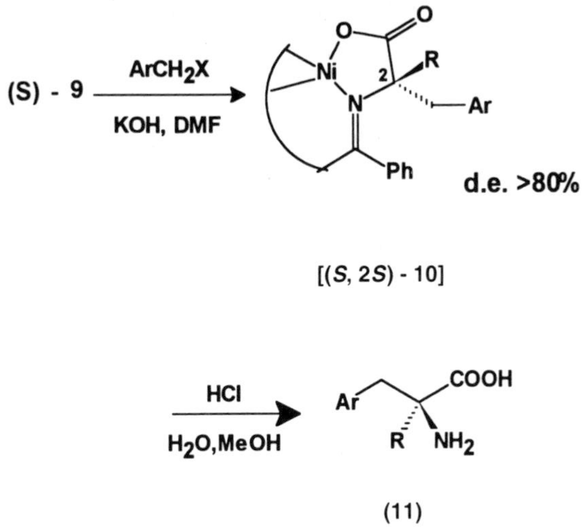

$$\text{d.e.} > 80\%$$

$$[(S, 2S) - 10]$$

(5.4)

(11)

Ar = 2-FC$_6$H$_4$; 3-FC$_6$H$_4$; 4-FC$_6$H$_4$; 3-CF$_3$C$_6$H$_4$; 4-Br-2-FC$_6$H$_3$;

6-Cl-2-FC$_6$H$_3$; C$_6$F$_5$

R = H, Me

As a consequence of the puckering of the whole chelate ring, the phenyl group adjacent to the C=N bond probably adopts a conformation which shields the Re side of the intermediate enolate. Moreover, in the alkylation product **10** a significant steric non-bonding interaction occurs between this phenyl ring and the alkyl residue introduced on C-2 of the glycine moiety. This interaction is minimized in $S,2S$-isomers **10a**, which are, in fact, thermo-dynamically more stable than the $S,2R$-epimers. This accounts for the high diastereoselection of the alkylation process under equilibrating conditions. Further, some X-ray analyses of model compounds show that the conversion of the sp^2 carbon of the intermediate enolate (C-2 of the glycine moiety) to the sp^3 carbon of the alkylation products **10a** brings minor changes in the whole conformation of the chelate ring when the $S,2S$-diastereoisomers are formed, whereas the same conversion results in significant distortions when the alkylated products have the $S,2R$ configuration. In other words, intermediate enolates are already biased towards the $S,2S$ products.

Reasonably, the steric hindrance exerted by the phenyl group is also the main reason why monoalkylated products can be obtained starting from glycine complex 9a. Finally, the steric shielding on the Re face of the enolate of the alanine complex (S)-9b by the phenyl residue is probably responsible for the preferential alkylation from the opposite Si face to give ($S,2S$)-2-methyl-2-alkyl complexes 10b.

Whereas the asymmetric synthesis of fluoro-substituted phenylalanines using Ni(II) complexes 9a and b gives results comparable to those obtained through the other approaches described above, the asymmetric synthesis of fluorinated β-hydroxy-α-amino acids was revealed to be particularly effective. In order to optimize the chemical and optical yields in the desired β-hydroxy-α-amino acids and taking into account the different reactivities of aliphatic and aromatic fluorinated aldehydes, four different reaction procedures have been studied for the aldol condensation of complex 9a.

When triethylamine is used as a base in methanol solution (1:2 ratio), ring fluorinated aromatic aldehydes afford a mixture of the four possible stereoisomers [22, 23]. The two epimers syn- and $anti$-12 having the same S absolute configuration at the nitrogen-substituted stereocentre (C-2) and opposite configuration at the carbinol carbon (C-3) account for 80–95% of the product mixture (equation 5.5). The ratio of these two stereoisomers is nearly

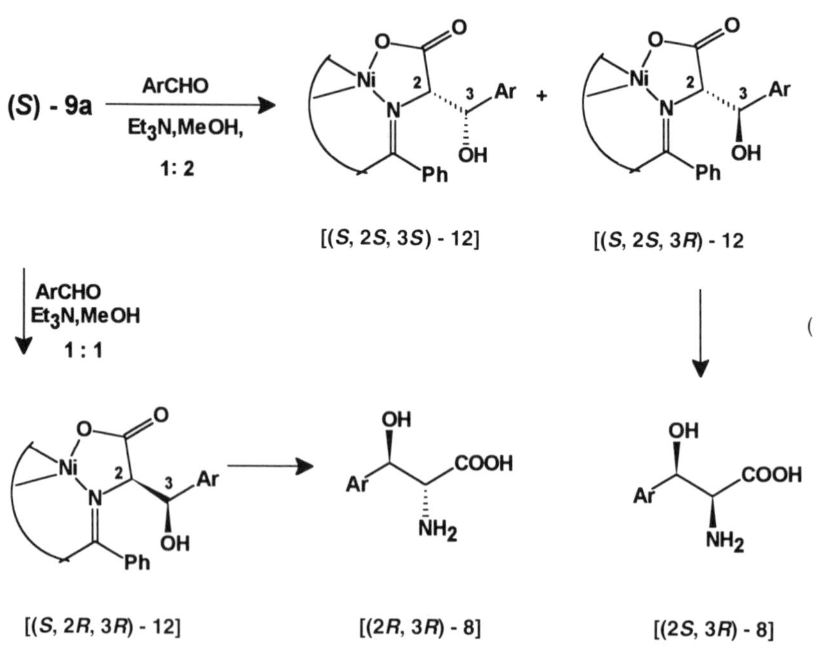

$$(S) - 9a \xrightarrow[\substack{Et_3N,MeOH, \\ 1:2}]{ArCHO}$$

[(S, 2S, 3S) - 12]   [(S, 2S, 3R) - 12]

ArCHO
Et₃N,MeOH
1:1

(5.5)

[(S, 2R, 3R) - 12]   [(2R, 3R) - 8]   [(2S, 3R) - 8]

Ar = 2-FC$_6$H$_4$; 3-FC$_6$H$_4$; 4-FC$_6$H$_4$; 2-CF$_3$C$_6$H$_4$; 2-F$_2$CHOC$_6$H$_4$;

4-F$_2$CHOC$_6$H4

1 : 1 in most cases, but the presence of a sterically demanding substituent in the *ortho* position of the benzaldehyde molecule increases this ratio up to 10 : 1, the *syn* isomer $(S,2S,3R)$-**12** being formed with great preference over the $(S,2S,3S)$-**12** epimer. This shows that the diastereoface selectivity for the entry on the aldehyde electrophile is low. In contrast, the *Si* face of the intermediate enolate is involved in the reaction with great preference over the *Re* face, in strict similarity to what has already been observed in the alkylation reaction. Reasonably, factors controlling the two processes are the same.

The increase in the concentration of triethylamine in the solution does not influence the ratios of condensation products **12** formed at room temperature. However, the solubility of different diastereoisomeric products is greatly affected and the *anti* isomer having the $S,2R,3R$ configuration precipitates from the reaction mixture when the triethylamine/methanol ratio is 1 : 1 (equation 5.5). Owing to the product equilibration during the reaction, all starting material can be transformed into the complex $(S,2R,3R)$-**12** which is isolated in enantiomerically and diastereoisomerically pure form in 57–65% yield.

With perfluorinated aldehydes, Dabco in chloroform solution gives better results than the triethylamine/methanol system described above. The stereoselectivity of the process is strictly similar to that observed in the triethylamine-catalysed condensation of aromatic aldehydes (equation 5.5) as complexes *anti*-$(S,2S,3R)$-**13** and *syn*-$(S,2S,3S)$-**13** are formed in nearly equimolar amounts (equation 5.6). The *syn*- and *anti*-β-hydroxy-α-amino acids

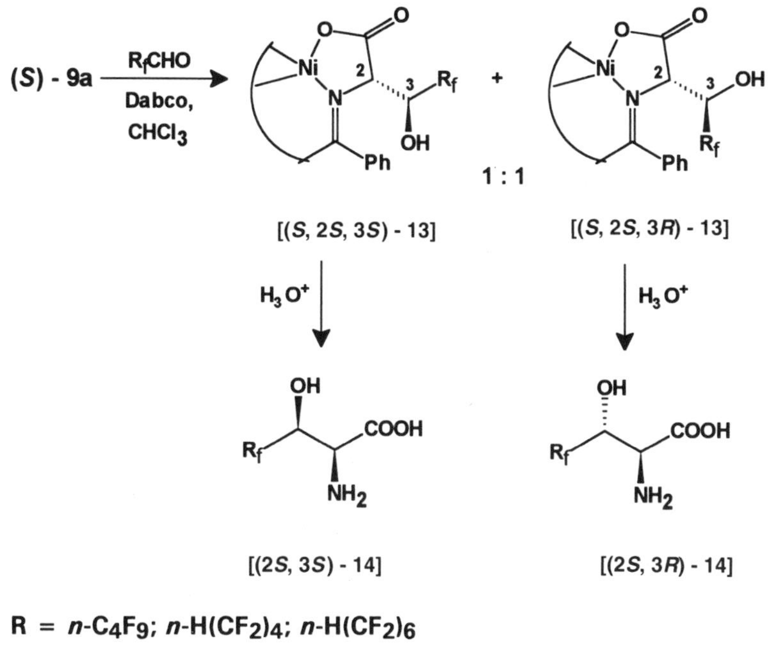

$$(S) - \textbf{9a} \xrightarrow[\substack{\text{Dabco,} \\ \text{CHCl}_3}]{\text{R}_f\text{CHO}}$$

[$(S, 2S, 3S)$ - **13**]   [$(S, 2S, 3R)$ - **13**]

[$(2S, 3S)$ - **14**]   [$(2S, 3R)$ - **14**]

(5.6)

$R = n\text{-}C_4F_9;\ n\text{-}H(CF_2)_4;\ n\text{-}H(CF_2)_6$

14, carrying a long perfluorinated chain on the β-carbon, are obtained in enantiomerically and diastereoisomerically pure form after hydrolyses of single complexes under standard conditions.

When fluoroalkylaldehydes are reacted with the glycine complex (S)-**9a** at room temperature with a 2.25 M solution of sodium methoxide in methanol, the aldol condensation reaction occurs slowly, probably as a consequence of the formation of unreactive hemiacetals. However, brief heating affords the diastereoisomeric complexes syn-(S,2S,3S)-**13** and syn-(S,2R,3R)-**13** in high chemical yields and in a 95:5 ratio (equation 5.7). The same results are obtained when corresponding hydrates or hemiacetals of fluoroalkyl aldehydes are used. Neither prolonged heating of the reaction mixture nor leaving it at room temperature has any effect on the ratio of the diastereoisomers, although the formation of some by-products is observed [24–26]. Similarly to previous cases, the aldehyde enters the enolate preferentially from the *Si* face which is not shielded by the phenyl. It is interesting to observe that when non-fluorinated aliphatic aldehydes are used, *syn* complexes having the *S,2R,3S* absolute configuration are formed preferentially (d.e. 40–50%). The reasons for this inversion of diastereoselection moving from fluorinated to hydrocarbon aldehydes will be discussed later.

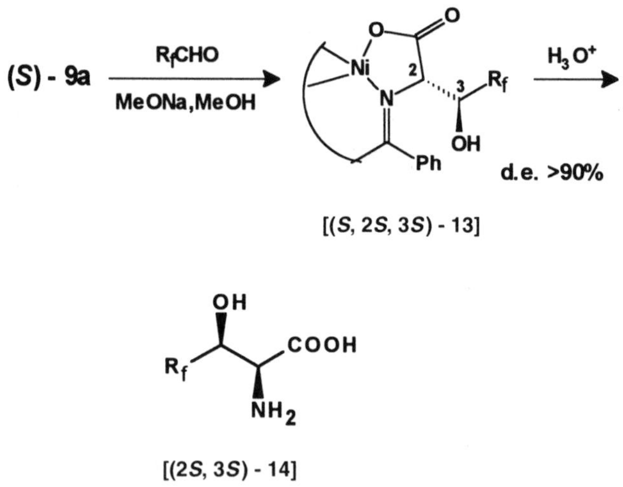

[(S, 2S, 3S) - 13]

(5.7)

[(2S, 3S) - 14]

$$R_f = CF_3 \; ; \; n\text{-}C_4F_9; \; n\text{-}H(CF_2)_2; \; n\text{-}H(CF_2)_4; \; n\text{-}C_7F_{15}$$

Under the same reaction conditions (sodium methoxide/methanol), trifluoroacetone affords exclusively the complex (S,2S,3S)-**15** (equation 5.8). A stereochemical preference strictly similar to that observed for perfluorinated aldehydes (equation 5.7) is shown, proving that substitution of methyl for hydrogen in the electrophile molecule has minor effects on the parameters controlling the reaction.

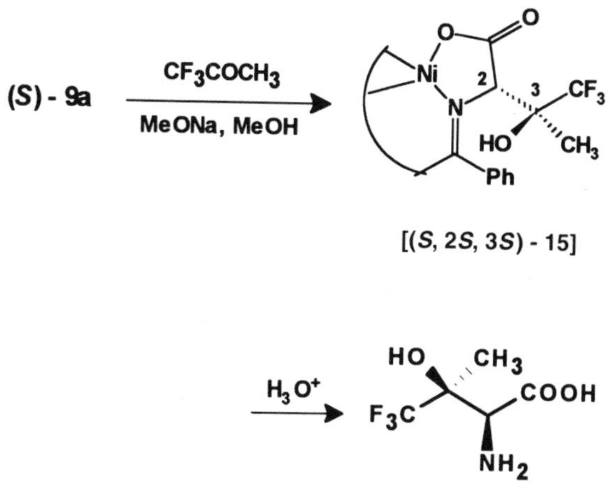

$$(S)-9a \xrightarrow[\text{MeONa, MeOH}]{\text{CF}_3\text{COCH}_3}$$

[(S, 2S, 3S) - 15]

(5.8)

$$\xrightarrow{\text{H}_3\text{O}^+}$$

HO CH$_3$
F$_3$C—C—C—COOH
       NH$_2$

[(2S, 3S) - 16]

In contrast, when sodium methoxide in methanol at room temperature is employed for the aldol condensation with fluoro-substituted benzaldehydes, the two *anti* complexes (S,2S,3S)-12 and (S,2R,3R)-12 are either not detected or are formed in minute amounts and the *syn* isomer 12 having the S,2R,3S absolute configuration greatly prevails over the other *syn* S,2S,3R-isomer (d.e. > 90%) (equation 5.9) [22, 24, 27]. This stereochemical result apparently contrasts with those reported in equations 5.4–5.7, but it can be rationalized through the same general principles. In detail, when sodium methoxide in methanol is used the thermodynamic product mixture is isolated as the formation of complexes 12 and their equilibration is essentially complete within a few minutes. At the high pH of the medium, the hydroxyl group becomes ionized and, as a consequence of its greater basicity, substitutes the carboxyl group in the coordination plane of complex 12 to give the alkoxide complex 17 (equation 5.10) [28]. Those steric factors which stabilize the S,2S configuration in carboxylic complexes 10 and 12 favour the orientation of the carboxyl group away from the phenyl adjacent to the C=N bond in the

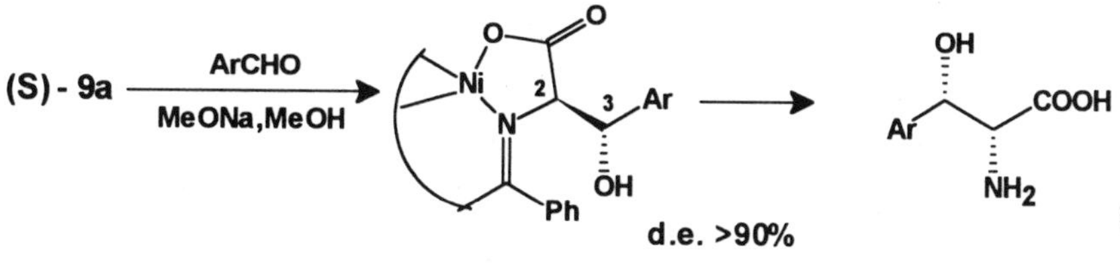

$$(S)-9a \xrightarrow[\text{MeONa,MeOH}]{\text{ArCHO}} \cdots \longrightarrow$$

d.e. >90%

[(S, 2R, 3S) - 12]   [(2R, 3S) - 8]

(5.9)

Ar = 2-FC$_6$H$_4$; 3-FC$_6$H$_4$; 4-FC$_6$H$_4$; 3-F-4-CH$_3$OC$_6$H$_3$; 2-CF$_3$C$_6$H$_4$;

2-F$_2$CHOC$_6$H$_4$; 4-F$_2$CHOC$_6$H$_4$; 4-CF$_3$OC$_6$H$_4$; 4-CH$_3$OC$_6$F$_4$; C$_6$F$_5$

$$\text{MeONa} \atop \overset{\longrightarrow}{\underset{\longleftarrow}{\text{AcOH}}} \quad (5.10)$$

[(S, 2R, 3S) - 12]   [(S, 2R, 3S) - 17]

alkoxide complex **17**. Such an orientation is allowed only by the $S,2R$ configuration which, therefore, under the adopted thermodynamic conditions, prevails in the reaction mixture. Further, in this latter complex **17** a *trans* disposition of the carboxylate on C-2 and of the aryl group on C-3 is favoured and leads to the observed *syn* configuration of final amino acids $(2R,3S)$-**8**.

When either the reaction conditions (equations 5.5 and 5.6) or the specific substrate employed (equations 5.7 and 5.8) do not allow the rearrangement of equation 5.10 to occur, the 'usual' complexes having the $S,2S$ configuration are formed preferentially. For instance, when pentafluorophenylaldehyde is used, the equilibration of diastereoisomeric condensation products **12** is less easy than for other aromatic aldehydes [22]. The thermodynamic mixture is now formed only after several hours at elevated temperature (probably as a consequence of the increased stability of the C-2—C-3 bond in the products) and kinetic products can be detected at an early stage of the reaction. As a further example, when the aldol condensation of aromatic aldehydes is performed by using triethylamine, the alkoxide anions of complexes **12** are not formed as the base employed is too weak to generate them. No rearrangement to the isomeric compounds **17** occurs and the 'usual' stereochemical course, leading to $S,2S$ products, is observed (equation 5.5). Further, as already reported, by using sodium methoxide in methanol the aldol condensation reaction of **9a** occurs with opposite diastereoface differentiation for hydrogenated and perfluorinated aldehydes. Specifically, the prevailing diastereoisomer in the product mixture has the $S,2R$ and $S,2S$ absolute configurations, respectively. A rationalization of this difference is that the intermediate perfluorinated alkoxides are relatively weak bases and therefore are incapable of substituting the carboxylate group in the coordination plane of complexes **12**. The 'usual' stereoselectivity due to the non-bonding interactions of the phenyl with the substituent on C-2 and resulting in the $S,2S$ configuration is thus observed.

As a summary, (S)-β-phenylalanines and (S)-α-methyl-β-phenylalanines (**11**) (equation 5.4), $(2S,3R)$-, or $(2R,3R)$- or $(2R,3S)$-β-fluorophenylserines (**8**) (equations 5.5 and 5.9), $(2S,3R)$- or $(2S,3S)$-β-fluoroalkylserines (**14**) (equations 5.6 and 5.7) and $(2S,3S)$-3-trifluoromethylthreonine (**16**) (equation 5.8) can be obtained starting from the same S-enantiomer of the chiral auxiliary BPB through the Ni(II) complexes **9a** and **b**. The synthetic usefulness of this auxiliary agent is thus definitively proved.

An attractive approach to fluorophenyl-substituted alanines rests on the alkylation of the *tert*-butylalanine Schiff base **18** under phase-transfer conditions (equation 5.11). When *N*-benzylcinchoninium chloride (**19a**) is used as a catalyst, the alkylation product **20** is isolated in high chemical yield, the isomer having the *R* absolute configuration being formed with medium enantiomeric excess (e.e. 50%) [29]. By using as a catalyst the 'pseudoenantiomer' *N*-benzylcinchonidinium chloride (**19b**), the enantiomeric alkylation product (*S*)-**20** is formed preferentially.

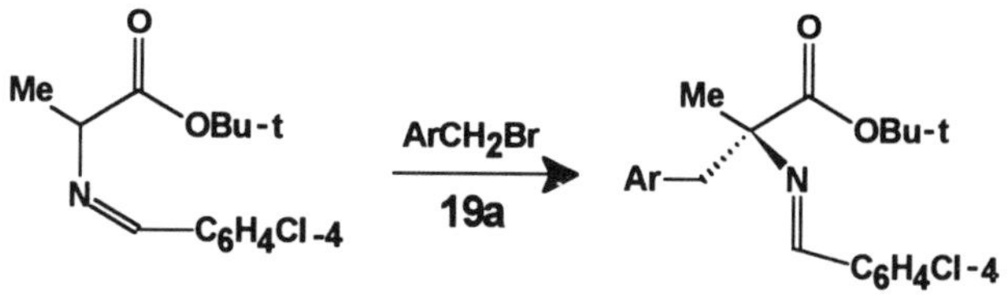

$$ \text{(18)} \qquad\qquad \text{[(R) - 20]} \tag{5.11}$$

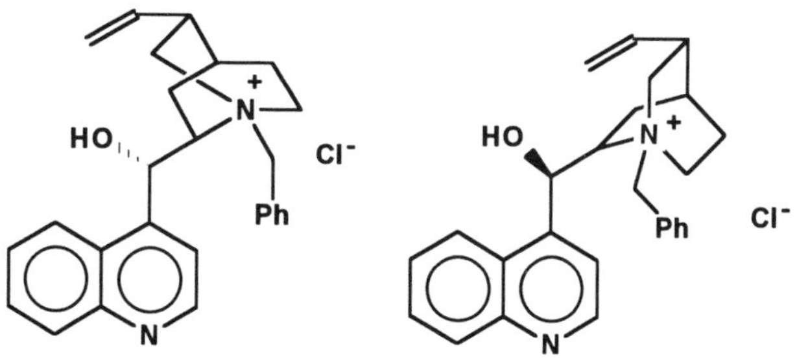

(19a)   (19b)

The diastereoselective alkylation of alanine and β-phenylalanine has been realized by using *O*-methylprolinol as a chiral auxiliary (equation 5.12). Interestingly, the lithium derivatives of amidines **21** can be alkylated with chlorodifluoromethane and corresponding α-difluoromethyl-α-amino acids **22** have been isolated in optically active form after acid hydrolyses. Neither the optical purity nor the absolute configuration of the products obtained has been established [30, 31]. In general, α-mono- and α-difluoromethylamino acids are particularly interesting compounds as they are selective and efficient mechanism-based inhibitors of pyridoxal phosphate-dependent enzymes.

A common feature of all the synthetic approaches described above is that the enolate of an α-amino acid (glycine, alanine, β-phenylalanine) is alkylated by fluorinated electrophiles (alkyl halides, aldehydes, ketones). In equation 5.13 a different approach is shown. The nucleophilic species is an α-sulphinyl carbanion which works as the synthetic equivalent of a chiral alkyl chain and

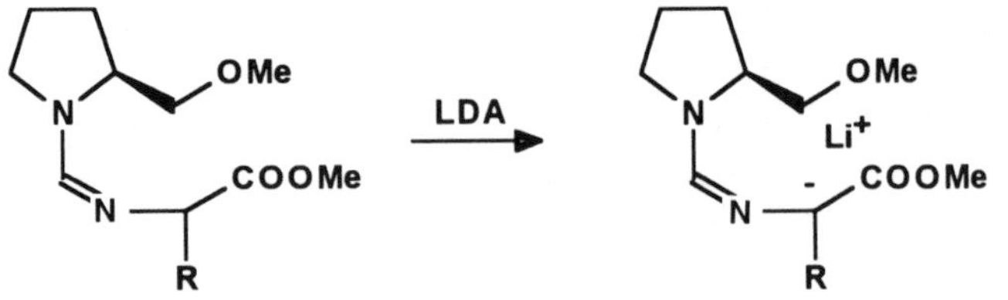

[(S) - 21]   (5.12)

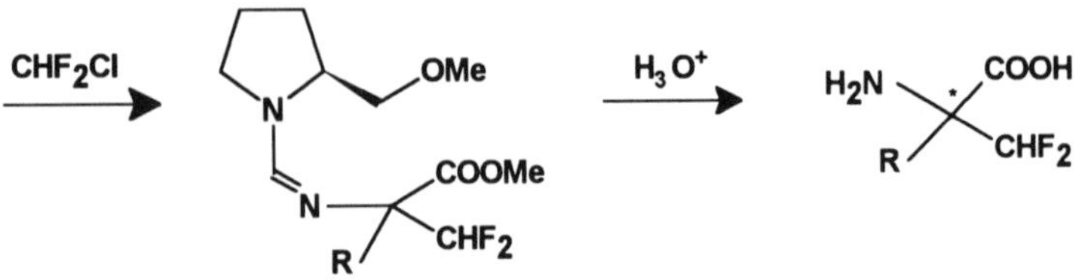

(22)

R = Me; C$_6$H$_5$CH$_2$

the electrophilic moiety is a fluorinated imine [32, 33]. Specifically, the lithium derivatives of (R)-alkyl 4-methylphenylsulfoxides attach the carbon–nitrogen double bond of the N-ethoxycarbonylimine of methyl trifluoropyruvate (23) with complete regio- but low diastereoselectivity. The two (R = H) or four (R = alkyl, aryl) adducts 24 are in fact formed. Single diastereoisomers can be easily obtained in pure form through flash chromatography and their reductive

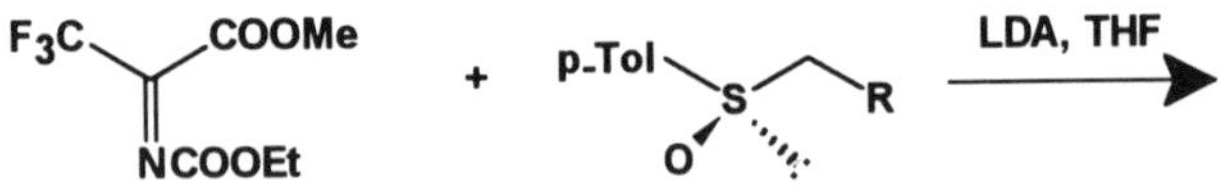

(23)   (R)   (5.13)

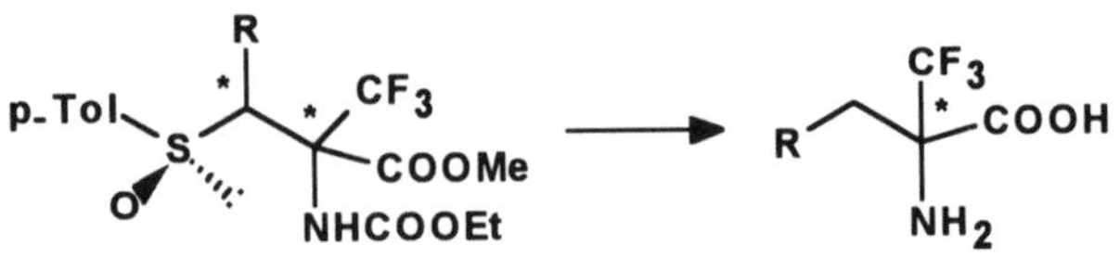

(24)   (25)

R = H; Me; Ph

desulfinylation affords both antipodes of the final α-trifluoromethyl-α-amino acids (25) in enantiomerically pure form.

Another approach ·in which the fluorinated compound works as the electrophilic species in the asymmetric C—C bond-forming reaction is the synthesis of 3-fluoro-β-lactams (27–29) (equation 5.14) [34]. When fluoroacetyl chloride is allowed to react with the optically active imine 26, the 3R,4S,4'S-diastereoisomer 27 of 3-fluoro-2-azetidone is formed in medium yield, complete diastereoselectivity and e.e. not less than 99%. The high stereoselectivity observed may be rationalized as a consequence of the *anti* addition of the imine 26 to a single face of *in situ*-generated fluoroketene. The so-formed intermediate zwitterionic species is postulated to collapse with 1,2-*lk,ul* topicity. This azetidone can be deprotonated with LDA and the enolate generated reacts with alkyl halides to give corresponding products 28 with no loss of stereochemical integrity at the fluorinated carbon. When the same enolate is condensed on aldehydes and ketones to give aldol products 29, complete retention of chirality at the fluorinated stereocentre is observed, but the stereoselectivity in the formation of the new chiral carbon in the side chain is poor and the two epimers at the hydroxylated atom are formed in nearly equal amounts.

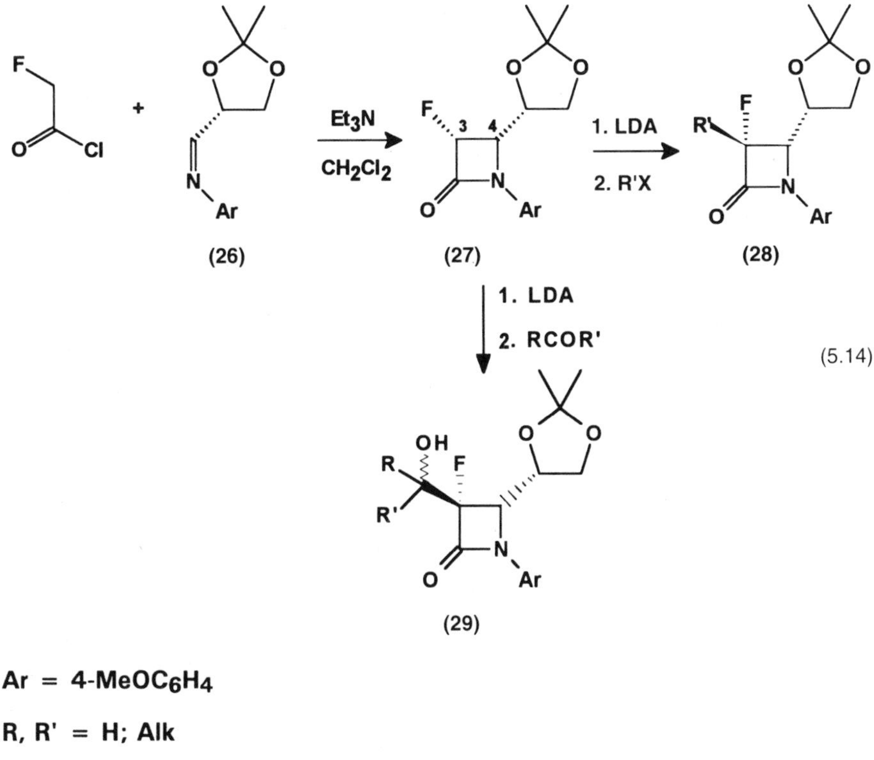

(5.14)

Ar = 4-MeOC$_6$H$_4$

R, R' = H; Alk

An asymmetric version of the classical Strecker synthesis of amino acids has been employed for the preparation of (2S,4R)-5,5,5-trifluoroleucine (32) (equation 5.15) [35]. The α-aminonitrile having the 2S absolute configuration is formed with the same diastereoselectivity (d.e. 60%) starting from the (4S,5S)-amine 30 and either the R- or the S-enantiomer of the fluorinated aldehyde. This shows that the configuration of the newly formed stereocentre is determined exclusively by the chirality of the amine. This observation is confirmed by the fact that in another synthesis of trifluoroleucine (32) starting from (3R)-3-trifluoromethyl-4-phenylthiobutanal, a 1 : 1 mixture of the two α-aminonitriles epimeric at the newly formed stereocentre was obtained when potassium cyanide and ammonia solution were employed [36].

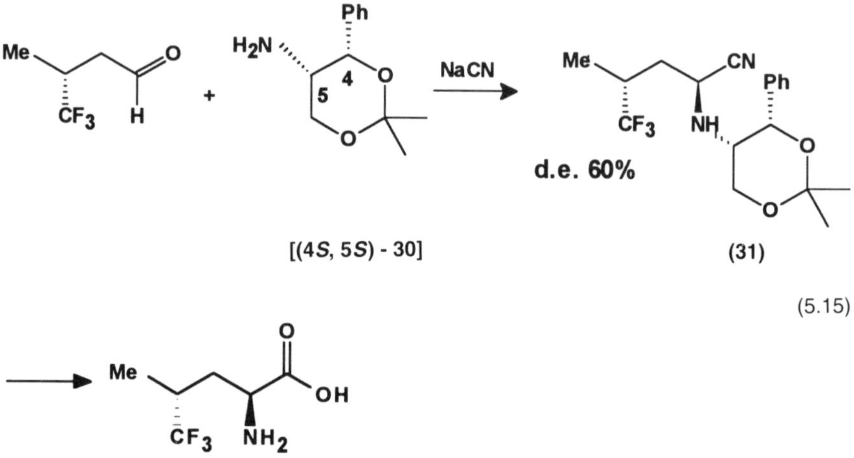

[(4S, 5S) - 30]   (31)

(5.15)

[(2S, 4R) - 32]

In all asymmetric C—C bond-forming reactions described above, fluorine is present on the molecule of the electrophile. In contrast, in the standard synthesis of the α,α-difluoro-β-hydroxy-γ-amino esters 33 and the α,α-difluoro-β-amino-γ-hydroxy esters 34, an enolate equivalent of ethyl difluoroacetate is condensed on α-aminoaldehydes [37–42] and α-hydroxyimines, respectively (equation 5.16). Usually, the zinc derivative of ethyl bromodifluoroacetate is employed, but also chloro- [43] and iododifluoroacetate have been used as starting materials and in some cases the best results have been obtained by resorting to the ketene silylacetal prepared in situ by adding the appropriate silyl chloride to the zinc reagent [44]. On condensation with both α-aminoaldehydes and α-alkoxyimines the formation of the syn products is favoured, probably as the organometallic reagent enters an α-chelated coordination species between the electrophile and the zinc halide. The α,α-difluoroamino esters 33 and 34 have been extensively used to prepare α,α-difluoroketone peptide analogues, which work as effective transition-state analogue inhibitors of several proteolytic enzymes (renin, porcine pancreatic elastase, human thrombin, etc.).

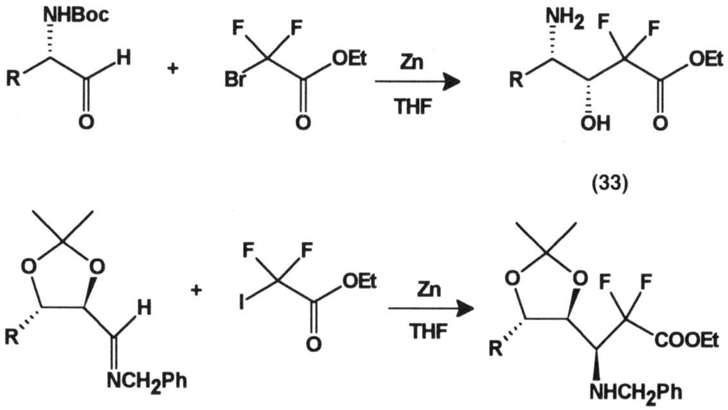

(33)

(5.16)

5.3 ENANTIOSELECTIVE C—H BOND-FORMING REACTIONS

Asymmetric hydrogenation is a particularly attractive field and the application of this process to α-amidoacrylic acid in order to prepare amino acid derivatives has reached industrial applications [45, 46].

The methodology has also been employed for the preparation of several β-fluoroaryl- and β-fluoroalkylalanines. For instance, (Z)-α-N-benzoylamino-β-fluorophenylacrylic acids and esters (35) have been hydrogenated with various catalysts. The best results were obtained by using the rhodium complex 36 and the enantioselectivity of the process (e.e. > 68%) is independent of the position of the fluorine substituent on the phenyl ring (equation 5.17). When the catalyst has the S absolute configuration, the (R)-amino acids are formed preferentially [47]. Optically pure compounds can be easily obtained through recrystallization of intermediate α-N-benzoylamino acids, whereas increasing the e.e. on α-N-benzoylamino esters through crystallization is more difficult. Both enantiomeric forms of the catalyst 36 are available, so that this method secures access to both antipodes of target β-fluorophenylamino acids in optically pure form.

The use of several catalysts has also been studied in the hydrogenation of (Z)-α-N-benzoylamino-6,6,6-trifluorohexenoate (37) and the highest e.e. (88%) has been obtained by using (R,R)-DiPAMP (38) (equation 5.18) [48]. (S)-Tri-fluoronorleucine (39) has been isolated after acid hydrolysis.

It has already been discussed how a *tert*-butyl residue on C-2 of an imidazolidin-4-one is very effective in controlling the stereochemical course of alkylation reactions at C-5 (equation 5.2). The same shielding effect which accounts for that diastereoselectivity can be used to rationalize the complete stereoselectivity in the hydrogenation of the 2-trifluoroethylideneimidazolidinone (R)-40 (equation 5.19). The 2,5-*cis* product is formed exclusively,

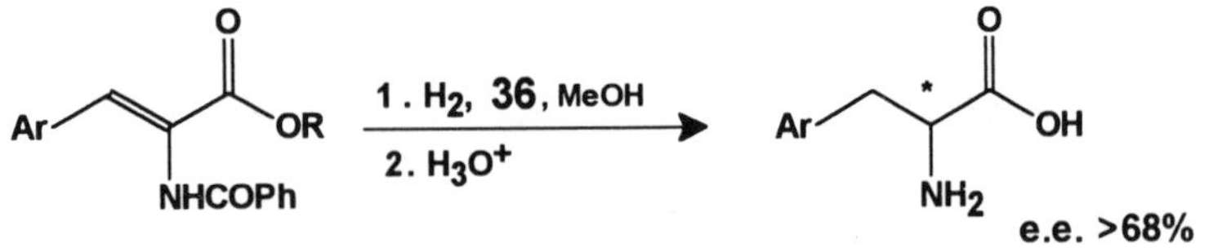

$$(5.17)$$

e.e. >68%

(35)   (4)

R = H; CH$_3$

Ar = 2-FC$_6$H$_4$; 3-FC$_6$H$_4$; 4-FC$_6$H$_4$; 4-CF$_3$C$_6$H$_4$; C$_6$F$_5$

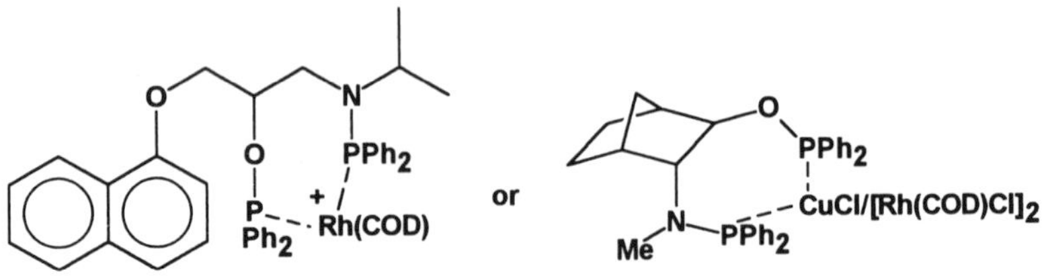

[(S) or (R) - 36]   or

COD =

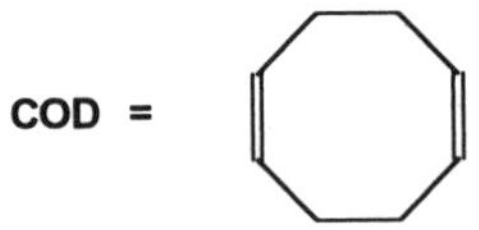

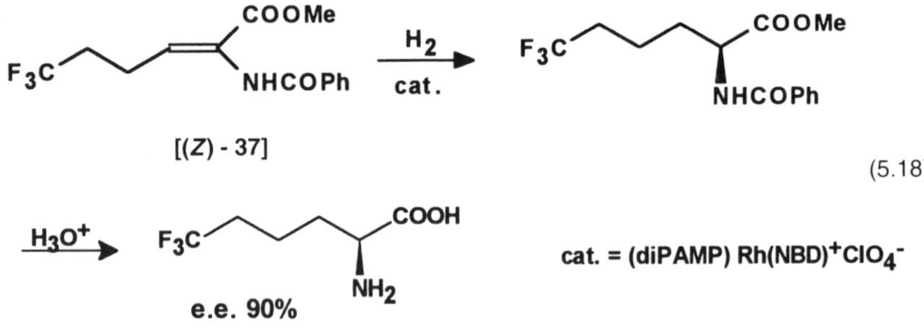

[(Z) - 37]

$$(5.18)$$

e.e. 90%

[(S) - 39]

cat. = (diPAMP) Rh(NBD)$^+$ClO$_4^-$

NBD = norbornadiene

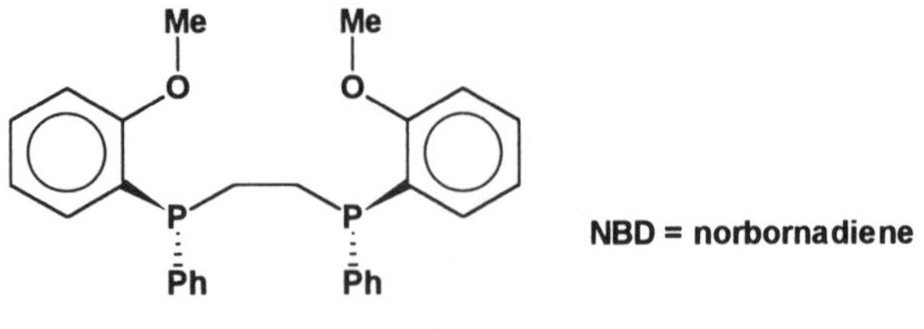

(38)   (R, R) - diPAMP

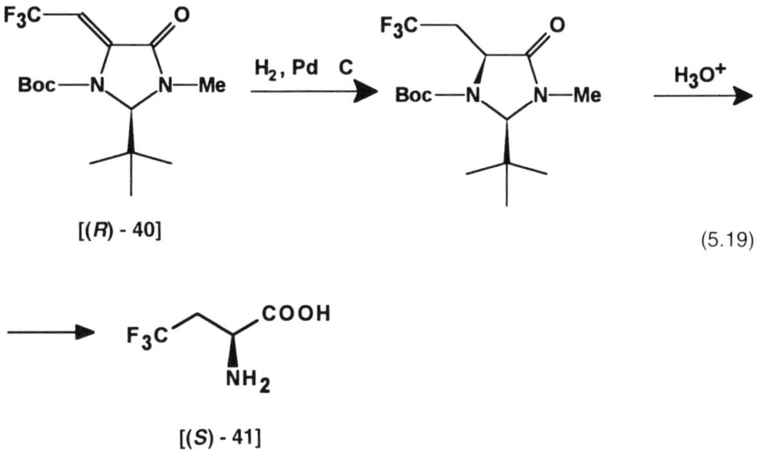

$$[(R) - 40] \qquad\qquad (5.19)$$

$$[(S) - 41]$$

through entry of hydrogen from the face opposite to the *tert*-butyl group [49] and (S)-3-trifluoromethylalanine (**41**) is obtained after acid hydrolysis. The starting trifluoroethylideneimidazolidinone **40** is obtained from the commercially available (R)-imidazolidinone and trifluoroacetoaldehyde so that, in general, the preparation of β-perfluoroalkylalanines through this method seems more convenient than through the direct alkylation of glycine enolate equivalents with α,α-dihydroperfluoroalkyl halides [50], which are poor alkylating species.

In the hydrogenation reaction chirality can be present on the catalyst (equations 5.17 and 5.18), on the substrate (equation 5.19) or in the reaction medium (equation 5.20). When benzylideneoxazolones **42** are treated with hydrogen in the presence of (S)-α-phenylethylamine, the (S,S)-amides **43** are formed with moderate diastereoselectivity (d.e. < 47%) through reduction of the carbon–carbon double bond and concomitant ring opening [51, 52]. As usual, the final, fluoro-substituted amino acids **4** are obtained through acid hydrolyses of the primary hydrogenation product.

In living organisms, the asymmetric 1,3-proton shift between two isomeric imines is often the key step in the synthesis of α-amino acids from α-keto acids. A similar process has been employed in an asymmetric and biomimetic synthesis of the β-polyfluoroalkyl-β-amino acids **46** (equation 5.21). Specifically, when (R)-3-polyfluoroalkyl-3-α-phenylethylamino acrylates (**44**) are heated in the presence of a trialkylamine, corresponding Schiff bases of β-amino acids (**45**) are formed in optically active form and the highest e.e. (75%) was obtained when Dabco was employed [53, 54]. Acid-catalysed removal of the protective groups at nitrogen and oxygen to give chiral and non-racemic β-amino acids (**46**) can be performed either in one or two synthetic steps.

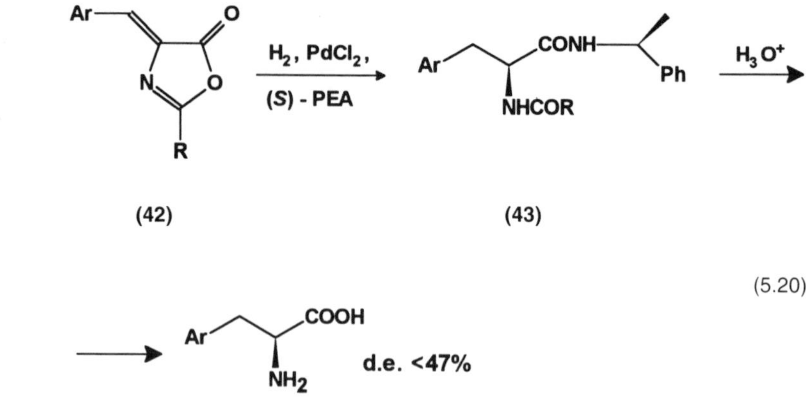

(42)   (43)

(5.20)

d.e. <47%

(4)

R = Me; Ph

Ar = 4-FC$_6$H$_4$; 4-F$_2$CHOC$_6$H$_4$

(S)-PEA = (S) - PhCH(NH$_2$)Me

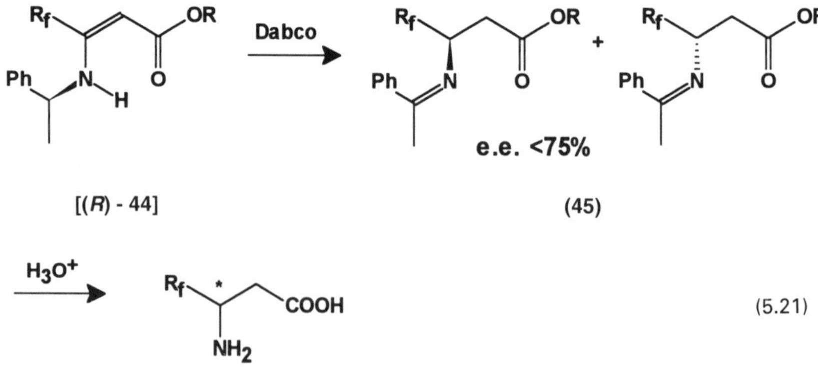

Dabco

e.e. <75%

[(R) - 44]   (45)

(5.21)

(46)

R$_f$ = CF$_3$; C$_2$F$_5$; n-C$_3$F$_7$; n-H(CF$_2$)$_2$; H(CF$_2$)$_4$

R = Me; Et

5.4 REFERENCES

1. R. Crossley, *Tetrahedron*, **48**, 8155 (1992).
2. E. J. Arien, *Eur. J. Clin. Pharmacol.*, **26**, 663 (1984).
3. B. Holmst, in H. Frank, B. Holmsted and B. Testa (Eds), *Chirality and Biological Activity*, Liss, New York, 1991, pp. 1–14.
4. *Guideline for Submitting Supporting Documentation in Drug Applications for the Manufacture of Drug Substances*, Office of Drug Evaluation and Research, Food and Drug Administration, Washington, DC, 1987, p. 3.
5. S. C. Stinton, *Chem. Eng. News*, September 28, 46 (1992).
6. G. M. R. Tombo and D. Bellus, *Angew. Chem., Int. Ed. Engl.*, **30**, 1193 (1991).
7. E. J. Ariens, J. J. S. van Rensen, W. Welling (Eds), *Stereoselectivity of Pesticides*, Elsevier, Amsterdam, 1988.
8. H. Frehse, E. Kesseler-Schmitz and S. Conway (Eds), *7th IUPAC Internatinal Congress of Pesticide Chemistry, Book of Abstracts*, Vol. I–III, Hamburg, 1990.
9. H. B. Kagan and J.-C. Fiaud, *Top. Stereochem.*, **10**, 175 (1978).
10. V. P. Kukhar and V. A. Soloshonok, *Russ. Chem Rev.*, **60**, 850 (1991).
11. J. T. Welch and S. Eswarakrisnan, *Fluorine in Bioorganic Chemistry*, Wiley, New York, 1991.
12. G. Resnati, *Tetrahedron*, **49**, 9385 (1993).
13. J. M. McIntosch, R. K. Leavitt, P. Mishra, K. C. Cassidy, J. E. Drake and R. Chadha, *J. Org. Chem.*, **53**, 1947 (1988).
14. R. Fitzi and D. Seebach, *Tetrahedron*, **44**, 5277 (1988).
15. D. Seebach, E. Dziadulewicz, L. Behrendt, S. Cantoreggi and R. Fitzi, *Liebigs Ann. Chem.*, 1215 (1989).
16. K. G. Grozinger, R. W. Kriwacki, S. F. Leonard and T. P. Pitner, *J. Org. Chem.*, **58**, 709 (1993).
17. D. Seebach, E. Juaristi, D. D. Miller, C. Schickli and T. Weber, *Helv. Chim. Acta*, **70**, 237 (1987).
18. D. Blaser and D. Seebach, *Liebigs Ann. Chem.*, 1067 (1991).
19. Yu. N. Belokon', *Janssen Chimi. Acta*, **10**, 4 (1992).
20. V. A. Soloshonok, Yu. N. Belokon', V. P. Kukhar', N. I. Chernoglazova, M. B. Saporovskay, V. I. Bakhmutov, M.T. Kolicheva and V. M. Belikov, *Izv. Akad. Nauk SSSR, Ser. Khim.*, 1630 (1990); *Chem. Abstr.*, **114**, 7135d (1991).
21. V. P. Kukhar', Yu. N. Belekon', V. A. Soloshonok, N. Yu. Svistunova, A. B. Rozhenko and N. A. Kuz'mina, *Synthesis*, 117 (1993).
22. V. A. Soloshonok, N. Yu. Svistunova, V. P. Kukhar', N. A. Kuz'mina and Yu. N. Belokon', *Izv. Akad. Nauk SSSR, Ser. Khim.*, 687 (1992);*Chem. Abstr.*, **117**, 212905h (1992).
23. V. A. Soloshonok, V. P. Kukhar', S. V. Galusko, A. B. Rozhenko, N. A. Kuz'mina, M. T. Kolycheva and Yu. N. Belokon', *Izv. Akad. Nauk SSSR, Ser. Khim.* , 1906 (1991); *Chem. Abstr.*, **116**, 21426x, (1992).
24. V. A. Soloshonok, V. P. Kukhar', S. V. Galushko, N. Yu. Svistunova, D. V. Avilov, N. A. Kuz'mina, N. I. Raevski, Yu. N. Struchkov, A. P. Pysarevsky and Yu. N. Belokon', *J. Chem. Soc., Perkin Trans. 1*, in press.
25. V. A. Soloshonok, V. P. Kukhar', A. S. Batsanov, M. A. Galakhov, Yu. N. Belokon' and Yu. T. Struchkov, *Izv. Akad. Nauk SSSR, Ser. Khim.*, 1548 (1991); *Chem. Abstr.*, **115**, 256590q (1991).
26. V. A. Soloshonok, N. Yu. Svistunova, V. P. Kukhar', A. B. Rozhenko and Yu. N. Belokon', *J. Fluorine Chem.*, **58**, 367 (1992).
27. V. A. Soloshonok, V. P. Kukhar', S. V. Galushko, M. T. Kolycheva, A. B.

Rozhenko and Yu. N. Belokon', *Izv. Akad. Nauk SSSR, Ser. Khim.*, 1166 (1991); *Chem. Abstr.*, **115**, 136682z (1991).

28. Yu. N. Belokon', A. G. Bulychev, S. V. Vitt, Yu. T. Struchkov, A. S. Batsanov, T. V. Timofeeva, V. A. Tsyryapkin, M. G. Ryzhov, L. A. Lysova, V. I. Bakhmutov and V. M. Belikov, *J. Am. Chem. Soc.*, **107**, 4252 (1985).

29. M. J. O'Donnell and S. Wu, *Tetrahedron: Asymmetry*, **3**, 591 (1992).

30. M. Kolb and J. Barth, *Tetrahedron Lett.*, 2999 (1979).

31. M. Kolb and J. Barth, *Liebigs Ann. Chem.*, 1668 (1983)

32. V. A. Soloshonok, I. I. Guerus, Yu. L. Yagupol'skii and V. P. Kukhar', *Zh. Org. Khim.*, **23**, 2308 (1987); *Chem. Abstr.*, **109**, 55185p (1988).

33. F. Viani, P. Bravo, V. A. Soloshonok and V. P. Kukhar', to be published.

34. J. T. Welch, K. Araki, R. Kawecki and J. A. Wichtowski, *J. Org. Chem.*, **58**, 2454 (1993).

35. K. Weinges and E. Kromm, *Liebigs Ann. Chem.*, 90 (1985).

36. T. Taguchi, A. Kawara, S. Watanabe, Y. Oki, H. Fuhushima, Y. Kobayashi, M. Okada, K. Ohta and Y. Iitaka, *Tetrahedron Lett.*, **27**, 5117 (1986).

37. S. Thaisrivongs, D. T. Pals, W. M. Kati, S. R. Turner and L. M. Jomasco, *J. Med. Chem.*, **28**, 1553 (1985).

38. S. Thaisrivongs, D. T. Pals, W. M. Kati, S. R. Turner, J. M. Jomasco and W. Watt, *J. Med. Chem.*, **29**, 2080 (1986).

39. K. Fearon, A. Spaltenstein, P. B. Hopkins and M. H. Gelb, *J. Med. Chem.*, **30**, 1617 (1987).

40. S. Thaisrivongs, H. J. Schostarez, D. T. Pals and S. R. Turner, *J. Med. Chem.*, **30**, 1837 (1987).

41. R. Shen, C. Priebe, C. Patel, L. Rubo, T. Su and M. Khan, *Tetrahedron Lett.*, **33**, 3417 (1992).

42. R. P. Robinson and K. M. Donahue, *J. Org. Chem.*, **57**, 7309 (1992).

43. R. W. Lang and B. Schaub, *Tetrahedron Lett.*, **29**, 2943 (1988).

44. T. Taguchi, O. Kitagawa, Y. Suda, S. Ohkawa, A. Hashimoto, Y. Iitaka and Y. Kobayashi, *Tetrahedron Lett.*, **29**, 5291 (1988).

45. W. S. Knowles, *Acc. Chem. Res.*, **16**, 106 (1983).

46. V. Caplar, G. Comisso and V. Sunjic, *Synthesis*, 85 (1981).

47. H.-W. Krause, H.-J. Kreuzfeld, C. Dobler and S. Taudien, *Tetrahedron: Asymmetry*, **3**, 555, (1992).

48. I. Ojima, K. Kato, K. Nakahaschi, T. Fuchikami and M. Fujita, *J. Org. Chem.*, **54**, 4511 (1989).

49. D. Seebach, H. M. Burger and C. P. Schickli, *Liebigs Ann. Chem.*, 669 (1991).

50. T. Tsushima, K. Kawada, S. Ishihara, N. Uchida, O. Shiratori, J. Higaki and M. Hirata, *Tetrahedron*, **44**, 5375 (1988).

51. L. F. Godunova, E. I. Karpeiskaya, E. S. Levitina, E. I. Klabunovskii, Yu. L. Yagupolskii and M. T. Kolycheva, *Izv. Akad. Nauk SSSR, Ser. Khim.*, 1359 (1987);*Chem. Abstr.*, **108**, 167905j (1988).

52. L. F. Godunova, E. S. Levitina, E. I. Karpeiskaya, E. I. Klabunovskii, Yu. L. Yagupol'skii and M. T. Kolycheva, *Izv. Akad. Nauk SSSR, Ser. Khim.*, 404 (1989); *Chem. Abstr.*, **111**, 134711b (1989).

53. V. A. Soloshonok, A. G. Kirilenko, V. P. Kukhar' and G. Resnati, *Amino Acids*, **4**, 130 (1993).

54. V. P. Kukhar', V. A. Soloshonok, S. V. Galushko and A. B. Rozhenko, *Dokl. Akad. Nauk SSSR*, **310**, 886 (1990); *Chem. Abstr.*, **113**, 78920w (1990).

6 Enzymatic Synthesis of Fluorine-containing Amino Acids

YASUSHI MATSUMURA and MASAHIRO URUSHIHARA
Asahi Glass Company Ltd, Research Center, Hazawa-cho, Kanagawa-ku, Yokohama 221, Japan

6.1 INTRODUCTION

The enzymatic synthesis of amino acids [1] has been one of the major areas in both academic and industrial research fields for over a century. The revolutionary progress during the last two decades in molecular biology and protein engineering has stimulated related studies with this field and revealed the almost infinite utility of naturally occurring or non-proteinogenic amino acids. The availability of chiral amino acids is also strongly appealing to the chemist to maximize their applicability in peptides, antibiotics and other biologically active molecules. In accord with strong demands in the food and pharmaceutical markets, the recent development of technology with purified or immobilized enzymes and whole cell system has accelerated rapidly.

The enzymatic synthesis of fluorine-containing amino acids has started to be exploited because of the interest in studying enzymatic reaction mechanisms and their biological activities. From the view point of the enantioselective synthesis of the amino acids, this enzymatic approach still seems to have several advantages in comparison with chemical synthesis in spite of great scientific endeavor and dramatic developments in asymmetric synthesis: (1) high enantiometric purity of the product; (2) mild reaction conditions; and (3) reasonable cost.

The efficiency of the synthetic process should be enhanced if the requirements such as ready availability of fluorinated substrates and easy operation, including purification, could be satisfied. Recently, several important processes for the bulk production of fluorine-containing amino acids have appeared and the most useful work seems often to result from a scientific basis in combination with chemical and biological methodology.

The biosynthesis of the amino acids can be classified into two types of strategy. The first approach is the enzymatic resolution of racemic fluorine-containing amino acid derivatives. The resolution seems rather classical, but the number and variety of fluorine-containing amino acids that have been obtained are testimony to the viability of this method. This topic will be covered in Chapter 7.

The second approach, discussed in this chapter, can be categorized as enzymatic bond formation with prochiral fluorine-containing substrates.

Fluorine-containing Amino Acids: Synthesis and Properties   Edited by V. P. Kukhar' and V. A. Soloshonok
© 1995 John Wiley & Sons Ltd

Currently, there are five reaction types (Table 6.1). It should be emphasized that the scope of these enzyme-catalyzed reactions has been greatly enlarged along with extensive screening of natural and recombinant biocatalysts to have broad substrate specificities. In the following sections, more details of the research over the last two decades are presented.

Table 6.1 Enzymatic synthesis of fluorine-containing amino acids

Amino acid	Biocatalyst	Yield $(\%)^a$	% e.e.	Ref.
1. Transamination				
Aspartic acid				
3-F	Glutamate–aspartate transaminase	Trace	—	2
3,3,-F$_2$	Glutamate–aspartate transaminase	Trace	—	3
3,3-F$_2$	Aspartate transaminase	Trace	—	4
Phenylalanine				
2'-F	Transaminase (*Alcaligenes faecalis*)	77.7	>99	5
3'-F	Transaminase (*Alcaligenes faecalis*)	51.7	>99	5
4'-F	Transaminase (*Alcaligenes faecalis*)	91.8	>99	5
4'-CF$_3$	Transaminase (*Alcaligenes faecalis*)	50	>99	5
2',3',4',5',6'-F$_5$	Transaminase (*Alcaligenes faecalis*)	60	>99	5
D-2'-F	D-Transaminase	17	>99	6
D-3'-F	D-Transaminase	34	>99	6
D-4'-F	D-Transaminase	27	>99	6
Valine				
4,4,4-F$_3$	Transaminase (*Alcaligenes faecalis*)	0	—	7
Trifluoroleucine				
5,5,5-F$_3$	Transaminase (*Alcaligenes faecalis*)	39	>99	8
D-5,5,5-F$_3$	D-Transaminase	27	>99	6
2. Reductive amination				
Glutamic acid				
(2R,3R)-3F	Glutamate dehydrogenase	95	100	9
(2R,3S)-3F	Glutamate dehydrogenase	95	100	9
Phenylalanine				
(2S)-4'-F	Phenylalanine dehydrogenase	>99^b	100	10
3. Condensation				
Tyrosine				
2'-F	Tyrosine phenol-lyase	—	100	11
3'-F	Tyrosine phenol-lyase	—	100	11
2'-F	Tyrosine phenol-lyase (*Erwinia herbicola* ATCC 21434)	90	>99	12
3'-F	Tyrosine phenol-lyase (*Erwinia herbicola* ATCC 21434)	85	>99	12
3',5'-F	Tyrosine phenol-lyase (*Erwinia herbicola* ATCC 21434)	Trace	—	12
Tryptophan				
4'-F	Tryptophanase (*E. coli* K-12)	100	>99	5
5'-F	Tryptophanase (*E. coli* K-12)	70	>99	5
6'-F	Tryptophanase (*E. coli* K-12)	72	>99	5

Table 6.1 *(continued)*

Amino acid	Biocatalyst	Yield $(\%)^a$	% e.e.	Ref.
4. Addition reactions				
Phenylalanine				
2'-F	Phenylalanine ammonia-lyase (*Rhodosporidium toruloides*)	85.8	>99	5
3'-F	Phenylalanine ammonia-lyase (*Rhodosporidium toruloides*)	90.3	>99	5
4'-F	Phenylalanine ammonia-lyase (*Rhodosporidium toruloides*)	25.8	>99	5
2'-CF$_3$	Phenylalanine ammonia-lyase (*Rhodosporidium toruloides*)	12.9	>99	5
4'-CF$_3$	Phenylalanine ammonia-lyase (*Rhodosporidium toruloides*)	4.5	>99	5
5. Unspecified biosynthesis				
Threonine				
4-F	(*Streptomyces cattleya*)	44	—	13

^{a} Isolated yield except as described elsewhere.
^{b} Determined by HPLC.

6.2 TRANSAMINATION

In 1960, Kun *et al.* [2] studied the mechanism of transamination by employing monofluorooxaloacetate as a substrate and an inhibitor. According to kinetic analysis, glutamic–aspartic transaminase of pig heart mitochondria proved to be inhibited by β-fluorooxaloacetate in competition with oxaloacetate. However, when β-fluorooxaloacetate was transaminated with aspartate, the resulting β-fluoroaspartate underwent subsequent dehydrofluorination and deamination to oxaloacetate.

The same group [3] reported later that the enzyme was also strongly inhibited by difluorooxaloacetate. They noted that this inhibition differed in certain important respects from the inhibitory effect of β-fluorooxaloacetate, since no measureable transamination could be detected with a relatively crude enzyme preparation. Only highly purified heart muscle enzyme catalyzed the transamination of difluorooxaloacetate at a rate approximately 10^{-4} times that of oxaloacetate.

Briley *et al.* [4] contributed to research on the transamination of the difluoro-α-keto acid **3** as the substrate of biosynthesis using aspartate transaminase. Ethyl difluorooxaloacetate (**2**) was prepared by replacement of the geminal hydrogen atoms on the β-carbon of diethyl oxaloacetate by fluorine by using perchloryl fluoride in 41% yield. The ester **2** was hydrolyzed by boiling in 3 M HCl under reflux for 1 h to afford the difluorooxaloacetic acid **3** after recrystallization from trifluoroacetic acid (equation 6.1).

The enzymatic reaction was carried out with cytoplasmic transaminase prepared in its aldimine form from fresh pig heart ventricles. They found that in steady-state experiments difluorooxaloacetate behaved as a competitive inhibitor of 2-oxoglutarate and as an uncompetitive inhibitor with respect to aspartate. Interestingly, ^{19}F NMR analysis demonstrated that difluoro-aspartate was produced by slow transamination of difluorooxaloacetate using cysteine sulfinate as an amino group donor in the presence of high concentrations of aspartate transaminase. In contrast to similar enzymatic transamination [2] of monofluorooxaloacetate, less than 5% decomposition of the transaminated amino acid was observed.

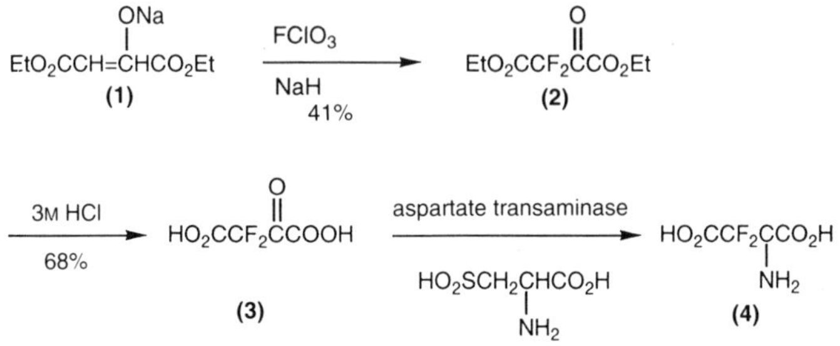

(6.1)

Uchida and Tanaka [5] at Asahi Glass have extensively studied transamination for the synthesis of various phenylalanine derivatives from the corresponding phenylpyruvic acids (equation 6.2).

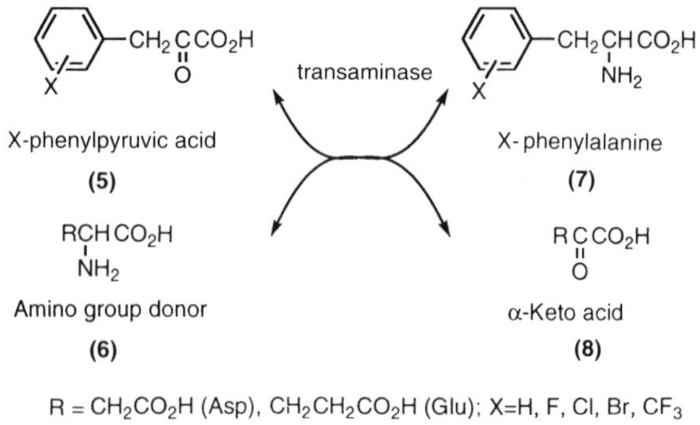

X-phenylpyruvic acid
(5)

transaminase

X-phenylalanine
(7)

RCHCO$_2$H
|
NH$_2$
Amino group donor
(6)

RCCO$_2$H
||
O
α-Keto acid
(8)

(6.2)

R = CH$_2$CO$_2$H (Asp), CH$_2$CH$_2$CO$_2$H (Glu); X=H, F, Cl, Br, CF$_3$

Synthesis of phenylpyruvic acid derivatives (13) is outlined in equation 6.3. Condensation of an aldehyde (9) and acetylglycine (10) provided the corresponding azlactones (11), which were hydrolyzed via 12 to afford the phenylpyruvic acids 13 in 20–50% yield.

(6.3)

X=F, Cl, Br, CF$_3$

It is well known that many living microorganisms contain the transaminase which converts α-ketocarboxylic acids into α-amino acids in the presence of an amino group donor. The authors have screened bacterial producers of fluorine-containing phenylalanine derivatives using *p*-fluorophenylpyruvic acid as a substrate. *Alcaligenes*, *Brevibacterium* and *Pseudomonas* bacteria showed high activities for transamination, whereas *Corynebacterium*, known as an amino acid producer, and *Rhodospovirudium*, a kind of yeast, showed weak or no activities. The authors found that *Alcaligenes faecalis* IAM1015 had the highest activity in the screening, which was selected for the enzymatic process.

The enzymatic reactions were carried out using the resting cells of *A. faecalis* IAM 1015 grown in nutrient medium (pH 7.0) in a shaking flask at 30 °C for 24–40 h. The cells were harvested by centrifugation and washed with a phosphate buffer to obtain wet cells for the transamination. Study of amino group donor of the transamination of phenylpyruvic acid is shown in Figure 6.1. In comparison with glutamic acid, aspartic acid had a far greater promoting effect on the reaction, resulting in a >80% production yield. The reaction mechanism involved decarboxylation of oxaloacetic acid (**17**), oxaloacetic acid decarboxylase (OAA decarboxylase) driving the equilibrium of the transamination towards the positive side to produce the phenylalanine **16** (equation 6.4). Enzyme assay of the decarboxylase showed a higher activity than the transaminase, supporting this supposition (L-transaminase, 20 mU mg^{-1} dry weight; OAA decarboxylase, 50 mU mg^{-1} dry weight).

Transaminase of *Alcaligenes faecalis* exhibited a remarkably broad spectrum of substrate specificities. Kinetic studies on *para*-substituted halophenylalanine derivatives are summarized in Figure 6.2. Fluorine substitution at the *para* position did not affect the reaction velocities, but substitution of a more bulky chlorine or bromine atom retarded the biosynthesis. Surprisingly, very electron-deficient derivatives, such as pentafluorophenylalanine or trifluoromethyl-substituted phenylalanines, were also obtained by slow transamination reactions in >50% yield. Table 6.2 gives the production yields of phenylalanine derivatives under optimized fermentation conditions.

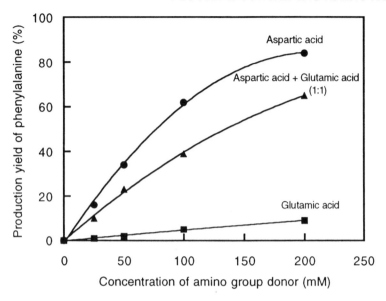

Figure 6.1 Effect of amino group donor on transamination of phenylpyruvic acid by *Alcaligenes faecalis*. Phenylpyruvic acid, 100 mM, pH 8.0, 30 °C, 24 h; cell, 4 mg dry weight ml^{-1}

Experiments on immobilized enzyme systems [14] provided promising data for the continuous production of *p*-fluorophenylalanine on a large scale. Entrapment of *Alcaligenes faecalis* IAM 1015 in calcium alginate gels was accomplished under mild conditions and the resulting beads of the immobilized cells were packed in a column reactor. The immobilization of the cells reduced the productivity to 50% of that achieved using the free cells, but remarkably improved the stability of the biocatalyst. After continuous reaction for 1000 h, 70% of the initial activity of the transaminase in the immobilized cells was found to be retained.

Asahi Glass researchers [7] extended their investigation on transamination to fluoro-aliphatic α-keto acids. Attempts at the biosynthesis of trifluorovaline were conducted by transamination of 4,4,4-trifluoro-3-methyl-2-oxobutanoic

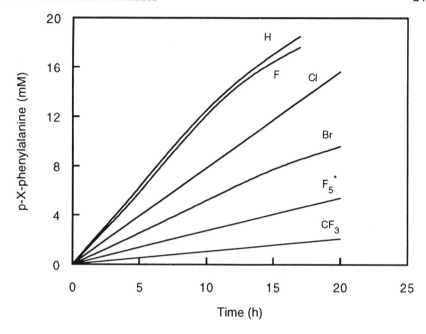

Figure 6.2 Effect of substituents in the *para*-position of phenylpyruvic acids on the transamination reaction. X-phenylpyruvic acid (X = H, Cl, Br, CF₃); F₅* = pentafluorophenylpyruvic acid, 50 mM; aspartate, 50 mM; pH 8.0, 30 °C

Table 6.2 Production yield of fluoro-phenylalanine with L-transaminase

Product	Yield (%)
L-Phenylalanine	77.0
L-*o*-Fluorophenylalanine	77.7
L-*m*-Fluorophenylalanine	51.7
L-*p*-Fluorophenylalanine	91.8

acid prepared by the condensation of trifluoroacetone with α-isocyanoacetate. However, owing to the strongly electron-withdrawing nature of the trifluoromethyl group neighboring the active site on the substrate, the enzyme could not catalyze the reaction at all.

More recently, successful application to an aliphatic system was reported by the same group [8], using 5,5,5-trifluoro-4-methyl-2-oxopentanoic acid (**23**) as a precursor of transamination for the synthesis of trifluoroleucine.

The preparation of **23** is shown in equation 6.5. Condensation of ethyl isocyanoacetate (**19**) with the known aldehyde [15] **20** afforded the oxazoline **21** in 75% yield, which was converted into **22** by a hydrolysis–dehydration sequence.

The ester **22** was directly obtained in 40% yield from **19** and **20** through iso-merization in the presence of lithium diisopropylamide by slowly raising the reaction temperature. Acid hydrolysis provided the desired keto acid **23** quantitatively.

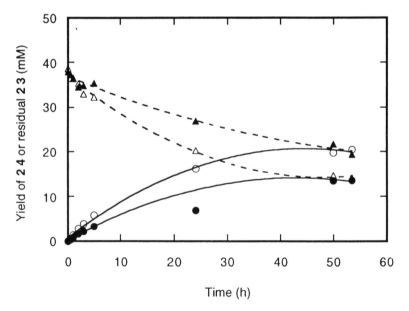

(6.5)

The enzymatic reactions were carried out using the resting cells of *Alcaligenes faecalis* IAM 1015 with L-glutamic acid or L-aspartic acid at 37 °C for 55 h. The results are shown in Figure 6.3. In contrast to the transamination

Figure 6.3 Production of trifluoroleucine. Glu as NH_2 donor: yield of **24** (○); residual **23** (△). Asp as NH_2 donor: yield of **24** (●); residual **23** (▲). **23**, 50 mM; aspartate or glutamate, 100 mM; pyridoxal 5-phosphate, 1 mM; pH 8.0, 37 °C

[5] of fluorinated phenylpyruvic acids. L-glutamic acid was more effective than L-aspartic acid as a donor of an amino group with respect to the reaction rate and selectivity. Whereas the reaction using L-aspartic acid produced L-alanine as a side-product, which caused serious problems in purification, no L-alanine was observed in the reaction mixture with L-glutamic acid. A larger scale production using L-glutamic acid gave **24** in 39% yield as a 1:1 mixture of (2S,4S)- and (2S,4R)-diastereomers, which was isolated after purification by ion-exchange chromatography.

D-Transaminase plays an essential role in the biosynthesis of D-amino acids, which are used as constituents of the peptidoglycan layer of bacterial cell walls. Using D-transaminase instead of L-transaminase, D-fluoroamino acids can apparently be produced, but optically pure D-amino acids could not be produced because of the L-transaminase present in whole cells. To overcome this problem, a few kinds of recombinant bacteria which have thermostable D-transaminase [16] screened from microorganisms in soil were developed that produce a variety of D-amino acids after inactivation of L-transaminase by heating.

Ohdo *et al.* [6] applied their enzymatic system for D-amino acids employing the cloned bacteria selected from their screening to the production of fluoro-D-phenylalanine derivatives **27a–c** and trifluoro-D-leucine (**27d**) (equation 6.6). When utilizing D-glutamic acid (**26**) as an amino group donor, **27a–d** were synthesized from the corresponding α-keto acids **25** with >99% optical purities in 17–34% yields. The optical purity of D-trifluoroleucine (**27d**) was as high as 97.4% even when DL-glutamate was used as the donor because L-transaminase activity was almost completely lost in the cloned bacteria.

RCH_2CCO_2H
O
α-Keto acid
(**25**)

D-Transaminase

RCH_2CHCO_2H
NH_2
D-Amino acid
(**27**)

(6.6)

$HO_2CCH_2CH_2 \cdot CHCO_2H$
NH_2
D-Glutamic acid
(**26**)

$HO_2CCH_2CH_2 \cdot C{-}CO_2H$
O
α-Keto acid
(**28**)

Yield (%)

		Yield (%)
27a	R= F—	27
27b	(F)	34
27c		17
27d	CF_3 / CH_3	27

6.3 REDUCTIVE AMINATION

In a related approach to transamination, NAD^+-dependent reductive amination has emerged as an important process for bulk production.

Vidal-Cros et al. [9] reported the use of glutamate dehydrogenase (GDH) for the synthesis of (2R,3R)- and (2R,3S)-3-fluoroglutamic acid (**35** and **36**). Preparation of sodium 2-oxo-3-fluoroglutarate (**34**), the substrate of the enzymatic reaction, is illustrated in equation 6.7. Condensation of diethyl fluorooxaloacetate sodium enolate (**31**), prepared by Claisen condensation of diethyl oxalate (**30**) and ethyl fluoroacetate (**29**), with ethyl bromoacetate furnished diethyl 2-oxo-3-carbethoxy-3-fluoroglutarate (**33**) accompanied by the O-alkylation product **32** in a ratio of 1 : 1. Decarboxylation of the corresponding α-keto acid after acidic hydrolysis of **33** in a mixture of hydrochloric acid and acetic acid occurred by simple heating to afford the desired 2-oxo-3-fluoroglutaric acid, which was isolated as the disodium salt **34**. According to the reductive amination of the racemic 2-oxo-3-fluoroglutarate **34** with

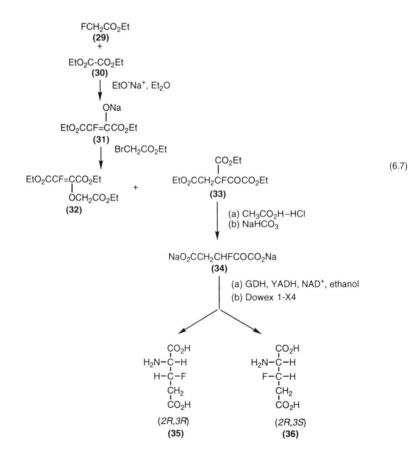

glutamate dehydrogenase, both $(2R,3R)$- and $(2R,3S)$-3-fluoroglutamic acid (**35** and **36**) were obtained in high chemical and optical yields. This enzyme catalyzes the initial formation of an α-iminoglutarate and subsequent reduction of the imine by NADH. The consumed NADH was regenerated *in situ* by using a yeast alcohol dehydrogenase (YADH)–ethanol system. In contrast to a pyridoxal 5'-phosphate (PLP)-dependent transaminase catalysis, undesired fluoride elimination is unlikely. The diastereomers **35** and **36** were separated by ion-exchange chromatography and proved to be optically pure by GC analysis on a chiral phase column after derivatization to the corresponding N-acetyldiisopropyl esters.

Asano *et al.* [10] reported studies on the substrate specificity of phenylalanine dehydrogenase from *Bacillus sphaericus* SCRC-R79a in the reductive amination reaction examined with various natural and synthetic analogues. They screened bacterial producers of the new enzyme, phenylalanine dehydrogenase (PheDH), and characterized the enzyme isolated from *B. sphaericus* SCRC-R79a [17]. The enzyme shows a higher stability and broader substrate specificity than other natural or recombinant enzymes screened. Various optically pure (S)-amino acids (**38**) were quantitatively synthesized using phenylalanine dehydrogenase with the regeneration of NADH by formate dehydrogenase (equation 3.8). Table 6.3 shows the relative rates in the reductive amination of a variety of 2-oxo acids (**37**). The enzyme utilized well phenylpyruvate analogues substituted at the phenyl ring, but substitution at the 3-position of pyruvic acid with a longer or bulkier group retarded the reaction. The enzymatic reaction employing (4-fluorophenyl)pyruvate as a substrate proceeded with 29% relative activity to provide optically pure 4-fluorophenylalanine quantitatively.

$$(6.8)$$

6.4 CONDENSATION

Enzyme-catalyzed condensation to produce fluorine-containing tyrosines was first reported by Nagasawa *et al.* [11]. They found that tyrosine phenollyase (β-tyrosinase) catalyzes the degradation of L-tyrosine to pyruvate, ammonia and phenol in the presence of pyridoxal 5'-phosphate, and at the same time the

Table 6.3 Substrate specificity of PheDH from *B. sphaericus* SCRC-R79a. (Reprinted with permission from Asano *et al.*, *J. Org. Chem.*, **55**, 5567. Copyright (1990) American Chemical Society)

Substrate (10 mM)	Relative activity[a]
(4-Hydroxyphenyl) pyruvate	100
Phenylpyruvate	74
(4-Vinylphenyl) pyruvate	38
(4-Fluorophenyl)pyruvate	29
2-Oxo-4-(methylthio)butyrate	8.1
2-Oxo-3-(R,S)-phenylvalerate	6.5
2-Oxoisocaproate	5.7
2-Oxobutyrate	4.6
2-Oxoisovalerate	4.0
2-Oxo-3-(R,S)-(4-fluorophenyl)butyrate	3.7
2-Oxo-3-(R)-phenylbutyrate	3.5
2-Oxo-4-phenylbutyrate	2.5
2-Oxo-3-(R,S)-methylvalerate	2.1
2-Oxo-5-phenylvalerate	2.0
2-Oxo-3-(R,S)-(3-methylphenyl)butyrate	0.74
2-Oxononanoate	0.54
2-Oxo-3-(2-naphthalenyl)propionate	0.34
2-Oxo-3-(R,S)-phenyl-4-methylvalerate	0.07

[a] The following compounds were inert as substrates: benzoylformate, ethyl phenylpyruvate, ethyl 2-oxo-4-phenylbutyrate, ethyl 3-oxo-4-phenylbutyrate, 3-oxo-4-phenylbutyrate, 2-oxo-phenylpropanol, 2-oxo-3-phenylnonanoate and 2-oxo-3-methyl-3-(4-propylphenyl)propionate.

reaction can be employed for the synthesis of L-tyrosine from them. The homogeneous enzyme was prepared in their laboratory from the cells of *Cinobacter intermedius* grown in a bouillon medium supplemented with L-tyrosine as an inducer. They studied the mechanism of the reversible reaction and applied the reaction to the alkylated or halogenated phenol derivatives for the synthesis of the corresponding tyrosines. The relative rates of synthesis of tyrosine analogues from ammonia, pyruvate and various phenol derivatives are given in Table 6.4. The reaction of a 3-substituted phenol proceeded much faster than that of 2-substituted phenol except for fluorophenol. From observations of the degradation reactions of tyrosine-related amino acids, the rate of degradation showed a tendency to be inversely proportional to the van der Waals radii (Figure 6.4). It was suggested that the steric size of the substituent could be an important factor in the interaction of the substrate with the enzyme. The presence of a bulky substituent such as an iodine atom adjacent to the hydroxy group on the benzene ring might strongly interfere with the interaction between the enzyme and the hydroxy group. However, the opposite trend in the reaction with fluorophenols cannot be explained by this simple theory, and it was noted that the electronic density of the substituents may contribute to the reaction rates.

Table 6.4 Relative rates of synthesis of tyrosine-related amino acids from ammonia, pyruvate and phenol derivatives by tyrosine phenol-lyase. (Reproduced by permission of Springer-Verlag from *European Journal of Biochemistry*, **117**, 33 (1981))

Substrate	Product	Relative rate
Phenol	L-Tyrosine	100
o-Fluorophenol	3-Fluoro-L-Tyrosine	66.3
m-Fluorophenol	2-Fluoro-L-Tyrosine	23.2
o-Chlorophenol	3-Chloro-L-Tyrosine	15.3
m-Chlorophenol	2-Chloro-L-Tyrosine	33.4
o-Bromophenol	3-Bromo-L-Tyrosine	2.0
m-Bromophenol	2-Bromo-L-Tyrosine	4.2
o-Iodophenol	3-Iodo-L-Tyrosine	0
m-Iodophenol	2-Iodo-L-Tyrosine	1.5
o-Methylphenol (*o*-cresol)	3-Methyl-L-Tyrosine	2.0
m-Methylphenol (*m*-cresol)	2-Methyl-L-Tyrosine	9.8
o-Ethylphenol	3-Ethyl-L-Tyrosine	0
m-Ethylphenol	2-Ethyl-L-Tyrosine	2.0
o-Methoxylphenol	3-Methoxy-L-Tyrosine	1.1
m-Methoxyphenol	2-Methoxy-L-Tyrosine	30.6

Tanaka *et al.* [12] reported the practical enzymatic synthesis of fluorotyrosines **41** from fluorophenols **39** and pyruvic acid (**40**) by using intact cells of *Erwinia herbicola* ATCC 21434 containing tyrosine phenol-lyase (equation 6.9).

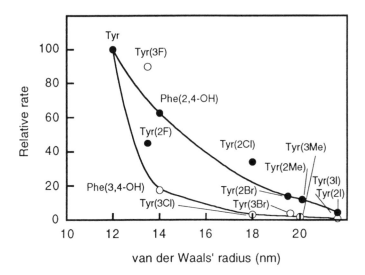

Figure 6.4 Effect of van der Waals radii of *ortho* (●) and *meta* substituents (○) on the rate of α, β-elimination reaction of tyrosine analogues. (Reproduced by permission of Springer-Verlag from *European Journal of Biochemistry*, **117**, 33 (1981))

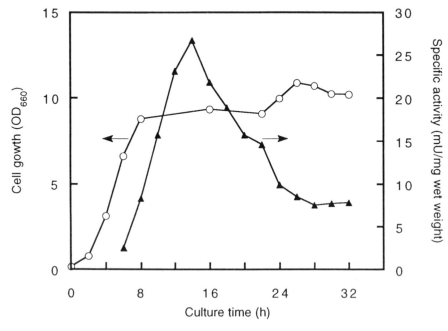

Tyrosine phenol-lyase is known as an inducible enzyme whose activity in microorganisms is induced by the addition of L-tyrosine to culture medium. However, the growth of *Erwinia helbicola* was partly repressed by the toxicity of phenol, which was formed from the degradation of L-tyrosine. Examination of various nutrition and control of the addition of L-tyrosine (0.05%) to the basal medium improved the cell growth. In the early stationary phase, maximum activity in growing cells was obtained, but the activity decreased quickly with the culture time (Figure 6.5). For the preparation of intact cells, the harvest time was fixed at the early stationary phase (12 h).

Under the optimized conditions, the reaction was carried out in the presence of various phenol derivatives as substrates for tyrosine phenol-lyase. As shown in

Figure 6.5 Time courses of cell growth and tyrosine phenollyase activity with *Erwinia herbicola* ATCC 21434. Cultivation was carried out at 30 °C in 3 l fermentation jar containing 2 l of medium with the pH controlled at 7.5 with KOH, and aeration at 0.25 vvm at an agitation speed of 400 rpm

Table 6.5 Formation of fluorotyrosines from fluorophenols[a]

Substrate	Product	Yield (%)
o-Fluorophenol	m-Fluorotyrosine	37
m-Fluorophenol	o-Fluorotyrosine	45
2,6-Difluorophenol	3,5-Difluorotyrosine	Trace

[a] Phenols, 50 mM; sodium pyruvate, 50 mM; ammonium acetate, 50 mM; pH 8.0, 30°C, 24 h.

Table 6.5, o-fluorophenol and m-fluorophenol were converted into m-fluorotyrosine and o-fluorotyrosine, respectively. However, 2,6-difluorophenol was hardly converted into 3,5-difluorotyrosine. Since fluorophenols are protein denaturing, which causes inactivation of the enzyme, the optimum concentration of fluorophenols in the reaction mixture was examined in order to obtain the maximum yield of fluorotyrosine. As shown in Table 6.6, fluorophenols showed higher substrate inhibition and lower K_m values than phenol. The trends in Table 6.6 may be explained by the high affinity of the fluorine-containing phenols for enzyme. After optimizing the concentrations of other substrates, pyruvate and ammonia, fluorotyrosines were formed in high yield with $> 99\%$ e.e.

Fluorotryptophans (**44**) were synthesized by the condensation of fluoro-indoles (**42**) and serine (**43**). *Achromobactor* and *E. coli* were used as a source of the enzyme tryptophanase, which catalyzed the production of tryptophan as shown in equation 6.10. Using *E. coli* K-12 ATCC 10768, the synthesis of various fluorotryptophans was reported by Uchida and Tanaka [5]. The effects of fluorine substitution on the indole ring are summarized in Table 6.7. It was found that 5- and 6-fluoroindole showed a small substrate inhibition, but when using 4-fluoroindole as a substrate the production rate of fluorotryptophan declined sharply with increase in the concentration of the substrate. Although

Table 6.6 Kinetic parameters for tyrosine phenol-lyase from intact cells with phenols

Substrate	V_{max}(mM min^{-1})	K_m(mM)	K_i(mM)
Phenol	0.47	23	150
m-Fluorophenol	0.23	20	60
o-Fluorophenol	0.16	20	66

(6.10)

Fluoroindole	Serine		Fluorotryptophan
(**42**)	(**43**)		(**44**)

Table 6.7 Kinetic parameters for tryptophanase with fluoroindoles

Indole	K_m(mM)	V_{max}(nmol min^{-1} OD^{-1})
Indole	2.0	40
4-Fluoroindole	2.0	60
5-Fluoroindole	2.0	50
6-Fluoroindole	2.0	20

the K_m values of fluoroindoles were not influenced by the position of substitution by fluorine in the indole ring, the maximum reaction rate of 4-fluoroindole was higher than those of other fluoroindoles. In order to prevent the substrate inhibition of 4-fluoroindole, the reaction was performed in a two-phase system (water and toluene). Because of the high solubility of 4-fluoroindole in toluene, 4-fluoroindole could be kept at a low concentration so as not to inhibit the desired enzymatic reaction. As shown in Figure 6.6, the yield of 4-fluorotryptophan increased dramatically from 20% to 100%. On the other hand, use of the two-phase system made little improvement to the production of 5- and 6-fluorotryptophan.

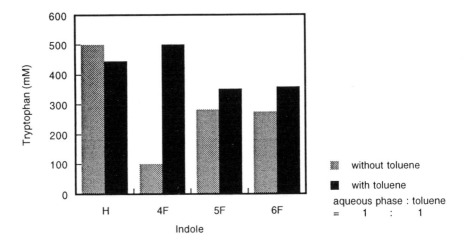

Figure 6.6 Effect of addition of toluene to the reaction mixture on the conversion of fluoroindoles into fluorotryptophans. Indoles, 500 mM; L-serine, 750 mM; sodium sulfite, 1.25 mg ml^{-1}; EDTA, 3.3 mg ml^{-1}; ammonium sulfate, 31.3 mg ml^{-1}; pH 7.5, 30 °C, 60 h; cell, 2 mg dry weight ml^{-1}

6.5 ADDITION REACTIONS

Biosynthesis employed with phenylalanine ammonia-lyase (PAL) is well known as an important industrial process for the production of L-phenylalanine from *trans*-cinnamic acid and ammonia. Uchida and Tanaka [5] at Asahi Glass reported the synthesis of fluorophenylalanines (**47**) from fluorocinnamic acids (**46**) by this strategy utilizing the enzyme originating from a kind of yeast, *Rhodosporidium toruloides* IFO 0559 (equation 6.11).

The fluorine-containing substrates **46** were prepared in good yields by Wittig–Horner reaction with fluorinated benzaldehyde derivatives (**45**) and subsequent hydrolysis. The enzymatic reaction is reversible and easily equilibrated, therefore an excess of ammonium ion was necessary to drive the reaction for the production of phenylalanine. PAL of *Rhodospridium toruloides* proved to be induced by L-phenylalanine and stabilized by the addition of isoleucine to the medium. Figure 6.7 illustrates the results demonstrating that the addition of isoleucine enhanced the cell growth and the enzyme activity.

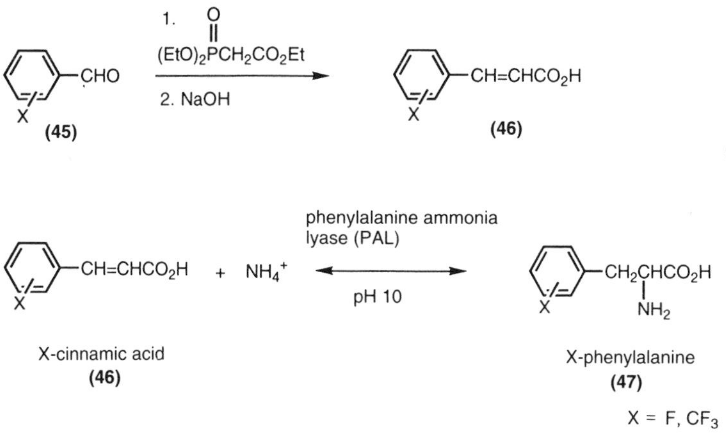

$$\text{(45)} \quad \xrightarrow[\text{2. NaOH}]{\substack{\text{1.} \quad \overset{O}{\underset{\|}{(EtO)_2PCH_2CO_2Et}}}} \quad \text{(46)}$$

(6.11)

X-cinnamic acid
(**46**)

X-phenylalanine
(**47**)

X = F, CF$_3$

Owing to the very unstable nature of PAL, which loses most of its enzyme activity after several hours, the practical production of phenylalanine requires a large amount of cells. As shown in Figure 6.8, 24 g of dry cells in 1 l of reaction solution could be sufficient to produce *o*-fluorophenylalanine from *o*-fluorocinnamic acid in over 90% yield. The efficiency of the reaction depended greatly on the position of the substituents in cinnamic acid (Figure 6.9). *p*-Fluorocinnamic acid gave *p*-fluorophenylalanine in poor yield, but *o*- and *m*-fluorocinnamic acid were superior as substrates, being transformed into the corresponding fluorophenylalanines in 85.8 and 90.3% yields, respectively. The enzyme also catalyzed the reaction with trifluoromethyl-substituted cinnamic acid derivatives to provide the corresponding phenylalanines in lower yields.

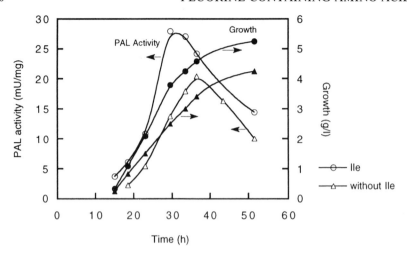

Figure 6.7 Time course of cell growth of *Rhodosporidium toruloides* IFO 0559 and PAL activity. Cultivation was carried out at 30 °C in a 10 l fermentation jar containing 2 l of medium with pH controlled at 6.0 with KOH, and aeration at 0.25 vvm at 400 rpm. Medium: yeast extract 1.0%, peptone 1.0%, NaCl 0.5%, L-Phe 0.05% (L-Ile 0.5%)

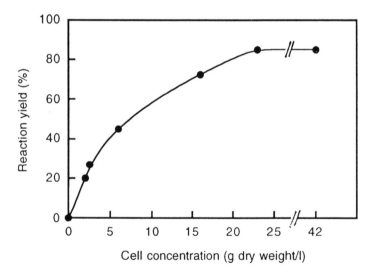

Figure 6.8 Influence of cell concentration on the yield of *o*-fluorophenylalanine. Cinnamic acid, 50 mM; ammonia, 8 M; pH 10.0, 30 °C, 22 h

6.6 UNSPECIFIED BIOSYNTHESIS

There are very few examples of naturally occurring fluorine-containing products [18]. In the course of collaborative studies to improve the production

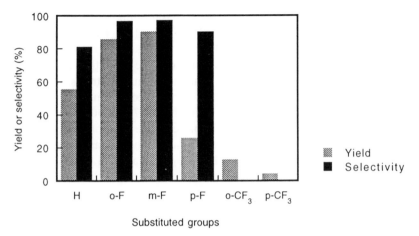

Figure 6.9 Effect of substitutuents on production of fluorophenylalanine from fluorocinnamic acid. Fluorocinnamic acid, 30 mM; ammonia, 8 M; pH 10.0, 30 °C, 40 h; cell, 20 g dry weight l^{-1}; initial PAL activity, 32 Ug^{-1} dry weight

of thienamycin by fermentation, Sanada *et al.* [13] at Banyu Pharmaceutical and Merck Sharp and Dohme discovered 4-fluorothreonine as an antimetabolite isolated from fermentation broths of *Streptomyces cattleya* in the presence of a fluorine-containing substrate or inorganic fluoride. The effects of halide compounds and soybean casein from various sources on the production of 4-fluorothreonine in fermentation medium are shown in Table 6.8.

Table 6.8 Effect of halide compounds on soybean casein 4-fluorothreonine production by *S. cattleya* (THC-3807E) in complex fermentation medium. (Reproduced by permission of the Japan Antibiotics Research Association)

Compound	Amount added (mM)	4-Fluorothreonine[a] ($\mu g\,ml^{-1}$)
None	—	0
KF	2	168
KF	10	134
NaF	2	144
NaCl	2	0
NaBr	2	0
KI	2	0
Soybean casein	0.3% (w/v)	
From Kishida Chemical		150
From Katayama Chemical		144
From Wako Chemical		0
m-Fluoro-DL-phenylalanine	2	160
p-Fluoro-DL-phenylalanine	2	12
Fluoroacetic acid	2	24

[a] Determined by bioassay.

4-Fluorothreonine was produced when the basal medium containing a fluoride source or soybean casein (from Kishida Chemical or Katayama Chemical) contained 0.7% fluorine as inorganic fluoride as measured by elemental analysis. The trial production of other halo-substituted organic derivatives by the addition of chloride, bromide and iodide failed. One of the mutant strains, MA 5176, an improved thienamycin producer, led to 4-fluorothreonine more efficiently than the original soil isolate (NRRL 8057), but MA 5617, the other mutant for the same purpose, had the opposite effect. Representative data for fermentation with resting cells of *S. cattleya* when incubated with fluorine-containing substrates are given in Figure 6.10. Fluoroacetate, 4-fluoroglutamate and potassium fluoride can be used as the fluoride source for the production of 4-fluorothreonine, but fluorocitrate was not suitable as the donor. The supernatant of each reaction mixture was examined by ^{19}F NMR spectrometry after incubation for 96 h (Table 6.9). Both fluoroglutamate and fluorothreonine were converted into fluoroacetate, whereas fluorocitrate appeared to be inert to metabolism by *S. cattleya*. A possible biosynthetic pathway was proposed via fluoroacetic acid as shown in equation 6.12.

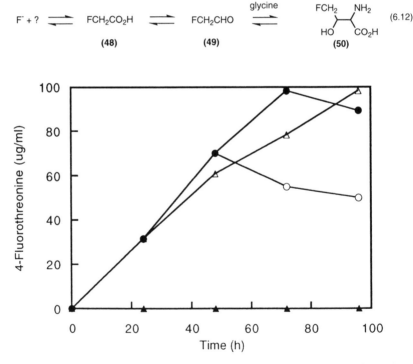

Figure 6.10 Production of 4-fluorothreonine with resting cells of *S. cattleya* (MA 5176). △, 4-Fluoroglutamate (12 mM); ▲, DL-fluorocitrate (2 mM); ○, sodium fluoroacetate (20 mM); ●, KF (2 mM). (Reproduced by permission of the Japan Antibiotics Research Association)

Table 6.9 Identification of fluorine compounds by ^{19}F NMR in resting cell reaction mixtures (MA 5176). (Reproduced by permission of the Japan Antiibiotics Research Association)

Fluorinated substrate	Organofluorine compound identified by ^{19}F NMR			
	F-acetate	F-citrate	F-glutamate	F-threonine
KF	F-acetate	None	None	F-threonine
Fluoroacetate	F-acetate	None	None	F-threonine
Fluorocitrate	None	F-citrate	None	None
4-Fluoroglutamate	F-acetate	None	F-glutamate	F-threonine
4-Fluorothreonine	F-acetate	None	None	F-threonine

6.7 CONCLUSION

This chapter on the enzymatic synthesis of fluorine-containing amino acids has highlighted the significant accomplishments which have been achieved in recent years. Certainly, these amino acids and physiologically important compounds containing them will reveal increasingly their fascinating aspects in all disciplines of biology and chemistry. In some instances novel discoveries have been emerging in events of great significance, such as the tumor invasion inhibitory activity [19] (potential anti-metastatic activity) of *p*-fluoro-phenylalanine (**51**) and the anti-leukemic activity of *p*-fluorophenylalanine-containing tripeptide (PTT-119) (**52**) [20].

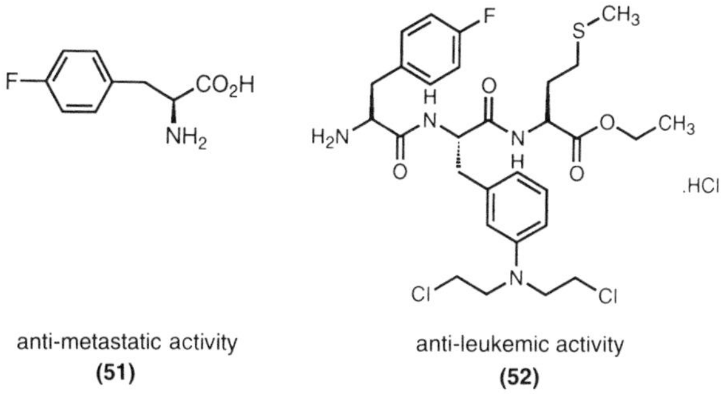

(6.13)

anti-metastatic activity
(**51**)

anti-leukemic activity
(**52**)

Considerable progress has been made towards exploiting practical bio-processes suitable for fluorine-containing amino acids. Nonetheless, the subject is still far from fully developed. For example, in contrast to the successful production of a variety of aromatic fluorine-containing amino acids, only a few aliphatic analogues have been prepared. For the design of the enzymatic system, there are two main issues of concern: prevention of the inhibitory effects of the fluorine-containing substrates, and extension of the reaction to a wide range of the derivatives. Although enzymatic mechanisms have been gradually elucidated in detail, major difficulties in resolving these problems often originate from the nature of enzymes. Other areas which have yet to be tackled involve approaches according to the technology based on molecular biology. Although a variety of biocatalysts isolated from living systems have been employed for the synthesis of fluorine-containing amino acids, it seems that there will be greater opportunities for using genetic methods to produce enzymes that are modified to affect some feature of the mode of action. These problems represent a fertile area for further studies.

6.8 REFERENCES

1. (a) R. M. Williams, *Synthesis of Optically Active α-Amino Acids*, Pergamon Press, Oxford, 1989; (b) Y. Izumi, I. Chibata and T. Itoh, *Angew. Chem., Int. Ed. Engl.*, **17**, 176 (1978); (c) I. Wagner and H. Musso, *Angew. Chem., Int. Ed. Engl.*, **22**, 816 (1983); (d) R. Breslow, J. Chemilelewski, D. Foley, B. Johnson, N. Kumabe, M. Varney and R. Mehra, *Tetrahedron*, **44**, 5515 (1988); (e) I. Chibata, in E. L. Eliel and S. Otsuka (Eds), *Asymmetric Reactions and Processes in Chemistry*, ACS Symposium Series, No. 185, American Chemical Society, Washington, DC, 1982, pp. 195–203.
2. E. Kun, D. W. Fanshier and D. R. Grassetti, *J. Biol. Chem.*, **235**, 416 (1960).
3. E. Kun, L. T. Gottwald, D. W. Fanshier and J. E. Ayling, *J. Biol. Chem.*, **238**, 1456 (1963).
4. P. A. Briley, R. Eisenthal, R. Harrison and G. D. Smith, *Biochem. J.*, **161**, 383 (1977).
5. K. Uchida and H. Tanaka, *J. Synth. Org. Chem. Jpn.*, **46**, 977 (1988).
6. S. Ohdo, Y. Ueno, M. Urushihara, H. Tanaka and K. Uchida, unpublished results.
7. Y. Matsumura, M. Urushihara, H. Tanaka, K. Uchida and A. Yasuda, unpublished results.
8. Y. Matsumura, M. Urushihara, H. Tanaka, K. Uchida and A. Yasuda, *Chem. Lett.*, 1255 (1993).
9. A. Vidal-Cros, M. Gaudry and A. Marquet, *J. Org. Chem.*, **54**, 498 (1989).
10. Y. Asano, A. Yamada, Y. Kato, K. Yamaguchi, Y. Hibino, K. Hirai and K. Kondo, *J. Org. Chem.*, **55**, 5567 (1990).
11. T. Nagasawa, T. Utagawa, J. Goto, C.-J. Kim, Y. Tani, H. Kumagai and H. Yamada, *Eur. J. Biochem.*, **117**, 33 (1981).
12. H. Tanaka, T. Ohdo, M. Urushihara and K. Uchida, *Asahi garasu Kenkyu Hokoku*, **38**, 123 (1988).
13. M. Sanada, T. Miyano, S. Iwadare, J. M. Williamson, B. H. Arison, J. L. Smith, A. W. Douglas, J. M. Liesch and E. Inamine, *J. Antibiot.*, **39**, 259 (1986).
14. (a) I. Chibata, *Immobilized Enzymes*, Kodansha, Tokyo, 1978; (b) K. Mosbach (Ed.), *Methods in Enzymology*, Vol. 44, Academic Press, New York, 1976.
15. I. Ojima, K. Hirai, M. Fujita and T. Fuchikami, *J. Organomet. Chem.*, **279**, 203 (1985).
16. K. Tanizawa, Y. Masu, S. Asano, H. Tanaka and K. Soda, *J. Biol. Chem.*, **264**, 2445 (1989).
17. Y. Asano, A. Nakazawa and K. Endo, *J. Biol. Chem.*, **262**, 10346 (1987).
18. Naturally occurring fluorine-containing products: (a) monofluoroacetic acid, M. M. Oliverira, *Experientia*, **19**, 586 (1963); (b) fluoroorganic acids, J. Y. O. Cheng,

M. H. Yu, G. W. Miller and G. W. Welkie, *Environ. Sci. Technol.*, **2**, 367 (1968); (c) nucleocidin, G. O. Morton, J. E. Lancaster, G. E. Van Lear, W. Fulmor and W. E. Meyer, *J. Am. Chem. Soc.*, **91**, 1535 (1969).

19. (a) *In vitro* invasion assay, K. Shinkai, M. Mukai and H. Akedo, *Cancer Lett.*, **32**, 7 (1986); (b) recent reports on antimetastatic agents, K. Hellmann, *Clin. Exp. Metastas.*, **2**, 1 (1984); A. Isoai, Y. Giga-Hama, K. Shinkai, M. Mukai, H. Akedo and H. Kumagai, *Jpn. J. Cancer Res.*, **81**, 909, (1990); G. Sava, L. Perissin, S. Zorzet, P. Piccini and T. Giraldi, *Clin. Exp. Metastas.*, **7**, 671 (1989).

20. M. J. Yagi, J. G. Bakesi, M. D. Daniel, J. F. Holland and A. De Barbieri, *Cancer Chemother. Pharmacol.*, **12**, 70 (1984).

7 Enzymatic Resolution of Racemic Fluorine-containing Amino Acids

TOSHIFUMI MIYAZAWA
Department of Chemistry, Faculty of Science, Konan University, Higashinada-ku, Kobe 658, Japan

7.1 INTRODUCTION

For the preparation of homochiral fluorine-containing amino acids, the following three routes are possible: resolution of the racemic amino acid derivatives, asymmetric synthesis and fluorination of the homochiral parent amino acids. Among these routes, the chemical synthesis of racemic forms followed by their optical resolution is still the preferable approach for the supply of this kind of uncommon amino acids in the required amounts. Besides chemical methods based on the formation of diastereomeric salts or derivatives, enzymatic methods are also available for the resolution of racemic amino acids. This chapter summarizes examples of the enzyme-based resolution of racemic fluorine-containing amino acids reported so far. These are subdivided into the following major alternative approaches:

1. Enzymatic resolution through amide bond formation.
2. Enzymatic resolution through deacylation of *N*-acylated derivatives.
3. Enzymatic resolution through hydrolysis of esters.
4. Enzymatic resolution through hydrolysis of leucine dipeptides.
5. Enzymatic resolution of fluorine-containing β-hydroxyamino acids as alcohols.

7.2 ENZYMATIC RESOLUTION THROUGH AMIDE BOND FORMATION

Since the first synthesis by Bergmann and Fraenkel-Conrat [1] of an amide bond between *N*-benzyloxycarbonyl (Z-)glycine and aniline using papain, the method based on this reaction has been utilized for the resolution of amino acids. As early as 1946, a fluorine-containing amino acid, 3-fluorotyrosine, was resolved using this enzyme [2]. The racemic *N*-benzoyl derivative **1** was incubated with aniline, papain and cysteine hydrochloride (as an activator) at pH 5.8 and 40 °C (equation 7.1). The precipitated L-anilide **2** was collected to give a total yield of 91% of the theoretical quantity after 2 weeks. It was converted into the free L-amino acid by

Fluorine-containing Amino Acids: Synthesis and Properties Edited by V. P. Kukhar' and V. A. Soloshonok

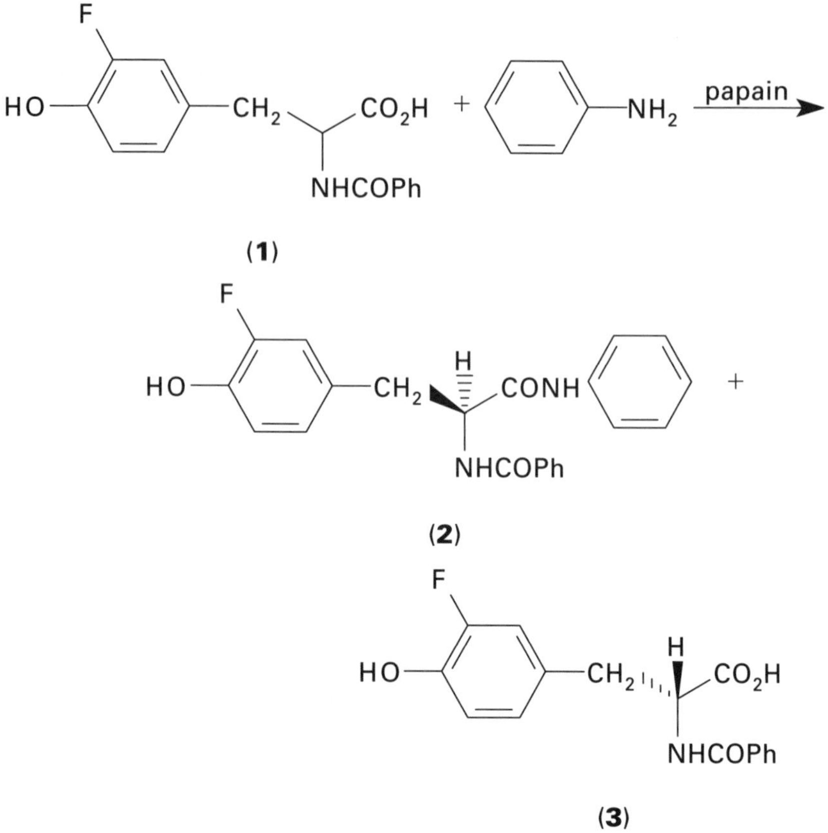

(1)

(2)

(7.1)

(3)

refluxing with hydrochloric acid. On the other hand, the D-amino acid was obtained from the filtrate remaining after the removal of the L-anilide by extraction with ethyl acetate at pH 1–2 and subsequent acidic hydrolysis followed by fractional recrystallization (31% overall yield).

Similarly, the three positional (o-, m- and p-) isomers of fluorophenylalanine were resolved using papain through conversion (at pH 4.6 and 40 °C) of the N-acetyl-L-amino acids into the corresponding L-phenylhydrazides (yields: o-, 87%; m-, 63%; p-, 31%) [3]. No indication of concomitant formation of the D-phenylhydrazides was obtained. The D-isomers were separated from the filtrates remaining after the removal of the L-phenylhydrazides. Changing the N-acyl group from acetyl to Z exerted a profound effect on the stereochemical course of the reaction. Thus, when N-Z-DL-2-fluorophenylalanine was treated with phenylhydrazine and papain, both the L- and D-phenylhydrazides were formed [4]. The phenylhydrazide of N-Z-L-2-fluorophenylalanine was separated with difficulty from the precipitated product by fractional recrystallization from toluene (25% yield).

Since the factors controlling the maximum production and precipitation of L-anilides or L-phenylhydrazides, e.g. the N-acyl group, pH and reaction time, are often different from amino acid to amino acid, it seems difficult to accept the papain-catalyzed synthesis as a general resolution procedure.

7.3 ENZYMATIC RESOLUTION THROUGH DEACYLATION OF N-ACYLATED DERIVATIVES

The most extensive studies of enzymatic resolution of amino acids have been carried out using such hydrolytic enzymes as acylase I and carboxypeptidase A. The procedure is based on the asymmetric removal of an N-acyl group of the racemic amino acid by enzymatic hydrolysis.

Carboxypeptidase A (from bovine pancreas) was employed for the resolution of ^{14}C-labeled 4-fluorophenylalanine prepared according to the Bücherer hydantoin synthesis [5]. Treatment of the neutral solution of the N-chloroacetyl derivative 4 (equation 7.2; labeled atoms are asterisked) with this enzyme afforded the carboxyl-labeled L-4-fluorophenylalanine (5) in 25% yield based on K^{14}CN used. The optical purity of the amino acid 5 was determined by employing D-amino acid oxidase to be 99.9%. The use of an acyl group which can be removed under milder conditions should be more desirable, because the acidic hydrolysis of the recovered N-chloroacetyl-D-

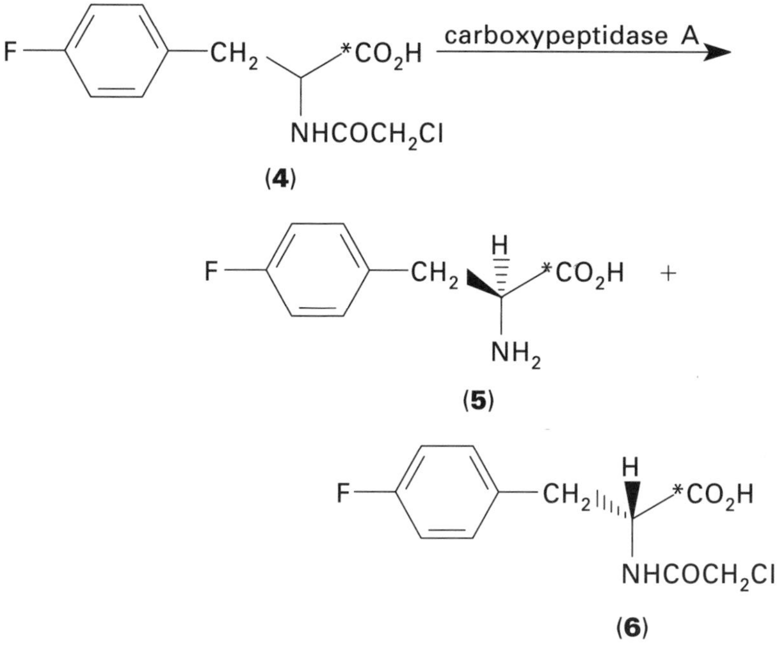

(7.2)

amino acid to the free amino acid may cause some racemization. The tri-fluoroacetyl group has been shown to be stereospecifically removed from phenylalanine by carboxypeptidase A at a rate much greater than any other N-acyl group [6]. Thus, 4-fluorophenylalanine was resolved as its N-tri-fluoroacetyl derivative using carboxypeptidase A [7]. The enzymatic hydrolysis was conducted at pH 7 and 37 °C for 4 h to give L-4-fluor-ophenylalanine (87% yield) and the unchanged D-isomer (87% yield). The latter was deprotected by hydrolysis with 2 equiv. of LiOH to afford D-4-fluorophenylalanine (97% yield).

Experimental protocols were developed for the synthesis and resolution using carboxypeptidase A and/or α-chymotrypsin (see below) of numerous ring-substituted phenylalanines, including 2-, 3- and 4-fluoro-phenylalanines and 4-(trifluoromethyl)phenylalanine, in half mole quantities [8]. The racemic N-trifluoroacetyl derivatives were treated with carboxypeptidase A at pH 7.2 and 37 °C overnight. The pH of the solution was maintained constant by delivering NaOH using an automatic titrator. In the case of 4-fluorophenylalanine, the free L-amino acid and the unhydrolyzed N-trifluoroacetyl D-isomer were separated in 85% and 88% yield, respectively.

The resolution of 7-fluorotryptophans, i.e. 7-fluoro- (**7a**), 4,7-difluoro- (**7b**) and 5,7-difluorotryptophans (**7c**), was also achieved by treatment of the N-tri-fluoroacetyl derivatives with carboxypeptidase A [9, 10]. The DL-amino acids (**7a–c**) were prepared by the Fischer indole cyclization. The enzymatic hydrolysis of the N-trifluoroacetyl derivatives was conducted at pH 7.5 and ambient temperature for 16 h to afford the free L-7-fluorotryptophans and the unchanged N-trifluoroacetyl-D-amino acids. The isolation of the L-amino acids was carried out by crystallization from a suitable solvent such as methanol–chloroform (70–80% yield). The treatment of the concentrated filtrate with 1 M aqueous piperidine, followed by recrystallization from methanol–chloroform, afforded the D-7-fluorotryptophans (60–70% overall yield). The optical purity of each enantiomer was measured by ligand-exchange HPLC to be >99%.

(**7**)

a, $R^1 = R^2 = H$; **b**, $R^1 = F$, $R^2 = H$; **c**, $R^1 = H$, $R^2 = F$

It is well known that N-acylated aromatic amino acids are susceptible to hydrolysis by renal acylase III or pancreatic carboxypeptidase A, while renal acylase I is active toward N-acylated aliphatic amino acids [11]. Nevertheless, porcine renal acylase I has been employed for the resolution of some aromatic fluorine-containing amino acids. The L-enantiomer of 2,3,4,5,6-penta-fluorophenylalanine was prepared through the hydrolysis of an N-acyl derivative with this enzyme [12]. Since the enzymatic attack on the N-acetyl derivative was too slow for synthetic purposes, the N-trifluoroacetyl derivative **8** was used. After 36 h of incubation at pH 7 and 38 °C, the free L-amino acid **9** was obtained in 52% yield (equation 7.3). The unhydrolyzed D-isomer **10** was not isolated, but it was racemized by treatment with trifluoroacetic anhydride for repeated resolution to obtain the L-amino acid.

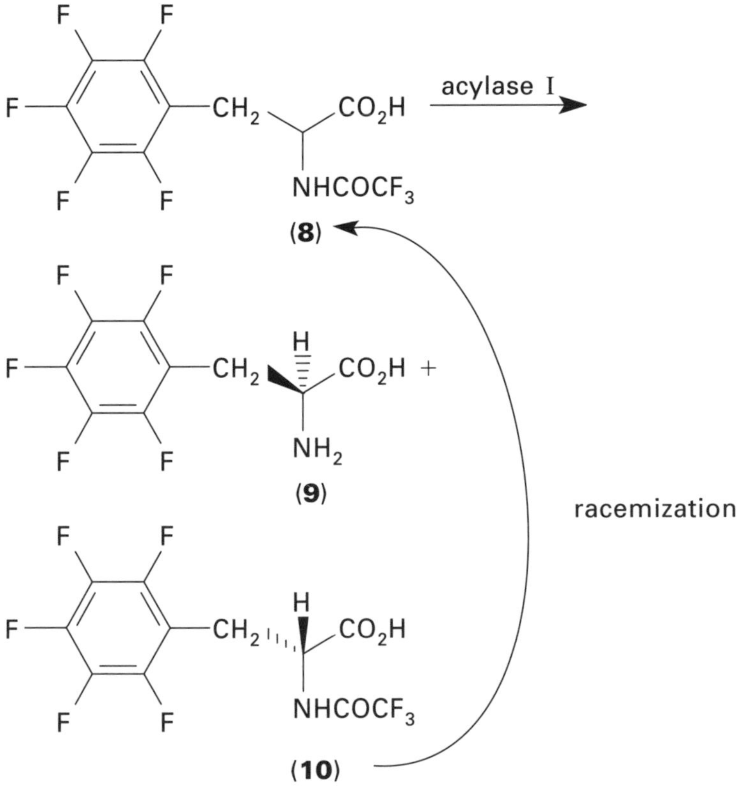

Similar resolution of racemates was possible also for [α, β-di-^{3}H]-2-fluoro-histidine [13]. Methyl α-benzoylamino-β-[4-(2-fluoroimidazolyl)]acrylate (**11**), synthesized via several steps from 2-fluoroimidazole, was reduced catalytically with tritium gas to give N-benzoyl-DL-2-fluorohistidine methyl ester (**12**) (equation 7.4). The methyl ester **12** was saponified and treated with porcine

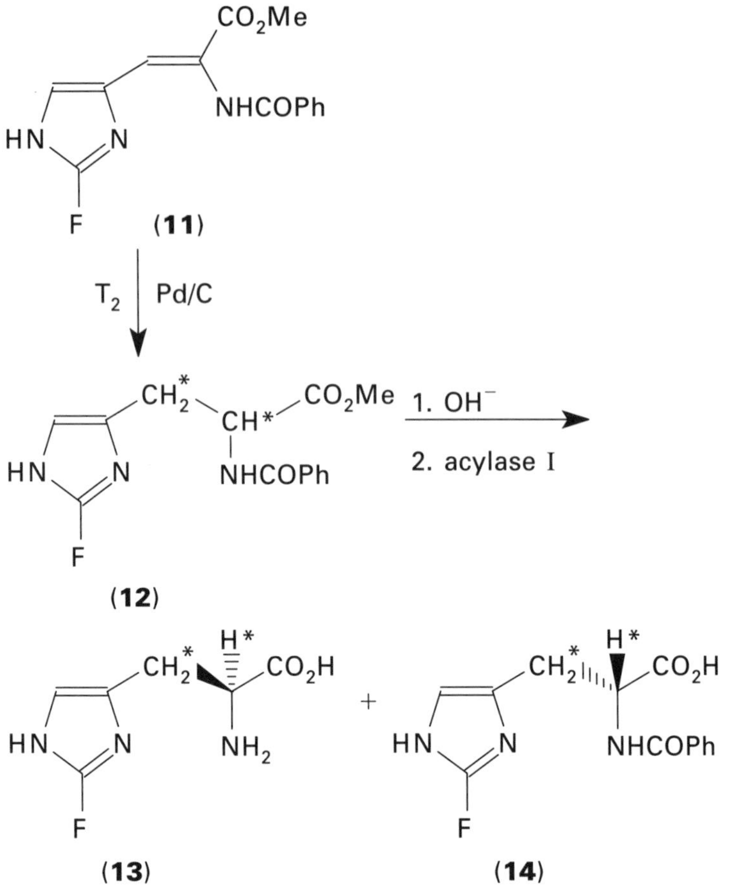

$$\text{(11)}$$

imidazole ring (HN–N, C–F) substituted with $CH=C(CO_2Me)(NHCOPh)$

$$T_2 \;\bigg|\; Pd/C \;\downarrow$$

$$\text{(12)}$$

imidazole ring substituted with $\overset{*}{C}H_2-\overset{*}{C}H(CO_2Me)(NHCOPh)$

$$\xrightarrow[\text{2. acylase I}]{\text{1. OH}^-}$$

$$\text{(13)} \qquad\qquad + \qquad\qquad \text{(14)} \qquad\qquad \text{(7.4)}$$

(13): imidazole with $\overset{*}{C}H_2-\overset{H^*}{C}(CO_2H)(NH_2)$

(14): imidazole with $\overset{*}{C}H_2-\overset{H^*}{C}(CO_2H)(NHCOPh)$

renal acylase I at pH 7 and 37 °C. L-2-Fluorohistidine (**13**) was separated from
N-benzoyl-D-2-fluorohistidine (**14**) by preparative HPLC.

Acylase I enzymes from microbial sources are known to show higher
activities than porcine renal acylase I toward aromatic amino acids [11, 14].
[18]F-labeled L-4-fluorophenylalanine (**18**) was prepared by treatment of the
racemic *N*-acetyl derivative **17** with acylase I from *Aspergillus oryzae* in the
presence of cobalt(II) chloride (equation 7.5) [15]. This microbial enzyme has
been used in the industrial-scale production of L-amino acids [16]. After
10 min of incubation at 50 °C, the labeled L-amino acid **18** was isolated from
the reaction mixture by gel permeation using polyamide gel. The malonic ester
derivative **15** was labeled by heterogeneous exchange [17]. The route based on
the malonic ester synthesis has an advantage that the *N*-acyl derivative as the
enzyme substrate can be obtained without an additional acylation step.

Another acylase I enzyme from *Aspergillus melleus* was employed for the
resolution of racemic trifluoromethyl-substituted phenylalanines (*o*-, *m*- and *p*-)

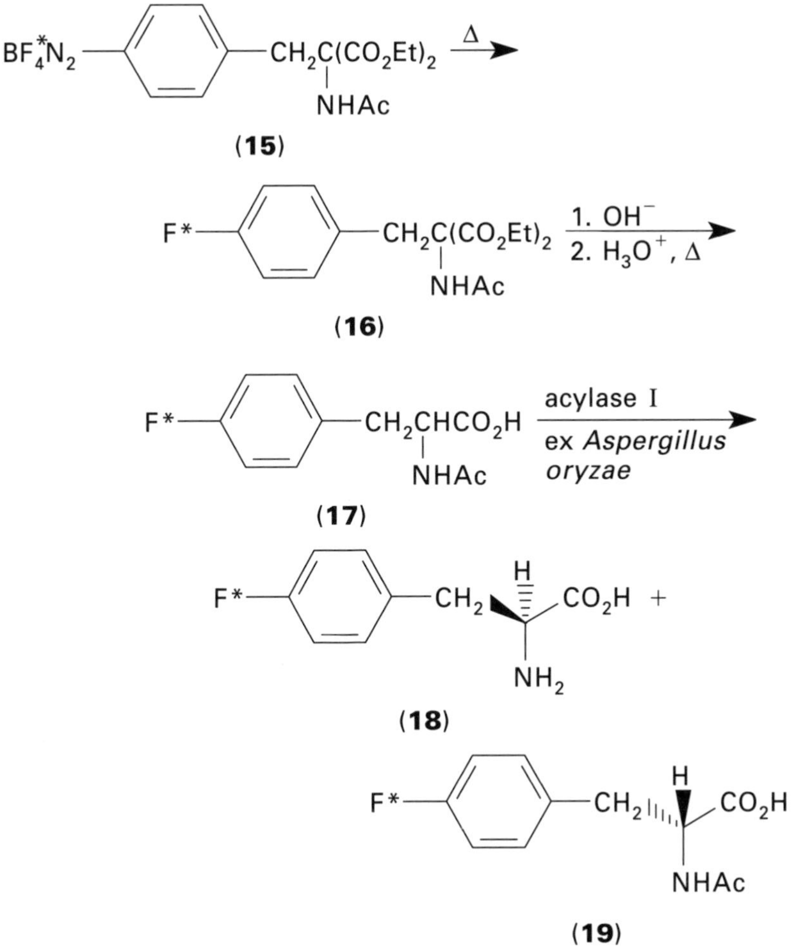

$$BF_4^*N_2 - \underset{}{\bigcirc} - CH_2C(CO_2Et)_2 \xrightarrow{\Delta}$$
$$\underset{NHAc}{|}$$

(15)

$$F^* - \underset{}{\bigcirc} - CH_2C(CO_2Et)_2 \xrightarrow[2.\ H_3O^+,\ \Delta]{1.\ OH^-}$$
$$\underset{NHAc}{|}$$

(16)

$$F^* - \underset{}{\bigcirc} - CH_2CHCO_2H \xrightarrow[\text{ex } \textit{Aspergillus}]{\text{acylase I}}$$
$$\underset{NHAc}{|} \quad \textit{oryzae}$$

(17)

$$F^* - \underset{}{\bigcirc} - CH_2 \overset{H}{\underset{NH_2}{\overset{||}{\diagdown}}} CO_2H \ +$$

(18)

$$F^* - \underset{}{\bigcirc} - CH_2 \overset{H}{\underset{NHAc}{\diagdown}} CO_2H \quad \textbf{(7.5)}$$

(19)

as their *N*-acetyl derivatives [18, 19]. After 24 h of hydrolysis in phosphate buffer (pH 7.5) at 37 °C, the L-amino acids were isolated by the use of a cation-exchange resin (yields: *o*-, 67%; *m*-, 79%; *p*-, 84%). Their optical purities were determined by ligand-exchange HPLC (on Daicel Chiralpak WH) to be >99%. Similarly, optically pure (>99.9% by HPLC) L-2,3,4,5,6-pentafluoro-phenylalanine was obtained through the enzymatic hydrolysis using this mold acylase I of the *N*-acetyl derivative in a 47% isolated yield. Interestingly, the pentafluorophenylalanine derivative, together with the 4-fluorophenylalanine derivative, was hydrolyzed more rapidly than the non-substituted phenylalanine derivative. In contrast, the (trifluoromethyl)phenylalanine derivatives were less susceptible to hydrolysis than the phenylalanine derivative. The lowest hydrolysis rate observed with the *ortho*-substituted phenylalanine derivative must be ascribed to the steric hindrance caused by the existence of an *o*-trifluoromethyl group near the reaction center.

Acylase I enzyme from *Streptoverticillium olivoreticuli* was also used for the resolution of racemic fluorophenylglycines (*o*- and *p*-) [20]. The enzymatic hydrolysis of their *N*-acetyl derivatives was conducted at pH 7.6–7.7 and ambient temperature, and the L-amino acids released were isolated with the aid of a cation-exchange resin (yields: *o*-, 86%; *p*-, 91%). Their optical purities were confirmed by chiral HPLC analysis [21].

A number of aliphatic fluorine-containing amino acids have been resolved using porcine renal acylase I in recent years. For example, L-*threo*- and L-*erythro*-3-fluoroglutamic acids were prepared according to Scheme 7.1 [22].

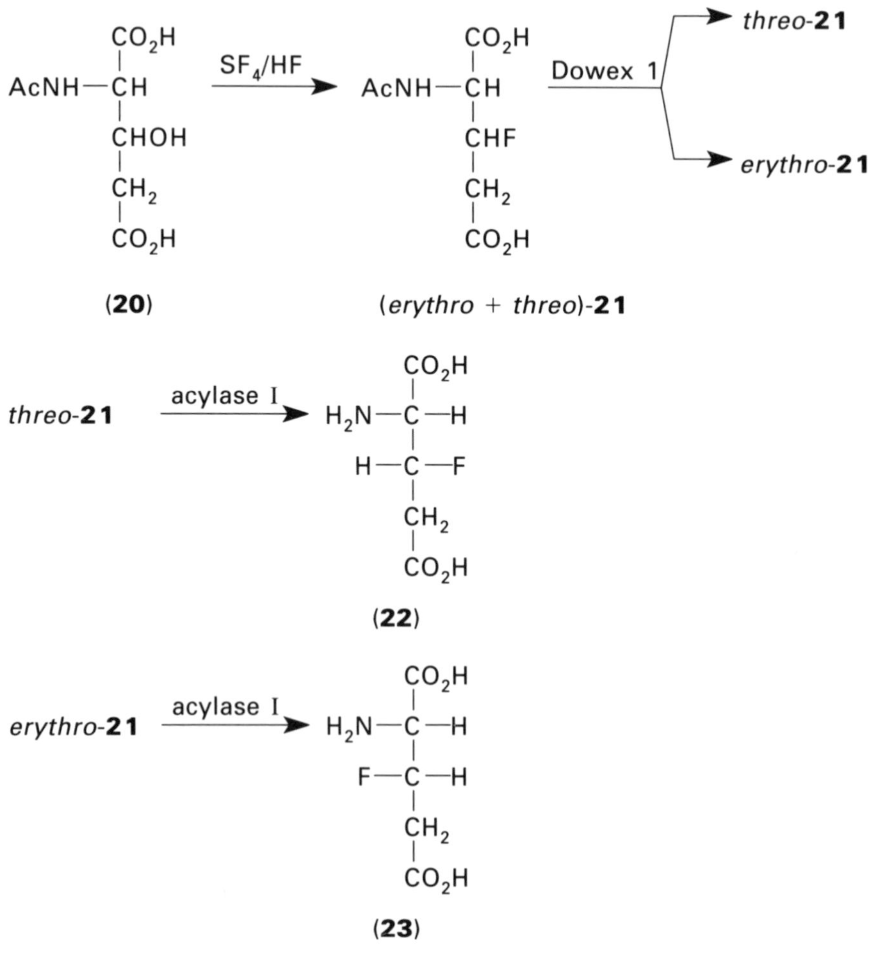

(20)   (*erythro* + *threo*)-**21**

(22)

(23)

Scheme 7.1

Fluorodehydroxylation of N-acetyl-3-hydroxyglutamic acid (20) according to Kollonitsch *et al.* [23] afforded almost quantitatively a mixture of *threo*- and *erythro-N*-acetyl-3-fluoroglutamic acids (21), which was then separated by ion-exchange chromatography on Dowex 1. Each diastereomer was deacylated with porcine renal acylase I in phosphate buffer (pH 7) at 25 °C for 26 h. The L-*threo*- and L-*erythro*-3-fluoroglutamic acids (22 and 23) were isolated by the use of the ion-exchange resin in 56% and 40% yield, respectively. The L-*threo*-amino acid 22 thus obtained proved to be optically pure by gas chromatography of the N-acetyl diisopropyl ester on a chiral column [24]. The assignment of the *threo* or *erythro* configuration was achieved by lactamization and determination of the *cis* or *trans* structure of the resulting 4-fluoro-2-pyrrolidone-5-carboxylic acid on the basis of ^{1}H NMR spectra.

An efficient enzymatic resolution using porcine renal acylase I was achieved with the N-acetyl derivatives of 2-amino-4,4,4-trifluorobutanoic acid (24a) and 6,6,6-trifluoronorleucine (24b), which were prepared via the direct alkylation of diethyl acetamidomalonate with the corresponding trifluoroalkyl trifluoromethanesulfonate (equation 7.6) [25]. The enzymatic hydrolysis was carried out at pH 7 and 36 °C (3.5 h reaction in the case of the N-acetyl derivative of 24b). The deacylated L-amino acids were separated by ion-exchange chromatography on Dowex 50W (yields: L-24a, 97%; L-24b, 98%); the L-24b obtained proved to be optically pure by HPLC analysis using a chiral solvent-generated phase [26]. The free D-amino acids were obtained by hydrolysis of the unchanged N-acetyl-D-amino acids with hydrochloric acid. Both the enantiomers of 5,5,5-trifluoroleucine were also obtained as the mixtures of (4R)- and (4S)-diastereomers via the hydrolysis with the same enzyme of the N-acetyl derivative in good yields [yields: L- or (2S)-enantiomer, 93%; D- or (2R)-enantiomer, 94%]. The HPLC analysis confirmed that they were enantiomerically pure. Likewise, L- and D-5,5,5-trifluoronorvalines with high optical purities (L-, >99%; D-, 95%) were obtained (L-, 80% yield; D-, 42% overall yield) through the enzymatic hydrolysis with porcine renal acylase I of the N-acetyl derivative 25, which was synthesized via the aminocarbonylation of 4,4,4-trifluorobutanal (equation 7.7) [27]. Their optical purities

$$\text{AcNHCH(CO}_2\text{Et)}_2 \xrightarrow{\text{CF}_3\text{(CH}_2)_n\text{OSO}_2\text{CF}_3} \underset{\underset{\text{NHAc}}{|}}{\text{CF}_3\text{(CH}_2)_n\text{C(CO}_2\text{Et)}_2}$$

$$\xrightarrow{\text{H}_3\text{O}^+} \underset{\underset{\text{NH}_2}{|}}{\text{CF}_3\text{(CH}_2)_n\text{CHCO}_2\text{H}} \qquad (7.6)$$

$$(24)$$

$$\textbf{a}, n = 1; \textbf{b}, n = 3$$

were determined on the basis of ^{19}F NMR analysis of (R)-(+)-α-methoxy-α-(trifluoromethyl)phenylacetyl [(R)-(+)-MTPA] methyl esters [28]. In contrast, the N-trifluoroacetyl [29] or N-acetyl [25] derivative of 4,4,4,4′,4′,4′-hexafluorovaline (26) totally resisted hydrolysis with this enzyme to give a complete recovery of the starting material.

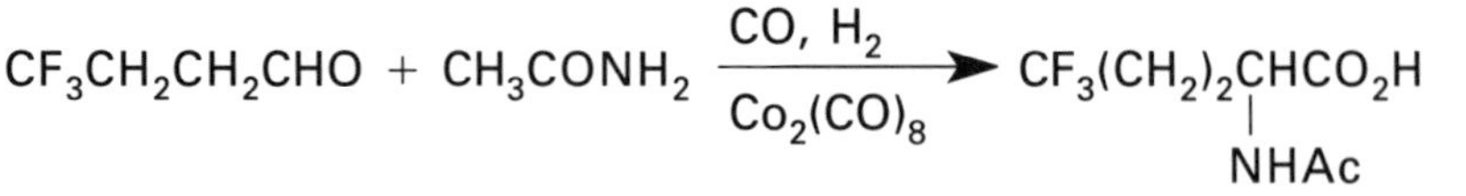

$$CF_3CH_2CH_2CHO + CH_3CONH_2 \xrightarrow[Co_2(CO)_8]{CO, H_2} CF_3(CH_2)_2\underset{\underset{NHAc}{|}}{C}HCO_2H \quad \textbf{(7.7)}$$

$$\textbf{(25)}$$

$$\underset{F_3C}{\overset{F_3C}{>}}CHCHCO_2H$$
$$\quad\quad\quad \underset{NH_2}{|}$$

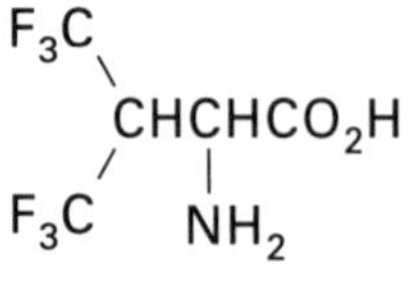

$$\textbf{(26)}$$

The L-form of 5-fluoroalloisoleucine (30) was obtained pure through the porcine renal acylase I-catalyzed hydrolysis of the N-acetyl derivative prepared according to equation 7.8 [30].

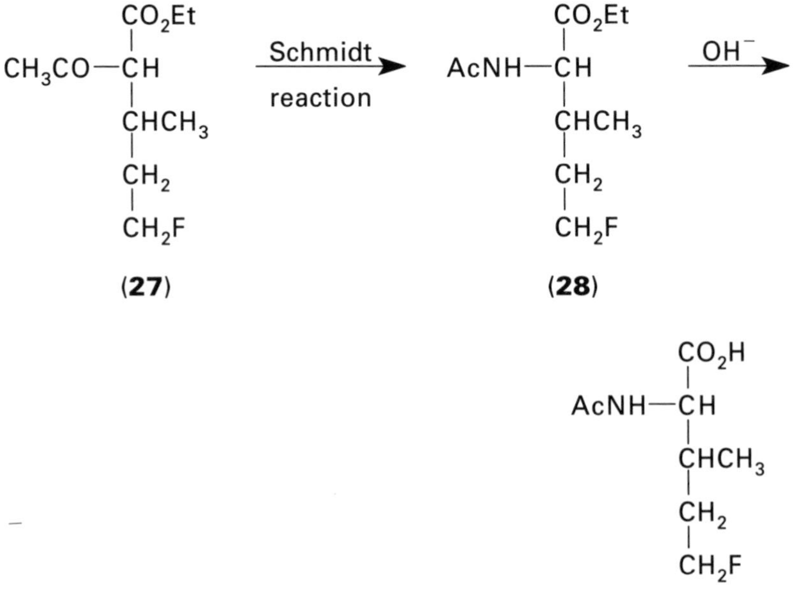

$$CH_3CO-\underset{\underset{\underset{\underset{CH_2F}{|}}{\underset{CH_2}{|}}}{\underset{CHCH_3}{|}}}{\overset{\overset{CO_2Et}{|}}{C}H} \xrightarrow[\text{reaction}]{\text{Schmidt}} AcNH-\underset{\underset{\underset{\underset{CH_2F}{|}}{\underset{CH_2}{|}}}{\underset{CHCH_3}{|}}}{\overset{\overset{CO_2Et}{|}}{C}H} \xrightarrow{OH^-}$$

$$\textbf{(27)} \quad\quad\quad\quad\quad\quad \textbf{(28)}$$

$$AcNH-\underset{\underset{\underset{\underset{CH_2F}{|}}{\underset{CH_2}{|}}}{\underset{CHCH_3}{|}}}{\overset{\overset{CO_2H}{|}}{C}H} \quad \textbf{(7.8)}$$

$$\text{DL-}(\textit{erythro} + \textit{threo})\text{-}\textbf{29}$$

Ethyl 2-acetamido-5-fluoro-3-methylpentanoate (**28**) was prepared by the Schmidt reaction at 0–10 °C of ethyl 2-acetyl-5-fluoro-3-methylpentanoate (**27**), which was obtained by the alkylation of ethyl acetoacetate with 3-bromo-1-fluorobutane. Alkaline hydrolysis of **28** afforded a mixture of *threo*- and *erythro*-2-acetamido-5-fluoro-3-methylpentanoic acids (**29**) (equation 7.8). The *threo* isomer was separated (18% yield) by systematic crystallization from water and subjected to enzymatic deacylation with the acylase (at pH 7.5 and 37 °C) to afford after 24 h the free L-*threo*-amino acid [5-fluoro-L-alloisoleucine (**30**)] (58% yield) and the unhydrolyzed *N*-acetyl-D-*threo*-amino acid (equation 7.9). Their separation was effected by crystallization, while the use of an ion exchanger converted the expected free L-amino acid **30** into a cyclization product with the replacement of the fluorine atom, i.e. *cis*-3-methyl-L-proline (**31**). This transformation provides unambiguous evidence for the stereochemistry of the compounds. Acidic hydrolysis of the corresponding *N*-acetyl-D-amino acid purified by passage through a cation-exchange resin afforded 5-fluoro-D-alloisoleucine (52% overall yield) contaminated with ca 5% of 5-hydroxy-D-alloisoleucine. On the other hand, *erythro*-2-acetamido-5-fluoro-3-methylpentanoic acid (DL-*erythro*-**29**) was not obtained entirely free of its *threo* isomer on crystallization, thus making the *erythro* series a much more difficult problem.

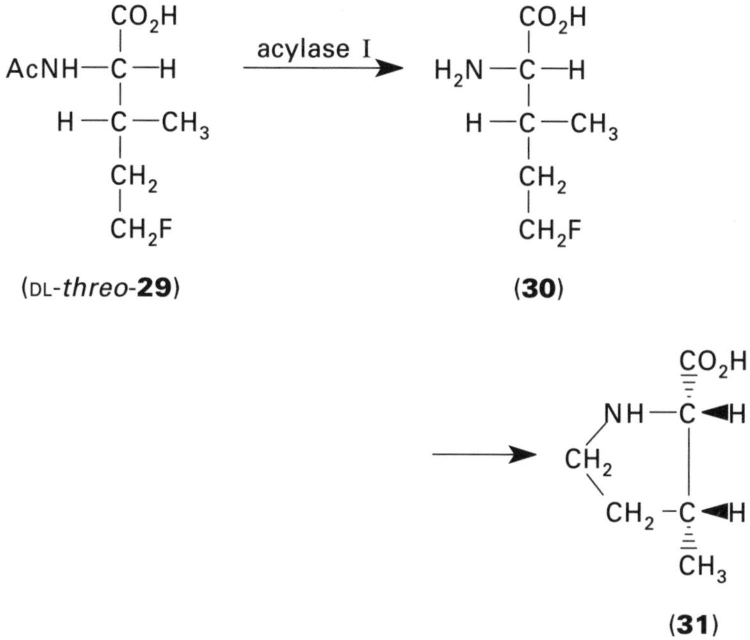

(DL-*threo*-**29**)   (**30**)

(**7.9**)

(**31**)

Fluorinated L-methionines (L-*S*-di- and L-*S*-trifluoromethylhomocysteines) were also prepared by treatment of the racemic *N*-acetyl derivatives (**33** and

34) with porcine renal acylase I at pH 7.5 (adjusted periodically by the addition of alkali) and 37 °C [31, 32]. The substrate, N-protected S-di- or S-trifluoromethylhomocysteine, was prepared by the reaction of chlorodifluoromethane or of iodotrifluoromethane under UV irradiation, respectively, with N-acetyl-DL-homocysteine which was prepared by *in situ* hydrolysis of N-acetylhomocysteine thiolactone (**32**) (equation 7.10). The free L-amino acids were separated from the unhydrolyzed N-acetyl derivatives and purified by the consecutive use of a cation-exchange column and a reversed-phase C_{18} column [86% yield from **33** (15 h of hydrolysis); 88% yield from **34** (18 h of hydrolysis)]. Their optical purities were confirmed by ^{1}H NMR analysis of the (R)-(+)-MTPA methyl esters [28].

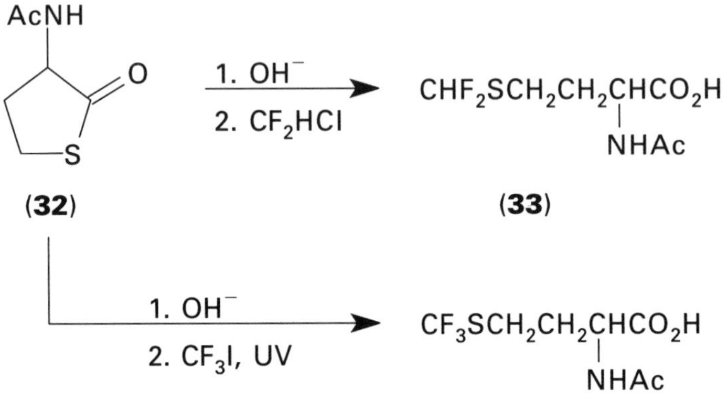

AcNH

$$\text{(32)} \xrightarrow[\text{2. CF}_2\text{HCl}]{\text{1. OH}^-} \quad \underset{\text{NHAc}}{\text{CHF}_2\text{SCH}_2\text{CH}_2\text{CHCO}_2\text{H}}$$

(32)   **(33)**   **(7.10)**

$$\xrightarrow[\text{2. CF}_3\text{I, UV}]{\text{1. OH}^-} \quad \underset{\text{NHAc}}{\text{CF}_3\text{SCH}_2\text{CH}_2\text{CHCO}_2\text{H}}$$

(34)

Enzymatic resolution of 2-trifluoromethylalanine was achieved by the porcine renal acylase I-catalyzed hydrolysis (pH 7.5, 25 °C) of the N-trifluoroacetyl derivative **35** (equation 7.11) [33]. The free (R)-amino acid (i.e. 2-trifluoromethyl-

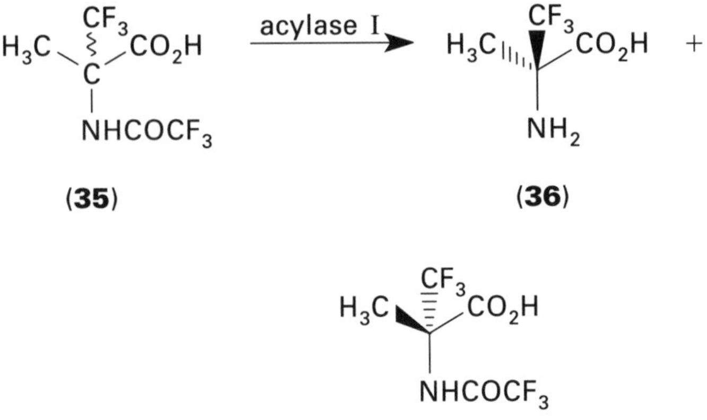

(35)   **(36)**

(7.11)

(37)

D-alanine, **36**) was isolated by ion-exchange chromatography on Dowex 50W (53% yield). The (S)-amino acid (i.e., 2-trifluoromethyl-L-alanine) was obtained after acidic hydrolysis of the unchanged amide **37** in 22% overall yield. HPLC analysis using a chiral solvent-generated phase [26] showed that the (R)- and (S)-amino acids thus obtained had optical purities of 98.2% and 97.0%, respectively. Minor enantiomeric components were ascribed to incomplete enantiospecificity by the enzyme, since the (R)-amino acid obtained from shorter reaction times or enzymatic hydrolysis of the more stable N-chloroacetyl derivative also contained ca 1% of the (S)-enantiomer. The absolute configuration of the (R)-amino acid was unambiguously determined by X-ray diffraction studies on its N-chloroacetyl derivative [34]. The enzymatic hydrolysis of **35** followed the generally observed stereochemical preference of the acylase, since the (R)-enantiomer in which the larger trifluoromethyl group occupies the pro-S position was preferentially hydrolyzed. However, the ability of the enzyme to discriminate between methyl groups and fluorinated methyl groups is limited, since no enantioselectivity was observed in the enzymatic hydrolysis of the N-trifluoroacetyl derivative of 2-(fluoromethyl)alanine (**38**).

$$H_3C \diagdown \overset{\overset{\displaystyle CH_2F}{\vert}}{\underset{\underset{\displaystyle NH_2}{\vert}}{C}} \diagup CO_2H$$

(38)

Attempts to resolve 4,4,4,4',4',4'-hexafluorovaline (**26**) by the enantio-selective hydrolysis with porcine renal acylase I of the N-trifluoroacetyl [29] or N-acetyl [25] derivative were unsuccessful, resulting in a complete recovery of the unchanged substrate.

Penicillin acylase from *Escherichia coli* was employed for the resolution of fluorophenylalanines [35, 36]. The enzymatic hydrolysis of the N-phenylacetyl derivatives **39a–e** was carried out at pH 7.6–7.7 and ambient temperature (equation 7.12). The progress of hydrolysis was controlled by the determination of free amino acids released by automatic titration with potassium hydroxide or spectrophotometrically using o-phthalaldehyde. When a conversion of 50% had been achieved (4–6 h), the unchanged phenylacetyl derivatives were extracted with ethyl acetate, and the free L-amino acids **40a–e** were isolated with the aid of a cation-exchange resin (yields: **40a**, 91%; **40b**, 89%; **40c**, 94%; **40d**, 95%; **40e**, 84%). The unchanged N-phenylacetyl derivatives of fluorophenylalanines **41a–e** were deprotected by hydrolysis with hydrochloric acid to afford free D-amino acids. The optical purity of each enantiomer was confirmed (>99% e.e.) by chiral HPLC analysis [21]. The position of the substituent did not influence substantially the catalytic properties of penicillin acylase.

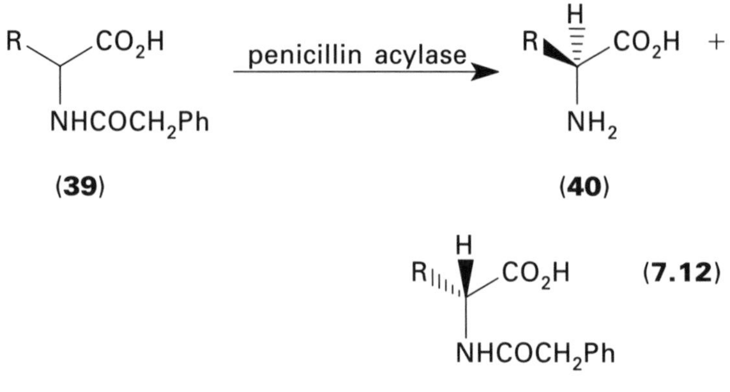

$$ \underset{\textbf{(39)}}{\underset{\text{NHCOCH}_2\text{Ph}}{\text{R}\diagup\overset{\text{CO}_2\text{H}}{|}}} \xrightarrow{\text{penicillin acylase}} \underset{\textbf{(40)}}{\underset{\text{NH}_2}{\text{R}\overset{\text{H}}{\diagdown}\overset{\text{CO}_2\text{H}}{|}}} + $$

$$ \underset{\textbf{(41)}}{\underset{\text{NHCOCH}_2\text{Ph}}{\text{R}_{\prime\prime\prime\prime}\overset{\text{H}}{\diagdown}\overset{\text{CO}_2\text{H}}{|}}} \qquad (7.12) $$

a, R = 2-FC$_6$H$_4$CH$_2$; **b**, R = 3-FC$_6$H$_4$CH$_2$;
c, R = 4-FC$_6$H$_4$CH$_2$; **d**, R = C$_6$F$_5$CH$_2$;
e, R = 4-FC$_6$H$_4$CH(OH) (*threo*)

Penicillin acylase from *E. coli* was employed also for the resolution of several fluorine-containing β-amino acids [37]. The enzymatic hydrolysis of the *N*-phenyl-acetyl derivatives **42a–e** was carried out at pH 6.5–7.5 (equation 7.13). The progress of hydrolysis was monitored by the consumption of ammonia solution. When a conversion of 50% had been achieved, the resulting (*R*)-amino acids **43a–e** were separated from the unchanged phenylacetyl derivatives **44a–e** by ion-exchange chromatography (yields: **43a**, 86%; **43b**, 78%; **43c**, 84%; **43d**, 67%; **43e**,

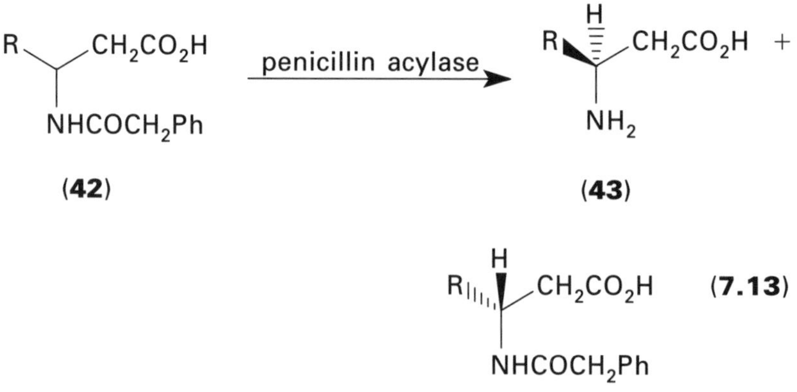

$$ \underset{\textbf{(42)}}{\underset{\text{NHCOCH}_2\text{Ph}}{\text{R}\diagup\overset{\text{CH}_2\text{CO}_2\text{H}}{|}}} \xrightarrow{\text{penicillin acylase}} \underset{\textbf{(43)}}{\underset{\text{NH}_2}{\text{R}\overset{\text{H}}{\diagdown}\overset{\text{CH}_2\text{CO}_2\text{H}}{|}}} + $$

$$ \underset{\textbf{(44)}}{\underset{\text{NHCOCH}_2\text{Ph}}{\text{R}_{\prime\prime\prime\prime}\overset{\text{H}}{\diagdown}\overset{\text{CH}_2\text{CO}_2\text{H}}{|}}} \qquad (7.13) $$

a, R = CF$_3$; **b**, R = C$_2$F$_5$; **c**, R = C$_3$F$_7$;
d, R = 2-FC$_6$H$_5$; **e**, R = 4-FC$_6$H$_5$

73%). Their optical purities were confirmed (>99% e.e.) by chiral HPLC analysis [21]. The absolute configuration of the free amino acid derived from **44a** was determined by X-ray analysis to be *S*.

7.4 ENZYMATIC RESOLUTION THROUGH HYDROLYSIS OF ESTERS

The esterase activity of proteases, exemplified by α-chymotrypsin, has been utilized for the resolution of racemic amino acids through the enantioselective hydrolysis of their *N*-unprotected and -protected esters.

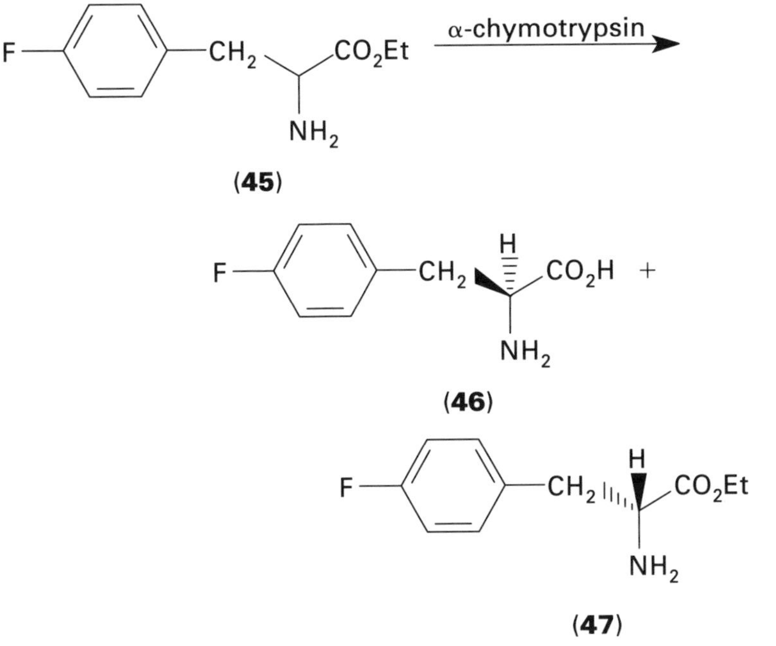

4-Fluorophenylalanine is the first example of fluorine-containing amino acids which were resolved by this procedure [38]. *N*-Unprotected 4-fluoro-phenylalanine ethyl ester (**45**) was treated with α-chymotrypsin at pH 5 (to eliminate the danger of spontaneous hydrolysis of the ester) and room temperature, the pH being kept constant by the automatic addition of LiOH from a pH-stat (equation 7.14). After 30 min of digestion, the free L-amino acid **46** was obtained by crystallization from the concentrated aqueous digest (60% yield). The unchanged D-amino acid ester **47** was isolated by extraction from a basic solution into ethyl acetate, and subsequently saponified at 45 °C to give the free D-amino acid (60% overall yield). The optical purity of the enantiomers was determined by chromatography on an amino acid analyzer of the diastereomeric dipeptides obtained by coupling with L-alanine *N*-carboxy-anhydride (L-, 100%; D-, >99.5%) [39]. Later this resolution procedure was repeated [40] and the optical purity of the L-4-fluorophenylalanine obtained was reported to be >99.5% by a gas chromatographic method [41].

L-2-Fluoro- and L-3-fluorophenylalanines were likewise prepared through the α-chymotrypsin-catalyzed hydrolysis of their racemic ethyl esters [42]. The resolution of α-methyl-4-fluorophenylalanine (48) was also tried using this protease [43]. The methyl ester of the racemic amino acid bearing an unacylated amino group was treated with α-chymotrypsin at pH 5 (pH-stat) and 37 °C. The hydrolysis rate was lowered by the introduction of an α-methyl group into the parent amino acid ester, but it was still tolerable for a practical method for resolution. After 35 h of digestion, the unhydrolyzed D-amino acid methyl ester was isolated in 78% yield (as the hydrochloride). Further, the L-form of 5-fluorotryptophan (49) was obtained through the α-chymotrypsin-catalyzed hydrolysis of the racemic methyl ester [44]. In contrast, the digestion of the ethyl ester of 4,4,4,4',4',4'-hexafluorovaline (26) with this enzyme resulted in the unhydrolyzed substrate only [29].

(48)   (49)

N-Protected fluorine-containing amino acid esters have also been employed as substrates for α-chymotrypsin. The N-acetyl methyl esters of 4-fluorophenylalanine, 4-(trifluoromethyl)phenylalanine and 5-fluorotryptophan (49) were resolved using this enzyme in water–acetonitrile at pH 7 [8]. Using a mixture of acetonitrile and water as the reaction solvent led to a faster hydrolysis than in the aqueous system and facile workup as compared with the use of other organic solvents such as dimethyl sulfoxide. When the hydrolysis was completed (within 4 h), acetonitrile was removed under reduced pressure. The N-acetyl L-isomer was water-soluble at pH 7, and the unhydrolyzed D-isomer could be extracted with organic solvents. The D- and L-isomers were hydrolyzed to free amino acids by refluxing with hydrochloric acid. On such sterically hindered substrates as tryptophans, the use of α-chymotrypsin on the N-acetyl ester gave a more convenient resolution system than that of carboxypeptidase A on the N-trifluoroacetyl derivative (see above), because the hydrolysis rate using the latter system was found to be too slow for a large-scale preparation.

L-threo-β-Fluorophenylalanine was also prepared by enzymatic hydrolysis of the N-protected DL-threo-β-fluorophenylalanine ester [45]. DL-threo-β-Fluorophenylalanine isopropyl ester (50) was prepared using the aziridine ring-opening reaction in hydrogen fluoride–pyridine [46]. The threo configuration of the product was determined by means of X-ray analysis of its N-acetyl derivative. The amino group was blocked by the *tert*-butoxycarbonyl (Boc) group and the resulting carbamate ester **51** was treated with α-chymotrypsin in water–*N,N*-dimethylformamide (DMF) (7 : 1) at pH 6.7–7.0 and ambient temperature. After 6 h of digestion, the hydrolysis product, *N*-Boc-L-*threo*-β-fluorophenylalanine (**52**), was separated (49% yield) from the unhydrolyzed D-isomer at pH 11, and deblocked by means of trifluoroacetic acid, followed by treatment with a cation-exchange resin, to give the free L-amino acid **53** (equation 7.15). The advantage of the enzymatic method in this case is that the conversion of the fluorine-containing amino acid ester into the free amino acid proceeds without loss of fluorine.

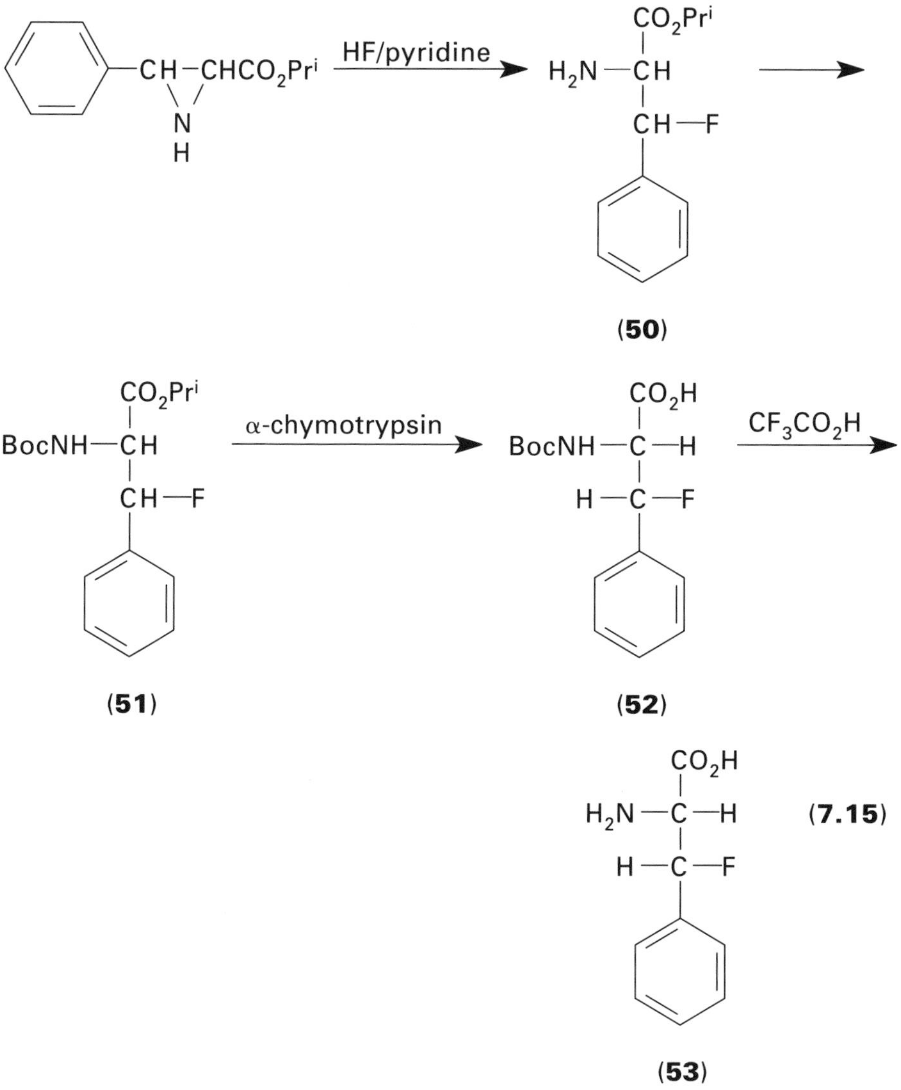

(**50**)

(**51**)   (**52**)

(**7.15**)

(**53**)

The Schiff bases (**54a** and **54b**) derived from the ethyl esters of 2- and 4-fluorophenylalanines and benzaldehyde were also employed as the substrates for enzymatic hydrolysis catalyzed by α-chymotrypsin [47, 48]. Another way to synthesize the Schiff bases **54** was to alkylate the Schiff base of ethyl glycinate and benzaldehyde with the corresponding fluorobenzyl bromide under phase-transfer conditions, although the yields were disappointingly low. The Schiff base **54** was dissolved in a mixture of acetonitrile and water (9 : 1) and treated with α-chymotrypsin at ambient temperature for 24 h with stirring (equation 7.16). The free L-amino acid **55** precipitated together with the undissolved α-chymotrypsin. It was separated from the enzyme by treatment with 5% aqueous trichloroacetic acid and subsequent ion-exchange treatment (yields: **55a**, 42%; **55b**, 78%). The remaining Schiff base **56** of the D-amino acid ester was hydrolyzed with hydrochloric acid to afford the free D-amino acid (*o*-, 60% overall yield; *p*-, yield not determined). The optical purity of the enantiomers was determined by chiral gas chromatography as their *N*-trifluoroacetyl isopropyl esters (L-*o*-, 98.4% e.e.; L-*p*-, 99.8% e.e.; D-*o*-, 49.7% e.e.). The yield and enantiomeric purity of both the L- and D-enantiomers have recently been greatly improved [48].

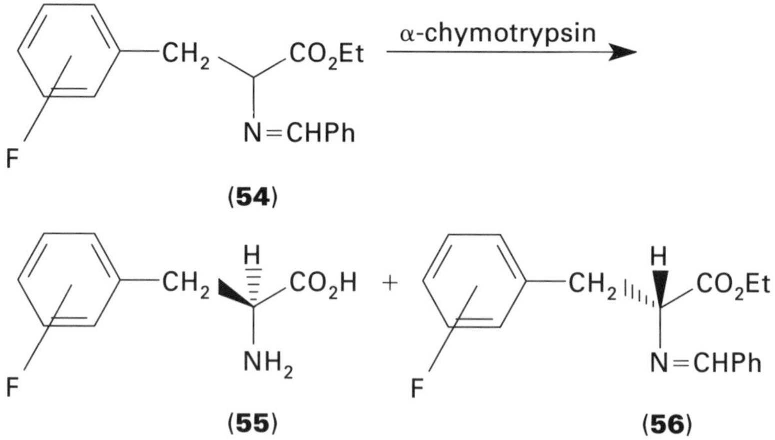

a, *o*-; **b**, *p*-

Another protease, subtilisin Carlsberg (from *Bacillus licheniformis*), was employed for the enantioselective hydrolysis of the *N*-Z methyl or ethyl esters (**57**) of 3-fluoro-, 4-fluoro- and 2,3,4,5,6-pentafluorophenylalanines, obtained by the malonate route (equation 7.17) [49]. The sparingly water-soluble substrates were suspended or emulsified in water or water–acetonitrile (9 : 11), and the enzymatic hydrolysis was conducted at pH 8 (pH-stat) and 25 °C. The hydrolysis products, *N*-Z-L-amino acids, were separated (yields: 78% from **57a**; 76% from **57b**; 42% from **57c**) from the hydrolysis-resistant esters of D-isomers at pH 8–9. *N*-Z-D-3-

fluorophenylalanine was obtained (58% overall yield) from the unhydrolyzed D-methyl ester by alkaline hydrolysis. These N-Z derivatives could be used directly for further peptide synthesis, while the free amino acids were obtained by removal of the N-Z group with HBr in acetic acid and subsequent isoelectric precipitation. Subtilisin Carlsberg was applied also to the resolution of 3- and 4-(trifluoromethyl)phenylalanines [50]. The racemic N-acetyl methyl esters were hydrolyzed with this enzyme in aqueous dimethyl sulfoxide at pH 7 (pH-stat). When the theoretical quantity of alkali had been consumed (ca 24 h), the unhydrolyzed N-acetyl-D-amino acid esters were extracted at pH 7 (yields: m-CF$_3$, 80%; p-CF$_3$, 76%). The D-amino acid hydrochlorides were obtained by refluxing the resolved esters in hydrochloric acid.

$$ArCH_2C(CO_2R)_2 \quad \xrightarrow[\text{(1 equiv.)}]{OH^-} \quad ArCH_2C \Big\langle \begin{matrix} CO_2R \\ CO_2H \end{matrix} \quad \xrightarrow{\Delta}$$
$$\underset{NHZ}{} \qquad\qquad\qquad \underset{NHZ}{}$$

(7.17)

$$ArCH_2CHCO_2R$$
$$\underset{NHZ}{|}$$

(57)

R = Me or Et
a, Ar = 3-FC$_6$H$_4$; **b**, Ar = 4-FC$_6$H$_4$; **c**, Ar = C$_6$F$_5$

Proteases from a variety of sources, i.e. plant, microbial and mammalian proteases, were tried for the enantioselective hydrolysis of racemic N-Z-4-fluorophenylalanine methyl ester [51–53]. The hydrolysis was conducted in phosphate buffer containing 20% DMF at pH 7 (maintained constant by the addition of NaOH) and 35 °C. After the desired degree of conversion (ca 40%), the liberated N-Z-L-amino acid was isolated and its enantiomeric excess was determined by HPLC analysis [51]. The results are shown in Table 7.1. In general, the hydrolysis rates were slow. The hydrolyses mediated by sulfhydryl proteases, i.e. papain, ficin and bromelain, were almost non-enantioselective or only moderately enantioselective. Of the microbial proteases examined, subtilisin Carlsberg from *B. licheniformis* showed only a moderate enantioselectivity under the hydrolysis conditions employed, while the proteases from *B. subtilis* exhibited good enantioselectivities and higher hydrolysis rates. The best result was obtained by using protease from *Rhizopus niveus*. The mammalian protease α-chymotrypsin also showed a good, though not outstandingly high, enantioselectivity. When the ester group was changed from methyl to 2,2,2-trifluoroethyl, the hydrolysis rates were accelerated to a certain extent with most proteases examined, generally at the expense of enantioselectivities. The best result with the trifluoroethyl ester was obtained by using

Table 7.1 Enantioselective hydrolysis of Z-DL-Phe(4F)-OMe using various
proteases[a]

Protease	Amount (mg)	Conversion (%)	Time (h)	e.e. (%)[b]
Papain[c]	100	43	29	0
Ficin[c]	200	26	175	2
Bromelain[c]	200	35	174	44
Ex *Aspergillus oryzae*[d]	200	8	336	87
Ex *Aspergillus oryzae*[e]	200	17	174	75
Ex *Aspergillus melleus*	200	25	194	66
Ex *Aspergillus sojae*	50	10	336	88
Ex *Bacillus polymyxa*	80	5	336	32
Ex *Bacillus subtilis*[f]	200	40	216	85
Ex *Bacillus subtilis*[g]	100	41	267	87
Ex *Bacillus licheniformis*	100	40	121	64
Ex *Rhizopus niveus*	200	40	195	93
α-Chymotrypsin	200	40	336	92
Pancreatin F[h]	100	40	336	70

[a] Reactions were carried out in a mixture (2 ml, pH 7) of 0.1 M phosphate buffer and DMF (20%)
at 35 °C using 1 mmol of the substrate, unless indicated otherwise.
[b] Enantiomeric excess of the liberated N-Z-amino acid.
[c] In the presence of L-cysteine.
[d] Amano M.
[e] Amano A.
[f] Amano N.
[g] Nagase neutral protease.
[h] Enzyme of Amano, at pH 8.

protease from *B. subtilis* (Nagase alkaline protease) under the same hydrolysis
conditions as in the case of the methyl ester [90% e.e. at 40% conversion
(23 h) using 200 mg of enzyme] [53].

In addition to proteases, lipases have recently been employed for the
enantioselective hydrolysis of N-protected amino acid esters [54, 55]. As
part of the investigation, the three positional (*o*-, *m*- and *p*-) isomers of
fluorophenylalanine were resolved using porcine pancreatic lipase [53, 55].
The N-Z 2,2,2-trifluoroethyl or 2-chloroethyl esters were treated with this
lipase at pH 7 (maintained constant) and 25 °C until ca 40% conversion was
achieved. The hydrolysis products, N-Z-L-amino acids, were separated and
their optical purities were determined by HPLC. The results are shown in
Table 7.2. All three positional isomers were resolved with good to excellent
enantioselectivities. The results with *ortho*- and *meta*-isomers were especially
good. The chloroethyl esters gave a slightly better result than the trifluoroethyl esters. Microbial lipases were also employed for the enantioselective hydrolysis of N-Z-4-fluorophenylalanine 2-chloroethyl ester under
the same reaction conditions [53]. Of the microbial enzymes tried, lipase
from *Aspergillus niger* [54] showed a good enantioselectivity (91% e.e. at
26% conversion).

Table 7.2 Enantioselective hydrolysis of Z-NHCH(R)CO$_2$R$'$ using porcine pancreatic lipase[a]

R	R$'$ = CF$_3$CH$_2$			R$'$ = ClCH$_2$CH$_2$		
	Conversion (%)	Time (h)	e.e. (%)[b]	Conversion (%)	Time (h)	e.e. (%)[b]
2-FC$_6$H$_4$CH$_2$	40	24	97	42	25	>99
3-FC$_6$H$_4$CH$_2$	40	23	95	40	19	98
4-FC$_6$H$_4$CH$_2$	40	117	90	38	133[c]	91

[a] Reactions were carried out using 1 mmol of the substrate and 300 mg of the enzyme in phosphate buffer (pH 7) at 25 °C, unless indicated otherwise.
[b] Enantiomeric excess of the liberated N-Z-amino acid.
[c] Using 400 mg of enzyme.

Lipase-catalyzed alcoholysis (transesterification) in organic solvents of the N-Z 2,2,2-trifluoroethyl esters **58** was also applied to the resolution of 3- and 4-fluorophenylalanines [56]. Only moderate enantioselectivities were obtained in the methanolysis of these aromatic amino acid derivatives in diisopropyl ether: **59a**, 73% e.e. at 38% conversion; **59b**, 70% e.e. at 41% conversion, using lipase from *R. javanicus*, for example (equation 7.18). In these alcoholysis reactions also the L-isomers reacted preferentially as in the corresponding hydrolysis reactions.

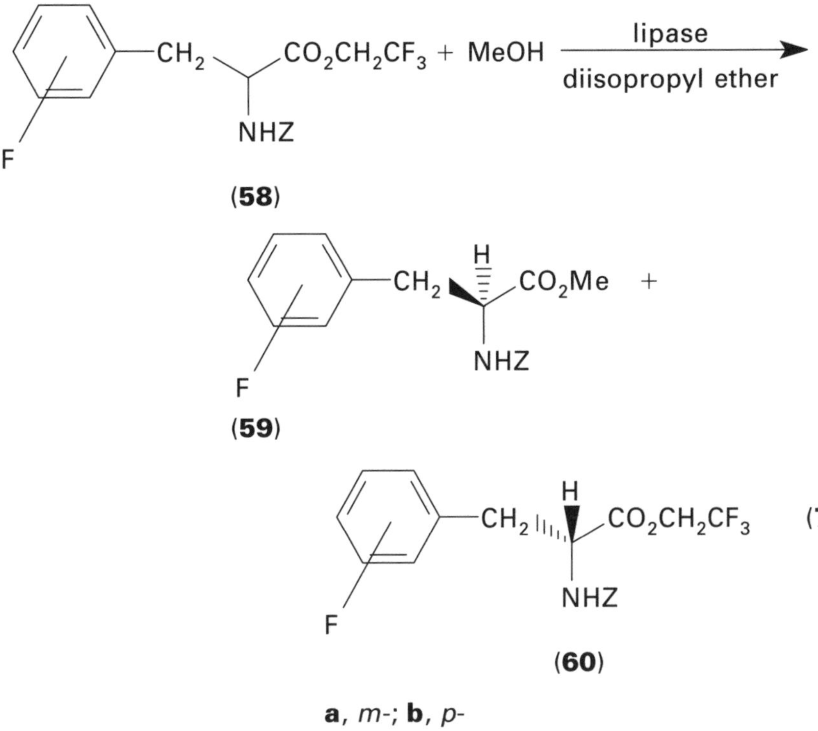

(58)

(59)

(60)   (7.18)

a, *m*-; **b**, *p*-

The hydrolytic properties of baker's yeast (*Saccharomyces cerevisiae* Hansen) have been employed for the resolution of *N*-acetylamino acid esters [57]. The active enzyme involved is thought to be an unspecific protease, from the close analogy of the hydrolytic behavior of *S. cerevisiae* Hansen with α-chymotrypsin. This microorganism was applied to the resolution of the three positional (*o*-, *m*- and *p*-) isomers of fluorophenylalanine [58]. Racemic *N*-acetylfluorophenylalanine ethyl esters were hydrolyzed in phosphate buffer (pH 7.5) at room temperature using lyophilized yeast. The pH optimum for lyophilized yeast for this reaction was found to be 7.5, while fermenting yeast itself showed a pH optimum between 3 and 4. The pH was maintained constant by addition of NaOH from an autoburette. The reactions stopped after 6–8 h at 52–53% conversion. No difference was observed in the hydrolysis rates of the three isomeric substrates. The unhydrolyzed *N*-acetyl-D-amino acid esters were obtained in good yields (yields: *o*-, 80%; *m*-, 76%; *p*-, 86%). Their enantiomeric purities were determined by ^{1}H NMR using Eu(hfc)$_3$ to be >96%. In addition, their optical rotations and melting points did not change on repeated recrystallizations.

7.5 ENZYMATIC RESOLUTION THROUGH HYDROLYSIS OF LEUCINE DIPEPTIDES

The enzymatic resolution of 4-fluoroglutamic acids has been achieved by the hydrolysis of the leucine dipeptides with leucine aminopeptidase.

A diastereomeric mixture of 4-fluoroglutamic acid (FGlu) was synthesized by condensation of diethyl fluoromalonate with ethyl α-acetamidoacrylate [59]. The *erythro* and *threo* racemates were then separated by ion-exchange chromatography on Dowex 1 (Scheme 7.2). The relative configuration of the two asymmetric carbon atoms was established by an unambiguous chemical method: on pyrolysis, *erythro*- or *threo*-4-fluoroglutamic acid cyclized to *trans*-or *cis*-3-fluoro-2-pyrrolidone-5-carboxylic acid, which was related to *trans*- or *cis*-4-fluoroproline, respectively. The racemate was converted into its *N*-L-leucyl derivative by reaction with the *N*-hydroxysuccinimide ester (ONSu) of *N*-Boc-L-leucine followed by removal of the Boc group with trifluoroacetic acid. The dipeptide was hydrolyzed at 37 °C in Tris–HCl buffer (pH 9) containing magnesium sulfate by leucine aminopeptidase from porcine kidney [60]. After 90 min of incubation, L-*erythro*-4-fluoroglutamic acid was isolated by ion-exchange chromatography in 77% yield. The L-*threo*-isomer was likewise obtained after 2 h of incubation in 92% yield.

The optical purity of the L-enantiomers obtained was examined by exhaustive treatment of each isomer with *E. coli* glutamate decarboxylase to be at least 98%. However, the optical rotations of these enantiomers were not reported. Later, this resolution procedure was re-examined [61]. In this case, a mixture of *erythro*- and *threo*-4-fluroglutamic acid was directly converted into

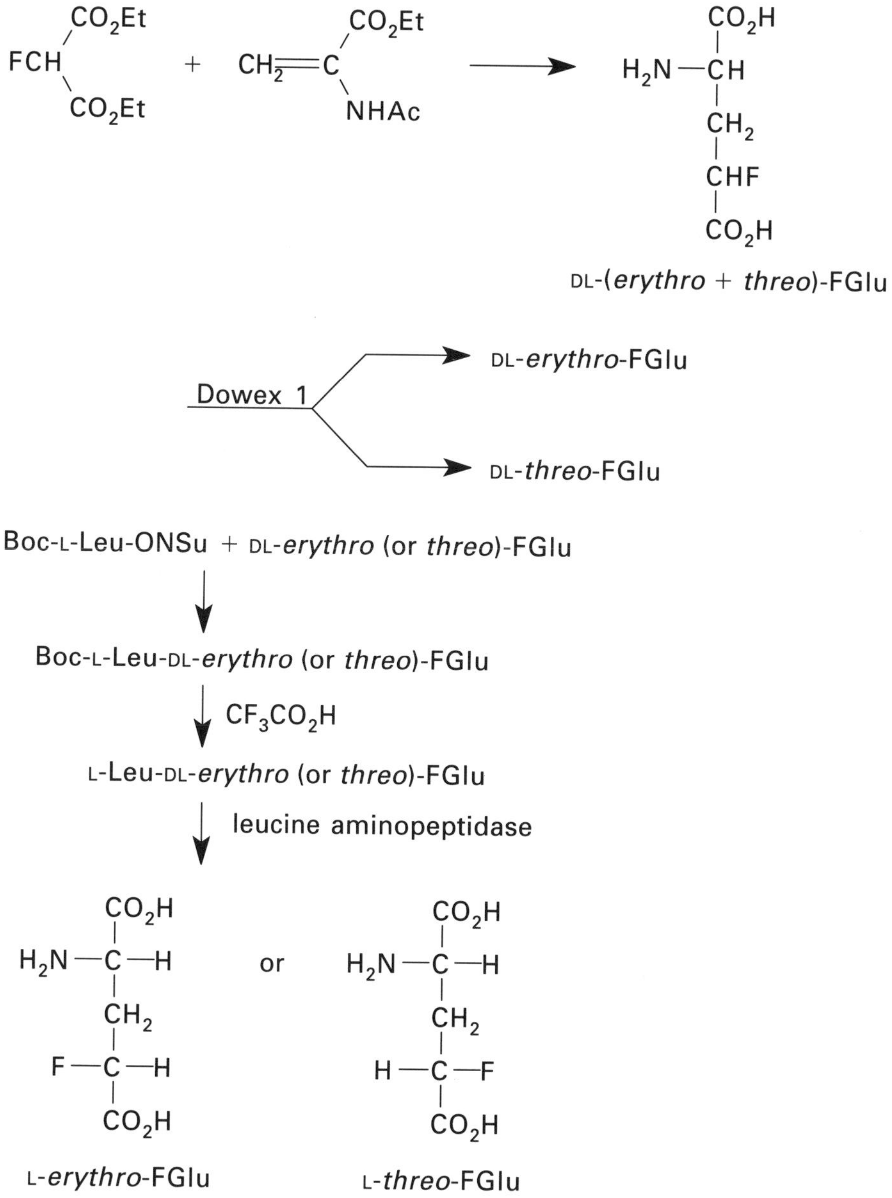

$$\underset{\underset{CO_2Et}{|}}{FCH}\!\!-\!\!CO_2Et \quad + \quad CH_2\!\!=\!\!\underset{\underset{NHAc}{|}}{C}\!\!-\!\!CO_2Et \quad \longrightarrow \quad H_2N\!-\!\underset{\underset{\underset{\underset{CO_2H}{|}}{CHF}}{\overset{\overset{CO_2H}{|}}{\underset{\underset{}{CH_2}}{\underset{|}{CH}}}}{}$$

DL-(*erythro* + *threo*)-FGlu

Dowex 1

DL-*erythro*-FGlu

DL-*threo*-FGlu

Boc-L-Leu-ONSu + DL-*erythro* (or *threo*)-FGlu

↓

Boc-L-Leu-DL-*erythro* (or *threo*)-FGlu

↓ CF_3CO_2H

L-Leu-DL-*erythro* (or *threo*)-FGlu

↓ leucine aminopeptidase

$$H_2N\!-\!\underset{\underset{\underset{\underset{CO_2H}{|}}{F\!-\!C\!-\!H}}{\overset{|}{CH_2}}}{\overset{\overset{CO_2H}{|}}{C}}\!\!-\!\!H \qquad or \qquad H_2N\!-\!\underset{\underset{\underset{\underset{CO_2H}{|}}{H\!-\!C\!-\!F}}{\overset{|}{CH_2}}}{\overset{\overset{CO_2H}{|}}{C}}\!\!-\!\!H$$

L-*erythro*-FGlu   L-*threo*-FGlu

Scheme 7.2

the *N*-L-leucyl derivative, which was hydrolyzed by the same enzyme at pH 9 and 36 °C for 3.5 h, and L-*erythro*- and L-*threo*-4-fluoroglutamic acids were isolated by ion-exchange chromatography. The latter isomer was determined to be optically pure by reversed-phase HPLC using *N,N*-dipropyl-L-alanine and copper(II) acetate in the aqueous mobile phase. However, the L-*erythro*-isomer was not optically pure (87% e.e.), indicating a lack of stereospecificity of this enzyme toward the *erythro*-isomers. This was also the case with the hydrolysis of L-leucyl-*erythro*-4-methylglutamic acid. Optically pure L-*erythro*-isomer was obtained through the leucine aminopeptidase-catalyzed hydrolysis of pure L-leucyl-L-*erythro*-4-fluoroglutamic acid thoroughly separated from its diastereomer by ion-exchange chromatography.

7.6 ENZYMATIC RESOLUTION OF FLUORINE-CONTAINING β-HYDROXYAMINO ACIDS AS ALCOHOLS

Fluorinated threonines have been resolved via the enantioselective hydrolysis of the *O*-acyl derivatives with hydrolytic enzymes.

A diastereomeric mixture of the 4,4,4-trifluorothreonine derivative **62** was prepared through the condensation of trifluoroacetaldehyde with the imine derived from benzaldehyde and ethyl glycinate, followed by diacetylation (Scheme 7.3) [62]. Column chromatography on silica gel separated the racemic *erythro* and *threo* isomers. The enzymatic hydrolysis of *threo*-**62** with lipase from *Candida rugosa* (Meito Sangyo, MY) at pH 7.3 and 40–41 °C gave the alcohol (2R,3R)-**63** with 86% e.e. at 37% conversion (96 h). This was reconverted into the acetate and hydrolyzed with the same enzyme to give (2R,3R)-**63** with >97% e.e. at 41% conversion, which gave free (2R,3R)-4,4,4-trifluorothreonine with >97% e.e. after acidic hydrolysis. From the recovered acetate, (2S,3S)-**63** with >93% e.e. was obtained by enzymatic hydrolysis with cellulase from *Trichoderma viride*. This was converted into the free (2S,3S)-amino acid with >93% e.e. On the other hand, the enzymatic hydrolysis of *erythro*-**62** to less than 25% conversion with lipase from *C. rugosa* afforded the alcohol (2S,3R)-**63** with 95% e.e., from which the free (2S,3R)-amino acid with >95% e.e. was obtained after acidic hydrolysis. When the lipase-catalyzed hydrolysis of *erythro*-**62** was allowed to proceed to 74% conversion, the recovered acetate gave (2R,3S)-**63** with 89% e.e. on treatment with cellulase. Its acidic hydrolysis afforded the free (2R,3S)-amino acid with >93% e.e. The absolute configuration of the (2S,3S)-amino acid was confirmed by transforming it into (S)-4,4,4-trifluorobutane-1,3-diol with known configuration [63].

Both enantiomers of *threo*-4,4-difluorothreonine were also prepared by a similar enzymatic resolution [64]. The condensation of ethyl N,N-dibenzylglycinate with ethyl difluoroacetate, followed by reduction with sodium tetrahydroborate, gave racemic N,N-dibenzyl-*threo*-4,4-difluorothreonine ethyl ester (**64**) (equation 7.19). Deprotection of the amino group and acetylation of both the amino and hydroxy groups yielded the substrate DL-*threo*-**65** for enzymatic resolution. The enzymatic hydrolysis of DL-*threo*-**65** with cellulase from *T. viride* to 76% conversion (41 h) afforded the alcohol (2S,3S)-**66** (62%

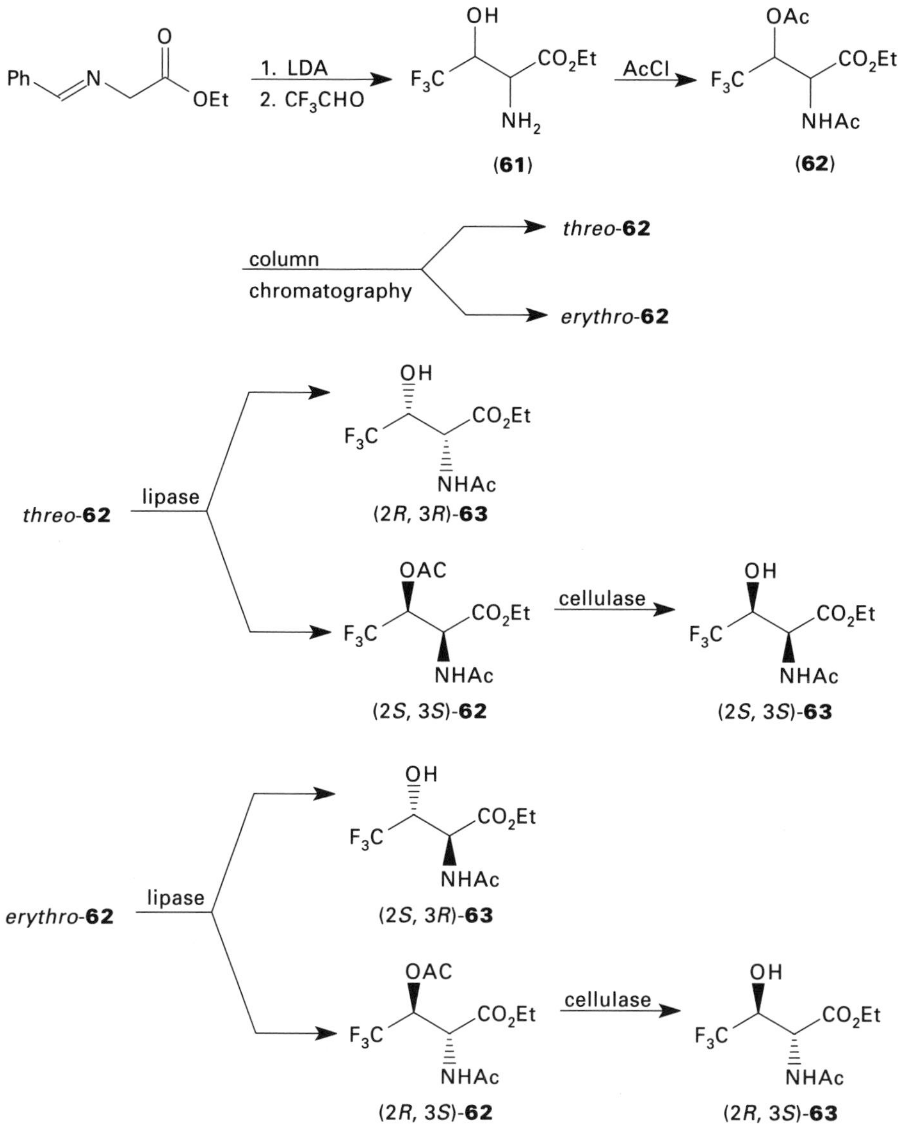

Scheme 7.3

yield) with the recovery of the acetate (2R,3R)-**65** with 82% e.e. (20% yield). To enhance the optical purity, these compounds were re-subjected to enzymatic processes. The alcohol (2S,3S)-**66** was re-converted into the acetate and hydrolyzed with the cellulase to 47% conversion (2.5 h) to give (2S,3S)-**66** with 71% e.e. (37% yield), which gave free (2S,3S)-4,4-difluorothreonine with 71% e.e. after acidic hydrolysis. On the other hand, the recovered acetate

(2*R*,3*R*)-**65** with 82% e.e. was hydrolyzed with lipase from *C. rugosa* to 69% conversion (5 days) to afford the alcohol (2*R*,3*R*)-**66** with 89% e.e. (55% yield). The free (2*R*,3*R*)-amino acid with 89% e.e. was obtained after usual acidic hydrolysis.

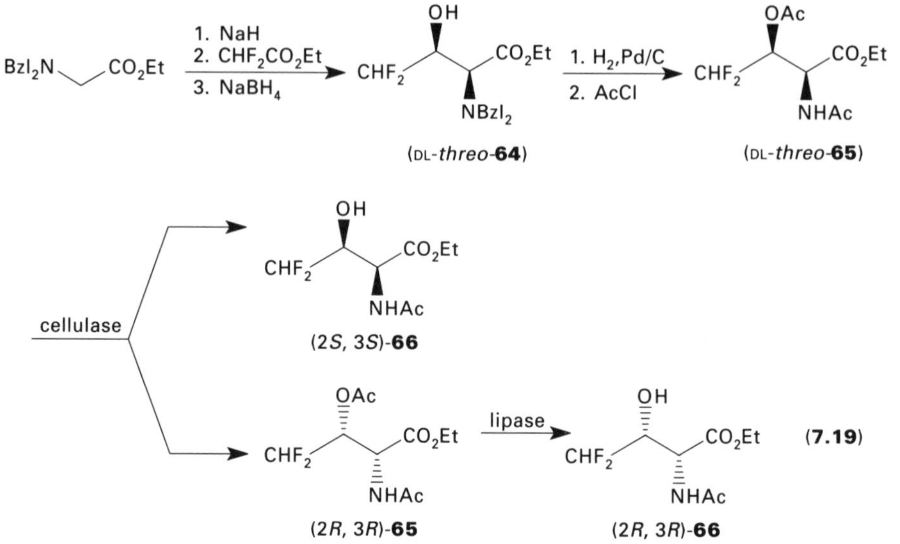

7.7 REFERENCES

1. M. Bergmann and H. Fraenkel-Conrat, *J. Biol. Chem.*, **119**, 707 (1937).
2. C. Niemann and M. M. Rapport, *J. Am. Chem. Soc.*, **68**, 1671 (1946).
3. E. L. Bennett and C. Niemann, *J. Am. Chem. Soc.*, **72**, 1800 (1950).
4. E. L. Bennett and C. Niemann, *J. Am. Chem. Soc.*, **70**, 2610 (1948).
5. R. B. Loftfield and E. A. Eigner, *Biochim. Biophys. Acta*, **130**, 449 (1966).
6. W. S. Fones and M. Lee, *J. Biol. Chem.*, **201**, 847 (1953).
7. W. H. Vine, D. A. Brueckner, P. Needleman and G. R. Marshall, *Biochemistry*, **12**, 1630 (1973).
8. J. Porter, J. Dykert and J. Rivier, *Int. J. Pept. Protein Res.*, **30**, 13 (1987).
9. M. Lee and R. S. Phillips, *Bioorg. Med. Chem. Lett.*, **1**, 477 (1991).
10. M. Lee, PhD Dissertation, University of Georgia (1992).
11. J. P. Greenstein and M. Winitz, *Chemistry of the Amino Acids*, Vol. 1, Wiley, New York, 1961, p. 728.
12. J.-L. Fauchère and R. Schwyzer, *Helv. Chim. Acta*, **54**, 2078 (1971).
13. K. Takahashi, K. L. Kirk and L. A. Cohen, *J. Labelled Compd. Radiopharm.*, **23**, 1 (1986).
14. H. K. Chenault, J. Dahmer and G. M. Whitesides, *J. Am. Chem. Soc.*, **111**, 6354 (1989).
15. R. W. Goulding and S. W. Gunasekera, *Int. J. Appl. Radiat. Isot.*, **26**, 561 (1975).
16. (a) I. Chibata, T. Tosa, T. Sato and T. Mori, *Methods Enzymol.*, **44**, 746 (1976); (b) I. Chibata, in E. L. Eliel and S. Otsuka (Eds), *Asymmetric Reactions and Processes in Chemistry*, American Chemical Society, Washington, DC, 1982, p. 195.

17. J. C. Clark, R. W. Goulding, M. Roman and A. J. Palmer, *Radiochem. Radioanal. Lett.*, **14**, 101 (1973).
18. (a) K. Uchida, *Bio Ind.*, **4**, 364 (1987); *Chem. Abstr.*, **107**, 196405f (1987); (b) K. Uchida and H. Tanaka, *Yuki Gosei Kagaku Kyokaishi*, **46**, 977 (1988); *Chem. Abstr.*, **110**, 75997d (1989).
19. M. Urushihara and Y. Matsumura, personal communication.
20. V. A. Soloshonok, I. Yu. Galaev, V. K. Švedas, E. V. Kozlova, N. V. Kotik, I. P. Shishkina, S. V. Galushko, A. B. Rozhenko and V. P. Kukhar', *Bioorg. Khim.*, **19**, 467 (1993).
21. S. V. Galushko, I. P. Shishkina and V. A. Soloshonok, *J. Chromatogr.*, **592**, 345 (1992).
22. A. Vidal-Cros, M. Gaudry and A. Marquet, *J. Org. Chem.*, **50**, 3163 (1985).
23. (a) J. Kollonitsch, *Isr. J. Chem.*, **17**, 53 (1978); (b) J. Kollonitsch, S. Marburg and L. M. Perkins, *J. Org. Chem.*, **44**, 771 (1979).
24. W. A. König, W. Franke and I. Benecke, *J. Chromatogr.*, **239**, 227 (1982).
25. T. Tsushima, K. Kawada, S. Ishihara, N. Uchida, O. Shiratori, J. Higaki and M. Hirata, *Tetrahedron*, **44**, 5375 (1988).
26. R. Wernicke, *J. Chromatogr. Sci.*, **23**, 39 (1985).
27. I. Ojima, K. Kato, K. Nakahashi, T. Fuchikami and M. Fujita, *J. Org. Chem.*, **54**, 4511 (1989).
28. J. A. Dale, D. L. Dull and H. S. Mosher, *J. Org. Chem.*, **34**, 2543 (1969).
29. W. H. Vine, K.-H. Hsieh and G. R. Marshall, *J. Med. Chem.*, **24**, 1043 (1981).
30. M. Hudlický, V. Jelínek, K. Eisler and J. Rudinger, *Collect. Czech. Chem. Commun.*, **35**, 498 (1970).
31. M. E. Houston, Jr, and J. F. Honek, *J. Chem. Soc., Chem. Commun.*, 761 (1989).
32. J. F. Honek, personal communication.
33. J. W. Keller and B. J. Hamilton, *Tetrahedron Lett.*, **27**, 1249 (1986).
34. J. W. Keller and C. S. Day, *Acta Crystallogr., Sect. C*, **40**, 1224 (1984).
35. V. A. Soloshonok, V. K. Švedas, V. P. Kukhar', I. Yu. Galaev, E. V. Kozlova and N. Yu. Svistunova, *Bioorg. Khim.*, **19**, 478 (1993).
36. V. A. Soloshonok, personal communication.
37. V. A. Soloshonok, V. K. Svedas, V. P. Kukhar, A. G. Kirilenko, A. V. Rybakova, V. A. Solodenko, N. A. Fokina, O. V. Kogut, I. Yu. Galaev, E. V. Kozlova, I. P. Shishkina and S. V. Galushko, *Synlett*, 339 (1993).
38. J. H. Tong, C. Petitclerc, A. D'Iorio and N. L. Benoiton, *Can. J. Biochem.*, **49**, 877 (1971).
39. J. R. Coggins and N. L. Benoiton, *J. Chromatogr.*, **52**, 251 (1970).
40. S. Weinstein, J. T. Durkin, W. R. Veatch and E. R. Blout, *Biochemistry*, **24**, 4374 (1985).
41. U. Beitler and B. Feibush, *J. Chromatogr.*, **123**, 149 (1976).
42. R. Sheardy, L. Liotta, E. Steinhart, R. Champion, J. Rinker, M. Planutis, J. Salinkas, T. Boyer and D. Carcanague, *J. Chem. Educ.*, **63**, 646 (1986).
43. G. M. Anantharamaiah and R. W. Roeske, *Tetrahedron Lett.*, **23**, 3335 (1982).
44. J. T. Gerig and J. C. Klinkenborg, *J. Am. Chem. Soc.*, **102**, 4267 (1980).
45. T. Tsushima, K. Kawada, J. Nishikawa, T. Sato, K. Tori, T. Tsuji and S. Misaki, *J. Org. Chem.*, **49**, 1163 (1984).
46. T. N. Wade, F. Gaymard and R. Guedj, *Tetrahedron Lett.*, 2681 (1979).
47. Yu. N. Belokon', K. A. Kochetkov, V. I. Tararov, T. F. Savel'eva, N. V. Fileva, N. S. Garbalinskaya, M. B. Saporovskaya, Z. B. Bakasova and A. G. Rait, *Bioorg. Khim.*, **17**, 773 (1991); *Chem. Abstr.*, **115**, 136693d (1991).
48. Yu. N. Belokon', personal communication.
49. H. R. Bosshard and A. Berger, *Helv. Chim. Acta*, **56**, 1838 (1973).

50. J. J. Nestor, Jr, T. L. Ho, R. A. Simpson, B. L. Horner, G. H. Jones, G. I. McRae and B. H. Vickery, *J. Med. Chem.*, **25**, 795 (1982).
51. T. Miyazawa, H. Iwanaga, T. Yamada and S. Kuwata, *Chirality*, **4**, 427 (1992).
52. T. Miyazawa, H. Iwanaga, T. Yamada and S. Kuwata, in N. Yanaihara (Ed.), *Peptide Chemistry 1992*, ESCOM, Leiden, 1993, p. 94.
53. T. Miyazawa, H. Iwanaga, T. Yamada and S. Kuwata, unpublished results.
54. T. Miyazawa, T. Takitani, S. Ueji, T. Yamada and S. Kuwata, *J. Chem. Soc., Chem. Commun.*, 1214 (1988).
55. T. Miyazawa, H. Iwanaga, S. Ueji, T. Yamada and S. Kuwata, *Chem. Lett.*, 2219 (1989).
56. T. Miyazawa, M. Mio, Y. Watanabe, T. Yamada and S. Kuwata, *Biotechnol. Lett.*, **14**, 789 (1992).
57. B. I. Glänzer, K. Faber and H. Griengl, *Tetrahedron*, **43**, 771 (1987).
58. R. Csuk and B. I. Glänzer, *J. Fluorine Chem.*, **39**, 99 (1988).
59. J. C. Unkeless and P. Goldman, *Mol. Pharmacol.*, **6**, 46 (1970).
60. J. C. Unkeless and P. Goldman, *Mol. Pharmacol.*, **7**, 293 (1971).
61. S. Bory, J. Dubois, M. Gaudry, A. Marquet, L. Lacombe and S. Weinstein, *J. Chem. Soc., Perkin Trans. 1*, 475 (1984).
62. T. Kitazume, J. T. Lin and T. Yamazaki, *Tetrahedron: Asymmetry*, **2**, 235 (1991).
63. J. T. Lin, T. Yamazaki and T. Kitazume, *J. Org. Chem.*, **52**, 3211 (1987).
64. T. Yamazaki, J. Haga and T. Kitazume, *Bioorg. Med. Chem. Lett.*, **1**, 271 (1991).

8 High-performance Liquid Chromatography of Fluorine-containing Amino Acids

S. V. GALUSHKO

Institute of Bioorganic Chemistry and Petrochemistry, Ukranian Academy of Sciences, 253660 Kiev, Ukraine

8.1 INTRODUCTION

High-performance liquid chromatography (HPLC) is widely used in bioorganic chemistry and the chemistry of fluorine-containing amino acids is no exception. However, some specific points arise in the HPLC of these compounds owing to the effect of fluorine atoms on the chromatographic behaviour of fluorine-containing analogues. The most important fields of application of HPLC in the chemistry of fluorine-containing amino acids are the following:

1. Quantitative and qualitative analysis.
2. Preparative separation of complex mixtures.
3. Determination of physico-chemical properties.

Some important chromatographic terms and definitions are defined here for readers who are not specialized in chromatography. Liquid chromatography is based on the distribution of solutes between two immiscible phases: a mobile phase (eluent) and a stationary phase (sorbent). As stationary phases, sorbents with either unmodified or modified surfaces can be used. Unmodified stationary phases (silica, alumina) as a rule are employed in low-performance liquid chromatography (flash chromatography, etc.). Sorbents with modified surfaces are now in widespread use in HPLC and most separations by reversed-phase HPLC (RP-HPLC) are performed with them. In this mode the sorbents are usually silica particles (3–20 μm in diameter) with chemically bonded alkyl (C_2, C_4, C_8, C_{18}) or aryl (C_6H_5) radicals. Methanol–water, acetonitrile–water and tetrahydrofuran–water binary mixtures and also ternary and quaternary mixtures of these solvents with additives of buffers, surfactants, complexones, etc., are mainly used as mobile phases in RP-HPLC.

In normal-phase HPLC, which is more suitable for preparative separations of mixtures in organic synthesis, unmodified silica and also chemically bonded phases such as those containing nitrile or amine functionalities are employed as stationary phases.

Fluorine-containing Amino Acids: Synthesis and Properties Edited by V. P. Kukhar' and V. A. Soloshonok
© 1995 John Wiley & Sons Ltd

The nature of the interactions that determine the distribution of solutes between two phases can be different: hydrophobic interactions, electrostatic forces, hydrogen bonds, π-bonds, metal–ligand coordination, etc.

A typical chromatogram that illustrates the basic quantitative characteristics of the chromatographic process is shown in Figure 8.1.

t_R and $t_0 =$ retention times of a retained and unretained substance, respectively;

$$k' = \frac{t_R - t_0}{t_0} = \text{capacity factor;}$$

$$\alpha = \frac{k'_2}{k'_1} = \text{selectivity of separation of two substances;}$$

$$w = \text{width of peak;}$$

$$N = 16\left(\frac{t_R}{w}\right) = \text{number of theoretical plates; } N \text{ is a}$$
measure of effectivity of chromatographic system;

$$R_S = (1/4)\,(\alpha - 1)\,N^{1/2}\left[\frac{k'}{(1 + k')}\right] = \text{resolution.}$$

For any column, the most important parameters are α and N. The greater are the α and N values, the easier is the separations. Modern HPLC systems allows the separation of substances with $\alpha = 1.03$–1.06.

8.2 REVERSED-PHASE HIGH-PERFORMANCE LIQUID CHROMATOGRAPHY

Organic chemists often deal with new substances. The optimum chromatographic conditions are unknown in such a case and it is desirable to find acceptable chromatographic conditions with the minimum time and effort. The selection of the optimum conditions in RP-HPLC is still heuristic and requires specific knowledge and expertise. To select the initial conditions for the reversed-phase separation of fluorine-containing amino acids and related compounds, some relationships between the structure and retention can be employed. Recently an approach has been developed that allows the retention in RP-HPLC on the basis of structural formula to be calculated [1, 2]. This approach has been realized in the commercial software CromDream [3, 4]. The software permits the prediction of acceptable RP-HPLC conditions (column and eluent) after drawing a structural formula of a compound on the PC screen. Prediction of acceptable conditions for mixtures of compounds and simulating chromatographic behaviour on different columns are also possible. The software can be used for prediction of acceptable eluent compositions and calculations of capacity factors for different classes of compounds. Results of the prediction of the eluent composition for some p-polyfluoroalkoxy(thio) derivatives of phenylalanine [5] are shown in Table 8.1.

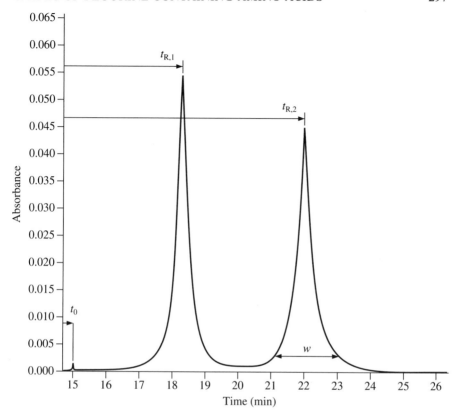

Figure 8.1 Typical chromatogram

Thus a computer-aided method for the selection of the initial conditions and calculation of the retention in RP-HPLC gives possibility of saving a lot of time and effort in experimental work.

An important point in the RP-HPLC of ionogenic substances is the efficiency of chromatographic separations. Ionizable groups in molecules of amino acids and their derivatives lead to broadened chromatographic peaks under RP-HPLC conditions. To provide a high efficiency of the separation of such compounds, different buffers are employed as the aqueous component of a mobile phase. To separate n-polyfluoroalkoxy(thio) derivatives of phenylalanine, methanol–0.02 M phosphate buffer (pH 2.9) mobile phases were used [5]. For studying the amination mechanism of α-bromo-γ-fluoro-α,β-unsaturated esters, which are intermediates in the synthesis of (E)-(β)-fluoromethylene-m-tyrosine, the compounds were separated on an octadecyl reversed-phase column with acetonitrile–0.05 M phosphate buffer (pH 6) as the mobile phase [6]. Both gradient and isocratic elution were employed for the

Table 8.1 Predicted and experimental concentrations of organic solvents in a mobile phase for *para*-substituted derivatives of phenylalanie in RP-HPLC.

$$X - \text{C}_6\text{H}_4 - CH_2\underset{\underset{NH_2}{|}}{C}HCOOH$$

Substituent X	π_x	Compound No.	Concentration (mol l^{-1})	
			Experimental	Predicted
H	—	1	10–25	8–20
F	0.14 [7]	2	10–30	8–24
	0.13 [8]			
	0.14 [5]			
CHF_2O	0.58 [7]	3	15–40	10–30
	0.57 [5]			
CF_3O	1.04 [7]	4	20–45	25–50
	1.21 [8]			
	1.06 [5]			
CHF_2S	0.95 [5]	5	20–50	15–40
CF_3S	1.44 [7]	6	35–60	40–65
	1.58 [8]			
	1.43 [5]			
CF_3CH_2O	0.95 [5]	7	30–50	35–55
CF_3CF_2O	1.80 [5]	8	40–65	30–45
CF_3CHFCF_2O	1.90 [5]	9	45–65	30–45
$CHClBrCF_2O$	0.69 [5]	10	40–55	30–45

determination of the physical properties and purity of the compounds and for kinetic studies [6].

In addition to the analysis of purity and reaction studies, RP-HPLC is also employed for the determination of different physico-chemical properties of compounds, such as hydrophobicity and pK values. Hydrophobicity is one of the most important features of bioactive compounds. The relationship between the hydrophobic properties of a substance and its biological activity has long been established. Introducing different fluoro substituents into a molecule allows the hydrophobicity of bioactive compounds to be regulated over a wide range. In drug design, the distribution coefficients of a substance in an octanol–water system as a measure of hydrophobicity are widely used. It should be noted that amino acids are highly polar ionogenic compounds that are difficult to study in an octanol–water system. In this case, the Leo–Hansch or Rekker methods for calculating the logarithm of the octan-1-ol–water partition coefficients (log P) are used [7, 8]. The hydrophobicity contributions π_x constants) have been determined only for several fluorine-containing substituents. The influence of different fluorine-containing groups on the retention parameters of amino acids in RP-HPLC and their correlation with the hydrophobicity constants of the substituents have been established [5]. The effect of different groups on the change in free energy of sorption (ΔG) of compounds on a hydrophobic surface is shown in Table 8.2. As can be seen, introducing an F atom in the *para*-position increases the ΔG value by $0.65\,\text{kJ mol}^{-1}$ (compounds **1** and **2**) and replacing a CH_2 group with a CF_2 group leads to an increase in ΔG of $4.96\,\text{kJ mol}^{-1}$ (compounds **7** and **8**). Introducing a CF_2 group increases the ΔG value by $4.16\,\text{kJ mol}^{-1}$ (compounds **4** and **8**) whereas introducing a CH_2 group increases ΔG only by $1.14\,\text{kJ mol}^{-1}$ (compounds **4** and **7**). From the results obtained, the hydrophobicity constants of fluorine-containing substituents were determined (Table 8.1).

As can be seen from Table 8.1, introducing fluorine-containing substituents allows the lipophilicity of amino acids to be varied over a wide range. It is interesting that a fairly good correlation also exists between the Hammett σ_p constants and the logarithm of the capacity factors of compounds [5]. The amino acids studied are in ionized state in pH range used in RP-HPLC (2–8). It is known that under RP-HPLC conditions even a slight change in the pK value of an ionogenic substance can appreciably affect its retention [9]. It seems that the influence of substituents on the pK values contributes to the retention of amino acids under the conditions of RP-HPLC.

8.3 LIGAND-EXCHANGE HIGH-PERFORMANCE LIQUID CHROMATOGRAPHY

HPLC is widely used to separate enantiomers of amino acids [10]. Atoms of fluorine affect considerably both the retention and selectivity in enantiomer separations, but some common approaches exist for separating amino acid enantiomers. The separations require the use of either a chiral stationary phase or an achiral stationary phase and a chiral mobile phase. Most papers have described the separation of amino acid enantiomers on chiral stationary phases, containing residues of amino acids chemically bonded to the surface of a silica or polymeric matrix (ligand A). A mobile phase in this case contains a

Table 8.2 Difference in free energy of sorption, $\Delta,(\Delta G)$, on octadecyl sorbent for derivatives of phenylalanine with 15% methanol in $0.02\,\text{M}$ phosphate buffer (pH 2.9) as eluent [5]

Compounds	$-\Delta(\Delta G)^a$ (kJ mol^{-1})	Compounds	$-\Delta(\Delta G)^a$ (kJ mol^{-1})
1–2	0.65	**1–3**	3.21
4–6	2.30	**3–4**	2.86
5–6	2.80	**8–9**	0.62
7–8	4.96	**4–7**	1.14

$^a -\Delta(\Delta G) = RT \ln\alpha$, where α is the selectivity.

constant concentration of methal ions (M). When a certain amount of another ligand (B) is introduced into the mobile phase, a mixed complex $(MAB)_S$ is formed on the surface. The capacity factor of the ligand B can be defined as follows [11]:

$$k' = \varphi \ \frac{[MAB]_S}{[B]_M + [MB]_M + \ldots + [MB_n]_M} \qquad (8.1)$$

where φ is the phase ratio in the column and M and S represent concentrations in the mobile phase and on the sorbent surface, respectively. The concentration of a metal ion in the mobile phase (C_M) is usually much higher than that of the ligand B and then equation 8.1 can be transformed into

$$K = Q/C_M \qquad (8.2)$$

where Q is constant for a given column and ligand B. Hence a plot of k' vs $1/C_M$ is a straight line of slope Q passing through the origin. A deflection from a straight line is observed only with a low concentration of metal ions in the mobile phase ($< 5 \times 10^{-4}$ M) [11]. The values of Q can be determined and the capacity factors of enantiomers can be calculated exactly with various concentrations of metal ions in the mobile phase. It has been shown that the enantiomers of fluoro derivatives of dihydroxyphenylalanine and tryptophan can be separated on a chiral column containing L-proline residues bonded to the surface of a silica matrix [12]. The mobile phase used in the work described [12] was 1 mM copper sulfate. The selectivity value ($\alpha = 4.0$ and 4.7) allows complete separation [12].

The chromatographic behaviour of α-trifluoromethyl-α-amino acids on chiral proline and hydroxyproline sorbents has been studied [11, 13]. The optimum concentrations of Cu^{2+} ions in the mobile phase were 1–5 mM. For poorly retained 3,3,3-trifluoroalanine and α-trifluoromethylalanine, it is preferable to use a mobile phase with a low Cu^{2+} concentration, whereas for α-CF_3Phe it is necessary to use a mobile phase with a Cu^{2+} concentration not lower than 5 mM. The chromatographic behaviour of α-trifluoromethylphenylglycine differs greatly from that of other CF_3-containing amino acids. The retention and selectivity of separation of enantiomers are so great that it is necessary to use a mobile phase with a higher concentration of H^+ ions and the separation should be carried out at 45–55 °C. An eluent of composition A–B $= 75:25$ or $50:50$ where A is 5 mM $CuSO_4$ containing ethanol (10%, v/v) and B is $0.005 M$ $CuSO_4$ (pH 3.0, HCl) are optimum for the separation of enantiomers of α-CF_3-phenylglycine. The selectivity of separation (α) is about 3.5, i.e. the conditions are suitable for both enantiomeric analysis and preparative work. The α-CF_3 group exerts a considerable influence on both the retention and the selectivity of the separation of amino acid enantiomers. The retention of α-CF_3-amino acids is much higher than that of natural amino acids. The change in the free energy of sorption when a CF_3 group is introduced is 1.2–2.5 kJ mol^{-1} on the proline column and 0.8–

$4.4\,kJ\,mol^{-1}$ on the hydroxyproline column. The difference in the selectivity of
the separation of enantiomers on proline and hydroxyproline columns is considerable (Table 8.3). On the L-Procolumn the selectivity of separation of α-
CF_3-amino acid enantiomers is much higher than that of natural amino acids.

The effects of replacing hydrogen atoms in molecules of amino acids with
fluorine atoms on the retention and selectivity of the separation of enantiomers
on chiral sorbents containing residues of L-Pro-OH and L-Val have been
studied in detail for fluorine-containing derivatives of alanine [13] and phenylalanine [14]. There are considerable differences in the changes in the retention
and the selectivity of enantiomer separations for the sorbents studied when H
atoms are successively replaced with F atoms in the molecule of alanine. For a
Nucleosil Chiral-1 HyPro column the introduction of each F atom into the
molecule of alanine leads to an increase in the selectivity whereas the
introduction of only the first F atom significantly affects the selectivity of
separation on Pro and Val columns. In addition, the introduction of F atoms
affects the retention of L- and D-isomers differently [13]. It is also interesting
that on Pro and Val columns the order of elution of D- and L-isomers is the
reverse of that with a Nucleosil L-Pro-OH column.

The introduction of fluorine atoms in different positions on the aromatic
ring of phenylalanine and phenylglycine also affects the separation of the
enantiomers differently [15]. It has been shown that the selectivity of the
separation of the enantiomers of phenylalanine and phenylglycine derivatives
depends substantially on the structures both of the bonded ligand and of the
enantiomers being separated (Table 8.4). The introduction of a fluorine atom

Table 8.3 Separation of amino acid enantiomers on 1 Chiral Pro Cu Si 100 and
Nucleosil Chiral-1 (Pro-OH) columns

Compound	1 Chiral Pro Cu Si 100		Nucleosil Chiral-1	
	k'_L	α	k'_L	α
Phe	16.8	2.0	3.7	1.8
α-CF$_3$Phe	28.0	6.0	5.2	1.3
Leu	5.8	1.0	1.4	2.1
α-CF$_3$Leu	14.3	1.6	4.7	2.0
Ala	4.3	1.0	0.5	1.7
αCF$_3$Ala	8.3	1.8	1.6	1.0
Nle	7.3	1.0	3.8	1.9
α-CF$_3$Nle	14.3	1.5	7.9	1.9
Nva	4.2	1.0	1.6	2.4
α-CF$_3$Nva	11.7	1.4	4.8	1.8
Asp	2.5	1.0	0.5	1.0
α-CF$_3$Asp	5.7	1.6	3.0	1.2
3,3,3-Trifluoroalanine	5.1	1.3	1.2	2.4
α-Trifluoroaminobutyric acid	7.6	1.5	3.2	1.6

Table 8.4 Separation of aromatic fluorine-containing amino acids on 1 Chiral Pro Cu Si 100, 1 Chiral Val Cu Si 100 (2) and Nucleosil Chiral-1 Pro-OH (3) columns

Compound	Column 1		Column 2		Column 3	
	k'_L	α	k'_L	α	k'_L	α
4-F-DL-α-PhGly	2.1	1.0	2.5	1.2	0.9	1.8
3-F-DL-α-PhGly	1.9	1.0	2.2	1.2	0.9	1.8
2-F-DL-α-PhGly	2.4	1.3	2.7	1.4	0.9	2.8
DL-α-PhGly	2.3	1.1	2.7	1.2	0.8	2.4
4-F-DL-α-Phe	4.8	2.3	4.6	1.5	1.3	1.6
3-F-DL-α-Phe	5.1	2.0	4.6	1.4	1.5	1.6
2-F-DL-α-Phe	5.3	2.0	4.6	1.3	1.6	1.5
DL-α-Phe	4.6	2.3	3.6	1.3	1.4	1.6
3-CF$_3$-DL-α-Phe	6.2	2.2	4.8	1.4	2.3	2.0
3,4-F$_2$-DL-α-Phe	4.4	1.9	3.7	1.7	1.7	1.7
1,2,3,4,5-F$_5$-DL-α-Phe	2.3	1.2	2.7	1.2	1.5	1.6

into the *ortho*-position of the aromatic ring of phenylglycine increases the selectivity of enantiomer separation on the chiral sorbents studied. For phenylalanine derivatives the replacement of one hydrogen atom with a fluorine one in any position on the phenyl ring hardly affects the selectivity of separation of enantiomers (Table 8.4). The introduction of two and five atoms of fluorine into the aromatic ring results in a decreased retention of enantiomers on L-Pro and L-Val columns. An increase in the number of fluorine atoms on the aromatic ring affects the selectivity of the separation of enantiomers differently. For a column with L-Pro residues on the surface the introduction of each F atom into the molecule of phenylalanine leads to a decrease in the selectivity of separation of 0.2, whereas for an L-Pro-OH column no effect was observed.

It has been found that a CF$_3$ group in the α-position of Phe exerts a considerable influence on both the retention and the selectivity of the separation of enantiomers [11]. Introduction of a CF$_3$ group into the phenyl ring of Phe has little effect on retention and selectivity. It seems that only a slight interaction occurs between a CF$_3$ group in the *meta*-position on the phenyl ring of Phe and the surface ligand of the sorbent.

Methanol in the mobile phase hardly affects the selectivity and retention, but increases the efficiency substantially. A mobile phase containing 2.5–5 mM copper sulfate and 0–30% (v/v) methanol is optimum for the separation of fluoro derivatives of phenylalanine and phenylglycyne [15, 26]. Chiral Pro Cu Si 100 is to be preferred for the separation of monosubstituted derivatives of phenylalanine. A Nucleosil Chiral-1 hydroxyproline column is to be preferred for the separation of monosubstituted fluorophenylglycine and pentafluoro derivatives of phenylalanine [15].

It should be noted that most papers describe the separation of enantiomers of amino acids that differ in the configuration of only one carbon atom. The separation of enantiomers of amino acids that contain two chiral centres is more sophisticated. Several methods for the synthesis of such amino acids have been developed [16–18] and the chromatographic separation of two enantiomer pairs can be very useful for both analytical and preparative purposes.

Galushko *et al.* [19] reported the possibility of the separation of all four [2*S*,3*S*(L,L); 2*R*,3*R*(D,D); 2*R*,3*S*(D,L); 2*S*,3*R*(D,D)] stereoisomers of threonine, phenylserine and its *o*- and *p*-fluoro derivatives by ligand-exchange HPLC. In searching for optimum conditions for the separation of the isomers, use was made of sorbents that contained residues of L-proline, L-hydroxyproline and L-valine bonded to the surface of a silica matrix. The experiments showed that an L-proline column permits the complete separation of all isomers of the amino acids studied with the retention times increasing in the order 2*R*,3*R* < 2*R*,3*S* < 2*S*,3*R* < 2*S*,3*S* (Figure 8.2). The conclusion was drawn

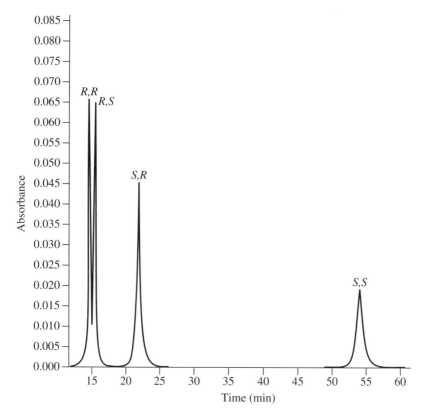

Figure 8.2 Separation of isomers of *p*-fluorophenylserine. Column, Chiral ProCu Si 100, 5 μm (250 × 4.6 mm i.d.) (Serva, Heidelberg, Germany); eluent, 5 mM CuSO₄; flow rate, 0.75 ml min⁻¹; temperature, 35 °C; detection wavelength, 225 nm.

that coordination interactions make a considerable contribution to the selectivity of separation of the isomers studied.

The introduction of atoms of fluorine into the o- and p-positions of the phenyl ring has little effect on the capacity factors (Table 8.5) and the selectivity of separation of isomers (Table 8.6). The threonine isomers have a much lower retention than phenylserine isomers. Mobile phases containing 5 mM and 1.25–2.5 mM copper sulfate are optimum for the separation of the isomers of phenylserine and threonine, respectively. The order of elution of the isomers studied on the L-valine and L-proline columns is the same whereas the selectivity of separation of enantiomers on the valine column is much less than that on the proline column but nevertheless the separation of four isomers of aromatic acids can be achieved.

It should be noted that a high column performance is needed for the separation of $2R,3R$–$2R,3S$ pairs on chiral proline and valine columns. It is also interesting that the decrease in the selectivity of separation of enantiomers on the valine column as compared with the proline column is the result of the decrease in retention mainly of $2S,3R$- and $2S,3S$-isomers (Table 8.5 and 8.6). The optimum concentrations of copper sulfate in the eluent are 4–5 mM for separation of isomers of phenylserine on a valine column. The retention and selectivity of the

Table 8.5 Retention (k') of isomers of α-amino-β-hydroxycarboxylic acids on chiral columns with 5 mM CuSO$_4$ solution as eluent

Compound	$2R,3R$			$2R,3S$			$2S,3R$			$2S,3S$		
	1^a	2^a	3^a	1	2	3	1	2	3	1	2	3
o-F-PhSer	1.5	1.5	1.6	1.7	1.8	0.9	3.4	2.7	0.85	8.3	4.1	0.9
o-F-PhSer	1.6	1.5	1.3	1.8	1.7	0.9	3.4	2.6	0.88	8.3	4.0	0.85
PhSer	1.4	1.6	1.5	1.6	2.0	1.1	3.4	2.5	1.05	8.6	4.0	1.1

^{a} Columns 1, 2 and 3 as in Table 8.4.

Table 8.6 Selectivity of separation of α-amino-β-hydroxycarboxylic isomers on chiral columns with 5 mM CuSO$_4$ solution as eluent

Compound	o-F-PhSer			p-F-PhSer			PhSer		
	1^a	2^a	3^a	1	2	3	1	2	3
R,R–R,S^b	1.1	1.2	1.8	1.1	1.1	1.4	1.1	1.2	1.4
R,R–S,R	2.3	1.8	1.9	2.1	1.7	1.5	2.4	1.7	1.4
R,R–S,S	5.5	2.7	1.8	5.2	2.7	1.6	6.1	2.5	1.4
R,S–S,R	2.0	1.5	1.1	1.9	1.5	1.0	2.1	1.3	1.0
R,S–S,S	4.9	2.3	1.0	4.6	2.3	1.0	5.3	1.9	1.0
S,R–S,S	2.4	1.5	1.1	2.4	1.5	1.0	2.5	1.6	1.0

^{a} Columns as in Table 8.4.
^{b} The pairs being separated.

separation of threonine isomers on this column were not high enough for complete resolution to be obtained. The conclusion was drawn that the ChiralProCu Si 100 column (Serva, Heidelberg, Germany) is an excellent sorbent for separating simultaneously all four isomers of the amino acids studied that contain two chiral centres (Figure 8.2). ChiralValCu Si 100 has similar possibilities for the separation of isomers of aromatic α-amino-β-hydroxy acids. A Nucleosil Chiral-1 hydroxyproline column (Macherey–Nagel, Düren, Germany) is more preferable for the separation of 2R,3R–2R,3S pair. Ligand-exchange HPLC on chiral columns was employed for controlling the enantiomeric purity of different fluoro derivatives of phenylserine [20].

It should be noted that the possibility of separating all four isomers of such amino acids with two chiral centres has some other useful applications in addition to analytical and preparative purposes. Since the capacity factors for enantiomers on ChiralProCu differ greatly and are weakly dependent on the substituents on the phenyl ring, ligand-exchange chromatography may be effective in establishing the absolute configuration of the molecules. In many cases this problem is difficult to solve and needs expensive instrumentation and much effort.

As has been mentioned, for separating enantiomers either a chiral stationary phase or a chiral mobile phase should be used. The use of a chiral eluent, with an amino acid–copper complex and an achiral reversed-phase stationary phase, allowed the separation of different fluoro derivatives of amino acids. Resolution of enantiomers of various fluorine-containing α-substituted ornithine and lysine analogues on reversed-phase columns with an L-proline–copper and N,N-di(n-propyl)-L-alanine–copper (L-DPA) mobile phase has been described [21]. Ornithine gives no separation with the eluents used, but excellent resolution has been obtained for various α-alkyl, α-halomethyl-, α-vinyl- and α-ethynyl-substituted ornithines, and also for dehydroornithine and lysine analogues (Tables 8.7–8.9). The eluate was monitored with a fluorescence detector after post-column derivatization of amino acids with o-phthalaldehyde and mercaptoethanol.

The different factors affecting the resolution, i.e. the nature and concentration of the chiral ligand, the pH and ionic strength of the eluent and the temperature of the column, have been studied. An increase in the ionic strength of the mobile phase leads to decreases in both the retention and the selectivity of the separation of enantiomers. The use of higher concentrations of the chiral additive resulted in an increase in the retention but had only a slight effect on the selectivity. For the ornithine analogues, the best resolutions were obtained with an eluent containing 8 mM L-Pro and 4 mM $CuSO_4$ (pH 5.5). An N,N-di(n-propyl)-L-alanine–copper (L-DPA) mobile phase yielded a markedly smaller selectivity than L-Pro–copper. The lower selectivity in separation with L-DPA versus L-Pro has also been observed for other α-substituted amino acids and β-alanine analogues [22]. The selectivity of the separation of enantiomers (α) increases with increasing size of the substituent in the order CH_3, CH_2F, CHF_2, CH_2Cl (Tables 8.7 and 8.9) for both ornithine and lysine analogues. Chlorofluoromethylornithine, containing two

Table 8.7 Separation of enantiomers of ornithine analogues by RP-HPLC with L-proline–copper as chiral mobile phase

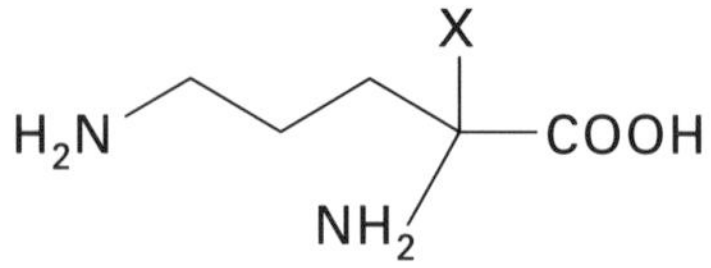

Substituent X	t_R (min)		
	I	II	α
H (DL-ornithine)	5.25	—	1.00
H (L-ornithine)	5.25	—	—
CH$_3$	5.60	8.50	2.38
CH$_2$F	5.80	10.70	3.09
CHF$_2$	6.50	13.85	3.42
CH$_2$Cl	8.20	26.25	4.82
CH$_2$OH	5.05	5.45	1.24
CH$_2$OCH$_3$	9.97	23.69	3.12
CH=CH$_2$	7.83	30.86	6.32
C≡CH	5.80	8.45	2.18
CHClF (pair I)	9.25	24.15	3.59
CHClF (pair II)	9.95	33.41	4.69

Table 8.8 Separation of enantiomers of the α-substituted dehydroornithine analogues by RP-HPLC with L-proline–copper as chiral mobile phase

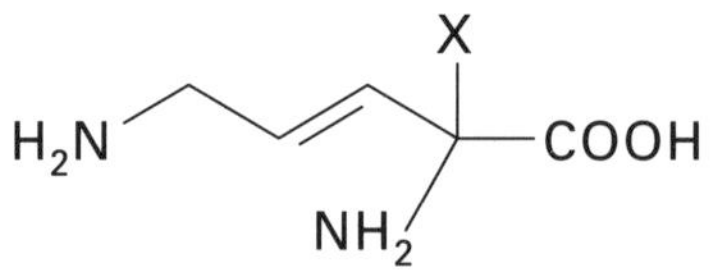

Substituent X	t_R (min)		
	I	II	α
CH$_3$	5.62	8.35	2.29
CH$_2$OH	5.02	5.33	1.21
CH$_2$F	5.97	11.47	3.22
CHF$_2$	6.40	12.47	3.09

chiral carbon atoms, was separated into four peaks. The excellent separations obtained (Table 8.7) allowed the semipreparative-scale resolution of enantiomers. Up to 4–5 mg of amino acids were separated per run on a μBondapack C$_{18}$ column (300 × 7.8 mm i.d., particle size 10 μm) with an L-Pro–copper eluent. The

Table 8.9 Separation of enantiomers of lysine analogues by RP-HPLC with L-proline–copper as chiral mobile phase

Substituent X	t_R (min)		
	I	II	α
H	5.93	6.45	1.22
CH$_3$	6.90	10.89	2.17
CH$_2$F	7.16	13.43	2.71
CHF$_2$	8.00	17.65	3.16
CCF$_2$Cl	10.50	32.74	4.18

method described was employed for controlling enantiomeric purity and establishing the absolute configuration of the enantiomers of α-chlorofluoromethylornithine [21]. RP-HPLC with an L-Pro–Cu(II) eluent also allows the complete separation of different dehydroornithine analogues (Table 8.8). The semipreparative resolution of the two enantiomers of fluoromethyldehydroornithine was performed under the conditions described for difluoromethylornithine. It should be noted that the enantiomers of the methyl ester derivative of fluoromethyldehydroornithine could not be separated under these conditions.

RP-HPLC with chiral eluents has been used for the separation of underivatized enantiomers of polyfluoro-2,2-dialkylglycines and several chiral fluorine-containing amino acids [23] (Table 8.10). In this case a 2 mM L-Phe–Cu(II) eluent containing 5 or 10% acetonitrile was employed. Increases in the concentration of acetonitrile in the eluent decreased the retention. As it can be seen from Table 8.10, amino acids with more fluorine atoms or larger alkyl groups produced later-eluting pairs of peaks. Such an effect was also described for the separation of the enantiomers of 3,3,3-trifluoroalanine versus the enantiomers of alanine with an L-Phe–Cu(II) eluent on a reversed-phase column [24].

An L-Phe–Cu(II) eluent was used for controlling the enantiomeric purity of α-trifluoromethylalanine by RP-HPLC after stereoselective enzymatic hydrolysis of *N*-acyl derivatives [25]. Milligram-scale separations of optical isomers of 2-pentafluoroethylalanine (PFA) and 2-trifluoromethylalanine (TMA) by RP-HPLC have been described [24]. The eluent was 2 mM L-Phe, 2 mM copper(II) acetate, 3 mM potassium acetate adjusted to pH 4.4 with acetic acid, to which was added 2% (v/v) (TMA) or 8% (v/v) (PEA), acetonitrile. Changes in the eluate composition were studied that allowed the calculation of the ligand concentration in the copper-saturated column. The ligand concentration in the packed column was

Table 8.10 Resolution of 2,2-dialkylglycines by RP-HPLC with L-phenylalanine–copper as chiral mobile phase

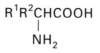

Compounds	R_1	R_2	5% CH$_3$CN		10% CH$_3$CN	
			k'_2	α	k'_2	α
2-Difluoromethylalanine	CHF$_2$	CH$_3$	1.9	1.2		
2-Trifluoromethylalanine	CF$_3$	CH$_3$	3.9	1.1	3.4	1.1
2-Amino-2-tri-fluoromethylbutyric acid	CF$_3$	CH$_2$CH$_3$	6.6	1.2	5.4	1.2
2-Pentafluoroethylalanine	CF$_2$CF$_3$	CH$_3$	16.5	1.4	12.6	1.2

86 mM, which is significantly lower than that of amino acid ligands covalently linked to silica (typically 350–400 mM) [26]. The PFA isomers were more strongly retained than the TMA isomers. It should be noted that the efficiency of the separations with L-Phe–Cu(II) eluents was as a rule significantly less than that when L-Pro–Cu(II) eluents were used [27].

8.4 CONCLUSION

HPLC is a powerful standard analytical tool for controlling the synthesis and determination of the enantiomeric purity of different fluorine-containing amino acids. The method can be easily scaled up with routine equipment for semipreparative resolutions.

Retention values in RP-HPLC, at least to a first approximation, are easily predictable and correlate well with the hydrophobic properties of compounds. Computer-aided methods for the selection of the optimum conditions in RP-HPLC gives the possibility of saving a lot of time and effort in experimental work.

Ligand-exchange HPLC with either a chiral stationary phase or a chiral mobile phase enable enantiomers of different fluorine-containing amino acids to be separated in a short time.

8.5 REFERENCES

1. S. V. Galushko, *J. Chromatogr.*, **55**, 91 (1991).
2. S. V. Galushko, *Chromatographia*, **36**, 39 (1993).
3. S. V. Galushko, A. A. Kamenchuk and G. L. Pit, *J. Chromatogr.*, **660**, 47 (1994).

4. *ChromDream*, Knauer, Berlin.
5. S. V. Galushko, M. V. Grigorieva, M. T. Kolicheva, I. I. Gerus, Yu. L. Yagupolsky and V. P. Kukhar', *Bioorg. Khim.*, **16**, 1024 (1990).
6. D. Shirlin, J. B. Ducep, S. Baltzer, *et al.*, *J. Chem. Soc., Perkin Trans. 1*, 1053 (1992).
7. C. Hansch and A. Leo, *Substituent Constants for Correlation Analysis in Chemistry and Biology*, Wiley, New York, 1979.
8. R. F. Rekker, *The Hydrophobic Fragmental Constants*, Elsevier, Amsterdam, 1977.
9. C. Horvath, W. Melander and I. Molnar, *Anal. Chem.*, **49**, 142 (1977).
10. V. A. Davankov, J. D. Navratil and H. F. Walton, *Ligand-Exchange Chromatography*, CRC Press, Boca Raton, FL, 1988.
11. S. V. Galushko, I. P. Shishkina, V. A. Soloshonok and V. P. Kukhar', *J. Chromatogr.*, **511**, 115 (1990).
12. J. R. Gerson and M. I. Adam. *J. Chromatogr.*, **325**, 103 (1985).
13. V. P. Kukhar' and S. P. Kobzev, *Zh. Anal. Khim.*,
14. S. V. Galushko, I. P. Shishkina, I. I. Gerus and M. T. Kolycheva, *J. Chromatogr.*, **600**, 83 (1992).
15. S. V. Galushko, I. P. Shishkina and V. A. Soloshonok, *J. Chromatogr.*, **661**, 51 (1984).
16. Yu. N. Belokon, A. G. Bulycheve, S. V. Vitt, *et al.*, **107**, 4252 (1985).
17. *Synthesis of Amino Acids, Tetrahedron*, Symp. Vol. **44** (1988).
18. D. Pons, M. Savignac and J. P. Genet, *Tetrahedron Lett.*, **31**, 5023 (1990).
19. S. V. Galushko, I. P. Shishkina and V. A. Soloshonok, *J. Chromatogr.*, **592**, 345 (1992).
20. V. A. Soloshonok, V. P. Kukhar' and S. V. Galushko, *Izv. Akad. Nauk SSSR, Ser. Khim.*, **8**, 1906 (1991).
21. J. Wadner, C. Gaget, B. Heintzelmann and E. Wolf, *Anal. Biochem.*, **164**, 102 (1987).
22. J. Wadner, E. Wolf, B. Heintzelmann and C. Gaget, *J. Chromatogr.*, **392**, 211 (1987).
23. J. W. Keller and K. O. Dick, *J. Chromatogr.*, **367**, 187 (1986).
24. J. W. Keller and K. Niwa,, *J. Chromatogr.*, **469**, 434 (1989).
25. J. W. Keller and B. J. Hamilton, *Tetrahedron Lett.*, **27**, 1249 (1986).
26. P. Roumeliotis, K. K. Unger, A. A. Kurganov and V. A. Davankov, *J. Chromatogr.*, **255**, 51 (1983).
27. V. A. Soloshonok, I. Yu. Galaev, V. K. Svedas, *et al.*, *Bioorg. Khim.*, **19**, 467 (1993).

9 General Features of Biological Activity of Fluorinated Amino Acids: Design, Pharmacology and Biochemistry

J. T. WELCH and A. GYENES
Department of Chemistry, University at Albany, Albany, NY, USA

M. J. JUNG
Marion Merrell Dow Research Institute, Cincinnati, OH, USA

9.1 INTRODUCTION

The crucial role of amino acids in biological functions is well established. Selective fluorination, generally involving replacement of a hydrogen or a hydroxyl group by fluorine, may modify the fundamental characteristics of those amino acids, significantly improving or altering their biological activity. These effects may be derived from the effect of fluorination on the overall reactivity and stability of the molecule, which in turn has an important impact on the biological activity. These modifications not only have biological effects on the amino acids, but also influence the chemical reactivity of peptides into which they are incorporated.

9.1.1 BINDING EFFECTS

Typical biological effects are the enhancement of binding to the natural substrate to an enzyme on replacement of hydrogen or a hydroxyl group by fluorine. The comparable sizes of fluorine and hydrogen precludes undesirable steric inhibition of binding at the active site of the receptor. In the special case of amino acids, fluorine substitution lowers the pK_a of the amine function by roughly one pK unit per fluorine atom, [1] while the effect on the acidity is less marked [2]. In addition, often fluorinated amino acids cannot be metabolized properly by the enzyme on binding.

Fluorinated analogs of amino acids have been synthesized as potent agonists and antagonists of natural amino acids [3–5]. To be effective the new inhibitors must have high selectivity for the enzymes with low effective doses (ED_{50} values) and limited toxicity (high LD_{50} values) (ED_{50} is the dose that is effective in 50% of test subjects and LD_{50} is the dose that is

Fluorine-containing Amino Acids: Synthesis and Properties Edited by V. P. Kukhar' and V. A. Soloshonok
© 1995 John Wiley & Sons Ltd

lethal in 50% of test subjects). The antiproliferative effects of amino acid antagonists, such as 4- and 6-fluorotryptophan, against *E. coli* B have been tested [6]. Protein biosynthesis seems to be an ideal target for antagonists of all 20 physiological amino acids; antiproliferative effects are therefore an integral part of the chemotherapy of bacterial [7] or malignant [8] diseases. Target enzymes are typically the aminoacyl-tRNA synthetases that catalyze the chemical activation and the transfer of natural amino acids (AA) to their cognate tRNA^{AA}s [6]. Amino acid antagonists may inhibit the synthesis of functional proteins by a number of different mechanisms. For example, 4-fluorotryptophan mimics the natural amino acids so perfectly that it becomes activated, then transferred to tRNA. When released as an aminoacyl-tRNA, misacylation results. The fluorinated material is therefore incorporated into non-functional proteins [9].

9.1.2 FLUORINE AS A LEAVING GROUP

Fluorine is the smallest of the halogens. It is also the least reactive in an S_N2 reaction with the consequence that β-fluorinated amines are reasonably stable to aqueous hydrolysis. In basia media, however, β-fluorinated amines may cyclize to the corresponding aziridine on heating [10]. The major interest in fluorine substitution arises from its leaving group ability when a negative charge has developed in the β-position as occurs in some enzymatic transformations. The reactivity of the halogens in $E1cB$ or $E2$ reactions is reversed from that found in S_N2-type processes [11].

Fluorine has also been employed as a leaving group in addition–elimination reactions where its superior leaving group ability relative to hydrogen is important. Some β-fluorine-substituted α-amino acids are potent irreversible inhibitors of pyridoxal phosphate-dependent enzymes, as illustrated in equation 9.1 [12, 13].

Typically, α-monofluoromethyl- and α-difluoromethylamino acids are potent enzyme-activated irreversible inhibitors of α-amino acid decarboxylases [12]. The enzymatic decarboxylation of amino acids, an important biosynthetic pathway, is specifically inhibited by these fluoromethylamino acid analogs. These fluorinated amino acids are therefore especially useful in the study of the mechanism of these enzymes.

Typical of the physiologically important amines formed by decarboxylation are dopamine, 5-hydroxytryptamine (serotonin), histamine, tyramine and γ-aminobutyric acid (GABA) [14]. The catecholamines are important in peripheral and central control of blood pressure [15]. Elevated histamine levels are observed in diseases such as allergies, hypersensitivity, gastric ulcers and inflammation [16]. High putrescine levels are associated with rapid cell development, including tumor growth [17].

These enzyme inhibitors are extremely selective and have great practical application in understanding the physiological roles of specific enzymes.

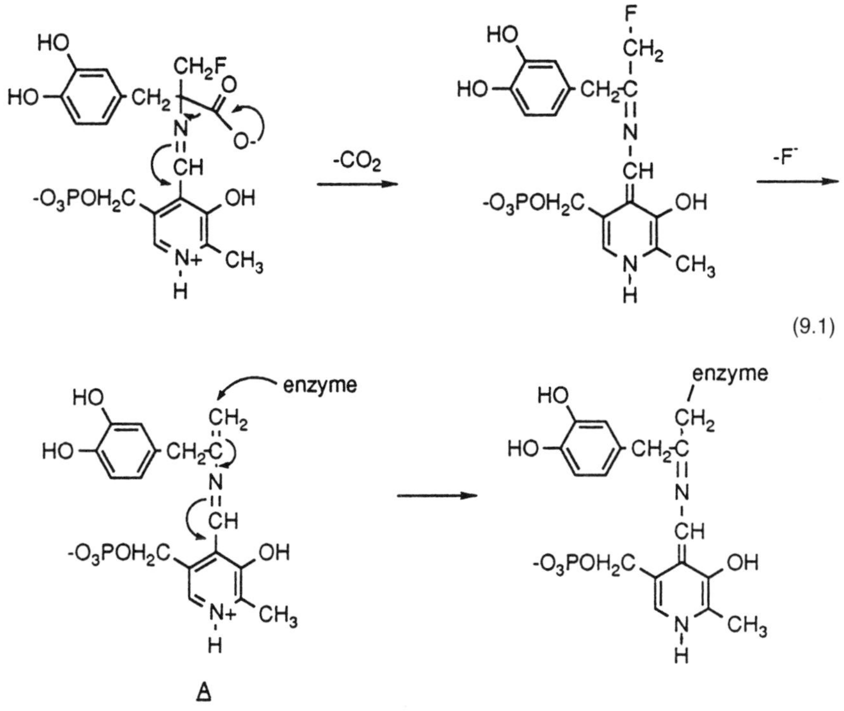

(9.1)

9.1.3 FLUORINATION AND METABOLISM

Fluorine substitution in appropriate positions has also been rationalized and exploited to transform a number of other amino acids into inhibitors of their catabolic enzymes [18]. Amino acids containing a trifluoromethyl group have proven therapeutic value as antimetabolites [19]. The relative non-toxicity and stability of trifluoromethyl compounds compared with mono- and difluoromethyl analogs make these molecules especially attractive. Uniquely, relative to mono- or difluoromethyl groups, the trifluoromethyl group presents a uniform sphere of electron-rich fluorines to the enzyme. This sphere potentially presents an electronic surface for the formation of strong hydrogen bonds with the enzyme. The enzyme is thus effectively prohibited from forming a complex with the natural substrate. The presence of a trifluoromethyl group in the α-position in amino acids strongly influences, both sterically and electronically, the reactivity of the carboxylic function of the amino acid. If the trifluoromethyl group is at the C-3 position, it may also cause a significant decrease in the nucleophilicity of the amino group and increase the steric hindrance when the modified amino acid is employed in standard peptide coupling transformations [20].

9.1.4 FLUORINATED AMINO ACIDS AS METABOLIC PRODUCTS

α-Fluorinated β-amino acids are relatively little known, but these compounds also have some potential biological utility. For example, α-fluoro-β-alanine is formed on metabolism of the antitumour agent 5-fluorouracil [21].

Some examples of fluorinated amino acids which have been synthesized and studied as potential enzyme inhibitors and therapeutic agents in the last few years are given in the following section to illustrate these principles.

9.2 BIOLOGICAL ACTIVITIES OF FLUORINATED AMINO ACIDS

Fluorinated amino acids exhibit a broad range of biological function, for example, demonstrating antibacterial, cancerostatic, cytotoxic or other inhibitory activities [22]. Some of the known fluorinated amino acids can also be useful in the treatment of disorders of the central nervous system.

9.2.1 α-FLUOROMETHYLATED AMINO ACIDS

It has been found that α-fluoromethyl and α-difluoromethyl substitution of glutamic acid caused significant changes in chemical and biological properties [23]. Mono- and difluoromethylgaluamic acid (1 and 2) are time-dependent irreversible inhibitors of bacterial glutamate decarboxylase [24, 25]. For the monofluoro derivative 1 it could be shown that the inhibited enzyme is alkylated stoichiometrically by the substrate analog during the inhibition process. However, no significant antibiotic effect have been reported.

Both tri- and monofluoromethylglutamic acid have been reported to inhibit glutamic acid decarboxylase, the enzyme that catalyzes the formation of the inhibitory neurotransmitter GABA [14, 26].

$$
\begin{array}{c}
\text{CHXF} \\
|
\end{array}
$$

$$\text{H}_2\text{N}-\overset{\displaystyle |}{\underset{\displaystyle |}{\text{C}}}-\text{CH}_2\text{-CH}_2\text{CO}_2\text{H}$$

$$\text{CO}_2\text{H}$$

1   X=H   α-monofluoromethylglutamic acid
2   X=F   α-difluoromethylglutamic acid

α-(Difluoromethyl)ornithine (3) is an inhibitor of ornithine decarboxylase, and can be used for the treatment of African sleeping sickness [27] and of *Pneumocystis carinii* pneumonia, the frequent opportunistic infection associated with acquired immune deficiency syndrome (AIDS) [28]. Ornithine decarboxylase [29] is an enzyme that catalyzes the first step in the biosynthesis of polyamines.

CHF$_2$
|
H$_2$N C CH$_2$CH$_2$CH$_2$NH$_2$
|
CO$_2$H

(3)
α-(difluoromethyl)ornithine

In vitro studies have also been accomplished on aromatic amino acid de-carboxylase and glutamic acid decarboxylase, both of which are involved in the biosynthesis of neurotransmitters, dopamine and γ-aminobutyric acid, respectively.

In a comparison of α-chlorofluoromethyl-α-amino acids with α-mono- or α-difluoromethyl-α-amino acids, the potency and selectivity of the inhibitors was determined [30]. Three types of α-chlorofluoromethyl-α-amino acids, bearing either a neutral, acidic or basic side-chain, have been prepared for development of a new family of putative irreversible inhibitors of pyridoxal phosphate-dependent decarboxylases. The presence of two asymmetric centers provides this category of α-alkylated α-amino acids with new properties in terms of selectivity and efficacy of inactivation of their corresponding dec-arboxylases.

Study of the kinetic parameters for the *in vitro* inactivation (diastereo-selectivity) of their corresponding decarboxylases shows that α–chloro-fluoromethyl-α-amino acids are less potent irreversible inhibitors of ornithine decarboxylase and of aromatic amino acid decarboxylase than α-mono- or α-difluoromethyl-α-amino acids. The efficacy of α-chlorofluoromethyl-α-amino acids is generally less than that of the corresponding α-difluoromethyl-α-amino acids, suggesting either a steric effect during the recognition process and/or the generation of less reactive alkylating agents during the inactivation process. All four diastereomers of α-(chlorofluoromethyl)ornithine (4) have been found to show activity, demonstrating the lack of enantiospecificity and moderate stereoselectivity of ornithine decarboxylase, an observation made previously by others [31] in the case of the two enantiomers of dehydromono- and difluoromethylornithine.

CHCIF
.
H$_2$N C CH$_2$ CH$_2$CH$_2$NH$_2$
CO$_2$H

(4)
α-(chlorofluoromethyl)ornithine

α-(Difluoromethyl)ornithine [30] has been found to have antigestational, antitrypanosomal, anticoccidial and antitumor activities [26]. As an inhibitor of ornithine decarboxylase, the enzyme that provides the polyamines needed for growth and development, α-(difluoromethyl)ornithine is also fungicidal with plant protection efficacy [32]. α-(Difluoromethyl)ornithine inhibits cell proliferation [33]. The cytostatic effect of α-(difluoromethyl)ornithine is due to a decline in intracellular levels of putrescine and spermidine. Mono- and difluoromethyldehydroornithine [34] are more potent suicide inhibitors of ornithine decarboxylase than the corresponding saturated analogs. α-(Difluoromethyl)dehydroornithine is an irreversible inhibitor of ornithine decarboxylase [34].

Histidine, a basic amino acid, is a precursor for histamine which shares many biological properties with the aforementioned biogenic amines. Indeed, histamine is a neurotransmitter in the central nervous system, in gastric acid secretion, an autocoid in inflammation and allergy and possibly a control agent in cell growth and replication. The only compound under discussion in this section is monofluoromethylhistidine (5). Since the first report on its activity, there has been a continuous flux of publications on its biological effects. Although the compound has been studied in phase one clinical trials, its present status is not clear.

(5)

α-fluoromethyl-histidine

Kollonitsch *et al.* [35] published the first report on the time-dependent inhibition of mammalian histidine decarboxylase by monofluoromethylhistidine. A number of publications subsequently appeared trying to explain the detailed chemistry of enzyme inactivation. It seems confirmed now that both mammalian and pyridoxal-dependent bacterial histidine decarboxylase are inactivated by monofluoromethylhistidine not through alkylation of the active site as proposed in similar cases but by formation of a stable adduct between the modified inhibitor and the cofactor [36]. The first step of the reaction scheme is common to both proposals: binding to the enzyme as a pyridoxal Schiff base, followed by decarboxylation with loss of fluorine to generate the Schiff base enamine which then reacts as described above. Mammalian histidine decarboxylases from fetal rat liver, gastric mucosa, CNS or peripheral blood leukocytes all respond similarly to monofluoromethylhistidine, explaining the diversity of biological effects. Bacterial histidine decarboxylases fall into two categories: pyridoxal-dependent in Gram-negative bacteria which are inhibited by monofluoromethylhistidine and pyruvoyl-dependent enzymes in Gram-positive bacteria which are insensitive to the inhibitor. However, no report on antibiotic activity even in sensitive strains could be found.

Histidine decarboxylase is similar to ornithine decarboxylase, a rapidly inducible enzyme with a short half-life. Therefore the maximum effect of an inhibitor, even a potent and irreversible one such as monofluoromethylhistidine, can be obtained only if the tissue concentration of the inhibitor is undergoing minimum fluctuation. Thus, continuous infusion of monofluoromethylhistidine via an osmotic minipump in rats at a rate of 5 mg kg^{-1} h^{-1} for 3 weeks depletes histamine in all tissues examined: hypothalamus (−94%), gastric mucosa (−75%), thymus (−56%), liver (−44%) and heart (38%). This treatment did not produce obvious behavioral changes [37]. Effects of single treatments at higher doses have also been reported in various organs, although the effects are usually less dramatic and of short duration [38].

In pylorus ligated rats, monofluoromethylhistidine inhibits gastric mucosal histidine decarboxylase in a dose-dependent manner, but does not modify the histamine content or the acid secretion. However, coadministration with cimetidine results in prolonged antisecretory effect of the H2-antagonist [39]. In the case of ulceration induced by physical restraint in the cold, mono-fluoromethylhistidine blocks the induction of histidine decarboxylase in the mucosa and the hypothalamus and protects against ulceration to a level comparable to cimetidine [40]. This effect was not confirmed, however, in another study using a different mode of administration [41].

Single doses of monofluoromethylhistidine (20–100 mg kg^{-1}) block totally histidine decarboxylase in cerebral cortex and hypothalamus and prevent new histamine synthesis without depleting existing pools of histamine [42]. A dose of 20 mgkg^{-1} induces an increase in deep, slow wave sleep, but not of paradoxical sleep [43]. Monofluoromethylhistidine potentiates the effect of tetrahydrocannabinol on thiopental sleeping time by blocking new histamine synthesis [44]. Similar doses reduce the muricidal behavior of thiamine-deficient rats and prevent the increase of aggressiveness due to histidine injection [45]. Monofluoromethylhistidine is reported to have orexigenic activity in cats at 25 mgkg^{-1} [46] and to reduce slightly motion sickness in an experimental model [47]. It also has an unfortunate effect on copulatory behavior in rats by increasing the post-ejaculatory intromission latency [48].

In a model of Lewis lung carcinoma in mice, monofluoromethylhistidine reduced both the growth of the main tumor and the number of metastases in lungs [49]. In an animal model of gastric carcinoid in an African rodent, a tumor known to overproduce histamine, monofluoromethylhistidine reduces the output of histamine and the appearance of duodenal ulcers without blocking tumor progression [50]. As expected, continuous subcutaneous infusion is far superior to intraperitoneal administration. Chronic admin-istration of monofluoromethylhistidine to gravid mice has no effect on implantation, fetal development or parturition [51]. Fetuses have markedly depleted histamine; however, this does not affect growth, but results in a deficit in norepinephrine in the sympathetic system on the third postnatal of week which disappears later on [52].

Monofluoromethylhistidine blocks histamine synthesis in 2H3 basophils without affecting growth or replication [53]. The reduction of lymphocyte blastogenesis after concavalin-A treatment seen at concentrations of mono-fluoromethylhistidine ten times higher than those needed to achieve maximum histidine decarboxylase inhibition may not be specific [54]. Although mono-fluoromethylhistidine reduced the secretion of corticosterone due to *E. coli* lipopolysaccharide administration, it did not prolong the survival time [55]. Patent literature claims antiasthmatic effects for monofluoromethylhistidine [56]. The discrepancy among the different studies may originate from the method of administration, i.e. injection as opposed to infusion as highlighted before. The potential drug has been studied in healthy volunteers and was found to be excreted in urine unchanged [57]. As it is a hydrophilic compound it would probably be absorbed through the skin and application via skin patches may represent the optimum mode of administration and could possibly revive interest in this non-toxic means of blocking histamine synthesis [58].

α-Mono- and difluoromethylphenylalanine (6 and 7) have been synthesized and used in model reactions with pyridoxal phosphate. For instance, difluoro-methylphenylalanine on heating to 80 °C in phosphate buffer (pH 6.3) in the presence of a catalytic amount of pyridoxal phosphate yields almost quanti-tatively 1-fluoro-3-phenylacetone [59].

$$\text{C}_6\text{H}_5\text{-CH}_2\text{-}\underset{\text{NH}_2}{\overset{\text{CHXF}}{\text{C}}}\text{-CO}_2\text{H}$$

(6) X=H   α-monofluorophenylalanine
(7) X=F   α–difluorophenylalanine

Fluorine substitution in the β-position would yield unstable compounds for *o*- and *p*-tyrosine through elimination of fluorine to the methylenequinone. This was exploited to synthesize potential inhibitors of tyrosine hydroxylase by introduction of a fluoromethyl group in the *ortho*-position on *p*-tyrosine [60]. Preliminary results seemed indeed to indicate that 0-monofluoromethyl-*p*-tyrosine (8) is a time-dependent inhibitor of rat adrenal tyrosine hydroxylase and that the inhibition needed molecular oxygen.

$$\text{HO-}\underset{}{\overset{\text{CH}_2\text{F}}{\text{C}_6\text{H}_3}}\text{-CH}_2\underset{\text{NH}_2}{\text{CHCO}_2\text{H}}$$

(8)

o-monofluoromethyltyrosine

Introduction of α-fluoromethyl groups in *o*-, *m*- or *p*-tyrosine has been more productive in generating interesting biological activities. Usually difluoro-methyl groups yielded lower activity and will not further be discussed in this section. α-Fluoromethyl-*o*- and -*m*-tyrosine are powerful inhibitors of aromatic amino acid decarboxylase (AADC), and in most characteristics resemble monofluoromethyl-DOPA in their effects on biogenic amine synthesis [61]. The *para* analog, however, deserves special attention [62]. This compound was designed as a bioprecursor of monofluoromethyl-DOPA: *p*-tyrosine is the substrate of tyrosine hydroxylase and is not decarboxylated by aromatic amino acid decarboxylase. The α-fluoromethyl analog is hydroxylated by purified tyrosine hydroxylase at about 1% the rate of tyrosine and is not a time-dependent inhibitor of the decarboxylase by itself. The methyl ester of monofluoromethyl *p*-tyrosine, when administered to rats or mice, gives a time- and dose-dependent decrease in decarboxylase activity in the brain and a selective decrease in dopamine and noradrenaline levels [63]. The intensity of the effects can be modulated by chemical or physical factors known to influence tyrosine hydroxylase activity *in situ* [64]. The compound could have pharmacological effects in the control of stress or on the onset of the beneficial action of neuroleptics, but is not under development at the moment. Difluoro-methyltyrosine has been found an effective hypnotizing agent [65].

Again for the sake of completeness, another fluorinated analog of tyrosine needs to be mentioned here: (*E*)-fluoromethylene-*m*-tyrosine (**9**). The (*E*)-fluoro-methylenephenethylamines are powerful irreversible inhibitors of monoamine oxidase (MAO) A or B, depending on the substitution of the aromatic ring [66]. (*E*)-Fluoromethylene-*m*-tyrosine was designed as a bioprecursor of an inhibitor with selectivity for amine terminal MAO and a lesser propensity to affect the sympathetic aminergic system when given with a peripheral ODPA decarboxylase inhibitor [67]. Indeed, activation of the amino acid precursor occurs neatly on incubation with AADC *in vitro* and *in vivo*, yielding long-lasting inhibition of MAO in the brain of rats or mice [68]. This compound may stimulate a renewal of interest in MAO inhibitors for the treatment of depression.

(9)

(*E*)-fluoromethylene-*m*-tyrosine

Difluoromethyl-DOPA (**10**) is an enzyme-activated inhibitor of aromatic amino acid decarboxylase. After systemic administration to rats or mice, it selectively blocks the decarboxylase in the periphery [69]. It does not influence the endogeneous synthesis of biogenic amines. However, difluoromethyl-DOPA inhibits the decarboxylation of exogeneously supplied DOPA or 5-hydroxytryptophan in the periphery [69]. Thus it could substitute for carbidopa in the treatment of Parkinson's disease in combination with DOPA and peripheral decarboxylase inhibitor. It reduces also the kidney damage due to 5-hydroxytryptophan administration to rats [70].

HO
HO—⟨benzene ring⟩—CH₂—|—CO₂H with CHXF above and NH₂ below

$$\text{HO} \quad \text{CHXF}$$
$$\text{HO}-\langle\text{ring}\rangle-\text{CH}_2\overset{\text{CHXF}}{\underset{\text{NH}_2}{|}}\text{CO}_2\text{H}$$

(10) X=F   α,α-difluoromethyl-DOPA
(11) X=H   α–fluoromethyl-DOPA

α-Monofluoromethyl-DOPA (MFMD) **(11)** caused rapid, time-dependent irreversible inactivation of AADC [71]. The inhibition is stereospecific as only the *S*-isomer shows activity. The inhibition occurs with loss of the carboxyl group and of fluorine with incorporation of one equivalent of modified inhibitor into the inactivated enzyme. When given to rats or mice it produces a progressive loss of decarboxylase activity in all peripheral organs and also in the CNS [72]. Depending on the dose, the enzyme activity can be reduced to zero or at least to a level where decarboxylation becomes rate limiting in the biosynthesis of biogenic amines. Therefore, especially after repeated administration, tissue catecholamine or indoleamine levels are depressed in a dose- and time-dependent way. The effect is even more marked on the metabolites of catecholamines and their precursors [73]. Indeed, fluoromethyl-DOPA has been used to measure turnover rates [74] and to determine the localization of catecholaminergic neurons (radiolabeled MFMD) [75]. Not unexpectedly, the pharmacological result is a profound depression of the peripheral sympathetic system which is reversed by infusion of dopamine illustrating the selectivity of action [76]. Blood pressure in spontaneously hypertensive rats was reduced by oral administration of $25\,\mathrm{mgkg^{-1}}$ of MFMD once a day for 3 days. At lower doses, a preferential peripheral effect was found which potentiates the central action of exogenous DOPA similar to effect observed with the difluoromethyl analog [77]. In normotensive animals the biochemical effects are maintained, but there is only a marginal decrease in blood pressure [78]. The consequence of central inhibition of biogenic amine synthesis is less evident at doses which do not produce heavy sedation and hypothermia. α-Difluoromethyl-DOPA has a selective peripheral activity whereas α-monofluoromethyl-DOPA inhibits both central and peripheral activities.

α-Mono- and difluoromethyl-5-hydroxytryptophan analogs **12** and **13** are powerful time-dependent inhibitors of AADC in every respect comparable to the DOPA derivatives *in vitro* [79]. In an attempt to duplicate the increased selectivity and modulability of action of monofluoromethyltyrosine over the DOPA analog (see above), the corresponding tryptophan derivatives were synthesized and tested *in vitro* and *in vivo*. Monofluoromethyltryptophan is hydroxylated by tryptophan hydroxylase partially purified from rat brain stem with an efficiency of about 0.5–1% of tryptophan. Oral administration of monofluoromethyltryuptophan to rats produced a slight decrease of brain serotonin and an accumulation of 5-hydroxytryptophan but no behavioral changes. The difluoromethyl analog was an even poorer substrate of tryptophan hydroxylase and had no effect *in vivo*. Recently, a stereospecific synthesis of (*S*)-mono-fluoromethyltryptophan was reported and the compound was shown to reduce the serotonin producing capacity of carcinoid tumors [80].

    (12) X=H α-monofluoromethyl-5-hydroxytryptophan
    (13) X=F α–difluoromethyl-5-hydroxytryptophan

9.2.2 *n*-*β*-FLUORINATED α-AMINO ACIDS

Amino acids that bear fluorine at a *β*-carbon can act as mechanism-based irreversible inactivators of certain enzymes and can block important metabolic pathways [81, 82] in a manner similar to that described above. Alanine racemase, for example, is an enzyme that provides D-alanine for bacterial cell wall formation and can be inactivated by *β*-fluoroalanine and its analogs [83, 84]. The chemotherapeutic importance of these compounds motivated the development of synthetic methods for preparing α-amino acids containing one or two fluorines at the *β*-position, yet retaining the parent side-chain.

Aspartic acid and asparagine building blocks in the biosynthesis of purinergic bases, are incorporated into proteins and serve as handle for glycosylation when incorporated in proteins. This explains the long-standing interest in aspartic acid analogs and the search for a convenient synthesis of *β*-fluoro and *β*,*β*-difluoro derivatives (e.g. **14–17**) [85–87]. Profound effects on enzymes acting both on the *β*-carboxyl function and the α-amino acid can be expected.

$$\underset{\text{YCO CXF CH CO}_2\text{H}}{\overset{\overset{\displaystyle\text{NH}_2}{|}}{}}$$

    (14) X=H, Y=OH β-monofluoroaspartic acid
    (15) X=F, Y=OH β–difluoro aspartic acid
    (16) X=H, Y=NH₂ β-monofluoroasparagine
    (17) X=F, Y=NH₂ β–difluoroasparagine

In contrast to the monofluoro derivative, which undergoes extensive fluoride ion elimination in the presence of aspartic acid aminotransferase, the difluoro derivative is slowly transaminated to difluoroaspartate with minimal loss of fluoride [86]. In the reverse reaction, only competitive inhibition of aspartic acid transamination has been reported for difluoroaspartic acid. Fluoro-fumaric acid is a very poor substrate of 3-methylaspartate ammonia lyase [88]. Fluoroaspartic acid (**14**) is a relatively good substrate for adenylosuccinate synthetase, transforming inosine monophosphate into fluoroadenyl succinate. This compound is a powerful inhibitor of adenylosuccinate lyase, the next and last enzyme in adenosine monophosphate synthesis. Fluoroaspartic acid is also a substrate of the enzyme which catalyzes the coupling of aspartic acid with 5-aminoimidazo 4-carboxylic acid ribonucleoside 5′-phosphate, an earlier enzyme in the biosynthesis of adenosine phosphate. In H4 cells treated with radioactive *threo*-fluoroaspartic acid, there is accumulation of the two fluorinated analogs. This may explain the inhibition of purine synthesis in these cells and the cytotoxicity of *threo*-fluoroaspartic acid. *erythro*-Fluoroaspartic acid has no effect on the growth of H4 cells and, in general, is much less cytotoxic than the *threo* analog [89]. It was also found that *threo*-fluoroaspartic acid is incorporated into proteins at a rate comparable to that for aspartic acid itself. The result on enzyme structure and activity is not known, however. The asparagine analogs behave in the same way: the *threo* analog is cytotoxic although less than fluoroaspartic acid, but the *erythro* analog has no significant activity. Thus, *threo*-fluoroasparagine at doses of $250\,\mathrm{mgkg}^{-1}$ increases the life span of mice bearing L1210 tumors by 60% [90].

Difluoroaspartic acid (15) and asparagine (17), which had been synthesized earlier, show limited cytostatic effects on 3T3-F cells at $10–100\,\mathrm{mgml}^{-1}$ but no effect on leukemia in mice at nontoxic doses [85]. Aspartic acid or asparagine is also essential for the biosynthesis of pyrimidine bases, and it could be expected that mono- or difluoroaspartic acid or asparagine would affect carbamoylation or that the corresponding carbamoyl derivatives would influence the other enzymes in the pathway such as dihydroorotase, dihydroorotate dehydrogenase, the phosphoribosyl transferase and the decarboxylase. However, no references on this type of work could be found.

Glycosylation of proteins on an asparagine site occurs most probably through an S_N2 displacement of dolichol pyrophosphate by the amide of asparagine from oligosaccharidyldolichol pyrophosphate. In a cell-free translation system from Krebs ascite tumor cells, *threo*- (but not *erythro*-) fluoroasparagine competes with asparagine for incorporation into proteins [91]. A precursor of human chorionic gonadotropin in which one or two asparagine residues were substituted by fluoroasparagine was synthesized. In the presence of membranes, this pre-hCG is normally glycosylated at two asparagine sites, yet in the fluoroasparagine pre-hCG there is only one or no site of glycosylation [91]. In a culture of mouse pituitary cells *threo*-fluoroasparagine inhibits the glycosylation of pro-opiomelanocortin, the precursor of ACTH and β-LPH, most probably by incorporation of fluoroasparagine in the place of asparagine [92].

Fluorinated glutamates have been described with the fluorine in the chain either in β- or γ-positions or in an α-methyl substituent.

$$\mathrm{NH_2}$$
$$\mathrm{HO_2C\ CH_2\ CXFCHCO_2H}$$

(18) X=H, Y=OH β-monofluoroglutamic acid
(19) X=F, Y=OH β,β–difluoroglutamic acid

β-Fluoroglutamate (18) can be synthesized chemically by fluorodehydroxylation [93] or by the reaction of glutamate dehydrogenase on 2-keto-3-fluoroglutamate [94]. Chain fluorination in a β-position, in contrast to the introduction of a fluoromethyl group, yields substrates of bacterial glutamate decarboxylase. The fate of the fluorine substituent differs according to the geometry of substitution [95]. The L-*threo* compound is decarboxylated with retention of fluorine to form optically active 4-amino-3-fluorobutyric acid. The L-*erythro* compound loses fluorine in the decarboxylation but without noticeable enzyme inactivation. The D-isomers could be of interest in bacterial cell wall synthesis, but no work in this field has been reported, to our knowledge.

β,β-Difluoroglutamate (19) is a potent inhibitor of polyglutamylation of folic acid, as was shown using tritium-labeled glutamate as substrate. In fact, it substitutes for the normal substrate and promotes chain elongation at increased rates [96]. The usefulness of folic acid conjugates with difluoroglutamate as antineoplastic agents has been claimed in the patent literature [97].

A number of syntheses of esters of β-fluorophenylalanine (20) have been reported. Chemical hydrolysis of the ester results in a massive loss of fluorine, but enzyme-catalyzed hydrolysis is feasible [98].

$$\text{CHFCHCO}_2\text{H}$$

with $\overset{|}{\text{NH}_2}$ group

(20)

β-fluorophenylalanine

Recently, a convenient synthesis based on iodotrimethylsilane-promoted cleavage of a benzyl ester allowed access to *erythro/threo*-fluorophenylalanine in gram amounts [99]. Structural studies highlighted the strong interaction between F and NH_2. Although a number of enzymes come to mind, e.g. phenylalanine transaminase, phenylalanine hydroxylase and tRNA ligase, the authors are not aware of biochemical studies with *erythro*- or *threo*-fluorophenylalanine.

9.2.3 TRIFLUOROMETHYLATED AMINO ACIDS

Trifluoroalanine [100, 104] acts as a suicide inhibitor for a number of pyridoxal enzymes such as γ-cystathionase [102, 103], alanine racemase [103], tryptophanase [100], tryptophan synthetase [100], β-cystathionase [100], and pyruvate–glutamate transaminase [100]. All these enzymes can perform elimination reactions, in addition to their catalytic process. β-Trifluoromethyl-β-alanine also has antibacterial properties [104].

Dipeptides with different trifluoromethylamino acids have been synthesized in high yields by aminolysis of 21 with amino acid esters (equation 9.2) [105] (Table 9.1).

Table 9.1 Synthesis of dipeptides according to equation 9.2

Compound	R^1	R^2	R^3	Yield (%)
23a	C_6H_5	H	$CH_2C_6H_5$	92
23b	$CH_2CH(CH_3)_2$	H	$CH_2C_6H_5$	97
23c	CH_3	$CH_2C_6H_5$	$C(CH_3)_3$	62
23d	$CH_2CH(CH_3)_2$	$CH_2C_6H_5$	$C(CH_3)_3$	74

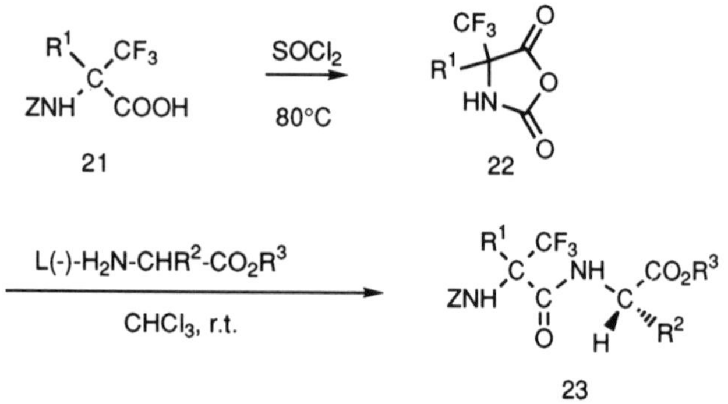

(9.2)

Trifluoromethyl Leuchs anhydrides (**22**) are therefore the derivatives of choice for the introduction of a wide variety of trifluoromethylamino acids into the N-terminal position of a peptide chain. Only a few dipeptides containing D,L-difluoroalanine have so far been described [106]. Trifluoromethionine (**24**) inhibits the growth of some microorganisms [107] and is able to form *in vivo* the corresponding adenosine derivative, preventing the demethylation of S-adenosylmethionine [108].

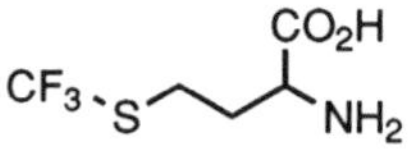

24

Trifluoromethionine

The α-amino-α-trifluoromethyl compounds **25–27**, can be used for regulation of the noradrenaline level.

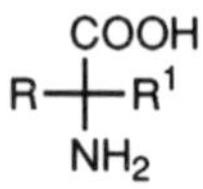

cmpd	R	R^1
25	CF_3	$4\text{-HOC}_6\text{H}_4\text{CH}_2$
26	CF_3	$3,4\text{-(HO)}_2\text{C}_6\text{H}_3\text{CH}_2$
27	CF_3	$3,4,5\text{-(HO)}_3\text{C}_6\text{H}_2\text{CH}_2$

9.2.4 ω-FLUORINATED α-AMINO ACIDS

Methotrexate (MTX, **28**) has wide potential utility as an antitumor agent [109]. Introduction of fluorine into the amino acid moiety (**28a, b**) enhances the acidity of the ω-carboxylic acid group and so diminishes the toxicity of methotrexate, enabling these compounds to be used in high-dose treatment regimens. A pronounced biological effect is obtained by incorporating a γ-fluorinated glutamic acid moiety into the parent molecule **28**.

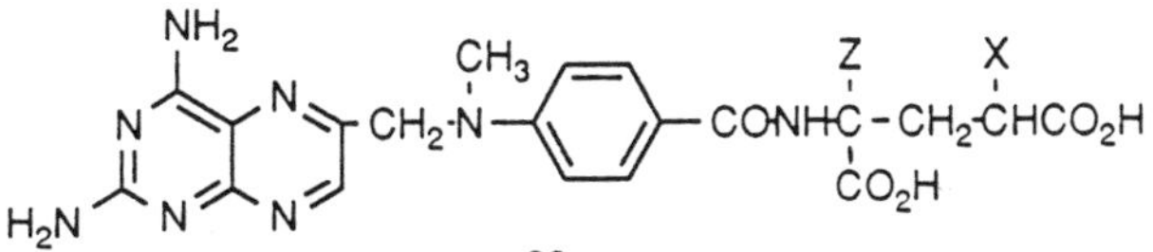

MTX : X = H   Z = H
α : X = F   Z = H
γ : X = H   Z = CHF$_2$

This compound has also been evaluated for *in vitro* dihydrofolate reductase inhibition. Relative binding affinities were determined, and they showed similar potency to that of methotrexate. Among several compounds bearing different groups such as fluoro, methyl, hydroxyl and methylthio at the γ-position, the fluoro compound showed the highest chemotherapeutic index (lowest effective dose to highest dose for toxicity) of all the compounds examined. It also showed an increase in life span (ILS) value close to that of methotrexate at its maximally tolerated dose, making **28** of interest for high-dose treatment of methotrexate-resistant cancers.

4-Fluoroglutamic acid (**29**) is also a substrate of polyglutamate synthetase [96]. The positional isomers of 4-fluoro-L-glutamate were also studied for their effect on bacterial glutamate mutase, a cobalamin-dependent enzyme catalyzing the reversible rearrangement of (*S*)-glutamate and (2*S*,3*S*)-3-methylaspartic acid. The 4-*S* analog is the most potent inhibitor reported ($K_i = 70\,\mu$M), whereas the 4-*R* isomer gives no inhibition at 4 mM. It is not known whether this inhibition could have biological consequences [110].

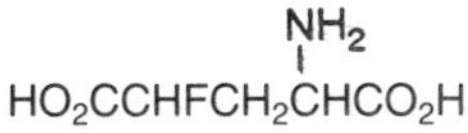

HO$_2$CCHFCH$_2$CHCO$_2$H
(with NH$_2$ above)

(29)

4-fluoro-glutamic acid

β-Fluoromethyleneglutamate (**30**) had been conceived as a potential bioprecursor of β-fluoromethylene-GABA, a good inhibitor of GABA transaminase. It was found, however to be only a weak competitive inhibitor of mammalian glutamate decarboxylase [111]. The effect on the bacterial enzyme was not tested.

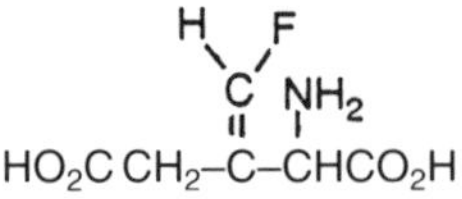

$$\begin{array}{c} H \diagdown \diagup F \\ \overset{|}{C} \overset{}{NH_2} \\ \underset{\parallel}{} \overset{|}{} \\ HO_2C\ CH_2-C-CHCO_2H \end{array}$$

(30)

β-fluoromethyleneglutamic acid

9.2.5 PEPTIDES CONTAINING FLUORINATED AMINO ACIDS

Peptides used as therapeutic agents are degraded rapidly by peptidases, explaining the vigorous search conducted for better non-natural amino acids with which to substitute natural amino acids. One strategy for the stabilization of peptide bonds can be the introduction of fluorine into peptides to improve the interactions with the receptor site and increase peptide activity [112]. This often produces compounds with reduced metabolism and higher transport rates *in vivo* [113].

4-Fluoroglutamic acid (29) was used for synthesis of fluorinated derivatives of 28 (X = F). Compound 28 (X = F) is methotrexate and is a known anticancer agent, as mentioned previously [109, 114–116].

9.2.6 α- AND β-FLUORINATED ω-AMINO ACIDS

Aminoglycoside antibiotics have been used clinically for the treatment of a wide range of Gram-negative infections. Their potential nephrotoxicity limits the doses that can be administered, so antibiotics exhibiting at least the same level of antibacterial activity as amikacin but with lowered nephrotoxicity would be desirable. Of the various fluorinated aminoglycosides which showed a similar level of antibacterial activity [117–119] or reduced toxicity [120], most effective were those that were fluorinated on the sugar moiety. Amikacin analogs with an ω-amino-α-fluoroalkanoyl side-chain have been prepared and were found to have almost the same biological activity as amikacin [121]. For instance, 1-*N*-[(*S*)-4-amino-2-fluorobutyryl]kanamycin A (31) (the 2‴-fluoro analog of amikacin) showed the best activity profile of the fluoro derivatives of

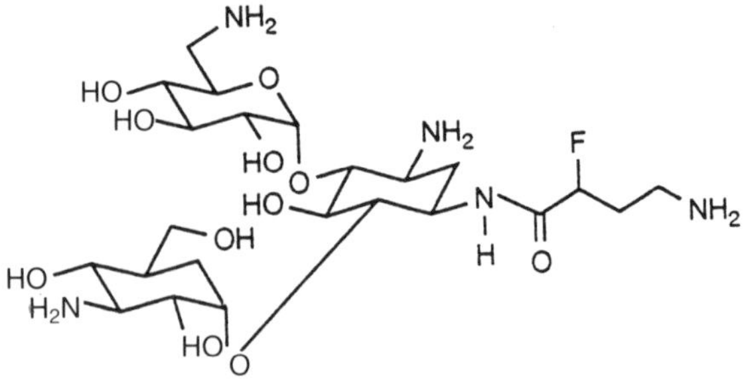

(31)

kanamycin A prepared to date. It was found to be nearly as active as amikacin against all the organism groups. ω-Amino-β-fluoro acids showed decreased biological activity, which was probably due to the decreased basicity caused by the fluorine on the β-carbon.

9.3 CONCLUSION

It is just over 100 years since the characterization of elementary fluorine by Henri Moisaan and it took another 70 years for chemists to learn to introduce fluorine into organic molecules. This has been a fruitful and rewarding exercise, as shown in this and other chapters. Replacement of hydrogen with fluorine in other classes of compounds (nucleosides, sugars, alcohols, fatty acids, etc.) has similarly yielded products with unique biological and pharmacological properties. There is no doubt that this will continue in the future, thanks to the development of creative and innovative chemistry as illustrated in other parts of this volume.

9.4 REFERENCES

1. C. Danzin, P. Bey, D. Schirlin and N. Claveric, *Biochem. Pharmacol.*, **31**, 387 (1982).
2. S. P. Kobser, V. A. Soloshonok, S. V. Galushko, Yu. I. Yagupolskii and V. P. Kukhar', *Zh. Obshch. Khim.*, **59**, 309 (1989).
3. (a) P. Goldman, *Science*, **164**, 1123 (1969); (b) F. A. Smith, *Chem. Tech.*, 422 (1973); (c) R. Filler, *Chem. Tech.*, 752 (1974); (d) *Ciba Foundation Symposium, carbon–Fluorine Compounds. Chemistry, Biochemistry and Biological Activities*, Elsevier, New York, 1972; (e) R. Filler, in R. F. Banks (Ed.), *Organofluorine Chemicals and Their Industrial Applications*, Ellis Horwood, Chichester (1979); (f) R. Filler and Y. Kobayashi (Eds), *Biomedical Aspects of Fluorine Chemistry*, Kodansha, Tokyo, 1982; (g) R. Filler (Ed.), *Biochemistry Involving Carbon–Fluorine Bonds*, American Chemical Society, Washington, DC, 1976.
4. M. R. C. Gerstenberger and A. Hass, *Angew. Chem., Int. Ed. Engl.*, **20**, 647 (1981).
5. (a) V. P. Kukhar', Y. L. Yagupolski and V. A. Soloshonok, *Russ. Chem. Rev.*, **59**, 89 (1990); (b) V. P. Kukhar' and V. A. Soloshonok, *Russ. Chem. Rev.*, **60**, 850 (1991); (c) V. P. Kukhar', Y. T. Yagupolski, I. I. Gerus and M. T. Kolycheva, *Russ. Chem. Rev.*, **60**, 1050 (1991).
6. R. Laske, H. Schoenenberger and E. Holler, *Arch. Pharm. (Weinheim)*, **322**, 847 (1989).
7. F. von der Haar, H.-J. Gabios and F. Cramer, *Angew. Chem.*, **93**, 250 (1981).
8. R. Laske, H. Schoenenberger and E. Holler, *Arch. Pharm. (Weinheim)*, **322**, 857 (1989).
9. M. J. W8ilson and D. L. Hatfield, *Biochim. Biophys. Acta*, **781**, 205 (1984).
10. F. Gerhart, W. Higgins, C. Tardif and J. B. Ducep, *J. Med. Chem.*, **33**, 2157 (1990).
11. M. J. Jung and J. Koch-Weser, in T. P. Singer and R. N. Ondarza (Eds),

Molecular Basis of Drug Action, North-Holland, Amsterdam, 1981, pp. 135–150.

12. J. Kollonitisch, L. M. Perkins, A. A. Patchett, G. A. Doldouras, S. Marburg, D. E. Duggan, A. L. Maycock and S. D. Aster, *Nature (London)*, **274**, 906 (1978).
13. (a) R. R. Rando, *Enzyme Inhibitors*, **8**, 281 (1975); (b) P. Bey, *Ann. Chim. Fr.*, **9**, 695 (1984).
14. C. Walsh, *Tetrahedron*, **38**, 871 (1982).
15. D. H. G. Versteeg, M. Palkouts, J. Van der Gugten, H. J. L. M. Wijnen, G. W. M. Smeets and W. DeJong, *Prog. Brain Res.*, **47**, 111 (1977).
16. W. W. Douglas, in L. S. Goodman and A. Gilman (Eds), *The Pharmacological Basis of Therapeutics*, 5th edn, Macmillan, New York, 1975, pp. 590–629.
17. D. H. Russell, in D. H. Russell (Ed.) *Polyamines in Normal and Neoplastic Growth*, Raven Press, New York, 1973, pp. 1–13.
18. H. M. Walborsky, M. Barum and D. F. Loncrini, *J. Am. Chem. Soc.*, **77**, 3637 (1955).
19. N. Seiler, M. J. Jung and J. Koch Weser (Eds), *Enzyme-activated Irreversible Inhibitors*, North-Holland, Amsterdam, 1978, pp. 1–389.
20. F. Bambino, R. T. C. Browlene and F. C. K. Chin, *Tetrahedron Lett.*, **32**, 3407 (1991).
21. C. Heidelberg, *Cancer Res.*, **30**, 1549 (1970).
22. S. Loy and M. Hudlicky, *J. Fluorine Chem.*, **7**, 421 (1976).
23. T. Tsushima, K. Kawada, S. Ishihara, N. Uchida, O. Shiratori, J. Higaki and M. Hirata, *Tetrahedron*, **44**, 5375 (1988).
24. D. Kuo and R. R. Rando, *Biochemistry*, **20**, 506 (1981).
25. M. Jung, unpublished results.
26. B. W. Metcalf, *Annu. Rep. Med. Chem.*, **16**, 289 (1981).
27. C. Dibari, G. Pastore, G. Roscigno, P. J. Schechter and A. Sjoerdsma, *Ann. Intern. Med.*, **105**, 83 (1986).
28. D. Schirlin, J. B. Ducep, S. Baltzer, P. Bey, F. Piriou, J. Wagner, J. M. Hornsperger, J. G. Heydt, M. J. Jung, C. Danzin, R. Weiss, J. Fischer, A. Mitschler and A. De Cian, *J. Chem. Soc., Perkin Trans. 1*, 1053 (1992).
29. P. Bey, C. Danzin and M. J. Jung, in P. P. McCann, A. E. Pegg and S. Sjoerdsma (Eds), *Inhibition of Polyamine Metabolism: Biological Significance and Basis for New Therapies*, Academic Press, Orlando, FL, 1987, pp. 1–31.
30. C. J. Bacchi, H. C. Nathan, S. H. Hutner, P. P. McCann and A. Sjoerdsma, *Science*, **210**, 332 (1980).
31. C. Danzin, J. B. Ducep, D. Schirlin and J. Wagner, in T. Korpela and P. Christen (Eds), *Biochemistry of Vitamin B6*, Birkhäuser, Basle, 1987, p. 333.
32. M. V. Rajam, L. H. Weinstein and A. W. Galston, *Proc. Natl. Acad. Sci. USA*, **86**, 1192 (1979).
33. J. Seidenfeld and L. J. Marton, *Biochem. Biophys. Res. Commun.*, **86**, 1192 (1979).
34. P. Bey, F. Gerhart, V. Van Dorsselaer and C. Danzin, *J. Med. Chem.*, **26**, 1551 (1983).
35. J. Kollonitsch, A. A. Patchett, S. Marburg, A. Maya, L. M. Perkins, G. A. Doldouras, D. E. Duggan and S. D. Aster, *Nature (London)* **274**, 906–908 (1978).
36. M. K. Bhattacharjee and E. E. Snell, *J. Biol. Chem.*, **265**, 6664 (1990), and references cited therein.
37. M. Bouclier, M. J. Jung and F. Gerhart, *Experientia*, **39**, 1303 (1983).
38. D. E. Duggan, K. F. Hooke and A. L. Maycock, *Biochem. Pharmacol.*, **33**, 4003 (1984).
39. M. Bouclier, M. J. Jung and F. Gerhart, *Biochem. Pharmacol.*, **32**, 1553 (1983).
40: M. Bouclier, M. J. Jung and F. Gerhart, *Eur. J. Pharmacol.*, **90**, 129 (1983).

41. V. S. Westerberger and J. D. Geiger, *Pharmacol. Biochem. Behav.*, **28**, 419 (1987).
42. M. Garbaarg, G. Barbin, E. Rodergas and J. C. Schwartz, *J. Neurochem.*, **35**, 1045 (1980).
43. J. S. Lin, K. Sakai and M. Jouvet, *Neuropharmacology*, **27**, 111 (1988).
44. R. Oishi, Y. Itoh, M. Nishibori, K. Saeki and S. Ueki, *Psychopharmacology*, **95**, 77 (1988).
45. K. Onodera and Y. Yasum, *Yakubutsu Seichin Kodo*, **5**, 11 (1985).
46. H. M. Hanson, *US Pat.*, *4 522 823* (1984).
47. T. Kaji, H. Saito, S. Ueno, T. Yasuhara, T. Nakajima and N. Matsuki, *Aviat. Space Environm. Med.* **62**, 1054 (1991).
48. T. Shimura, T. Horio and M. Shimokochi, *Neurosciences (Kobe, Jpn)*, **11**, 194 (1985).
49. J. Bartholeyns and M. Bouclier, in *Proceedings of the 13th International Congress on Chemotherapy*, 1983, pp. 276–234.
50. S. Hososa, T. Saito, H. Kumazawa, T. Watanabe and H. Wada, *Biochem. Pharmacol.*, **34**, 4327 (1985).
51. J. Bartholeyns and M. Bouclier, *Contraception*, **26**, 535 (1982).
52. T. Slotkin, J. Bartolome and W. L. Whitmore, *Life Sci.*, **33**, 2137 (1983).
53. E. W. Mussie and M. E. Beaven, *Mol. Pharmacol.*, **28**, 191 (1985).
54. C. Oh, H. Okamoto and K. Nakano, *Agric. Biol. Chem.*, **53**, 377 (1989).
55. (a) S. Seiji and N. Kiwao, *Biochem. Pharmacol*, **35**, 3039 (1986); (b) E. Neugegauer, T. Beckwarts, W. Lorenz, D. Maroske, H. Merte, Ge. Horeseck and W. Dietz, *Agents Actions*, **18**, 23 (1986).
56. J. Kollonitsch, *US Pat.*, *5 030 545* (1990).
57. T. D. August, D. G. Musson, S. S. Hwang, D. E. Duggan, K. F. Hooke, I. J. Roman, R. J. Ferguson and W. F. Bayne, *J. Pharm. Sci.*, **74**, 871 (1985).
58. T. Watanabe, A. Yamatodani, K. Maeyama and H. Wada, *Trends Pharm. Sci.*, **11**, 363 (1990).
59. P. Bey, in N. Seiler, M. J. Jung and J. Koch Weser (Eds), *Enzyme-activated Irreversible Inhibitors*, North-Holland, Amsterdam, 1978, pp. 27–41.
60. I. A. McDonald, P. L. Nyce, M. J. Jung and J. S. Sabol, *Tetrahedron Lett.*, **32**, 887 (1991).
61. M. Jung, unpublished results.
62. M. J. Jung, J. M. Hornsperger, I. McDonald, J. R. Fozard and M. G. Palfreyman, in P. Buri and M. Gumma (Eds), *Drug Targeting*, North-Holland, Amsterdam, 1985, pp. 165–178.
63. M. J. Jung, J. M. Hornsperger, F. Gerhart and J. Wagner, *Biochem. Pharmacol.*, **33**, 327 (1984).
64. J. M. Hornsperger, *These d'Etat*, Université Louis Pasteur, Strasbourg (1984).
65. J. Kollonitsch, A. A. Patchett, S. Marburg, *Eur. Pat.*, 7600 (1980); *Chem Abstr.*, **93**, 114978 (1980).
66. P. Bey, J. Fozard, J. J. M. Lacoste, I. A. McDonald, M. Zreika and M. G. Palfreyman, *J. Med. Chem.*, **27**, 9 (1984).
67. I. A. McDonald, J. M. Lacoste, S. Bey, J. Wagner, M. Zreika and M. G. Palfreyman, *J. Am. Chem. Soc.*, **106**, 3354 (1984).
68. M. G. Palfreyman, I. A. McDonald, J. R. Fozard, Y. Rely, A. J. Sleight, M. Zreika, J. Wagner, P. Bey and P. I. Lewis, *J. Neurochem.*, 1850 (1985).
69. M. G. Palfreyman, C. Danzin, M. J. Jung, J. R. Fozard, J. Wagner, J. K. Woodward, M. Aubry, R. C. Dage and J. Koch-Weser, in N. Seiler, M. J. Jung and J. Koch-Weser (Eds), *Enzyme-activated Irreversible Inhibitors*, North-Holland, Amsterdam, 1978, pp. 221–233.
70. G. Zbinden and E. Braendle, *Toxicol. Lett.*, **5**, 125 (1980).

71. A. L. Maycock, S. D. Aster and A. A. Patchett, *Biochemistry*, **19**, 709 (1980).
72. M. J. Jung, M. G. Palfreyman, J. Wagner, P. Bey, G. Ribereau-Gayon, M. Zraika and J. Koch-Weser, *Life Sci.*, **24**, 1037 (1979).
73. P. Bey, M. J. Jung, J. Koch-Weser, M. G. Palfreyman, A. Sjoerdsma, J. Wagner and M. Zraika, *Br. J. Pharmacol.*, **70**, 571 (1980).
74. M. G. Palfreyman, S. Huot and J. Wagner, *J. Pharmacol. Methods*, **8**, 185 (1982).
75. R. Maneckjee and S. B. Baylin, *Biochemistry*, **22**, 6058 (1983).
76. J. R. Fozard, M. Spedding, M. G. Palfreyman, J. Wagner, J. Mohring and J. Koch-Weser, *J. Cardiovasc. Pharmacol.*, **2**, 229 (1980).
77. S. Johansson and M. Henning, *Naunyn-Schmiedeberg's Arch. Pharmacol.*, **321**, 28 (1982).
78. J. R. Fozard, J. Mohring, M. G. Palfreyman and J. Koch-Weser, *J. Cardiovasc. Pharmacol.*, **3**, 1038 (1981).
79. D. Schirlin, F. Gerhart, J. M. Hornsperger, M. Hamon, J. Wagner and M. J. Jung, *J. Med. Chem.*, **31**, 30 (1988).
80. D. Zembower, J. A. Gilbert and M. A. Ames, *J. Med. Chem.*, **36**, 305 (1993).
81. (a) J. Kollonitsch, in R. Filler and Y. Kobayashi (Eds), *Biomedicinal Aspects of Fluorine Chemistry*, Kodansha, Tokyo, 1982, pp. 93–125; (b) C. Walsh, *Adv. Enzymol. Relat. Areas Mol. Biol.*, **55**, 197 (1983); (c) C. Walsh, *Annu. Rev. Biochem.*, **53**, 493 (1984); (d) P. Bey, *Ann. Chem. Fr.*, **9**, 695 (1984); (e) P. K. Rathod, A. H. Tashjian, Jr, and R. H. Abeles, *J. Biol. Chem.*, **261**, 6461 (1986).
82. N. Seiler, M. J. Jung and J. Koch-Weser, *Pharmacol. Rev.*, **36**, 111 (1984).
83. N. Esaki and C. T. Walsh, *Biochemistry*, **25**, 3261 (1986).
84. G. A. Flynn, D. W. Beight, E. H. Bohme and B. W. Metcalf, *Tetrahedron Lett.*, **26**, 285 (1985).
85. J. J. M. Hageman, M. J. Wanner, G. K. Koomen and U. K. Pandit, *J. Med. Chem.*, **20**, 1677 (1977).
86. P. A. Briley, R. Eisenthal, R. Harrison and G. D. Smith, *Biochem. J.*, **16**, 383 (1977).
87. M. Hudlicky, *J. Fluorine Chem.*, **40**, 99 (1988).
88. M. Akhtar, N. S. Botting, M. A. Cohen and D. Gani, *Tetrahedron*, **43**, 5899 (1987).
89. P. J. Casey, R. H. Abeles and J. W. Lowenstein, *J. Biol. Chem.*, **26**, 16367 (1986).
90. A. Stern, R. H. Abeles and A. H. Tashjian, *Cancer Res.*, **44**, 5614 (1984).
91. G. Hortin, A. M. Stern, B. Miller, R. H. Abeles and I. Boime, *J. Biol. Chem.*, **258**, 4047 (1983).
92. M. A. Philips, A. M. Stern, R. H. Abeles and A. H. Tashjian, *J. Pharmacol. Exp. Ther.* **226**, 276 (1983).
93. A. Vidal-Cros, M. Gaudry and A. Marquet, *J. Org. Chem.*, **50**, 3163 (1985).
94. C. B. Grissom and W. W. Cleland, *Biochim. Biophys. Acta*, **916**, 437 (1988).
95. A. Vidal-Cros, M. Gaudry and A. Marquet, *Biochem. J.*, **229**, 675 (1985).
96. J. J. McGuire, W. H. Haile, P. Bey and J. K. Coward, *J. Biol. Chem.*, **265**, 14073 (1990).
97. P. Bey and M. H. Kolb, *Eur. Pat. Appl.*, EP 451 835 (1991), p. 13.
98. T. Tsushima, K. Kawada, J. Nishikawa, T. Sato, K. Tori, T. Tsuji and S. Misaki, *J. Org. Chem.*, **49**, 1163 (1984), and references cited therein.
99. M. J. O'Donnell, C. L. Barney and J. R. McCarthy, *Tetrahedron Lett.*, **26**, 3067 (1985).
100. R. H. Abeles and A. L. Maycock, *Acc. Chem. Res.*, **9**, 313 (1976).
101. P. F. Bevilacqua, D. D. Keith and J. L. Roberts, *J. Org. Chem.*, **49**, 1430 (1984).
102. C. W. Fearon, J. A. Rodkey and R. H. Abeles, *Biochemistry, 321*, 3790 (1982).
103. R. B. Silverman and R. H. Abeles, *Biochemistry*, **16**, 5515 (1977).

104. W. W. Doublass, in L. S. Goodman and A. Gilman (Eds), *The Pharmacological Basis of Therapeutics,* 5th edn, Macmillan, New York, 1975, pp. 590–629.
105. C. Schierlinger and K. Burger, *Tetrahedron Lett.,* **33**, 193 (1992).
106. F. Weygand, W. Steglich, W. Oettmeier, A. Maierhofer and R. S. Loy, *Angew. Chem.,* **78**, 640 (1966).
107. W. A. Zygmunt and P. A. Tavormina, *Can. J. Microbiol.,* **12**, 143 (1966).
108. J. A. Steko, *Transmethylation Methionine Biosyn.,* 231 (1965).
109. T. Tsushima, K. Kawada, S. Ishihara, N. Uchida, o. Shiratori, J. Higaki and M. Hirata, *Tetrahedron,* **444**, 5375 (1988).
110. U. Leutbecher, R. Bocher, D. Linder and W. Buckel, *Eur. J. Biochem.,* **205**, 759 (1992).
111. I. A. McDonald, M. G. Palfreyman, M. J. Jung and P. Bey, *Tetrahedron Lett.,* **26**, 4091 (1985).
112. A. F. Spatola, *Chemistry and Biochemistry of Amino Acids, Peptides and Proteins VIII,* 1983, pp. 267–357.
113. C. Walsh, *Tetrahedron,* **38**, 871 (1982).
114. R. L. Buchanan, F. H. Dean and F. L. M. Pattison, *Can. J. Chem.,* **40**, 1571 (1962).
115. T. Tsushima, K. Kawada, O. Shiratori, *et al., Heterocycles,* **23**, 45 (1985).
116. J. J. McGuire and J. K. Coward, *J. Biol. Chem.,* **260**, 6747 (1985).
117. R. Albert, K. Dax and A. E. Stütz, *J. Carbohydr. Chem.,* **3**, 267 (1984).
118. R. Albert, K. Dax and A. E. Stütz, *J. Antibiot.,* **38**, 275 (1985).
119. T. Tsuchiya, Y. Takahashi, Y. Kobayashi, S. Umezawa and H. Umezawa, *J. Antibiot.,* **38**, 1287 (1985).
120. T. Tsuchiya, T. Torii, H. Umezawa and S. Umezawa, *J. Antibiot.,* **35**, 1245 (1982).
121. H. Hoshi, S. Aburaki, S. Imura, T. Yamasaki, T. Naito and H. Kawaguchi, *J. Antibiot.,* **43**, 858 (1990).

10 Renin Inhibitors with Fluorine-containing Amino Acid Derivatives

HING L. SHAM

Pharmaceutical Discovery Division, D-47D; AP9A, Abbott Laboratories, Abbott Park, IL 60064-3500, USA

10.1 INTRODUCTION

The renin–angiotensin system has been successfully manipulated to control high blood pressure. Many angiotensin-converting enzyme (ACE) inhibitors have been reported to be effective drugs for the control of high blood pressure. Renin is an aspartic protease that catalyzes the first step in the angiotensinogen cascade to produce angiotensin II, a very potent vasoconstrictor. It is produced mainly in the juxtaglomerular apparatus of the kidney. It cleaves the circulating angiotensinogen to produce angiotensin I, which is in turn cleaved by ACE to produce angiotensin II. Renin is an enzyme of high substrate specificity and its inhibition has the potential of providing specific therapeutic intervention in the control of high blood pressure [1].

During the hydrolysis of the peptide bond of the substrate by an aspartic protease such as renin, the amide carbonyl group is postulated to be hydrated to form a tetrahedral transition state as shown in equation 10.1. The collapse of this intermediate gives rise to angiotensin i and the other fragment. Fluorine, because of its high electronegativity, has an important effect on its neighboring functional groups in a molecule. Fluoroketones, in which the carbonyl group is next to fluoro substituents, are usually fully hydrated (equation 10.2). As a result, fluoroketones are good transition-state mimics of the sp^3-hybridized hydrated carbonyl at the cleavage site of substrate peptides [2]. Renin inhibitors with fluorine-containing amino acids derivatives including fluoroketones will be described.

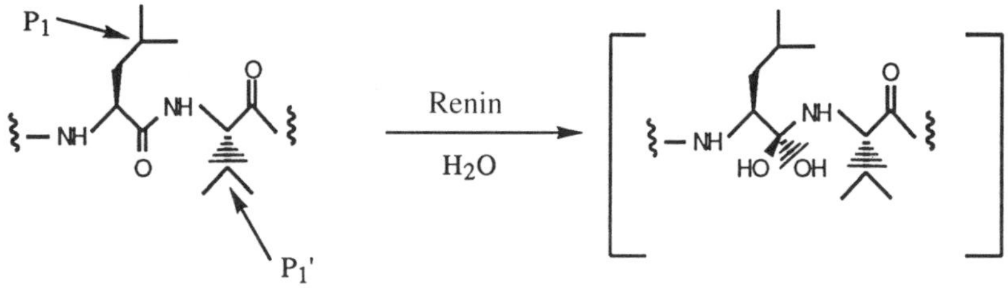

$$(10.1)$$

Fluorine-containing Amino Acids: Synthesis and Properties   Edited by V. P. Kukhar' and V. A. Soloshonok
© 1995 John Wiley & Sons Ltd

$$(10.2)$$

10.2 CHEMISTRY AND RENIN INHIBITORY ACTIVITIES

Highly potent renin inhibitors containing a difluorostatone or side-chain modified difluorostatone analogues have been reported by several groups. The synthesis of renin inhibitors with difluorostatine and difluorostatone and related analogues (equation 10.3) was first reported by Thaisrivongs *et al.* [3].

Compound **1**, which contains the difluorostatone unit, is a potent inhibitor of renin with an $IC_{50} = 0.52$ nM. Other difluorostatone-containing renin inhibitors such as **2**, **3** nd **4** have also been reported [4–6].

Fluoro ketones containing the retroamide-type bond at the normal carboxyl terminus has been synthesized [7, 8]. This series of compounds is represented by the prototypical compound **5**.

The difluorostatine ester was prepared as reported [3] and reaction with ammonia provided the primary amide which was reduced by borane dimethyl sulfide to provide the corresponding primary amine (equation 10.4). Further elaboration of this intermediate provided **5**, which is also a potent renin inhibitor. A related series of renin inhibitors which are highly selective and more soluble were derived by deprotection of the retroamide to provide a free amino group [9]. This series of compounds is represented by the prototypical compound **6**. The weakly basic amino alkyl group enhances the solubility of the compound in aqueous media.

Yet another interesting difluoro ketone-containing dipeptide isostere that maintains the normal sequence length of the substrate has been synthesized by a multi-step synthesis [10]. The difference between this dipeptide isostere, represented by **7** (equation 10.5), and the difluorostatone series is that the P_1' side-chain is restored.

Renin inhibitors containing an α,α-difluoro-β-aminodeoxystatine moiety had been reported by Thaisrivongs *et al.* [11]. The starting material for the synthesis is difluorostatine [3]. As shown in equation 10.6, the 4-methoxyphenyl-protected α,α-difluoro-β-aminodeoxystatine was synthesized in several steps from the sodium salt of difluorostatine. Incorporation of the protected difluoro-β-aminodeoxystatine in an appropriate peptide sequence followed by deprotection with cerium(IV) ammonium nitrate provided **8**, which is a relatively weak inhibitor of human plasma renin ($IC_{50} = 2800$ nM). Starting with the α-hydroxy isomer of difluorostatine, **9** with the (*R*)-amino group can be synthesized (equation 10.7). it is a better inhibitor of human plasma renin with an IC_{50} of 340 nM.

Renin inhibitors containing a tetrafluoro ketone moiety have been reported

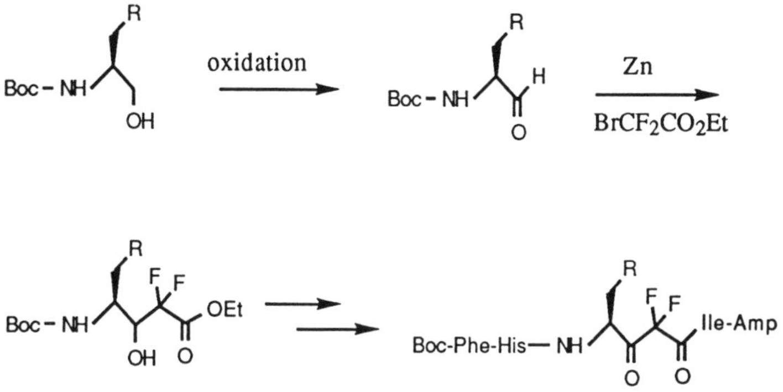

(10.3)

α- and β-OH

difluorostatine

(1) R = isopropyl
     $IC_{50} = 0.52$ nM

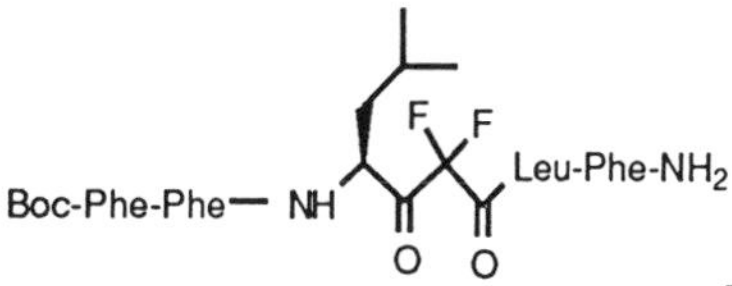

$IC_{50} = 7$ nM

(2)

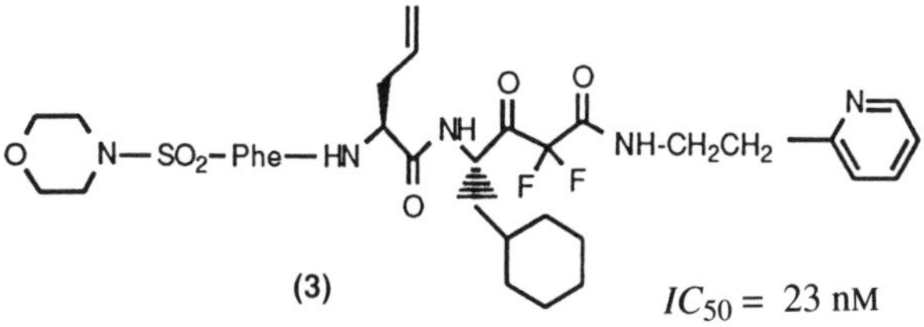

(3) $IC_{50} = 23$ nM

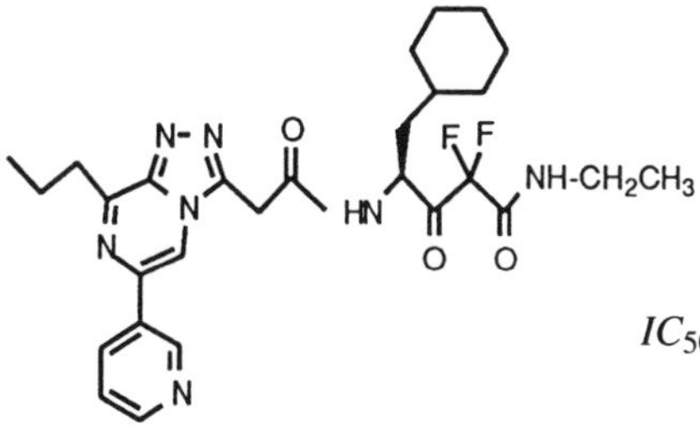

$IC_{50} = 1.4$ nM

(4)

[12]. Electrochemical reaction of an analogue of difluorostatine with monoethyl diifluoromalonate provided the tetrafluoro alcohol intermediate, which, when incorporated in a peptide sequence followed by oxidation, provided the tetrafluoro ketone **10** (equation 10.8). Compound **10** is an inhibitor of human renin with an IC_{50} of 23 nM.

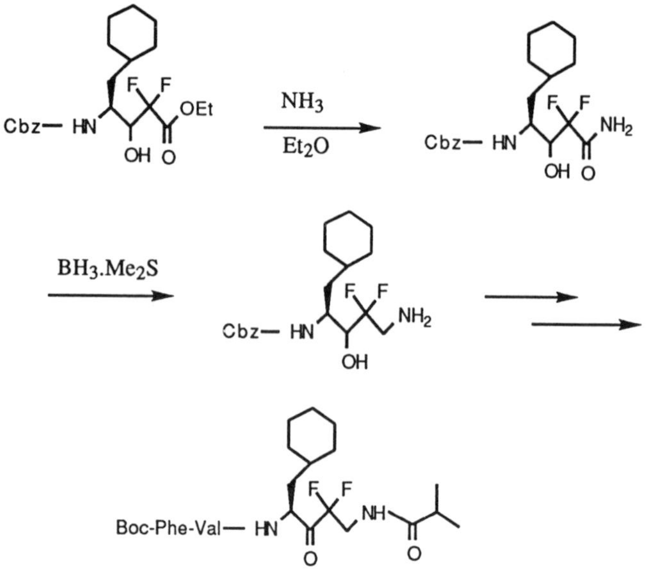

(10.4)

(5) $IC_{50} = 3.5$ nM

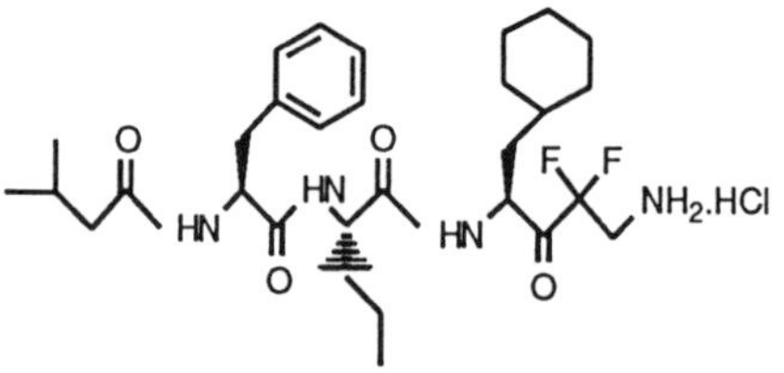

(6) $IC_{50} = 16$ nM

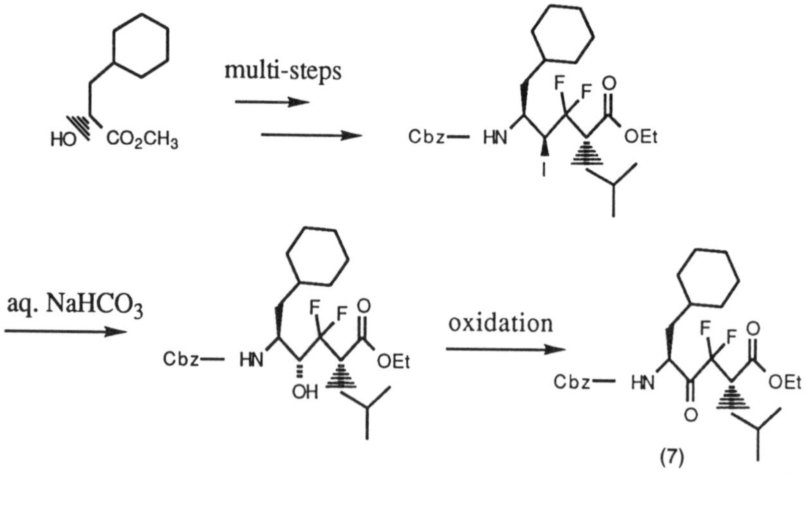

(10.5)

(7)

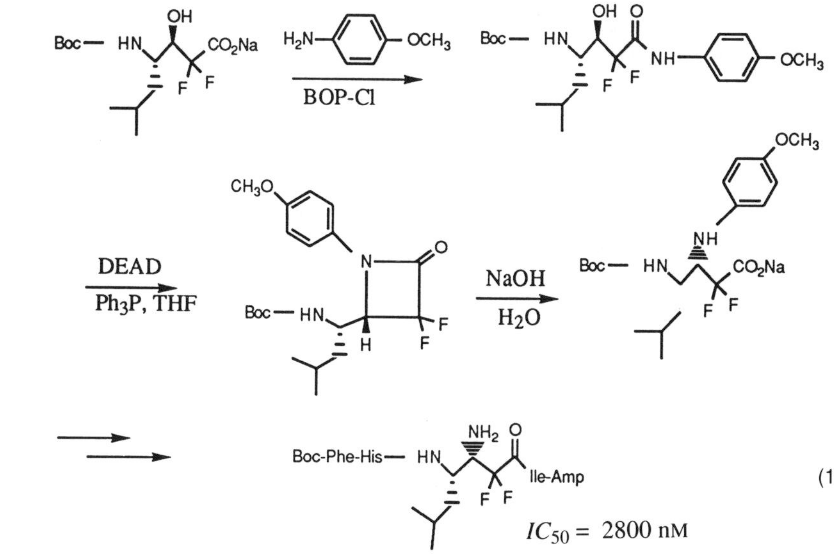

$$IC_{50} = 2800 \text{ nM}$$

(10.6)

(8)

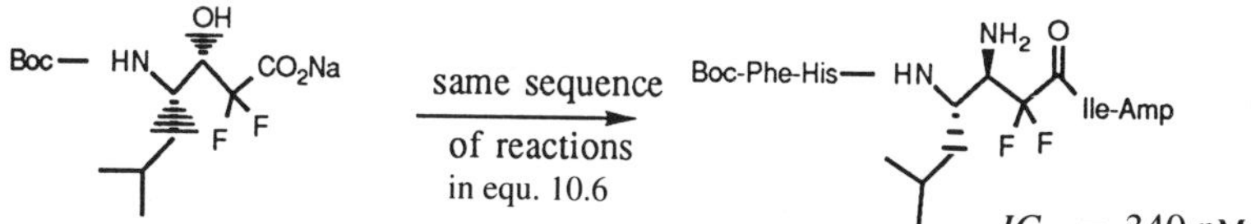

$$IC_{50} = 340 \text{ nM}$$

(10.7)

(9)

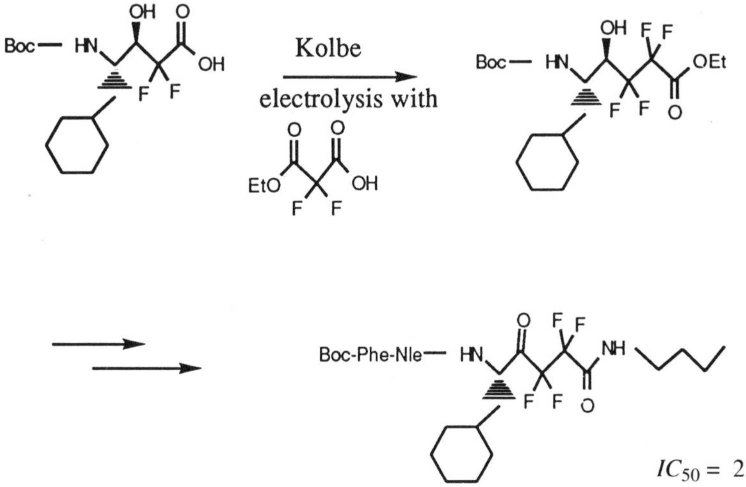

(10.8)

$$IC_{50} = 23 \text{ nM}$$

(10)

Inhibitors containing a difluoro alcohol or ketone with an alkyl sulfone terminus have been synthesized [13] starting with a derivative of a difluoro-statine analogue (equation 10.9). Oxidation of the difluoro alcohol **11** provided the corresponding difluoro ketone **12**, which is a potent inhibitor of human renin with an IC_{50} of 1 nM.

Several synthesis of renin inhibitors incorporating trifluoromethyl alcohol and trifluoromethyl ketones at the C-terminus of peptide sequences have been reported. One of these reported syntheses [14] starts with 2-cyclohexyl-propionic acid (equation 10.10). Lithiation and aldol condensation with tri-

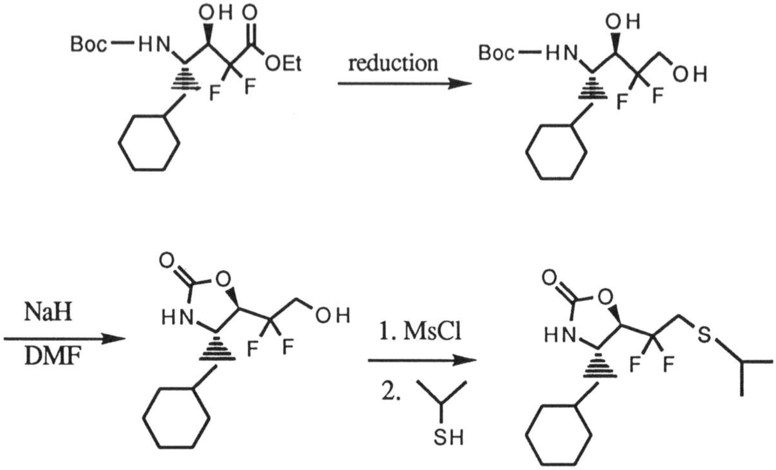

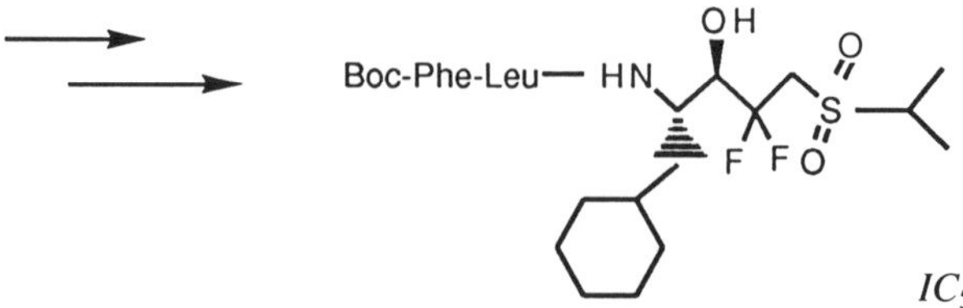

$IC_{50} = 10$ nM

(11)

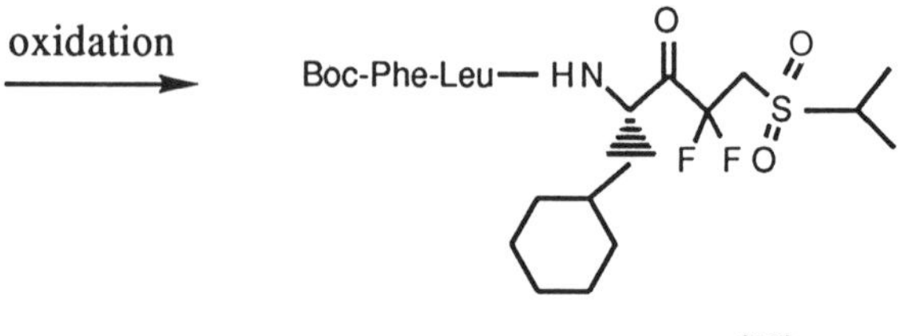

$IC_{50} = 1$ nM

(12)

(10.9)

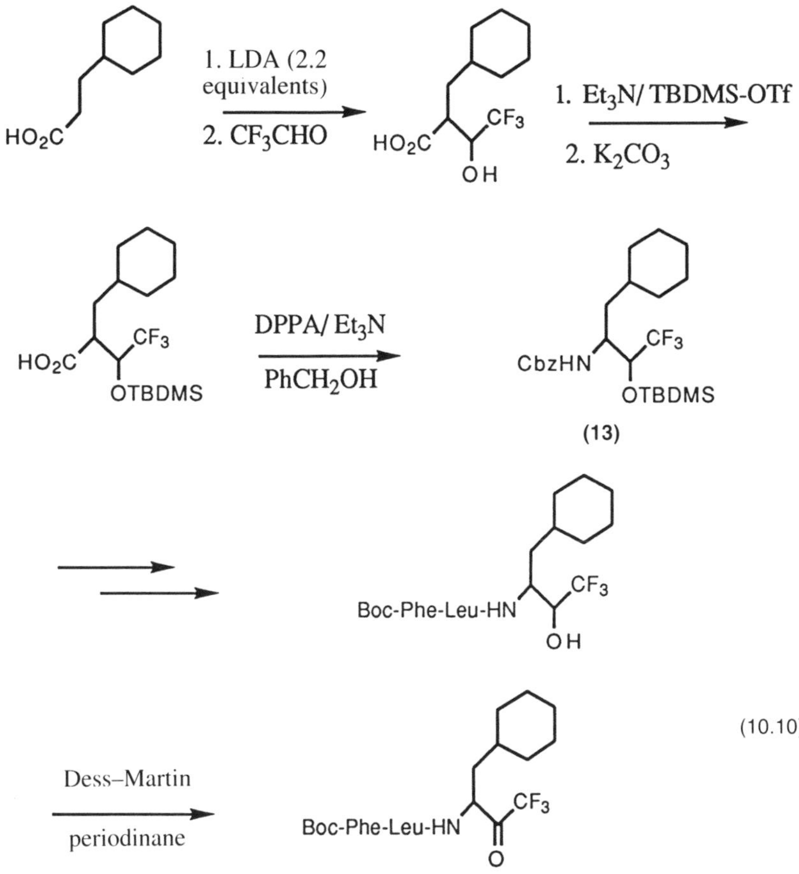

(13)

(10.10)

(14) $IC_{50} = 250$ nM

fluoromethylacetaldehyde provided the trifluoromethyl alcohol. Protection of the hydroxy function as the silyl ether and Curtius rearrangement with diphenylphosphoryl azide in the presence of benzyl alcohol produced the key intermediate 13. Incorporation of this intermediate in an appropriate peptide sequence followed by oxidation provided the trifluoromethyl ketone 14, which has an IC_{50} of 250 nM against human renin.

Another alternative synthesis of trifluoromethyl alcohols and ketones starts with 2-phenyl-1-nitroethane (equation 10.11). Condensation with trifluoroacetaldehyde provided the corresponding trifluoromethyl alcohol. Reduction of the nitro group to the amino group is accomplished by hydrogenation using Raney nickel as catalyst. Incorporation of the resulting amine 15 into the C-terminus of a peptide sequence followed by oxidation provided the trifluoromethyl ketone [15].

A facile synthesis of renin inhibitors containing perfluoroalkyl ketones at the

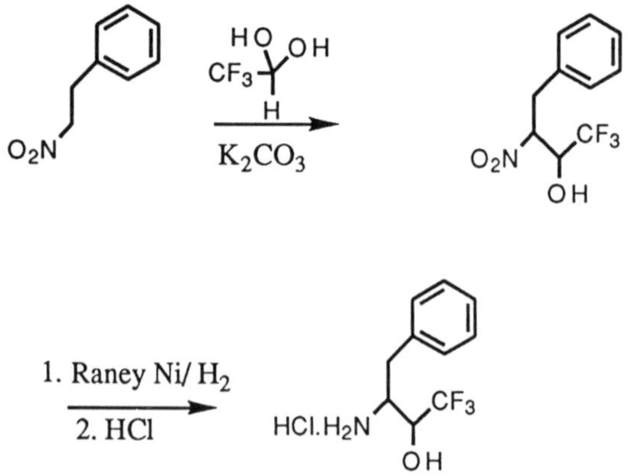

(10.11)

(15)

C-terminus of a peptide sequence has been reported [16]. Reaction of per-fluoroalkyllithium at low temperature with an aldehyde derived from an N-terminal protected amino acid produced the corresponding perfluoroalkyl alcohols (equation 10.12). Incorporation of the perfluoroalkyl alcohols at the

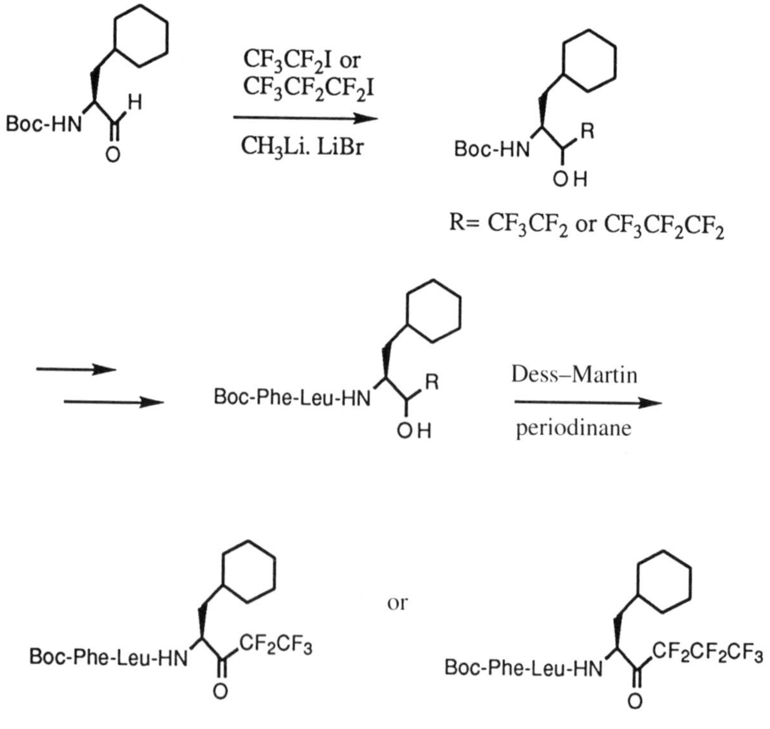

$R = CF_3CF_2$ or $CF_3CF_2CF_2$

(10.12)

or

(16) $IC_{50} = 13$ nM

(17) $IC_{50} = 3$ nM

C-terminus of a peptide sequence followed by oxidation provided perfluoro-alkyl ketones such as **16** and **17**, which are potent inhibitors of human renin.

A closely related synthesis of perfluoroalkyl ketones has been reported [17]. Instead of starting with an N-terminal protected α-aminoaldehyde, the synthesis started with N-terminal protected amino acids. Coupling with N-methyl-N-methoxyamine hydrochloride provided the Weinreb amide [18]. Reaction of the Weinreb amide with perfluoroalkyllithium generated *in situ* from perfluoroalkyl iodide and methyllithium provided the perfluoroalkyl ketones such as **18**. Incorporation of the perfluoroalkyl ketones in a peptide sequence similar to those in equation 10.12 should provide inhibitors of human renin.

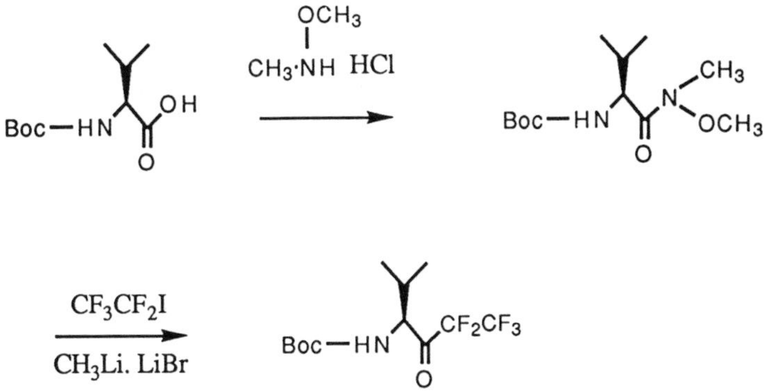

(10.13)

(18)

10.3 CONCLUSION

The synthesis of a wide variety of fluoro ketones has been described. The carbonyl group next to the highly electronegative fluorines showed a very high degree of hydration. The tetrahedral hydrated carbonyl mimics very well the tetrahedral hydrated amide formed during the hydrolysis of peptide bonds catalyzed by aspartyl proteases such as renin. Most of the fluoro ketones described are potent inhibitors of renin, confirming the usefulness of the transition state analogue concept in the design of potent enzyme inhibitors.

10.4 REFERENCES

1. W. J. Greenlee, *Med. Res. Rev.*, **10**, 173 (1990).
2. M. H. Gelb, J. P. Svaren and R. H. Abeles, *Biochemistry*, **24**, 1813 (1985).
3. S. Thaisrivongs, D. T. Pals, W. M. Kati, S. R. Turner, L. M. Thomasco and W. Watt, *J. Med. Chem.*, **29**, 2080 (1986).
4. K. Fearon, A. Spaltenstein, P. B. Hopkins and H. Gelv, *J. Med. Chem.*, **30**, 1617 (1987).

5. A. M. Doherty, I. Sircar, B. E. Kornberg, J. Quin, R. T. Winters, J. S. Kaltenbronn, M. D. Taylor, B. L. Batley, S. R. Rapundalo, M. J. Ryan and C. A. Painchaud, *J. Med. Chem.,* **35**, 2 (1992).

6. R. H. Bradbury and J. E. Rivett, *J. Med. Chem.,* **34**, 151 (1991).

7. D. Schirlin, S. Baltzer and J. M. Altenburger, *Tetrahedron Lett.,* **29**, 3687 (1988).

8. C. Tarnus, M. J. Jung, J. M. Remy, S. Raltzer and D. Schirlin, *FEBS Lett.,* **249**, 47 (1989).

9. D. Schirlin, C. Tarnus, S. Baltzer and J. M. Remy, *Bioorg. Med. Chem. Lett.* **2**, 651 (1992).

10. D. J. Hoover and D. B. Damon, *J. Am. Chem. Soc.,* **112**, 6439 (1990).

11. S. Thaisrivongs, H. J. Schostarez, D. J. Pal and S. R. Turner, *J. Med. Chem.,* **30**, 1837 (1987).

12. E. Pfenninger and A. Weidmann, *Br. Pat. Appl.,* 86-19182 (1986).

13. S. Rosenberg, *US Pat.,* 4 857 507 (1989).

14. D. V. Patel, K. R. Gauvin and D. E. Ryono, *Tetrahedron Lett.,* **29**, 4665 (1988).

15. R. H. Abeles and B. Imperiali, *Tetrahedron Lett.,* **27**, 135 (1986).

16. H. L. Shain, H. Stein, C. A. Rempel, J. Cohen and J. J. Plattner, *FEBS Lett.,* **220**, 299 (1987).

17. M. R. Angelastro, J. P. Burkhart, P. Bey and N. P. Peet, *Tetrahedron Lett.,* **23**, 3265 (1992).

18. S. M. Weinreb and S. Nahm, *Tetrahedron Lett.,* **22**, 3815 (1981).

11 Synthesis and Biochemical Applications of Fluorine-containing Peptides and Proteins

KENNETH L. KIRK

Laboratory of Bioorganic Chemistry, National Institute of Diabetes and Digestive and Kidney Diseases, National Institutes of Health, Bethesda, MD 20892, USA

11.1 INTRODUCTION

Fluorine-containing peptides have been used as pharmacological agents, potential medicinal agents, biochemical tracers and mechanistic probes. Fluorine also has been used as a probe to study the structures and functional properties of many proteins. The preparation, biological properties and biochemical applications of fluorine-containing peptides and proteins are reviewed in this chapter.

11.1.1 BIOLOGICAL APPLICATIONS OF FLUORINE-CONTAINING PEPTIDES AND PROTEINS

In a peptide or protein, introduction of fluorine can represent a fairly minimal structural variation, and often little or no perturbation in biochemical properties is observed. This may not be the case if fluorine is present at a critical amino acid required, for example, for catalytic action. With respect to potential research value, however, either outcome can be advantageous. Thus, to the extent that the presence of fluorine has minimal effects on biological properties, ^{19}F NMR and other physical organic parameters can be used to assess the properties of the native molecule. On the other hand, alterations in biological properties that result from the presence of a fluorinated amino acid can be a useful measure of the function of the peptide or protein residue that has been labeled with fluorine. For example, strategies to alter beneficially the biological behavior of peptide hormones have included synthetic preparation of analogues containing fluorinated amino acids.

The NMR properties of ^{19}F make it a particularly useful probe for the study of the structure and function of fluorine-labeled biomolecules, including peptides and proteins. ^{19}F has spin $\frac{1}{2}$, has a high natural sensitivity, has a much higher chemical shift range than does ^{1}H, chemical shifts are particularly sensitive to environment, and, of special importance, there is no endogenous

Fluorine-containing Amino Acids: Synthesis and Properties   Edited by V. P. Kukhar' and V. A. Soloshonok
© 1995 John Wiley & Sons Ltd

signal present. There has been extensive use of fluorine-labeled peptides using ^{19}F NMR spectrometry to probe such parameters as local environments of individual amino acid residues, effects of ligands on chemical shifts in fluorine-labeled receptors, intramolecular communication, and others [1].

This review will be of sufficient breadth to cover areas of research wherein fluorine plays an important role, even though not present as a fluorinated natural amino acid. For example, much ^{19}F NMR research has been carried out on peptides and proteins that have been tagged with a fluorine-containing label. Another important area of current research in medicinal chemistry involves development of inhibitors targeted to proteolytic enzymes. Important contributions to this research include the development of active site-selective electrophilic alkylating agents, transition-state mimics or isosteric replacements for the peptide bond. The special properties of the carbon–fluorine bond have been used effectively in each of these approaches. In this strategy, fluorine usually is introduced into an inhibitor at sites proximal to the corresponding position of the scissile bond of the substrate.

Toxic amino acids have been incorporated into peptides for targeted delivery into cells using peptide transport systems. Intracellular peptidases subsequently release the toxic moiety to produce the desired chemotherapeutic effect. Use of toxic fluorinated amino acids in this strategy represents a third application, more limited to date, of fluorine-containing peptides.

Finally, the use of ^{18}F-labeled proteins and peptides as scanning agents for positron emission tomography (PET) has received some attention, and will undoubtedly increase in importance.

11.1.2 PREPARATION OF FLUORINE-CONTAINING PEPTIDES AND PROTEINS

11.1.2.1 Attachment of fluorinated prosthetic groups to peptides and proteins

The potential of using ^{19}F NMR to study the dynamic properties of proteins prompted early interest in the synthesis of fluorine-labeled proteins and peptides. Many early ^{19}F NMR studies were carried out on semisynthetic material having fluorine-labeled probes attached to the peptide chain. Although other routes to fluorine-labeled peptides and proteins are now available, attachment of fluorinated probes remains an important and efficient strategy for preparing substrates for NMR and other structural and functional studies. Selected examples of this approach will be given in this chapter. Reviews that are cited contain other examples and more detailed descriptions of this work. For example, Gerig [1b] has reviewed reagents used in experiments to study fluorinated ligands in equilibrium with enzymes (type I experiment) and experiments wherein the fluorinated molecule becomes covalently attached to (or dissociates very slowly from) the enzyme (type II experiment). Attachment of fluorinated prosthetic groups to peptides has also been used to prepare potent inhibitors of proteolytic enzymes.

11.1.2.2 Total synthesis of fluorinated peptides

The development of many procedures for the synthesis of fluorinated molecules, together with the recognition of the special advantages of fluorine substitution, have made available an impressive inventory of fluorinated amino acids. Parallel advances in peptide chemistry, especially the development of solid-phase syntheses and rapid procedures for sequencing polypeptides, have made total synthesis a powerful technique for producing biologically active peptides and their analogues. Fluorinated amino acids have been incorporated into a growing list of biologically active peptides by total synthesis. These have been used for ^{19}F NMR studies, and biological sequelae have included altered potency and/or biological half life.

11.1.2.3 Biosynthetic incorporation of fluorinated amino acids into proteins

As with other classes of fluorinated biomolecules, the diverse biological properties of fluorinated amino acids have produced a wealth of useful biochemical and pharmacological tools. For example, fluorinated amino acids can function as inhibitors of specific enzymes and often serve as biological precursors for other fluorinated biologically critical molecules, such as aminergic neurotransmitters. An additional dimension to the biological activity of fluorinated amino acids is their potential as substrates for the biochemical machinery responsible for protein biosynthesis. The survival of an organism requires transfer of genetic information to produce proteins with maximum fidelity of peptide bond formation. The minimal steric alterations present in fluorinated amino acids allow many of these to pass stringent biochemical proofreading mechanisms present in peptide bond assembly. Accordingly, biosynthetic incorporation of fluorinated amino acids has been exploited effectively for a variety of research purposes, including, for example, the study of protein biosynthetic mechanisms as determined by analogue incorporation, the effect of analogue incorporation on cellular functions and, of major impact, the study of the effects of incorporation on protein structure and function, with special emphasis on the use of ^{19}F NMR. Examples of the consequences of incorporation of fluorinated amino acids into proteins include the production of fraudulent enzymes, inactive regulatory proteins and altered structural proteins. Early work has been reviewed by Richmond [2] and Kirk [3a] has provided additional discussion.

Following the synthetic work in the 1930s that made accessible fluorinated analogues, first of phenylalanine and then of tyrosine, much early research was focused on the effects of these analogues on the growth of microorganisms. Several examples of inhibited or altered growth patterns were published. In the late 1950s and early 1960s, there appeared several reports of the isolation of protein fractions from several sources into which had been incorporated fluorinated amino acids [4]. Microorganisms soon became a valuable source of several classes of fluorine-labeled proteins (enzymes, structural proteins, transport proteins, etc.) that have had many applications, particularly for use in ^{19}F NMR studies. The recent availability of the powerful tools of molecular biology, together with advances in NMR techniques, have made this a particularly active and productive area of research.

11.1.2.4 Direct fluorination of peptides

Biological incorporation and total syntheses represent the two general methods for producing peptides containing fluorinated analogues of natural amino acids. Recently, direct, electrophilic fluorination and photochemical trifluoromethylation of biologically active peptides have been shown to be alternative approaches [5, 6].

11.2 PEPTIDES CONTAINING FLUORINATED PROSTHETIC GROUPS AS NMR AND MECHANISTIC PROBES

Nucleophilic sites of peptides and proteins have been derivatized with a wide variety of fluorine-containing small molecules. The fact that the amino acid residues possessing these nucleophilic side-chains often are at critical functional sites of the proteins has made the fluorine-labeled analogues even more useful for probing functional and structural parameters of the protein. Selected examples of this are given below. The reviews by Gerig [1a, b] contain further discussions of this approach.

11.2.1 RIBONUCLEASE

In an early example of the labeling of proteins with fluorinated NMR probes, Huestis and Raftery [7] acylated the Lys[1] and Lys[7] residues of ribonuclease with trifluoroacetyl groups. ^{19}F NMR was used to measure conformational changes induced by the presence of inhibitors. Gerig [1a] has reviewed several additional examples of this strategy.

11.2.2 CYTOCHROME c

Treatment of cytochrome c with limiting amounts of ethyl thiotrifluoroacetate followed by ion-exchange chromatography gave proteins modified at Lys[22] or Lys[25]. Exhaustive trifluoromethylation with the same reagent followed by incubation at pH 10.7 (15 h) afforded a derivative that retained the more stable TFA groups by Lys[13], Lys[55] and Lys[99]. Results of kinetic experiments indicated that Lys[13] and Lys[25] are important for binding cytochrome oxidase [8].

11.2.3 INSULIN

Several N-trifluoroacetylated derivatives of beef insulin were prepared by the reaction of the hormone with ethylthiotrifluoroacetate [9] or phenyltrifluoroacetate [10]. Derivatives having the trifluoroacetyl group on the N-terminal Gly of the A-chain, on the N-terminal Phe of the B-chain and on a Lys residue near the C-terminus of the B-chain were isolated and shown to have high biological activity. These n-trifluoroacetylated analogues of insulin were used to study the aggregation properties of the A- and B-chains as a function of pH, using ^{19}F NMR together with optical rotatory dispersion–circular dichroism and sedimentation velocity experiments [1a, 11].

Recently, a prosthetic group approach was used to label the B-chain N-terminal Phe of insulin with an $[^{19}F]$- and an $[^{18}F]$({[N-(4-fluoromethyl)benzoyl]amino}butylamino)suberoyl group (Figure 11.1) [12]. In the fluorination step, reaction of N-[4-(bromomethy)benzoyl]-N'-(tert-butyloxycarbonyl)butane-1,4-diamine (1) produced the corresponding $[^{19}F]$- or $[^{18}F]$fluoromethyl compound (2). After deprotection (TFA), this was coupled to A^1,B^{29}-di-Boc-B^1-[(N-hydroxysuccinimidyl)suberoyl]insulin (3). Final deprotection gave the fluorinated insulin derivative (fluoroinsulin) (4). The overall process is rapid, suitable for work with the short-lived ($t_{\frac{1}{2}} = 110$ min) ^{18}F isotope. Fluoroinsulin was found to retain the essential biological properties of native insulin. $[^{18}F]$-fluoroinsulin has been used to study binding of insulin receptors in monkeys *in vivo* [13].

11.2.4 α-CHYMOTRYPSIN

A large number of analogues of chymotrypsin have been prepared wherein fluorine-containing reporter groups have been covalently attached at or near the active site [1a,b]. Several stable esters of the catalytically essential Ser^{195} have been prepared, including 4-fluorobenzoyl, 3,5-(bistrifluoromethyl)benzoyl, 4-(trifluoromethyl)benzenesulfonyl, 4-fluorobenzenesulfonyl and 3,5-(trifluoromethyl)benzenesulfonyl. Enzyme-induced chemical shift differences were determined for many of these to probe rotational flexibility and/ or conformational constraints of the aryl rings. Met^{192} has been alkylated with N-(2-, 3-, and 4-trifluoromethyl)phenyl-2-haloacetamide. From observed indole-dependent upfield chemical shift differences of the fluorine resonances, the conclusion was reached that alkyl groups attached to Met^{192} are normally near the indole binding site and are displaced to a solution-like environment in the presence of indole (for references to original articles and more details, see reviews by Gerig [1a, b]).

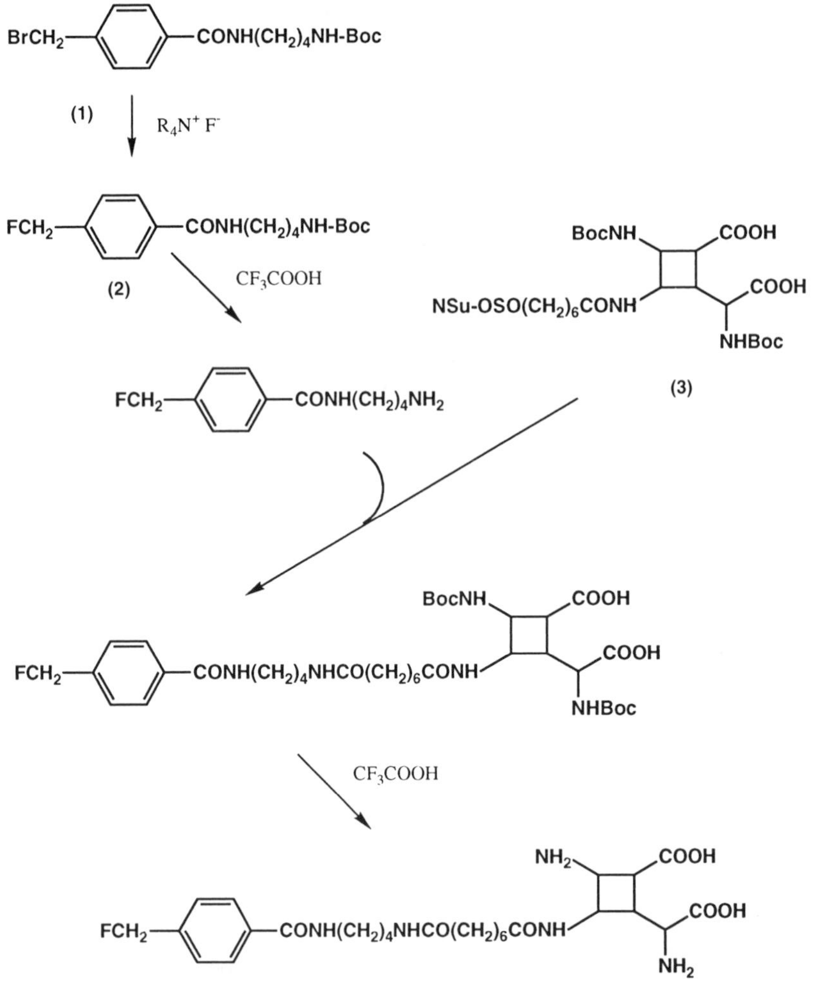

Figure 11.1 Schematic representation of the synthesis of a fluorinated analogue of insulin. The square represents the two peptide chains of insulin connected by the two disulfide bridges [12]

Fluorinated probes have been used effectively to study the hydrophobic binding domains of α-chymotrypsin. The enzyme is stoichiometrically inactivated by diphenylcarbamoyl chloride (**5a**) (Figure 11.2), presumably by reaction with Ser[195] at the active site. The fluorinated analogues (**5b–i**) also react stoichiometrically with the enzyme, and the fluorine-substituted diphenyl-carbamoyl-α-chymotrypsins so formed have been used effectively to probe the hydrophobic binding pocket of the enzyme. The ^{19}F NMR spectrum of *N,N-*

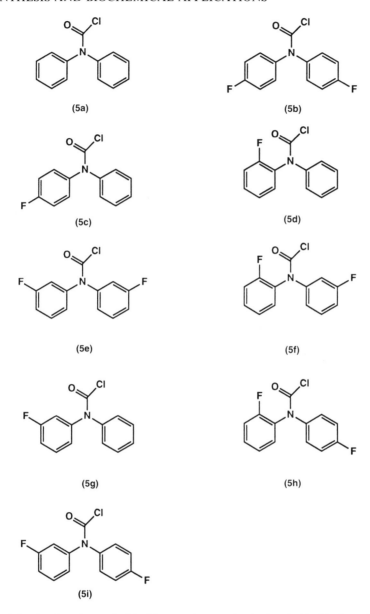

(5a)

(5b)

(5c)

(5d)

(5e)

(5f)

(5g)

(5h)

(5i)

Figure 11.2 Diaryl carbamoyl chlorides used to probe hydrophobic binding sites of chymotrypsin [17, and references cited therein]

bis(4-fluorophenyl)carbamoylchymotrypsin derived from the 4,4'-difluoro analogue **5b** with the enzyme shows two broad signals separated by 3.4 ppm at 25 °C. The appearance of two signals was attributed to restricted rotation about the carbonyl carbon–nitrogen bond, with the low-field signal assigned to the fluorophenyl group residing in the substrate specificity pocket [14]. The fluorophenyl ring of N-(4-fluorophenyl)-N-phenylcarbamoyl-α-chymotrypsin derived from (5c) also resides in two magnetically distinct environments. This protein was used to measure the competition of the fluorophenyl and phenyl rings for the hydrophobic pocket. ^{19}F NMR data indicated that the hydrophobic pocket more readily binds the fluorophenyl ring relative to the phenyl ring, a result consistent with the greater hydrophobicity of the fluorine-substituted aromatic ring [15]. N-(2-Fluorophenyl)-N-phenylcarbamoyl-α-chymotrypsin derived from 5d, asymmetric with regard to the C_1—C_4 axis of rotation, was used to study rotation about the N—C_{phenyl} bonds [16]. One of the fluoroaromatic groups of N,N-[bis-(2-fluorophenyl)]-α-chymotrypsin was shown to rotate rapidly while one rotated slowly at 25 °C. An extended family of fluorinated N,N-diphenylcarbamoyl-α-chymotrypsin derivatives (5e–i) were used to probe further aspects of the active site binding pockets. For example, chemical shift data suggest two binding sites for the aromatic rings of the diphenylcarbamoyl group, with the effect on chemical shift resulting from binding to either pocket dependent on the position of the fluorine. The extent to which a fluoroaromatic ring is found in one of the sites also is dependent on the position of the fluorine. Whereas a 3-fluoro substituent has little influence on the degree of occupancy, a 2-fluorophenyl ring essentially dominates the determination of site occupation [17].

11.2.5 PAPAIN

Papain is a cysteine protease isolated from papaya fruit. Alkylation of papain with 3-bromo-1,1,1-trifluoropropanone resulted in essentially complete loss of activity, along with the loss of titratable sulfhydryl groups. The relationship between the ^{19}F NMR chemical shift and pH below 5 gave a titration curve having a pK of 2.56. These data were used together with fluorescence data to assign a pK of about 4.5 to Asp[158] and 3.5 to His[159] [18].

11.3 PEPTIDES CONTAINING FLUORINATED ANALOGUES OF NATURAL AMINO ACIDS

11.3.1 SYNTHETIC OR SEMI-SYNTHETIC BIOLOGICALLY ACTIVE PEPTIDES CONTAINING FLUORINATED AMINO ACIDS

Rapid advances in protein synthesis methodology have led to the increasing use of total synthesis for the incorporation of amino acid analogues into peptides. Several examples of analogues of peptides containing fluorinated natural amino acids will be given in this section (Table 11.1).

Table 11.1 Synthetic or semisynthetic peptides containing fluorinated analogues of naturally occurring amino acids

Peptide	Analogue[a]	Activity	Method	Ref.
SRNase-S'	4-FPhe[8]	Full activity	Solid phase	19
	4-FHis[12]	Stable complex, no catalytic activity	Solid phase	20
Oxytocin	4-FPhe[2]	Full activity	Solution	22
Bradykinin	4-FPhe[4]	Increased activity (170%)	Merrifield, solution	23, 24
	4-FPhe[8]	Increased activity (150%)	Merrifield, solution	23, 24
	4-FPhe[4],4-FPhe[8]	Decreased activity (30%)	Merrifield	24
Physalaemin C-terminal hexapeptide	4-FPhe[3]	Full hypotensive activity	Solution	25
(Val[5])angiotensin II		Increased spasmogenic activity		
	4-FPhe[8]	Full activity	Solid phase	26
	4-FPhe[4]	Antagonist	Solid phase	26
	Hfv[5]	Increased (133%) agonist activity	Fragment condensation/solid phase	27
	Ac-Asn[1],Hfv[8]	Antagonist	Fragment condensation/solid phase	27
	Sar[1],Hfv[8]	Antagonist	Fragment condensation/solid phase	28
	Sar[1],D-Hfv[8]	Antagonist	Fragment condensation/solid phase	28
TRH	4-FHis[2]	Decreased receptor affinity; Comparable in vivo potency	Solution phase	29a,b
	(2-CF$_3$)His[2]	Decreased receptor affinity; Comparable in vivo potency	Solution phase; photochemical	29c,e / 6
	(4-CF$_3$)His[2]	Decreased receptor affinity; Comparable in vivo potency	Fragment condensation/solid phase	28
Dynorphin	[D-Ala[2],(F$_5$)Phe[4]]-Dynorphin$_{1-13}$-NH$_2$	Decreased (κ-selective) receptor affinity; increased in vivo activity; toxic	Solid phase	30.31
Leu[5]-enkephalin	2-FTyr[1]	No effect on activity	Solid phase	32
	3-FTyr[1]	No effect on activity		
	4-FPhe[1]	Reduced activity (μ and δ)		

(continued overleaf)

Table 11.1 (*continued*)

Peptide	Analogue[a]	Activity	Method	Ref.
Leu[5]-enkephalin	2-FPhe[4]	No effect on activity	Solid phase	32
	3-FPhe[4]	No effect on activity		
	4-FPhe[4]	Marked increase in potency (μ and δ)		
	(p-CF$_3$)Phe[4]	Decreased activity (μ and δ)	Solution phase	33
	[2R,4R]-trifluoro-Leu[5]	Decreased activity (μ and δ)		
	[2R,4S]-trifluoro-Leu[5]	Decreased μ and δ activity		
	[2R,4S]-trifluoro-Leu[5]	Decreased μ, increased δ activity		
	[2S,4S]-trifluoro-Leu[5]	Decreased μ and δ activity		
Tyr-D-Ala-Gly-Phe-D-Leu [(D-Ala[2],D-Leu[5])enkephalin]	2-FTyr[1]	Little effect on μ and δ activity	Solid phase	34
	3-FTyr[1]	Little effect on μ and δ activity		
	2-FPhe[4]	Little effect on μ and δ activity		
	3-FPhe[4]	Little effect on μ and δ activity		
	4-FPhe[4]	Little effect on μ and δ activity		
Tyr-D-Ala-Gly-Phe-MetNH$_2$[(D-Ala[2],Met[5])-enkephalin amide]]	4-FPhe[4]	15 times more potent	Solid phase	35a
	3-FTyr[1]	25–98% opiate activity; toxic	Solid phase	35b
	F$_5$Phe[4]	10 times more potent		36
Dermorphin analogues				
Tyr-D-Ala-Phe-Gly-NH$_2$ (*C*-terminal tetrapeptide from dermorphin)	3-FTyr[1]	Lower activity, comparable μ selectivity	Photochemical	5a
Tyr-D-Arg-Phe-Lys-NH$_2$	3-FTyr[1]	Lower activity, comparable μ selectivity	Photochemical	5b
	2-FPhe[3]	Small decrease in μ and δ affinities		
	3-FPhe[3]	Small decrease in μ and δ affinities		
	4-FPhe[3]	Marked decrease in μ and δ affinities		
Tyr-D-Arg-Phe-βAla-NH$_2$	2-FTyr[1]	Half μ affinity, comparable selectivity	Solid phase	38
	3-FTyr[1]	High μ affinity, μ selectivity		
	(3-FTyr[1],4-FPhe[3])	Marked decrease in μ affinity		

Peptide	Substitution	Property	Method	Ref.
Arg-Pro-Lys-Pro-Gln-Gln-Phe-Phe-Gly-Leu-Met-NH_2 (substance P)	3-FPhe^7	Full activity; (circular dichroism, ^{19}F NMR)	Solid phase	39
	3-FPhe^8	Full activity; (circular dichroism, ^{19}F NMR)		
	$(3\text{-FPhe}^7,3\text{-PFhe}^8)$	Full activity; (circular dichroism, ^{19}F NMR)		
Arg-Arg-Leu-Glu-Glu-Glu-Glu-Ala-Tyr-Gly (gastrin-related peptide)	3-FTyr^{10}	Lower K_m, (higher V_{max} (Tyr phosphorylation)	Solid phase	40
$N\alpha$-acetyl-$\text{TnI}_{104-115}$amide	4-FPhe^{106}	Interaction with TnC (^{19}F NMR)	Solid phase	41
N-Ac-*threo*-3-FAsn-Leu-Thr-$NHCH_3$	3-FAsn	Poor substrate for enzymatic glycosylation	Solution phase	49
N-Ac-*erythro*-3-FAsn-Leu-Thr-$NHCH_3$		Poor substrate for enzymatic glycosylation		
N-Benzoyl-*threo*-3-FAsn-Leu-Thr-$NHCH_3$		Poor substrate for enzymatic glycosylation		
N-Benzoyl-*erythro*-3-FAsn-Leu-Thr-$NHCH_3$		Poor substrate for enzymatic glycosylation		

11.3.1.1 Ribonuclease S

Semisynthetic ribonuclease-S' (SRNase-S') consists of a non-covalent complex made up of the synthetic polypeptide fragments (1–15) [corresponding to residues 1–15 of bovine ribonuclease A (RNase-A)] and RNase-S(21–124) (the native fragment of RNase-A containing residues 21–124). This semisynthetic complex is as active as native ribonuclease-S' (RNase-S'), which contains the NH_2-terminal eicosapeptide fragment of RNase-A instead of the synthetic pentadecapeptide [19].

The full activity of SRNase-S' has provided a convenient strategy for the investigation of the effects of variations in fragments 1–15 on ribonuclease structure and function. For example, using solid-phase synthesis, 4-FPhe and ^{13}C-enriched Phe were introduced as replacements for Phe^8 in synthetic (1–15) peptides [19]. After reconstitution and purification, the semisynthetic RNase-S' [(Phe^8-^{13}C) and (4-FPhe^8)] analogues were as active enzymatically as the normal complex. ^{13}C and ^{19}F NMR experiments gave evidence that conversions of the (Phe^8-^{13}C) and (4-FPhe^8)synthetic-(1–15) peptides to (Phe^8-^{13}C) and (4-FPhe^8)SRNase-S' analogues is accompanied by changes in environment at the aromatic ring of Phe^8. Likewise, NMR showed that addition of the competitive inhibitor 2'-cytidine monophosphate to the (Phe^8-^{13}C) and (4-FPhe^8)SRNase-S' analogues produced a change in environment at Phe^8.

The involvement of His^{12} in the catalytic process mediated by bovine pancreatic RNase has been well established. A synthetic pentadecapeptide was prepared having His^{12} replaced with 4-FHis [20]. (4-FHis^{12})synthetic-(1–15) peptide formed a stable non-covalent complex with native ribonuclease-S-(21–124) to give (4-FHis^{12})SRNase-S'. This complex, however, is devoid of catalytic activity, a result attributed to the drastically lowered imidazole pK_a (2.5) of 4-FHis now at the active site [21].

11.3.1.2 Oxytocin (Cys-Tyr-Ile-Asn-Cys-Pro-Leu-Gly-NH$_2$)

Oxytocin, elaborated by the posterior pituitary gland, functions as the principal hormone responsible for stimulation of lactation and uterus contraction. [4-FPhe^2]oxytocin (replacement of Tyr with 4-FPhe) was found to have uterotonic potency comparable to that of oxytocin. Similar ^{19}F NMR spectra of [4-FPhe^2]oxytocin and an S-substituted acyclic precursor suggested that the p-fluorophenyl side-chain of the fluoropeptide is freely exposed to solvent. The high biological activity of [4]FPhe^2]oxytocin makes it unlikely that hydrogen bonding of the Tyr hydroxyl of oxytocin is important in hormone–receptor interactions[22].

11.3.1.3 Bradykinin (Arg-Pro-Pro-Gly-Phe-Ser-Pro-Phe-Arg)

Replacement of Phe^5 with 4-FPhe produces an analogue [Arg-Pro-Pro-Gly-(4F)Phe-Ser-Pro-Phe-Arg] having 170% and 150% of the activity of bradykinin in bronchoconstriction and vasodilation assays, respectively [23]. Fisher *et al.* [24] confirmed these results and extended the series to include substitution of 4-FPhe at the 8-position [Arg-Pro-Pro-Gly-Phe-Ser-Pro-(4F)Phe-Arg] [(4-FPhe[8])bradykinin] and at both the 4- and 8-positions [Arg-Pro-Pro-Gly-(4F)Phe-Ser-Pro-(4-F)Phe-Arg] [(4-FPhe[4], 4-FPhe[8])bradykinin]. Whereas each of the singly substituted analogues was approximately 50% more active than native bradykinin, the disubstituted analogue had only 30% of the activity of bradykinin (rat uterus and guinea pig ileum).

11.3.1.4 Physalaemin (pGlu-Ala-Asp-Pro-Asn-Lys-Phe-Tyr-Gly-Leu-Met-NH₂)

Substitution of 4-FPhe for Tyr in the potent vasodilator and hypotensive *C*-terminal hexapeptide of physalaemin gives an analogue [Lys-Phe-(4F)Phe-Gly-Leu-Met-NH₂] with approximately 100%, 675% and 220% of the activity of physalaemin in hypotensive, large intestine and ileum assays, respectively [25].

11.3.1.5 Angiotensin II (Asp-Arg-Val-Tyr-Val-His-Pro-Phe) (AII)

Several fluorine-containing AII analogues have been made to investigate the effects of fluorine substitution on biological activity and metabolic stability and for ^{19}F NMR analysis. Whereas substitution of Phe[8] with 4-FPhe in AII produced an analogue [Asp-Arg-Val-Tyr-Val-His-Pro-(4F)Phe] essentially equipotent to AII, replacement of Tyr[4] with 4-FPhe [Asp-Arg-Val-(4F)Phe-Val-His-Pro-Phe] resulted in an antagonist of AII. ^{19}F NMR was used to study conformational properties of the fluorinated AII analogue in solution [26].

The natural (Val[5])AII agonist has Val in positions 3 and 5, while substitution of Val for Phe at position 8 of AII produces a potent AII antagonist. In order to increase the ^{19}F NMR sensitivity for conformational analyses of the hormone–receptor interactions, $\gamma, \gamma, \gamma, \gamma', \gamma', \gamma'$-hexafluorovaline (Hfv) was incorporated into AII as a replacement for Val[5] to give Asp-Arg-Val-Tyr-Hfv-His-Pro-Phe [(Hfv[5])AII] and at Phe[8] of (Ac-Asn[1])AII to give Ac-Asn-Arg-Val-Tyr-Val-His-Pro-Hfv [(Ac-Asn[1],Hfv[8])AII] [27]. (Hfv[5])AII was shown to have 133% of the activity of AII with respect to contraction of rat uterus, while (Ac-Asn[1],Hfv[8])AII was a potent inhibitor of AII *in vivo* and *in vitro*. Moreover, the fluorinated derivates were resistant to proteolytic digestion by carboxypeptidase A, hog renal acylase and α-chymotrypsin, an observation of considerable significance with respect to the design of orally effective drugs. Incorporation of Hfv into the 8-position of sarcosine[1] (Sar[1])-containing AII antagonists (solution phase coupling of Boc-Sar-Arg-Val-Tyr-Val-His-Pro with racemic Hfv benzyl ester followed by resolution) gave Sar-Arg-Val-Tyr-Val-His-Pro-Hfv (Sar[1],Hfv[8])AII and (Sar[1],D-Hfv[8])AII. Both (Sar[1],Hfv[8])AII and (Sar[1],D-Hfv[8])AII were significantly more potent *in vivo* than

(Sar[1],leu[8])AII as AII inhibitors, blocking the pressor activity of AII for more than 1 h. (Sar[1],Hfv[8])AII appears to be one of the most effective AII antagonists reported [28].

11.3.1.6 Thyrotropin-releasing factor (TRH) (pGlu-His-Pro-NH$_2$)

TRH is one of several hormonal-releasing factors elaborated by the hypothalamus. The principal biological role of TRH initially had been thought to be confined to stimulation and release of thyrotropin (TSH) and prolactin from the pituitary gland. More recent research has shown that TRH regulates a wide variety of biological functions, including activation of central noradrenergic neurons and related cardiovascular stimulation, actions unrelated to pituitary function. These diverse functions suggest the existence of multiple TRH receptors. To investigate structural requirements for biological activity, several TRH analogues have been made in which His has been replaced by a ring-substituted His. 4-FHis and 2-trifluoromethyl- and 4-trifluoromethyl[L-histidine (2- and 4-CF$_3$His) were incorporated into TRH using solution chemistry to give the analogues (4-FHis^2)TRH, (2-CF$_3$His^2)TRH and (4-CF$_3$His^2)TRH. (2-CF$_3$His^2)TRH and (4-CF$_3$His^2)TRH have also been prepared by direct photochemical trifluoromethylation [6]. The effects of all three analogues on blood pressure and heart rate in conscious rats were comparable [29]. In *in vitro* prolactin release assays, (2-FHis^2)TRH and (4-CF$_3$His^2)TRH showed no detectable displacement of [^{3}H]TRH from loaded rat pituitary cells [29a]. However, following central injection into the anterior hypothalamus, (4-FHis^2)TRH gave a twofold increase in plasma prolactin levels. Further, (2-CF$_3$His^2)TRH and (4-CF$_3$His^2)TRH gave a four- to fivefold increase in plasma prolactin levels following intra-arterial injection, an increase substantially higher than that seen following injection of TRH or (4-FHis^2)TRH. Receptors outside the pituitary and/or indirect action of these analogues may be responsible for these unexpected results.

11.3.1.7 Opioid peptides

Dynorphine A (Tyr-Gly-Gly-Phe-Leu-Arg-Arg-Ile-Arg-Pro-Lys-Leu-Lys-Trp-Asp-Asn-Gln)

This extremely potent neuropeptide contains Leu^5-enkephalin as the NH$_2$-terminal sequence. Intraventricular administration of dynorphin, in addition to enkephalins, produces little opiate effects, probably, at least in part, because these peptides undergo rapid enzymatic cleavage upon exposure to brain tissue. The fluorinated analogue [D-Ala^2,(F$_5$)Phe^4]-dynorphine$_{1-13}$-NH$_2$ retains the pharmacology of dynorphin, but exhibits marked *in vivo* potency [30]. Lethal effects observed in mice injected peripherally with the fluorinated peptide may be the consequence of actions on non-opiate systems [31].

Leu^5-enkephalin (Tyr-Gly-Gly-Phe-Leu)

Maeda *et al.* [32] used solid-phase synthesis to prepare Leu-enkephalin analogues containing fluorinated aromatic amino acids {[2-FTyr^1]-, [3-FTyr^1]-, [4-FPhe^1]-, [2-FPhe^4]-, [3-FPhe^4]-, [4-FPhe^4]- and [(4-CF$_3$)Phe^4]-Leu^5-enkephalin}. Fluorine in the *para*-position of Phe^4 potentiated μ and δ activity (inhibition of electrically induced contractions of guinea pig ilium and mouse vas deferens, respectively), whereas replacement of the *p*-OH of Tyr^1 with fluorine markedly reduced potency in both systems. *para*-Substitution of CF$_3$ on Phe^4 also reduced potency. Fluorine in the other positions of Tyr^1 or Phe^4 had little effect on activities relative to leu-enkephalin. Fluorine substitution also had no effect on *in vitro* enzymatic hydrolysis.

The enkephalin analogues [2R,4R]-, [2R,4S]-, [2S,4R]- and [2S,4S]-trifluoro-Leu^5-enkephalin (6) (Figure 11.3) were prepared by solution-phase synthesis. The inhibitory effects of these analogues on electrically induced contractions of guinea pig ilium and mouse vas deferens were studied. With respect to both assays, all analogues showed dose-dependent responses that were weaker relative to leu^5-enkephalin, except for the 2S,4R-isomer, which was more potent in the mouse vas deferens assay [33].

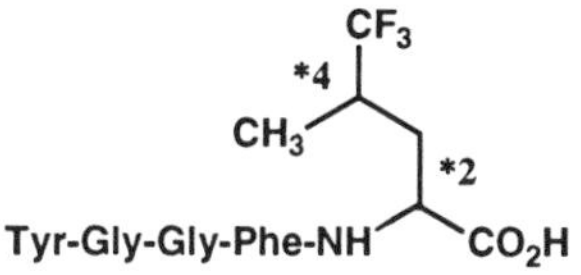

CF$_3$

*4

CH$_3$

*2

Tyr-Gly-Gly-Phe-NH   CO$_2$H

(6) (2R,4R)
    (2R,4S)
    .(2S,4S)
    (2S,4R)

Figure 11.3 Isomers of trifluoro-Leu^5 enkephalin

[D-Ala^2, D-Leu^5]enkephalin (Tyr-D-Ala-Gly-Phe-D-Leu)

Analogues of the δ-selective [D-Ala^2, D-Leu^5]enkephalin (DADLE) having 2-FPhe, 3-FPhe and 4-FPhe as replacements for Phe^4 and 2-FTyr and 3-FTyr as replacements for Tyr^1 were prepared by solid-phase synthesis [34]. Opioid receptor affinities were determined by competition for binding with receptor-specific radioligands to rat brain (μ and δ) and guinea pig brain (κ) membrane fractions. Affinity profiles of the five fluorinated analogues for μ- and δ-opioid receptors were comparable to the parent peptide and the affinity of all for κ-receptors was negligible.

[D-Ala^2]-methionine enkephalin amide (Tyr-D-Ala-Gly-Phe-Met-NH$_2$)

Tyr-D-Ala-Gly-Phe-MeMet-NH$_2$ (metkephamide) has shown promise as an analgesic agent in humans. As part of a structure–activity study, Gesellchen *et al.*

[35] prepared several analogues of metkephamide, including several containing fluorinated aromatic amino acids. For example, [4-FPhe[4], D-Ala[2], MeMet[5]]-enkephalin amide showed increased activity in mouse vas deferens and mouse hot-plate jump test [35a] whereas [3-FTyr[1], D-Ala[2], MeMet[5]]-enkephalin amide showed weaker activity in the same assays [35b]. 3-FTyr-containing analogues showed toxicity consistent with formation *in vivo* of toxic 3-FTyr.

In another study, [F$_5$-Phe[4], D-Ala[2], met[5]]-enkephalin amide was shown to be ten times more potent (central injection) and also longer lasting than the highly potent parent [D-Ala[2]]-methionine enkephalin amide [36].

Dermorphin (Tyr-D-Ala-Phe-Gly-Tyr-Pro-Ser-NH₂)

Dermorphin (Tyr-D-Ala-Phe-Gly-Tyr-Pro-Ser-NH$_2$)

Dermorphin is a μ-specific opioid peptide isolated from the frog skin of genus Phylomedusae. The *N*-terminal tetrapeptide amide (Tyr-D-Ala-Phe-Gly-NH$_2$) of dermorphin retains much of the activity of the parent heptapeptide. Structural modifications of this tetrapeptide have led to a number of potent and receptor-specific analogues [37]. This peptide was chosen as the initial substrate to explore the feasibility of direct fluorination of tyrosine-containing peptides with electrophilic fluorinating agents. Fluorination of Boc-Tyr(Bzl)-D-Ala-Phe-Gly-NH$_2$, Boc-Tyr-D-Ala-Phe-Gly-NH$_2$ or the completely un-protected Tyr-D-Ala-Phe-Gly-NH$_2$, with acetyl hypofluorite gave good yields of 3-FTyr-containing peptides (Figure 11.4) [5a]. Another dermorphin analogue, Tyr-D-Arg-Phe-Lys-NH$_2$, is one of the most μ-selective opioid agonists known. Direct fluorination of Tyr-D-Arg-Phe-Lys-NH$_2$, leaving the terminal amino group and the side-chains of Lys, Arg, and Tyr unprotected, gave a 50% yield of 3-FTyr-D-Arg-Phe-Lys-NH$_2$ [5b]. Whereas the affinities of the fluorinated peptides 3-FTyr-D-Arg-Phe-Lys-NH$_2$ and 3-FTyr-D-Ala-Phe-Lys-NH$_2$ are reduced two to ninefold relative to the non-fluorinated parents, their selectivity for μ-opioid receptors is not diminished.

Sasaki *et al.* [38] used solid-phase synthesis to prepare fluorinated analogues of Tyr-D-Arg-Phe-βAla-NH$_2$, a potent and extremely μ-selective opioid peptide. As in their work cited above, they used 2-FPhe, 3-FPhe and 4-FPhe as replacements for Phe[4] and 2-FTyr and 3-FTyr as replacements for Tyr[1]. In general, as in their previous work, incorporation of fluorinated aromatic amino acids had relatively little effect on receptor affinities. Among the FPhe[3] analogues, Tyr-D-Arg-4-FPhe-βAla-NH$_2$ showed a marked decrease in both μ- and δ-affinities, whereas Tyr-D-Arg-2-FPhe-βAla-NH$_2$ and Tyr-D-Arg-3-FPhe-βAla-NH$_2$ showed little change relative to the parent peptide. 3-FTyr-D-Arg-Phe-βAla-NH$_2$ retained full potency and μ-selectivity whereas 2-FTyr-D-Arg-Phe-β-Ala-NH$_2$ had a lower affinity, but retained μ-selectivity. A difluorinated analogue, 3-FTyr-D-Arg-4-FPhe-βAla-NH$_2$, had a greatly reduced μ-affinity. The potencies and selectivities of Tyr-D-Arg-Phe-βAla-NH$_2$ and the 2-FTyr- and 3-FTyr-containing analogues have added importance owing to the resistance of these peptides to enzymatic degradation.

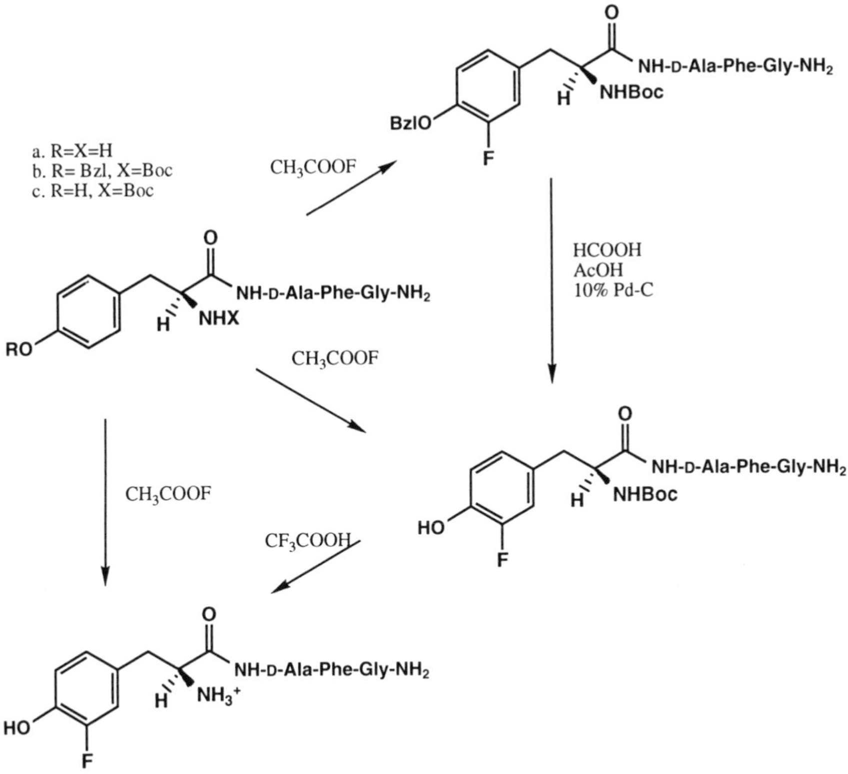

Figure 11.4 Electrophilic fluorination of dermorphin with acetyl hypofluorite [5a]

11.3.1.8 Substance P (Arg-Pro-Lys-Pro-Gln-Gln-Phe-Phe-Gly-Leu-Met-NH₂)

Using the Merrifield solid-phase method., Tanaka *et al.* [39] prepared three analogues of the neuropeptide substance P containing 3-FPhe as replacements for Phe[7], for Phe[8] and for both Phe[7] and Phe[8]. The three derivatives were equipotent with native substance P with respect to contraction of guinea pig ileum. Comparable circular dichroism spectra suggested that fluorine substitution had little effect on the solution structure of substance P. The ^{19}F NMR signals of the analogues were separated and assignable, indicating that the fluorinated substance P derivatives will be useful for studying interactions of substance P with its receptor.

11.3.1.9 3-FTyr[8] gastrin-related peptide (Arg-Arg-Leu-Glu-Glu-Glu-Glu-Glu-Ala-3-FTyr-Gly) as a substrate for tyrosine kinase

Tyrosine kinases play important roles in many aspects of cellular growth and transformation. As part of an investigation to determine the chemical features that are important in phosphorylation of the tyrosine residue, peptides derived from the sequence of gastrin were prepared and studied as substrates for insulin/tyrosin kinase. In the presence of insulin, the 3-FTyr-containing peptide (Arg-Arg-Leu-Glu-Glu-Glu-Glu-Glu-Ala-3-FTyr-Gly) was phosphorylated with approximately a three times lower K_m while the V_{max} for the reaction was more than eight times lower compared with the non-fluorinated peptide. Similar trends were seen during the less efficient phosphorylation in the absence of insulin. These data show that the substrate with the lower pK_a binds better but is phosphorylated less well. This suggests that the substrate becomes ionized on binding to the kinase, but that the more nucleophilic non-fluorinated phenolate anion is more reactive [40].

11.3.1.10 Tropinin I peptide

The regulatory proteins tropinin and tropomyosin are required for Ca^{2+}-mediated regulation of contraction of vertebrate skeletal muscle. Tropinin consists of three subunits, including tropinin C (TnC) that binds Ca^{2+}, tropinin I (TnI) that binds actin and inhibits actomyosin ATPase and tropinin T (TnT) that binds to tropomyosin. An important part of the activation of contraction of skeletal muscle is a Ca^{2+}-dependent interaction of TnC and TnI. Formation of a 1:1 complex between TnI and Ca^{2+}-saturated TnC blocks inhibition of ATPase, leading ultimately to muscle contraction. Several studies have shown that residues 104–115 make up the minimum sequence that is required for inhibition of ATPase, and binding of Ca^{2+} saturated TnC to this region relieves inhibition of actomyosin ATPase. Campbell et al. [41] prepared the synthetic peptide N_α-acetyl[4-FPhe^{106}]TnI(104–115)amide and studied its interaction with turkey skeletal TnC in the presence and absence of Ca^{2+} using ^{19}F NMR. ^{19}F NMR chemical shift differences were observed for the peptide bound to apo- and Ca^{2+}-saturated protein, indicative of different environments. Dissociation constants for each interaction were obtained.

11.3.1.11 Folyl-poly-γ-Glu

Among the several important biological functions of γ-glutamic acid (γ-Glu) is a role in folate-dependent one-carbon biochemistry. The predominant forms of intracellular folate coenzymes exist as pteroyl γ-glutamyl polypeptides. The formation of these γ-Glu peptides is catalyzed by folyl-poly-γ-glutamyl synthase, an enzyme with narrow substrate specificities, while the hydrolysis of pteroyl γ-glutamyl peptides is catalyzed by folyl-poly-γ-glutamyl hydrolase. With either tetrahydrofolic acid (H$_4$PteGlu) (7) or methotrexate (MTX) (8) (Figure 11.5) as substrate, folyl-poly-γ-glutamyl synthase catalyzes the incorporation of threo-4-FGlu into pteroyl γ-Glu peptides. With MTX, the principal product is 4-NH$_2$-10-CH$_3$PteGlu-threo-4-FGlu (9). threo-4-FGlu thus serves as a substrate and chain-terminating

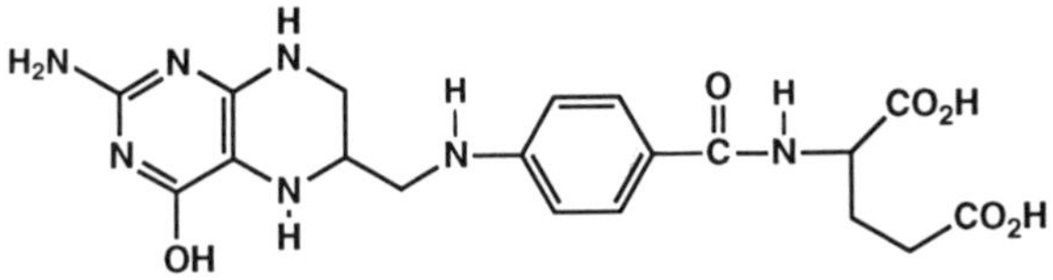

(7) H₄PteGlu

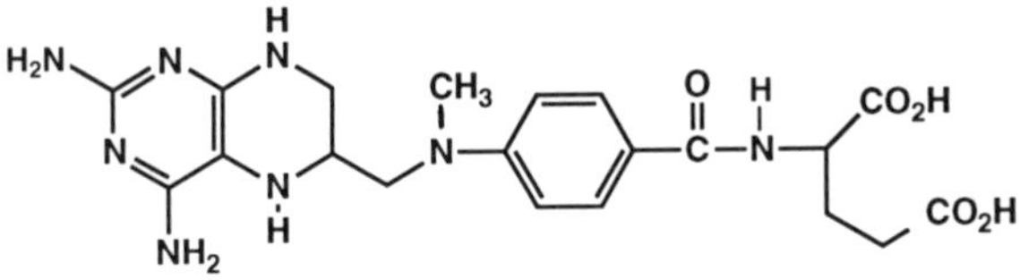

(8) MTX

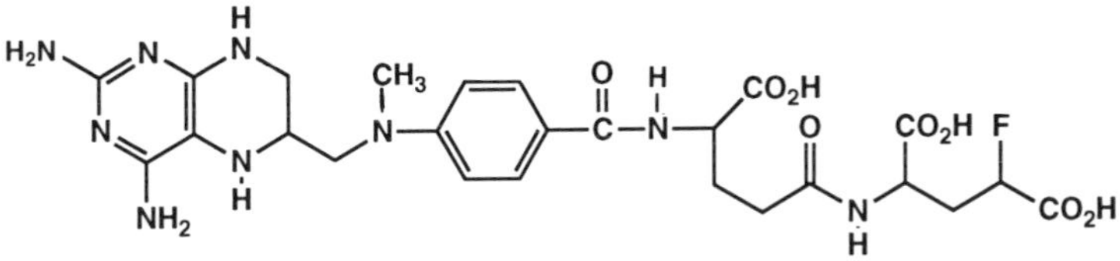

(9)

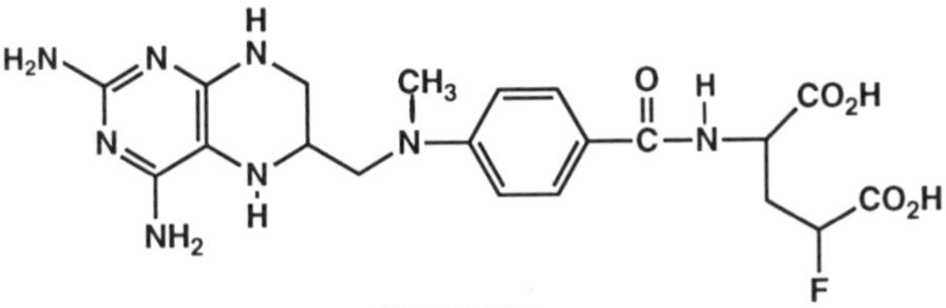

(10) FMTX

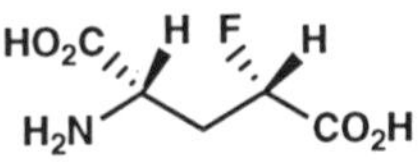

threo-4-F-Glu (2S isomer)

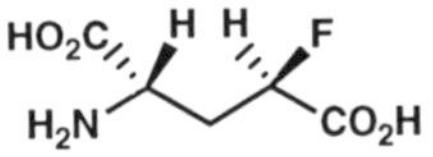

erythro-F-Glu (2S isomer)

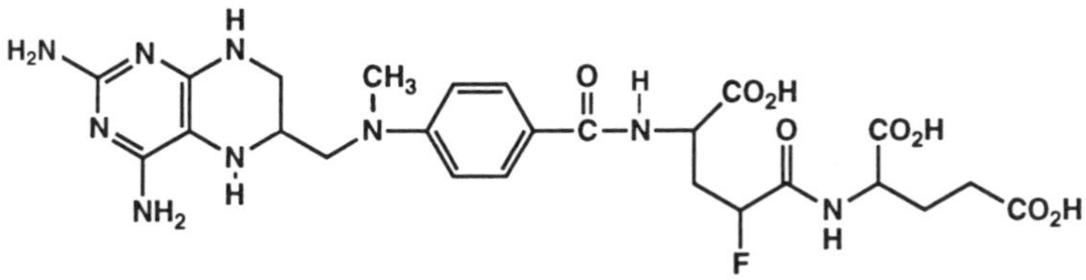

(11) FMTX-γ-Glu

Figure 11.5 Folates and antifolates containing fluorinated analogues of glutamic acid [46, and references cited therein]

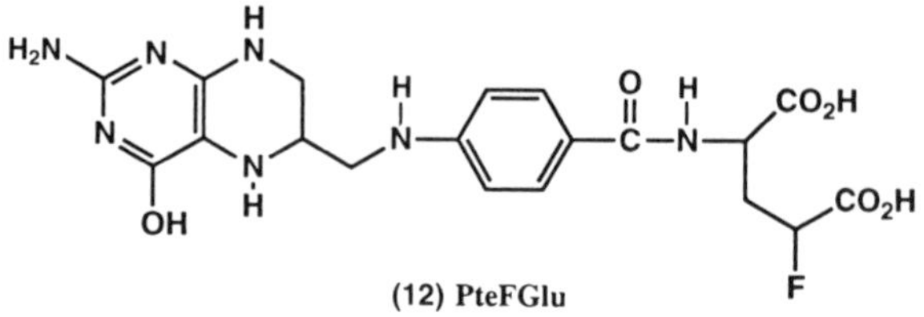

(12) PteFGlu

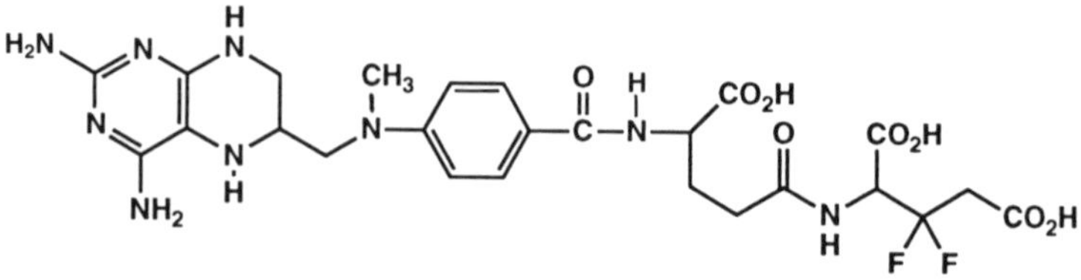

(13)

Figure 11.5 *(continued)*

inhibitor of folyl-poly-γ-Glu peptide synthesis [42]. However, this analogue had no effect on glutamylation of MTX by hepatic cells *in vitro* [43]. Toxic effects and failure to accumulate polyglutamated MTX appear to be related to effects on cell membranes.

These studies were extended to the synthesis of folates and anti-folates having fluorine at the γ-position of the glutamate moiety. γ-Fluoro-methotrexate (FMTX) (10) was prepared and found to be a potent inhibitor of dihydrofolate reductase, a key enzyme in one-carbon metabolism [44]. Of therapeutic relevance is the fact that FMTX is a poor substrate for folyl-poly-γ-glutamyl synthase, the enzyme responsible for the accumulation of highly retained, cytotoxic MTX polyglutamates. FMTX having the *erythro*- and *threo*-γ-Glu configuration (see Figure 11.5) appear to have similar biochemical properties [45]. *N*-[*N*-(4-Deoxy-4-amino-10-methylpteroyl)-4-fluoroglutamyl]-γ-glutamate (FMTX-γ-Glu) (11) was synthesized in the anticipation that this analogue would be a good substrate for folyl-poly-γ-glutamyl synthase and γ-glutamyl hydrolase. Whereas FMTX-γ-Glu was a good substrate for the synthetase, it unexpectedly was a poor substrate for the hydrolase. The use of fluoroglutamate-containing peptides as hydrolase-resistant folates or anti-folates was suggested by the above results.

Based on the results obtained with FMTX, the corresponding FGlu-containing analogue of folic acid, pteroyl(4-fluoro)glutamate (PteFGlu, 12), was prepared [46]. As expected, based on the behavior of FMTX-γ-Glu, both *erythro*- and *threo*-PteFGlu were extremely poor substrates for γ-glutamyl synthase. The reduced derivatives of 12 appear to function as substrates for several folate-dependent enzymes, including dihydrofolate reductase, thymidylate synthase and 10-formyltetrahydrofolate synthase.

3,3-Difluoroglutamic acid (3,3-DiFGlu) was examined as a substrate for folyl-poly-γ-Glu synthase. The absence of the second chiral centre was expected to simplify interpretation of the results and, with a pK_a comparable to that of 4-FGlu, 3,3-diFGlu was expected to have similar chain-terminating properties. In fact, with MTX as the pteroyl substrate, 3,3-diFGlu was found to be an excellent substrate for the synthase, and actually stimulates the addition of several more Glu or 3,3-diFGlu to the growing Glu-γ-diFGlu chain of the product 13 [46].

The behaviour of fluorinated glutamic acids and the folates and antifolates conjugated with fluorinated γ-Glu-containing peptides has provided information regarding the substrate specificities and mechanism of action of the enzymes responsible for maintaining the balance of conjugated and free tetrahydrofolate. In addition, the ability of fluorine substitution to render certain folates and antifolates resistant to the action of γ-glutamyl hydrolase, together with the potentially useful behavior of 3,3,-diFGlu as a stimulator of polyglutamylation, suggests that this line of research may produce new potent and selective chemotherapeutic agents.

11.3.1.12 N-Acetyl-*threo*-β-asparaginyl peptides as inhibitors of N-linked glycosylation

threo-β-Fluoroasparagine (*threo*-3-FAsn, 14) is highly toxic to certain mammalian cells, a fact that has therapeutic significance in that some lymphoma and leukemia cells require asparagine owing to a deficiency in asparagine synthase [47]. Asparagine-linked oligosaccharides are an essential structural unit of glycoproteins, and interference with protein glycosylation has been considered likely to contribute to the toxicity of *threo*-3-FAsn. An Asn-sensitive inhibition of protein glycosylation was demonstrated in a cell free translation system prepared from Krebs ascites II tumor cells, and this was attributed to replacement of Asn by *threo*-3-FAsn in protein biosynthesis [48]. At least two factors could contribute to the inability of *threo*-3-FAsn-containing protein to undergo glycosylation: abnormal secondary structures or, alternatively, substitution of fluorine on Asn residues could interfere directly with the catalytic process between the protein substrate and the oligosaccharyl transferase. To investigate this problem directly, short synthetic

(14)

peptides containing N-acetyl- or N-benzoyl-3-FAsn were prepared and studied as substrates for protein glycosylation in a cell-free system. N-Acetyl-Asn-Leu-Thr-NHCH$_3$ and N-benzoyl-Asn-Leu-Thr-NHCH$_3$ are excellent substrates for glycosylation, with the latter having greater activity. In contrast, N-acetyl-*threo*-3-FAsn-Leu-Thr-NHCH$_3$,   N-acetyl-*erythro*-3-FAsn-Leu-Thr-NHCH$_3$, N-benzoyl-*threo*-3-FAsn-Leu-Thr-NHCH$_3$ and N-benzoyl-*erythro*-3-FAsn-Leu-Thr-NHCH$_3$ all fail to undergo glycosylation, except at very high substrate concentrations. The N-benzoyl-*threo*-3-FAsn-containing peptide had a 100-fold lower V_{max}/K_m than the corresponding N-benzoyl-Asn-containing peptide, and *threo*- and *erythro*-3-FAsn-containing peptides were N-glycosylated at the same rate. These data give further evidence that the toxicity of *threo*-3-FAsn and the accompanying inhibition of N-glycosylation result from incorporation of the analogue into cellular protein. Since both *threo*- and *erythro*-3-FAsn-containing peptides show comparable weak binding to the oligosaccharyltransferase, the lack of toxicity of the *erythro*-3-FAsn apparently is caused by the failure of this isomer to be incorporated into protein, or by more rapid metabolic degradation of this isomer [49].

11.3.2 DRUG DELIVERY (TOXIC FLUORINATED AMINO ACIDS)

Clinical applications of amino acid analogues have been complicated by problems of transport and host toxicity. For example, trifluoroalanine is a potent inhibitor of alanine racemase, an essential enzyme for bacterial cell wall biosynthesis, but has little antibacterial activity, probably because of poor transport as a result of the low amine pK_a (5.8). Incorporation of antimicrobial amino acids into peptides has been used effectively to circumvent problems associated with poor transport, and also to decrease toxicity by providing a mechanism for targeted drug delivery [50].

3-FPhe is a substrate for phenylalanine hydroxylase. The product of this hydroxylation, 3-fluorotyrosine, is acutely toxic to animals as a result of its further metabolism to fluorocitrate, a potent inhibitor of mitochondrial tricarboxylic acid transport and a reversible inhibitor of aconitase [3b]. Despite this potential lethality, 3-FPhe is only weakly toxic to fungal cells. However, a series of di- and tripeptides containing 3-FPhe were effective inhibitors of the growth of *Candida albicans*. Evidence was obtained that indicated a peptide transport system delivered the 3-FPhe-containing peptides into *C. ablicans* where peptidases catalyzed the release of 3-FPhe. Ultimate toxicity may result from intracellular conversion of 3-FPhe to fluorocitrate, and/or biosynthesis of fraudulent and malfunctioning 3-FPhe-containing proteins. Peptides containing D-F-Phe were inactive. The active peptides included 3-FPhe-Ala, 3-FPhe-Ala-Ala, 3-FPhe-Met-Met and Ala-Ala-3-FPhe-Ala [51]. Against *Pythium ultimum*, the tripeptide 3-FPhe-Ala-Ala was much more fungitoxic than the dipeptide 3-FPhe-L-Ala or 3-FPhe [52].

11.4 BIOLOGICAL INCORPORATION OF FLUORINATED ANALOGUES OF NATURALLY OCCURRING AMINO ACIDS INTO PEPTIDES AND PROTEINS

11.4.1 INTRODUCTION

In 1959, Munier and Cohen [53] reported the incorporation of 2-FPhe and 3-FPhe into the protein of *Escherichia coli*. Two years later, Westhead and Boyer [54] published their work on the incorporation of 4-FPhe into rabbit enzymes and other proteins. Since these and other similar reports, the biochemical machineries of many cell types have been used to prepare a large number of fluorine-containing proteins for a variety of research purposes. Applications include the use of fluorinated amino acids to study protein biosynthesis, the use of fluorine-containing peptide products to study the effects of analogue incorporation on structure and function and the use of fluorinated proteins as mechanistic probes using ^{19}F NMR spectrometry. As this field has advanced, several strategies have been developed to enhance the degree of incorporation of amino acid analogues and, more recently, to define the precise position of the incorporated analogue. The use of auxotrophic bacteria, often combined with induced protein synthesis, has been extensively adopted. More recently, the powerful techniques of molecular biology have led to the insertion of amino acid analogues at precise locations into cloned proteins. Illustrative examples of biochemical procedures that have been used to incorporate fluorinated amino acids into bacterial protein will be given, along with brief discussions of the research purposes to which the fluorinated proteins have been applied. More comprehensive discussions can be found in reviews by Gerig [1a] and Sykes and Weiner [55]. Selected examples of the biological incorporation of fluorinated amino acids into mammalian protein will also be given, along with discussions of the research carried out with the fluorinated products.

11.4.2 INCORPORATION OF FLUORINATED AMINO ACIDS INTO BACTERIAL PROTEIN

The general inhibitory action of most fluorinated amino acids on bacterial cell growth was an obstacle to be overcome in order to achieve efficient incorporation into bacterial protein. In one important strategy to enhance incorporation of an amino acid analogue, auxotrophic bacteria (bacteria unable to synthesize a particular amino acid) are starved for that amino acid and subsequently fed an analogue that can be utilized in its place in protein synthesis. Although high levels of incorporation often can be realized, the overall protein yield often is low owing to decreased cell viability [55]. An additional strategy used to increase the degree of incorporation and overall yield is to induce the synthesis of the desired protein at the same time as the analogue is added. Early examples of this strategy are found in the preparation of several fluoroamino acid-containing enzymes from *E.*

coli, including alkaline phosphatase [56], β-galactocidase [57] and lactose permease [57]. These early results are discussed briefly below and these and additional examples are given in Table 11.2. More detailed reports can be found in the review by Sykes and Weiner [55].

11.4.2.1 (3-FTyr)(4-FTrp) alkaline phosphatase

In an example of enzyme induction in an auxotroph, Sykes *et al.* [56a] developed conditions for growing a Tyr auxotroph of *E. coli* such that the culture was depleted of Tyr and phosphate simultaneously. Addition 3-FTyr together with the low phosphate concentration led to the induction of alkaline phosphatase with a 10% introduction of 3-FTyr into the enzyme. After purification of the protein, ^{19}F NMR revealed fluorine resonances corresponding to 11 tyrosines in chemically different environments. Similar studies were carried out by Browne and Otvos [56b], who prepared 4-FTrp-containing alkaline phosphatase and reported a higher yield preparation of 3-FTyr-containing alkaline phosphatase.

11.4.2.2 (3-FTyr, 6-FTrp)dihydrofolate reductase from *Lactobacillus casei*

A strain of *L. casei* auxotrophic to Tyr, Trp and Phe grown in the presence of $1.0 \, \mu\text{gmL}^{-1}$ of methotrexate and optimum concentrations of 3-FTyr or 6-FTrp produced 3-FTyr- and 6-FTrp-containing dihydrofolate reductase. ^{19}F NMR spectra of the analogues displayed well resolved resonances for each of the five Tyr and five Trp present in the protein. Binding of methotrexate (inhibitor), dihydrofolate (substrate) or NADPH (cofactor) to the enzyme caused changes in chemical shifts, the most notable being a large downfield shift of the resonances of one 3-FTyr and one 6-FTrp on methotrexate binding. Folate binding gave a much smaller effect, indicating different binding modes for substrate and inhibitor [58].

11.4.2.3 (3-FTyr)aspartate transcarbamoylase

Aspartate transcarbamoylase catalyzes the first committed step in pyrimidine biosynthesis in *E. coli*. Using a similar strategy as in the above example, a Tyr auxotroph that overproduces this enzyme on uracil starvation was developed. Addition of 3-FTyr to uracil-starved cells resulted in an 85% incorporation of the analogue [59]. In spite of this high level of incorporation, enzyme activity was retained. ^{19}F NMR was used to study the 'communication' between the catalytic and regulatory subunits [60].

11.4.2.4 (3-FTyr)gene 5 protein

To control viral DNA replication in bacteriophages, phage-coded gene 5 protein forms a stoichiometric complex with single-stranded DNA from

Table 11.2 Examles of fluorine-containing proteins prepared by biosynthetic incorporation of fluorinated amino acids into bacterial protein

Protein	Source	Analogue	Applications	Ref.
Alkaline phosphatase	E. coli	3-FTyr 4-FTrp	NMR: identification of ^{19}F resonances	56a 56b
Dihydrofolate reductase	Lactobacillus casei	3-FTyr 6-FTrp	NMR: effects of substrates and inhibitors on ^{19}F-resonances; fluoroenzymes active, less stable	58
Aspartate transcarbamylase	E. coli 3-FTyr		Fluoroenzyme (85% replacement) fully active. ^{19}F NMR used to study communication between subunits	59, 60
Gene 5 protein	Phage-infected E. coli	3-FTyr	^{19}F NMR study of Tyr environments; fluoroprotein has normal DNA binding	62
Gene 8 protein	Phage-infected E. coli	3-FTyr	^{19}F NMR signals for two conformations used to study conformational equilibria	63, 64 65
Ribosomal proteins	B. subtilis	4-FTrp	NMR, fluorescence; auxotroph favors 4-FTrp over Trp	66 67
lac Repressor	Prophage E. coli	3-FTyr 5-FTrp	^{19}F NMR; point mutations used to assign positions of fluorinated residues	68 69 70
β-Galactosidase	E. coli	3-FTyr 4-FTrp 5-FTrp 6-FTrp	^{19}F NMR; 3-FTyr located at or near active site	71 57 57 57
D-Lactate dehydrogenase	E. coli	4-FTrp 5-FTrp 6-FTrp	Five Trp (^{19}F NMR) residues assigned by point mutations. F-Trp substituted for Phe, Tyr and Leu; full activity	72 73 74
Histidine-binding J protein	Salmonella typhimurium	5-FTrp	70% incorporation; fully active. ^{19}F NMR study of communication between J and P proteins	76 76

(continued overleaf)

Table 11.2 (*continued*)

Protein	Source	Analogue	Applications	Ref.
Chromatophoric proteins	*Rhodospirullum rubum* G[9]	3-FPhe 4-FPhe 6-FTrp 3-FTyr	NMR ddemonstrated incorporation; 3-FPhe immobilized in membrane	75
Glutamine-binding protein	*E. coli*	6-FTrp	No effect on glutamine binding	77
Rat cellular retinol-binding protein II	Induced in *E. coli*	6-FTrp	^{19}F NMR resonances for four Trp residues studied as function of ligand binding	78, 79
Rat cellular retinol-binding protein	Induced in *E. coli*	6-FTrp	^{19}F NMR study of transfer of ligands between CRBP and CRBP-II	80
Rat intestinal fatty acid-binding protein	Induced in *E. coli*	6-FTrp	^{19}F NMR of 6-FTrp[6]; 6-FTrp[82] used to study behavior during folding	81
Cyclic AMP receptor protein	*E. coli*	3-FPhe 3-FTyr 5-FTrp	^{19}F NMR study of conformational changes during binding of cAMP	82a 82b,c
D-Galactose chemosensory protein	*E. coli*	5-FTrp 3-FPhe	^{19}F NMR study of conformational changes during binding of sugar, Ca^{2+} and cleft opening	83–85
Phosphoglycerate kinase	*Saccharomyces cerevisiae*	5-FTrp	5-FTrp-containing enzyme detected by ^{19}F NMR in intact cell	86

daughter virions, preventing their functioning as templates. Bacteriophage-coded proteins have been labeled with fluorinated amino acids by simultaneous initiation of bacteriophage infection and addition of the analogue to the bacterial growth medium. An *E. coli* Tyr auxotroph was grown in a medium containing limiting Tyr and optimum concentrations of the other 19 natural amino acids until the early logarithmic phase. Addition of 3-FTyr followed 10 min later by bacteriophage fd led to efficient incorporation of the analogue. ^{19}F NMR of isolated gene 5 protein revealed the presence of three surface and two buried 3-FTyr residues [61]. This 3-FTyr-containing protein was used in subsequent NMR studies of gene 5 protein–oligonucleotide complexes [62].

11.4.2.5 (3-FTyr)bacteriophage M13 gene 8 coat protein

Gene 8 protein, the major coat protein of bacteriophage M13, is a 50 amino acid polypeptide of known sequence. The protein is easily prepared from intact phage, which can be recovered from the *E. coli* host growth medium. The two Tyr residues of this protein occur at positions 21 and 24, in the most hydrophobic region. 3-FTyr-containing M13 phage has been grown, and gene 8 protein having the analogue incorporated has been isolated and used for several ^{19}F NMR studies [63]. For example, NMR was used to demonstrate that the fluorotyrosyl residues are buried in the hydrophobic portion of the bilayer [64]. In experiments to study conformational equilibria, gene 8 protein having 76% replacement of Tyr was isolated and purified. Four ^{19}F resonances were observed, corresponding to two F-Tyr for each of two conformations of the protein. Using ^{19}F NMR, the equilibrium between the two conformations was studied as a function of pH, temperature and presence of detergents [65].

11.4.2.6 (4-FTrp)*Bacillus subtilis* ribosomal proteins: an example of an altered genetic code

The use of auxotrophic bacteria is one strategy used to overcome the bias that favors the incorporation of the natural amino acid over incorporation of a structural analogue, a bias that accompanies membership in the ensemble of amino acids recognized by the genetic code. Serial mutation of a Trp auxotroph of *B. subtilis* produced a mutant that was able to survive much more effectively on 4-FTrp than on Trp [66]. The change in fitness to the genetic code was attributed to the ability of the new strain to function better with proteins containing 4-FTrp than with normal Trp-containing proteins. This organism provides a convenient source of fluorinated analogues of proteins endogenous to, or clonable to, *B. subtilis,* including proteins that are vital to cell growth. This was illustrated by a study of *B. subtilis* ribosomal proteins using 4-FTrp-related decreases in Trp fluorescence [67].

11.4.2.7 (3-FTyr) and (5-FTrp)*lac* repressor

3-FTyr and 5-FTrp have been incorporated into the *lac* repressor, a macro-molecular component of the well studied lactose operon. In this case, induction and replication of a prophage in the cell line greatly increased production of labelled *lac* repressor protein. 3-FTyr initially was incorporated and ^{19}F NMR was used to study the number and environment of Tyr residues in the protein [68]. Chain-terminating (non-sense) mutations of the *lac* repressor gene at positions corresponding to the eight Tyr and two Trp residues in the *lac* repressor subsequently were used to study systematically the site of incorporation of the fluorinated analogues [69]. Using the fluorine-labeled analogues, binding of both inducers and anti-inducers to the *lac* operon were studied using ^{19}F NMR [70].

11.4.2.8 (4-, 5-, 6-FTrp)(3-FTyr)β-galactosidase

The biosynthesis of β-galactosidase by *E. coli* is controlled by the *lac* operon. Induction of the *lac* operon in the presence of 4-FTrp caused an induction of β-galactosidase to a specific activity of 60% of control (Trp). In the presence of 5- or 6-FTrp, induction of activity was 10% of control [57]. An 81.4% incorporation of 3-FTyr into β-galactosidase was achieved by growing a Tyr auxotroph in Tyr-limited medium followed by simultaneous induction of the *lac* operon and addition of 3-FTyr [71]. 3-FTyr-containing β-galactosidase had a phenolic pK_a about 1.5 pH units lower than the native enzyme. These and other data indicated that the enzyme has a Tyr residue located near or at the active site where the phenolic OH can function in general acid–base catalytic processes.

11.4.2.9 (4-, 5-, and 6-FTrp)D-lactate dehydrogenase

^{19}F NMR has been used in several studies of the structure and dynamics of membrane-bound proteins containing fluorinated amino acid residues [72]. For example, the membrane-bound *E. coli* enzyme D-lactate dehydrogenase (LDH) was labeled with 4-, 5- and 6-FTrp. To increase the yield of labeled enzyme, a DNA plasmid was used that, on temperature induction, produced 300 times more LDH than did wild-type cells. Five resolved ^{19}F NMR resonances were obtained with each analogue. These signals were assigned by generating mutant enzymes having each labeled amino acid replaced by a different residue [73].

The utility of ^{19}F NMR to probe the structure and function of D-lactate dehydrogenase was expanded by the introduction of ^{19}F signals (as 5-FTrp) at amino acid positions not corresponding to Trp [74]. Using oligonucleotide-directed site specific mutagenesis, 5-FTrp was substituted for nine Phe, Tyr and Leu residues in the enzyme with no loss of activity. These new signals were used to study sensitivity to substrate, accessibility to solvent and proximity to a lipid-bound spin-label in order to gain more information regarding the catalytic and cofactor-binding domains. Evidence was obtained for the presence of a membrane-binding third domain situated between the catalytic and cofactor binding domains.

11.4.2.10 (3-FPhe, 4-FPhe, 6-FTrp, 3-FTyr)chromatophoric proteins of *Rhodospirillum rubrum* G9$^+$

3-FPhe, 4-FPhe, 6-FTrp and 3-FTyr were incorporated into polypeptide chains of the chromatophore membranes of the photosynthetic bacterium *Rhodospirillum rubrum* [75]. The 3-FPhe-containing protein was purified and ^{19}F NMR showed the 3-FPhe to be immobilized within the membrane.

11.4.2.11 (5-FTrp)histidine-binding protein J (*Salmonella typhimurium*)

The high-affinity histidine transport of *S. typhimurium* requires communication between at least two proteins, the histidine-binding J protein and the membrane-bound P protein. Genetic evidence had shown that the histidine-binding site of the J protein is distinct from the site responsible for interaction with the P protein. In order to study histidine-induced conformational changes that could result in J–P interactions, a mutant strain of *S. typhimurium* that overproduces the J protein was used to achieve a 70% incorporation of 5-FTrp into the protein. The 5-FTrp-containing protein was fully active and was used to study the His-dependent communication between the J and P proteins [76].

11.4.2.12 (6-FTrp)glutamine-binding protein (*E. coli*)

Communication between a substrate recognition protein and membrane-bound components of the glutamine active transport system also was studied by ^{19}F NMR. As part of an investigation into the role of the two Trp residues (positions 32 and 220) in glutamine-binding protein (GlnBP), wild-type and mutant strains of *E. coli* were produced in which Trp residues were replaced by 6-FTrp. As part of the Gln transport system of *E. coli*, GlnBP binds Gln with high affinity. Association of GlnBP with membrane-bound components of the transport system is an integral part of the Gln active transport. ^{19}F NMR experiments show that both Trp residues have limited mobility and are located away from the surface of the protein. Replacement of Trp with 6-FTrp had no effect on binding of Gln or on association of GlnBP with membrane-bound proteins. However, replacement of Trp with Phe or Tyr significantly decreased ligand transport activity in comparison with wild-type, suggesting that either or both Trp residues are involved in ligand translocation [77].

11.4.2.13 (6-FTrp)rat cellular retinol-binding protein II (produced in E. coli)

Rat cellular retinol binding protein II (CRBP II) is a 15.6 kDa intestinal protein that binds all-*trans*-retinol and all-*trans*-retinal. Efficient (93%) incorporation of 6-FTrp into CRBP II was accomplished by inducing the expression of the protein in a Trp auxotroph of *E. coli* and growing the bacteria in the presence of the analogue. ^{19}F NMR resonances corresponding to two of the four Trp residues underwent large downfield shifts as a consequence of ligand binding. Resonances corresponding to the two other Trp residues underwent only minor shifts associated with binding [78]. To identify the positions of the Trp residues that undergo the large chemical shifts, four CRBP II mutants were constructed wherein Trp^9, Trp^{89}, Trp^{107} and Trp^{110} were replaced by other hydrophobic residues (Phe or Ile). Examination of effects of binding on the ^{19}F NMR spectra of each mutant in comparison with wild-type Trp CRBP II revealed that Trp^{107} experiences the largest change in chemical shift (2.0 ppm downfield), consistent with the predicted location of this residue at the binding site. Despite being remote from the binding site, Trp^9 also undergoes a 0.5 ppm downfield shift [79].

These studies were extended by using a similar strategy to incorporate 6-FTrp into the homologous protein, rat cellular binding protein (CRBP). Clearly distinguished ^{19}F NMR resonances associated with 6-FTrp-apoCRBP, 6-FTrp-CRBP-retinol (or retinal), 6-FTrp-apoCRBPII, and 6-FTrp-CRBPII-retinol (or retinal) were used to monitor the transfer of all-*trans*-retinol and all-*trans*-retinal between CRBP and CRBP II. The K_d for CRBP-retinol was about 110 times less than that for CRBP II-retinol, whereas retinal had comparable affinities for each protein [80].

11.4.2.14 (6-FTrp)rat intestinal fatty acid-binding protein (produced in E. coli)

^{19}F NMR spectrometry has been used to study the folding properties of intestinal fatty acid-binding protein (IFABP). IFABP is a member of a family of intracellular hydrophobic ligand-binding proteins. The protein consists of two orthogonal β-sheets of five strands each enclosing a large cavity that binds the fatty acid. Using techniques similar to those used in the synthesis of CRBPII described above [79], 6-FTrp-containing IFABP was produced [81]. The two 6-FTrp ^{19}F NMR resonances corresponding to the two Trp residues of IFABP were found to have different behaviors during denaturation experiments. Whereas the 6-FTrp^6 resonance was in slow chemical exchange between resonance frequencies for native and completely unfolded protein, the 6-FTrp^{82} resonance gave evidence for one or more intermediates, resembling the unfolded form, in equilibrium with native and unfolded protein.

11.4.2.15 (3-FTyr, 5-FTrp, 3-FPhe)cyclic AMP receptor protein from
E. coli

In bacteria, cylic AMP receptor protein (CRP) regulates several catabolite-sensitive genes by binding to specific DNA sites in the presence of cylic AMP. CRP is a dimer consisting of two identical subunits, each having a DNA- and cylic AMP-binding site. 3-FTyr, 5-FTrp and 3-FPhe-containing analogues of CRP were prepared using biosynthetic incorporation using a strain of *E. coli* auxotrophic for Phe, Trp and Tyr [82]. ^{19}F NMR studies with 3-FTyr- and 5-FTrp-containing CRP indicated that cyclic AMP induces conformational changes both in the binding domain and in the hinge region connecting the cyclic AMP binding domain and the DNA-binding region [82c].

11.4.2.16 (5-FTrp)(3-FPhe)D-galactose chemosensory receptor from
E. coli

The *E. coli* aqueous receptor for glucose and galactose possesses a hinge mechanism that regulates access to the substrate binding site. Common to many such proteins, the substrate site consists of a deep cleft between two protein domains connected by strands of polypeptides (the hinge). A simple model has been proposed wherein the hinge is flexible while the domains are essentially rigid. In the absence of substrate, the cleft exists in a stable open form. As substrate binds, the cleft closes, trapping the substrate inside. However, an alternative model would permit substrate-induced structural changes to be transmitted to other, remote regions of the protein. To explore these conformational characteristics of the *E. coli* galactose receptor, fluorine labels were introduced into widely distributed positions in the protein. 5-FTrp was incorporated (65% efficiency) into five Trp sites and 3-FPhe was incorporated (20% efficiency) into seven Phe sites, using Trp and Phe auxotrophs. Site-directed mutagenesis was used to assign ^{19}F resonances. Sugar binding caused chemical shift changes at 10 of the 12 labeled sites. These data indicate that substrate binding produces a global structural change in the protein [83]. The *E. coli* D-galactose receptor also possesses a Ca^{2+}-binding site. In contrast to the effects of sugar binding, Ca^{+2} binding affects the chemical shift of only two labeled sites, 5-FTrp^{127} and 5-FTrp^{183}, indicative of a local structural change [84]. Trp^{183} is located inside the binding cleft of the galactose receptor. 5-FTrp^{183} was used as a probe to detect opening and closing of the cleft. In the presence of saturating substrate, the cleft is closed and ^{19}F NMR confirmed that 5-FTrp^{183} was inaccessible to an aqueous paramagnetic probe. In the absence of substrate, the 5-FTrp^{183} resonance was broadened, indicating that the cleft is open, at least transiently, to solvent [85]. These and other data from this research illustrate the effective use of ^{19}F NMR to probe functionally significant conformational changes in proteins.

11.4.2.17 (5-FTrp)phosphoglycerate kinase in *Saccharomyces cerevisiae*

5-FTrp-labeled phosphoglycerate kinase (PGK) was synthesized in the yeast *S. cerevisiae* by inducing enzyme synthesis by a Trp auxotroph in the presence of 5-FTrp. To induce enzyme synthesis, the cells were transformed with a plasmid in which the PGK coding sequence was under the control of a galactose-inducible promoter. Incubation in the presence of 5-FTrp and galactose produced fluorine-labeled PGK, as detected by ^{19}F NMR of the intact cell. Two resonances corresponding to the two Trp residues were detected. Effects of metabolic activity of the cells on chemical shifts also were investigated [86].

11.4.3 INCORPORATION OF FLUORINATED AMINO ACIDS INTO MAMMALIAN PROTEIN

As the above brief discussions suggest, many microorganisms can be used effectively to prepare protein samples selectively labeled with fluorinated amino acids. Low levels of fluorinated amino acids have also been incorporated directly into mammalian protein, both *in vitro* and *in vivo*, for a variety of research purposes. (As discussed above, fluorinated mammalian proteins can also be prepared by incorporating the appropriate expression vectors in bacteria and inducing protein synthesis in the presence of the fluorinated amino acid analogue.) Competitive inhibition of uptake of an amino acid by its fluorinated analogue has been used as indirect evidence for incorporation, while detection of radioactivity in protein fractions due to incorporation of a radiolabeled fluorinated amino acid, or detection of the analogue following digestion of the protein, provide more direct evidence. In certain cases, fluoroamino acid-containing proteins have been purified, and ^{19}F NMR has been used to verify incorporation, and to assign the position of the analogue. Examples of these approaches will be discussed below, with emphasis given to studies wherein individual proteins have been isolated and studied.

11.4.3.1 Effects of incorporation of fluorinated aminoacids on cellular function

Several examples of alteration of cellular functions caused by incorporation of amino acid analogues, including fluorinated amino acids, have been published. For example, Wheatley and Henderson [87] found that entry of HeLa cells into mitosis is rapidly and almost completely inhibited in the presence of 4-FPhe. Sunkara *et al.* [88] found evidence that incorporation of 4-FPhe into proteins necessary for mitosis results in an inhibition of chromosome condensation. These results have provided a tool for studying cells in the G_2 (gap) phase, since a pre-synchronized cell culture can be held in this phase for several hours by treatment with 4-FPhe [89].

If an amino acid critical to protein function is replaced with an analogue, significant biological consequences may be expected. For example, histidine often has an important role in acid–base and nucleophilic processes required for the catalytic activity of many enzymes. Thus, as discussed above, replacement of His with 4-FHis was shown to abolish catalytic activity of semisynthetic ribonuclease S. In this regard, 2-FHis, but not 4-FHis, is incorporated into mammalian protein, in a cell free system, in cell cultures [90] and *in vivo* [91]. One consequence of this incorporation is the inhibition of induction of several enzymes as illustrated by the inhibition of pineal gland *N*-acetyltransferase activity, in cell culture and *in vivo*. Inhibition of enzyme activity was accompanied by a His- and cycloheximide-sensitive incorporation of 2-FHis into cellular protein, as verified by the isolation of [^{3}H]-2-FHis following pronase digestion. A possible mechanism for inhibition of *N*-acetyl-transferase is the formation of defective or inactive enzyme containing 2-FHis as a replacement for His [92].

Incorporation of fluorinated amino acids can affect the normal processing of proteins. For example, as discussed above, *threo*-3-FAsn inhibits protein glycosylation in a cell-free translation system. Replacement of Asn with *threo*-3-FAsn inhibits glycosylation of synthetic peptides, providing additional evidence that biosynthetic incorporation of *threo*-3-FAsn also is responsible for inhibition of glycosylation, and presumably plays a part in the toxicity of *threo*-3-FAsn [47].

Several Pro analogues have been incorporated into collagen, a protein that has an unusually high Gly, hydroxyproline and Pro content. Several Pro analogues, including *cis*-4-FPro, have been incorporated *in vitro* into collagen. The presence of the analogues in collagen appears to prevent normal triple helical conformations from forming, resulting in more rapid degradation [93].

In an early example of *in vivo* incorporation of a fluorinated amino acid into mammalian protein, Westhead and Boyer [54] reported that significant amounts of 4-FPhe could be incorporated into rabbit protein by maintaining the animal on a diet containing the analogue. In subsequent research, Gerig *et al.* [94] isolated and purified carbonmonoxyhemoglobin from rabbits maintained on a diet containing 0.3% 4-FPhe. Approximately 1 mol of 4-FPhe was incorporated for each mole of $\alpha_2\beta_2$ dimer. The ^{19}F NMR spectrum of the cyanomet form of the protein contained 16 fluorine resonances, one for each Phe residue present in α- and β-globins. Since the genetic techniques used to identify positions of fluorinated residues in bacterial protein were not applicable to these studies, several indirect methods were used to make tentative assignments of these resonances. In order to assign resonance to either the α- or β-chain, hybrid hemoglobins were prepared wherein only one of the two chains contained fluorine. In addition, the effects on NMR signals of complexation with oxygen, temperature and pH dependence on chemical shifts, spin–lattice relaxation times, ^{19}F{^{1}H} nuclear Overhauser effects and the effects of modification of β-93 sulfhydryl with a nitroxide spin label were used in making tentative assignments of most of the 16 resonances [94, 95]. Similar studies have been carried out with primate (chimpanzee) hemoglobin [96] and with rabbit carbonic anhydrase [97].

11.5 FLUORINE-CONTAINING PEPTIDES AS ACTIVE SITE-DIRECTED INHIBITORS OF HYDROLYTIC ENZYMES

11.5.1 INTRODUCTION

Proteolytic enzymes are important in the regulation of many biological processes. Examples include the renin–angiotensin system that plays a critical role in the control of blood pressure and the elastases, responsible for the hydrolytic degradation of elastin, a structural protein that otherwise is generally resistant to proteolysis. Imbalances of extracellular elastase levels have been associated with pathogenesis of certain diseases, including cystic fibrosis and smoking-induced emphysema. Proteases also are required for normal post-translational processing of precursor proteins.

The therapeutic potential of protease inhibitors in such areas as hypertension, emphysema and HIV infection continues to spur research activity toward the development of potent and selective inhibitors. A strategy that has been employed extensively involves design of a peptide analogue that closely resembles the natural substrate, but which contains an electrophilic center as a replacement for, or in close proximity to, the scissile bond. This reactive center may be designed to react reversibly, or irreversibly, with a functional group (thiol, hydroxyl, carboxylic acid) responsible for carrying out substrate bond scission. Inhibitors also are designed to mimic the tetrahedral intermediate that is formed during peptide bond cleavage.

11.5.2 FLUORO KETONE-CONTAINING PEPTIDES AS IRREVERSIBLE INHIBITORS OF SERINE PROTEASES

An early success using halomethyl ketones to alkylate enzymes was demonstrated by the work of Schoellmann and Shaw [98] in experiments to characterize the active site of chymotrypsin, a serine protease. Stoichiometric inactivation by the substrate analogue, chloromethyl (1-N-tosylamido-2-phenyl)ethyl ketone (TPCK, **15**) was accompanied by loss of one histidine residue. This provided the first direct evidence for the presence of histidine at the active site of chymotrypsin. TPCK was also shown to be an effective reagent for lysine residues [99]. A large number of peptidyl bromo- and chloromethyl ketones have been synthesized and used effectively to determine active site residues and binding requirements of serine and cysteine proteases. Shaw [100] has provided an account of early developments that emphasized

$$\text{(structure of TPCK)}$$

(15) TPCK

the potential of using the selectivity of enzyme binding sites in the design of reagents for selective modifications of enzymes.

Whereas peptidyl chloromethyl ketones have proved to be extremely useful pharmacological tools, their potential therapeutic value is compromised by the high reactivity of chloromethyl ketones resulting in non-specific alkylation of non-target nucleophiles *in vivo*. These considerations have prompted interest in the development of fluoromethyl ketones as serine protease inhibitors. Fluoroalkyl compounds are much poorer alkylating agents than the corresponding chloroalkyl compounds, a fact that would be expected to decrease non-specific alkylation. While early progress to this end had been thwarted by difficulties in preparing peptidyl fluoromethyl ketones, recent synthetic procedures have made available for study several series of peptide analogues containing this moiety as a reactive group.

Imperiali and Abeles [101] prepared a series of peptide analogues incorporating the trifluoro-, difluoro- and fluoromethyl ketones (general structure **16**) using β-amino alcohols as the key building blocks (Figure 11.6). The β-amino alcohols were readily available by alkylation of nitroalkanes with an appropriately fluorinated aldehyde hydrate or hemiacetal followed by reduction. The P_1 residue is determined by the choice of nitromethane. Peptide coupling followed by oxidation produced the peptidyl fluoro ketones. Inhibitors were targeted to proteases of known specificity by replacing the amide of the scissile bond with a fluorinated ketone while retaining the determinants (P_1, P_2, etc.) of secondary binding interactions. A member of this series, AcLeu-*ambo*-PheCH$_2$F (**17**, R = CH$_2$F) functions as an irreversible inhibitor of chymotrypsin with a second-order rate constant of $1.7 \, \mathrm{l \, mol^{-1} \, s^{-1}}$, representing a rate of alkylation that is only 15 times lower than the rate of alkylation with the analogous chloromethyl ketone. Difluoromethyl and trifluoromethyl ketone analogues were found to be reversible inhibitors of chymotrypsin and elastase (see below) [102].

Another route to peptidyl fluoromethyl ketones developed by Shaw and co-workers [103] is shown in Figure 11.7. Cbz-Phe-CH$_2$F prepared by this route inactivates chymotrypsin and alkylates the active site histidine at about 2% of the rate of the corresponding chloromethyl ketone. This strategy was extended to the synthesis of lysylfluoromethanes. Ala-Phe-Lys-CH$_2$F was shown to be an active site-directed inhibitor of the serine proteases, plasmin and trypsin.

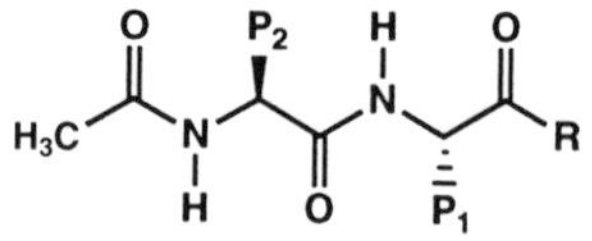

(16) R = CF₃, CHF₂, CH₂F)

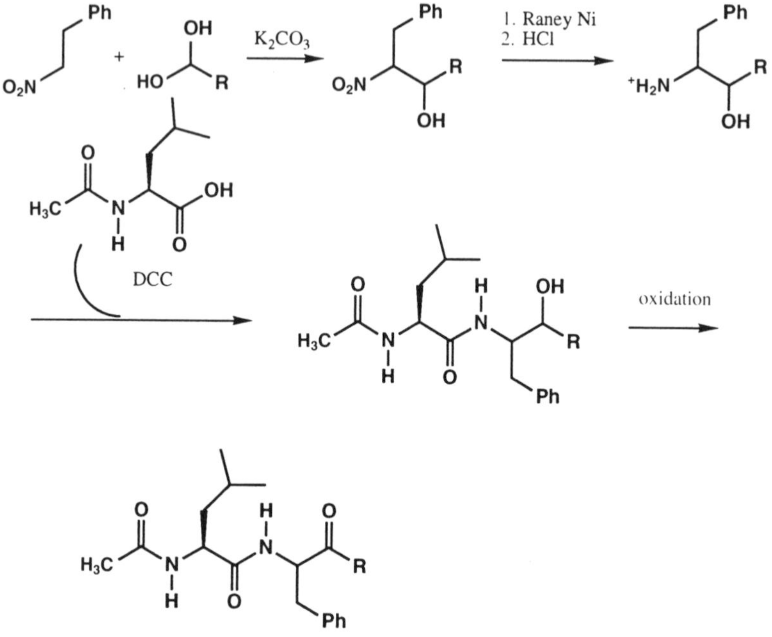

(17)

Figure 11.6 General route to fluoro-, difluoro- and trifluoromethyl ketone-containing peptide analogues [101]

The rate of alkylation was an order of magnitude lower compared with alkylation by the corresponding chloromethyl ketones. The fluoro derivative also was an extremely effective inhibitor of the cysteine protease cathepsin B (see below) [104].

11.5.3 N-TRIFLUOROACETYL PEPTIDES AS SERINE PROTEASE INHIBITORS

Reversible peptidyl serine protease inhibitors have been developed that contain N-trifluoroacetyl groups, trifluoromethyl ketones and, more recently, a com-

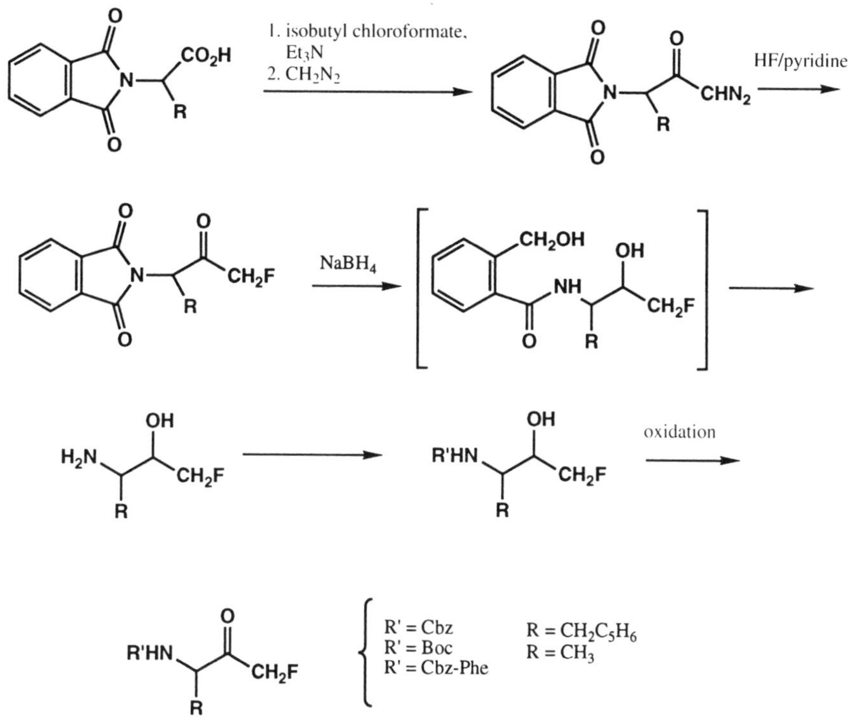

Figure 11.7 Synthesis of peptidyl fluoromethyl ketones [103]

bination of N-trifluoroacetyl and aryl trifluoromethyl groups [105]. N-Trifluoroacetyl (TFA) di- and tripeptides have been used as probes to study the active site and mechanism of action of elastase, and can function as inhibitors or substrates, depending on peptide length (Table 11.3). The trifluoroacetylated peptides were orders of magnitude better inhibitors than the corresponding acetylated peptides. The tightly binding reversible inhibitors act through a substrate–protease recognition mechanism and by hydrogen bonding in a crevice

Table 11.3 N-Trifluoroacetylated peptides as tightly binding reversible inhibitors of elastases [106b]

	K_i (mM)	
Inhibitor	Human leucocyte elastase	Porcine pancreatic elastase
CF$_3$CO-Ala-Ala	9	8.6
CF$_3$CO-Ala-Ala-Ala	0.17	0.0079
CF$_3$CO-Ala-Ala-Ala-Ala	0.1	0.032
CF$_3$CO-Gly-Leu-Tyr	0.57	0.052
CF$_3$CO-Val-Tyr-Val	0.5	0.00056

outside, or situated near, the active site. ^{19}F NMR studies of the inhibitor–elastase complex indicated that the trifluoroacetyl group is located at the S_4-binding site, while the free carboxyl terminus is near the active site [106].

Several irreversible inhibitors of elastases have been studied, including chloromethyl ketones. The efficient binding of the trifluoroacetyl peptides to porcine pancreatic elastase suggested that potent, irreversible inhibitors could result from incorporation of both chloromethyl ketone and N-trifluoroacetyl groups into the same peptide analogue. In addition to potential therapeutic use, such inhibitors would permit ^{19}F NMR studies of the interactions of these peptides with the enzyme. Accordingly, Renaud et al. [107] prepared a series of N-trifluoroacetyl di-, tri and tetrapeptidyl chloromethyl ketones (TFA-Ala-Ala-CH$_2$Cl, TFA-Ala-Ala-Ala-CH$_2$Cl, TFA-Ala-Ala-Ala-Ala-CH$_2$Cl) for study as pancreatic elastase inhibitors. Comparison of kinetic data obtained with this series with published data from the corresponding N-acetylated peptidyl chloromethyl ketones revealed that, whereas the latter were much better irreversible inhibitors, the N-trifluoroacetylated peptides had much higher affinity for the enzyme. The proposal was made that the TFA analogues can form non-productive complexes with the extended active center of elastase. ^{19}F NMR data provided supporting evidence for this hypothesis. Further, elastase that had been irreversibly inactivated by alkylation of Ser^{195} with phenylsulfonyl fluoride binds CF$_3$CO-Ala-containing anilides more tightly than native enzyme. NMR data indicated that binding of TFA peptides in the non-productive mode causes a conformational change in the enzyme that results in reduced catalytic activity [108].

Although human leukocyte elastase has a much lower affinity for the TFA peptides, these are again more tightly bound and less reactive than the corresponding acetylated peptides [109].

A more detailed description of the binding interactions of the TFA peptide inhibitors with porcine pancreatic elastase has been obtained through X-ray analyses. A series of dipeptides substituted at both ends by trifluoromethyl groups (CF$_3$CO-X-Ala-NH-C$_6$H$_4$-CF$_3$) have been shown to be potent reversible inhibitors of elastase activity (K_m in the range 10^{-5}–10^{-10} M) [108, 110]. Both ^{13}C and ^{19}F NMR indicate a 1 : 1 stoichiometry for the inhibitor + enzyme complex, with the inhibitor tightly bound to a single conformation state. A 1 : 1 complex of the inhibitor, CF$_3$CO-Leu-Ala-NH-C$_6$H$_4$-CF$_3$ ($K_m = 2.5 \times 10^{-8}$ M), was obtained for X-ray studies. Analysis of close contacts indicated the trifluoromethylamide bond, but not the Leu-Ala or Ala-anilide bonds, interacts with the active site.

11.5.4 FLUORO KETONE PEPTIDE ANALOGUES AS REVERSIBLE SERINE PROTEASE INHIBITORS

Carboxaldehyde-containing peptide analogues are effective inhibitors of serine and cysteine proteases. For example, papain is inhibited by Ac-Phe-Gly-CHO

($K_i = 5 \times 10^{-8}$ M) [111] and elastase is inhibited by Ac-Pro-Ala-Pro-Ala-CHO ($K_i = 8 \times 10^{-7}$ M) [112]. NMR data provide evidence that these analogues function as 'transition-state analogues' by forming tetrahedral complexes with active site nucleophiles [113]. Aldehydes are also inhibitors of metalloproteases. In this case, the hydrated form of the aldehyde appears to be the inhibitory species, functioning as a mimic of the tetrahedral intermediate that is formed during substrate bond scission [114]. In these analogues, a reactive center is provided by the electrophilic carbonyl group and selectivity is introduced by designing the aldehyde side-chain to resemble the peptide sequence of the natural substrate.

Gelb *et al.* [115] examined fluoro ketone analogues as inhibitors of a series of proteolytic enzymes, including acetylcholinesterase (a serine protease), carboxypeptidase A (a zinc metalloenzyme) and pepsin (an aspartyl protease) [115]. The series of acetyl choline analogues 18–20 clearly demonstrates the effectiveness of difluoro- and trifluoromethyl ketones as substrate analogues (Figure 11.8). Formation of a hemiacetal with an active site serine was assumed to be the mechanism of inhibition.

As noted above, difluoromethyl and trifluoromethyl ketone analogues prepared by the general β-amino alcohol route were found to be reversible inhibitors of chymotrypsin and elastase (Table 11.4) [102, 105]. These function as transition-state analogue inhibitors through the formation of a tetrahedral intermediate with the active site serine hydroxyl group, thereby mimicking the

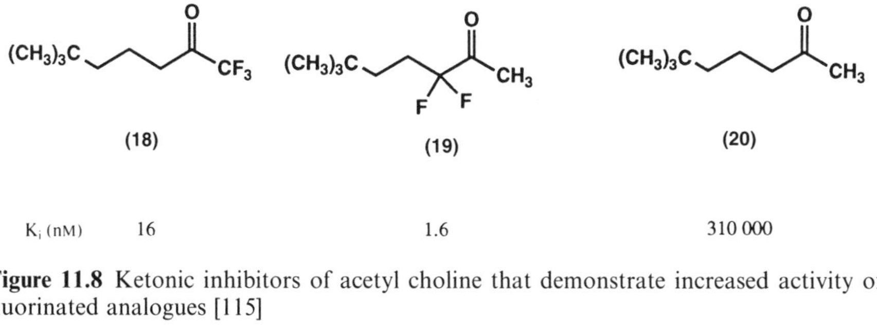

| K_i (nM) | 16 | 1.6 | 310 000 |

Figure 11.8 Ketonic inhibitors of acetyl choline that demonstrate increased activity of fluorinated analogues [115]

Table 11.4 Fluoro ketone-containing peptides as inhibitors of chymotrypsin and elastases [102, 105]

Protease	Inhibitor	K_i (μM)
Chymotrypsin	CH$_3$CO-*ambo*-PheCF$_3$	40
Chymotrypsin	CH$_3$CO-Leu-*ambo*-PheCF$_3$	1.2
Chymotrypsin	CH$_3$CO-Leu-*ambo*-PheCF$_2$H	25
Chymotrypsin	CH$_3$CO-Leu-*ambo*-PheCFH$_2$	200
PP elastase	CH$_3$CO-Pro-*ambo*-AlaCF$_3$	3000
PP elastase	CH$_3$CO-Ala-Ala-Pro-*ambo*-AlaCF$_3$	0.34
HL elastase	Z-Ala-Ala-Pro-*ambo*-Val-CF$_3$	0.001

transition state for amide bond cleavage. A potential advantage of the fluoro ketone inhibitors over peptidyl aldehydes would be greater *in vivo* stability, since the latter are readily oxidized under physiological conditions [105].

Using a synthetic strategy similar to that developed by Imperiali and Abeles [101, 102], Skiles and co-workers [116] prepared tripeptide trifluoromethyl ketones (general structure **21**) having non-naturally occurring N-substituted glycine residues, in place of Pro that is present in most HLE inhibitors, at the P_2 position (Figure 11.9) [116]. Several of these were potent and selective HLE inhibitors. The most active compounds spanned the P_5–P_1 subsites and contained a val-CF$_3$ residue at P_1. It is noteworthy that N-substitution of bulky lipophilic substituents on the P_2 residue had no adverse effect on activity. In contrast, the P_1 residue is very sensitive to modification, while deletion of amino acids at P_3 or at P_3 and P_2 also gave inactive compounds. The inhibitors were tested against all four classes of proteases (e.g. serine, cysteine, aspartate and metallo) and found to inhibit only HLE. The potent tripeptide **22** containing N-(2,3-dihydro-1H-inden-2-yl)glycine at the P_2 position was shown to halt the progression of HLE-induced emphysema-like lesions in the hamster [117].

The Merrell Dow research group has used a modified Dakin–West reaction to prepare N-benzoylated α-amino trifluoromethyl ketones (**23**) as building blocks for peptidyl trifluoromethyl ketones (Figure 11.10) [118]. Following N-acylation of the derived amino alcohols **24**, a series of peptide analogues were elaborated. This is illustrated in Figure 11.10 for the synthesis of N^α-(adamantylsulfonyl-N^ε-(MeO-Suc)Lys-Pro-Val-CF$_3$ (**25**), the most potent elastase inhibitor prepared in this study ($K_i = 0.58$ nM). In this study, a comparison was made between trifluoromethyl ketones and analogous peptidyl α-keto esters. In certain cases, the latter were more potent inhibitors.

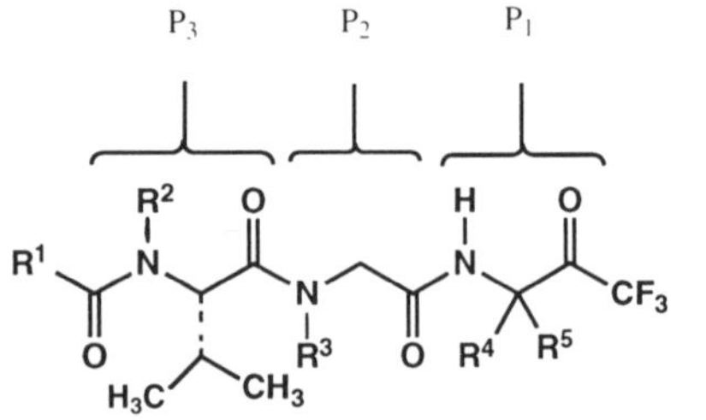

(21)

(22) $R_1 = p$-[p-Cl(C$_6$H$_4$)SO$_2$NHCO](C$_6$H$_4$)-, R^2 = H, R^3 = ⟨indan-2-yl⟩ — , R^4 = H, R^5 = CH(CH$_3$)$_2$

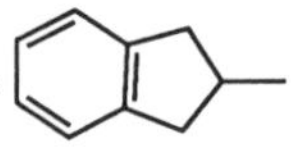

Figure 11.9 Potent elastase inhibitors. General structure (**21**) of tripeptidyl trifluoromethyl ketones having non-naturally occurring N-substituted glycine residues and the structure of the potent Val-CF$_3$-containing analogue [116]

The fluoro ketones inhibitors of serine proteases discussed to this point occupy maximally S_1–S_4 binding subsites of the enzyme. Crystallographic studies with serine proteases bound to naturally occurring inhibitors have shown that the S_1'-site represents the limit of the binding cleft. The ability of an inhibitor to occupy additional binding subsites on the leaving group side of the peptide (subsites S_1'–S_3') could be expected to reduce the K_i significantly. The demonstrated effectiveness of difluoromethyl ketones as reversible serine protease inhibitors suggested that substituted difluoromethyl ketones also would be sufficiently electrophilic to function as reversible P' inhibitors, with the added advantage of allowing for introduction of P_1'–P_3' residues. Reversible chymotrypsin inhibitors

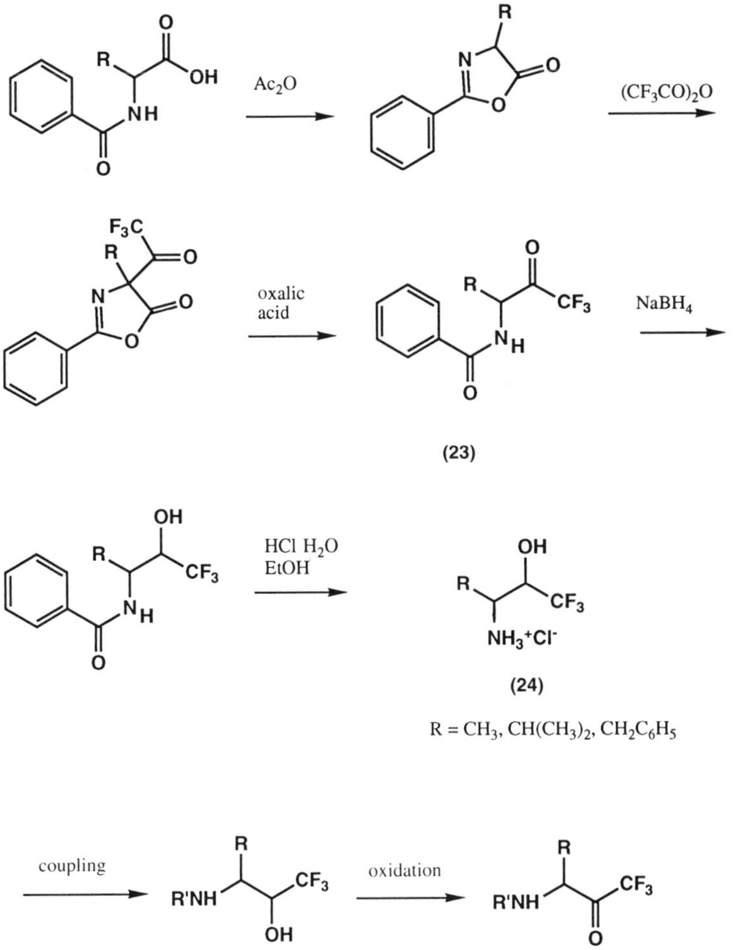

Figure 11.10 Synthesis of trifluoromethyl ketone elastase inhibitors based on Dakin–West preparation of amino ketones, illustrated by the synthesis of the potent analogue **25**

Figure 11.10 *(continued)*

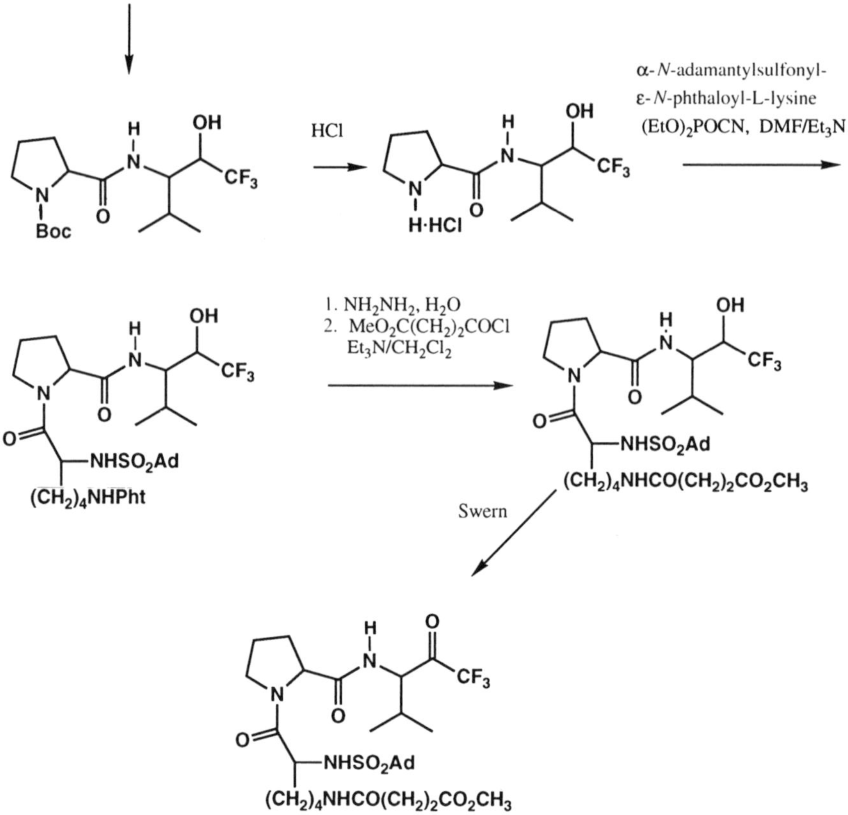

based on electrophilic internal difluoromethyl ketones were prepared according to the route shown in Figure 11.11, as illustrated for the synthesis of the potent inhibitor **26**. The synthetic strategy utilizes an initial aldol condensation between an α,α'-difluoroaldehyde and a substituted nitromethane, similar to the strategy used in the synthesis of the fluoromethyl ketone inhibitors [119]. The importance of S'-subsite interactions can be seen from a comparison of Ac-Leu-*ambo*-Phe-CF_2-CH_2CH_2-Leu-Val-OCH_2CH_3 (**27**), $K_i = 0.30\,\mu M$, having Val at P_3', and Ac-Leu-*ambo*-Phe-CF_2-CH_2CH_2-Leu-Arg-OCH_3 (**26**), $K_i = 0.009\,\mu M$, having Arg, a residue previously shown to favor highly interaction with S_3'.

Skiles *et al.* [120] used a similar rationale to design more potent HLE inhibitors based on previously prepared tripeptides having N-substitution at the P_2 position, as in **21** [120]. The structure was extended to take advantage of

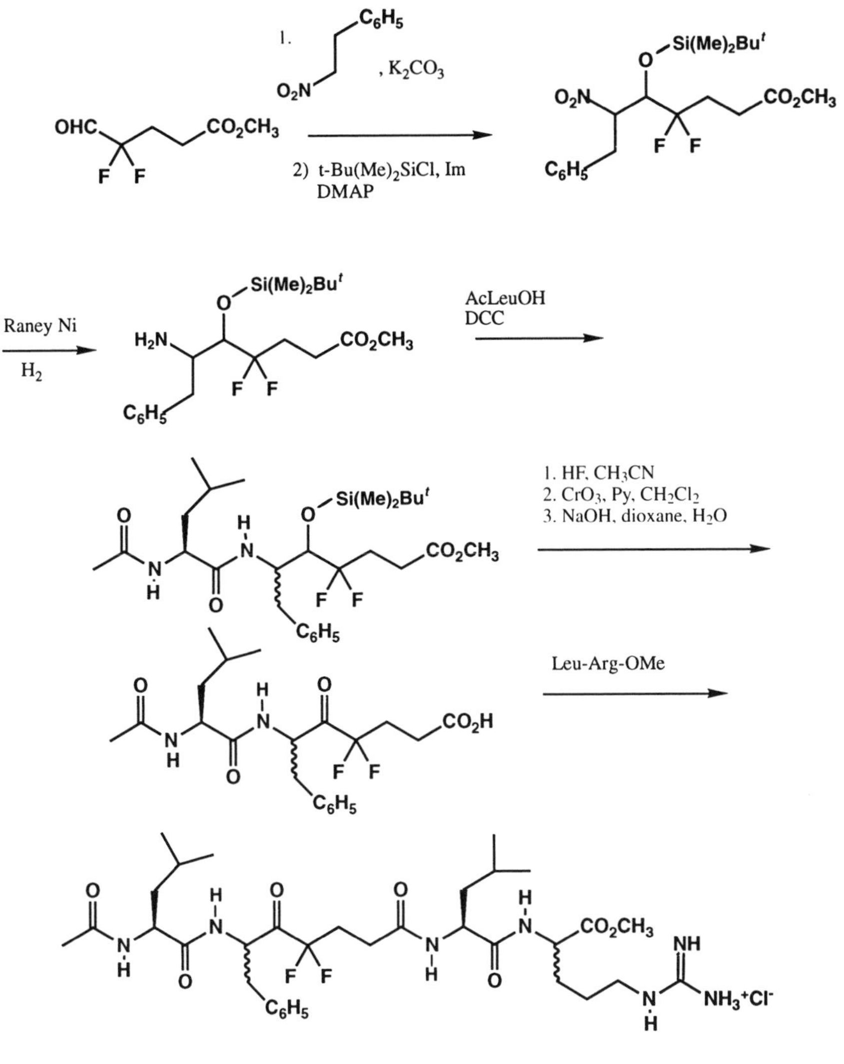

Figure 11.11 A series of elastase inhibitors capable of binding to the 'leaving group' side of the peptide were prepared by the route shown, illustrated for the potent member of the series **26** [119]

interactions on the 'leaving group' side of cleaved peptides by replacing one of the fluorine substituents with a side-chain designed to interact specifically with S' binding sites. In the preparation of a series of tripeptides containing the difluoro ketones, molecular modeling and X-ray crystallography were used to design a spacer group in order to optimize potential $P_3'-S_3'$ interactions. The synthetic approach is illustrated for the preparation of the particularly potent analogue **28** ($IC_5 = 0.057 \mu M$) containing an *N*-dihyroindenyl group at P$_1$ as a proline replacement (Figure 11.12). This peptide effectively blocked elastase-induced pulmonary hemorrhage in a dose-dependent manner. Hemiacetal formation between the α,α-difluoromethyl ketone and the active site Ser[195] has been proposed as a mechanism of inhibition.

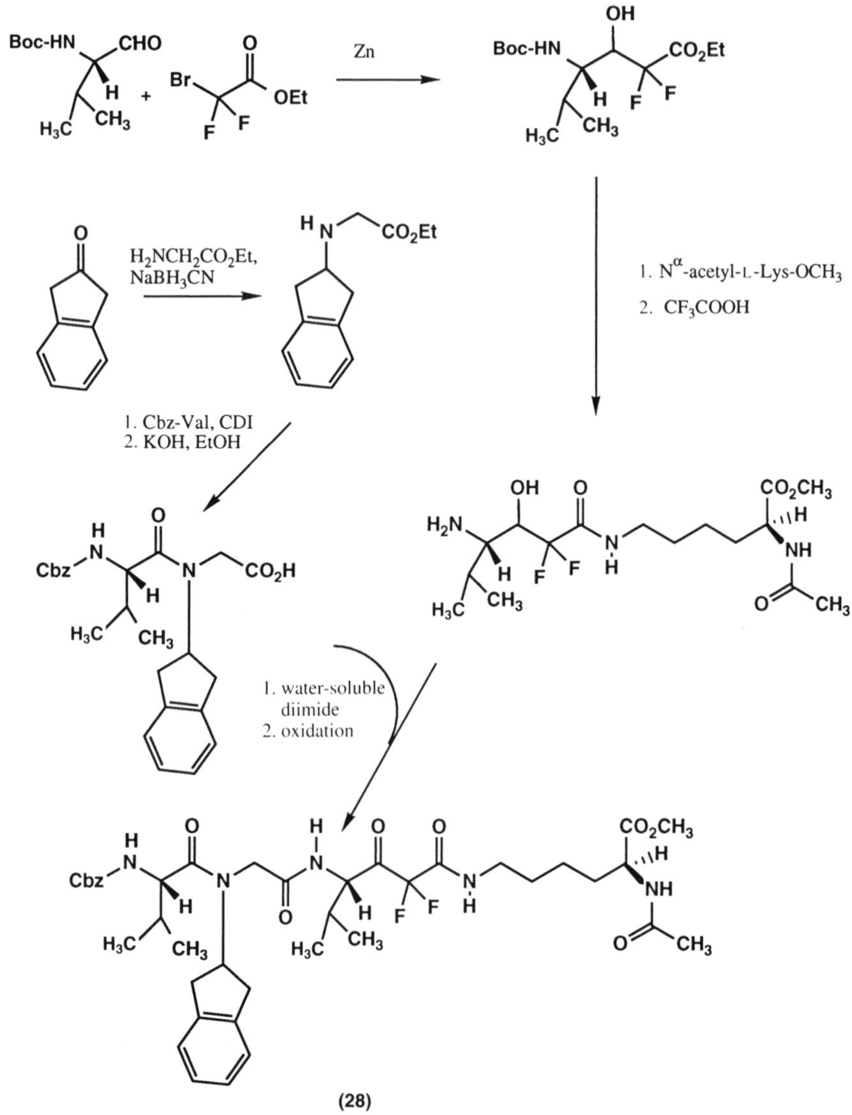

Figure 11.12 Example of the synthetic strategy used to prepare a series of difluoro ketone-containing peptides that function as potent HLE inhibitors [120]

11.5.5 INHIBITORS OF SULFHYDRYL PROTEASES

Peptide bond hydrolysis by serines and cysteine proteases both involve acyl-enzyme intermediates that are formed and hydrolyzed through tetrahedral intermediates. Thus, the design of inhibitors of cysteine proteases should overlap that for serine proteases. In fact, the first peptidyl fluoromethyl ketone synthesis was reported by Rasnik [121] and studied as an inhibitor of human cathepsin B. A modified Dakin–West synthesis was used to make CBz-Phe-AlaCH$_2$F. The fluoromethyl ketone was 30 times more potent than the corresponding diazomethane but less active than the chloromethyl ketone.

Cbz-Phe-Phe-CH$_2$F and Cbz-Phe-Ala-CH$_2$F, discussed above as serine protease inhibitors, also were investigated as inhibitors of cathepsin B [103]. The differences in reactivities between the fluoromethyl and chloromethyl ketones was less than that seen with serine proteases. Ala-Phe-Lys-CH$_2$F also was a very effective inactivator of cathepsin B [104].

11.5.6 INHIBITORS OF ASPARTYL PROTEASES

Renin is a highly specific aspartyl protease that cleaves the Leu–Val peptide bond of angiotensinogin to form angiotensin I, the biological precursor of the vasoconstricting peptide angiotensin II. There has been much recent interest in the development of inhibitors of renin as a strategy for the management of hypertension. The structure of pepstatin, a microbially produced inhibitor of pepsin, provided an important lead in this strategy (Figure 11.13). The structure of the novel hydroxy amino acid statine, present at a site corresponding to the scissile bond of renin, suggested that pepstatin could function by mimicking the tetrahedral intermediate formed during cleavage of the Leu–Val bond of renin. Accordingly, several potent renin inhibitors were made in which statine is present as a structural mimic of the Leu–Val sequence cleaved by the enzyme. Since the hydrated form of a ketone would more resemble a tetrahedral intermediate than would statine, the ketone analogue **29** of peptstatin was prepared [122]. Although NMR data suggested the enzyme-catalyzed formation of a hemiketal, the statone analogue **29** proved to be

Valeryl-Val-Val-Statyl-Ala-Statine

Pepstatin

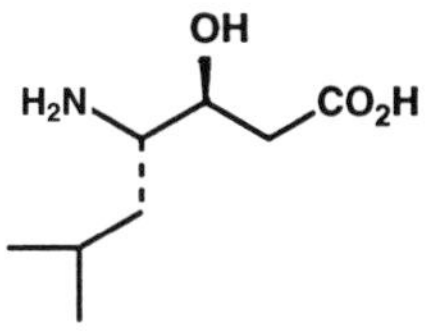

Statine

Figure 11.13 Structures of pepstatin, a naturally-occurring renin inhibitor, and statine, a critical component of pepstatin.

about 50 times less effective than the corresponding statine-containing peptide as a renin inhibitor [123]. In contrast, the corresponding difluorostatone (StoF$_2$)-containing analogue **30** was an extremely potent inhibitor (K_i = 0.06 nM) of renin, about 1000 times more potent than the Sto analogue. Rapid onset of inactivation implicated the hydrated form of the difluoro ketone as the inhibitory species [115].

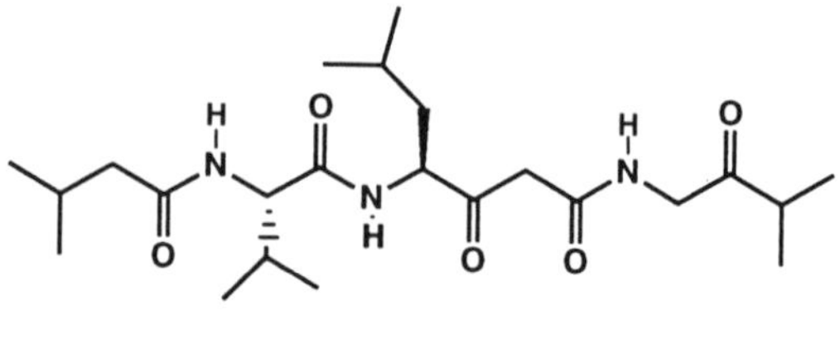

(29)

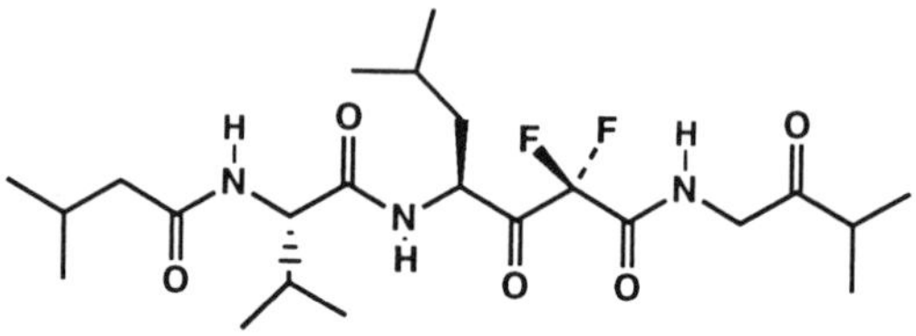

(30)

Fearon *et al.* [124] have made potent renin inhibitors by incorporating difluorostatone into a peptide sequence corresponding to the angiotensinogen sequence recognized by the enzyme. Comparison of the statine (**31**), statone (**32**) and difluorostatone (**33**) analogues reveal the effectiveness of the difluoroketone moiety as a transition state mimic (Figure 11.14) [124]. In a similar approach, Thaisrivongs *et al.* [125] demonstrated the greater inhibitory activity of the difluorostatone analogue **36** over the statine- (**34**) and statone-containing (**35**) analogues (Figure 11.15). Further, unlike peptstatin, **36** was highly specific for renin relative to pepsin, cathepsin D and angiotensin-converting enzyme [125]. More recently, Doherty *et al.* [126] prepared a series of peptides containing difluorocarbinol and difluoro ketone groups as an approach to potent, selective and orally active renin inhibitors. Compounds **37** (IC_{50} =1.20 nM) and **38** (IC_{50} =0.92 nM) showed best oral activity and good selectivity relative to cathepsin D [126].

Tightly binding fluorine-containing peptide inhibitors have been used effectively to study details of inhibitor and substrate binding. For example, isovaleryl-Val-Val-StaF$_2$NHCH$_3$ (**39**) and isovaleryl-Val-Val-StoF$_2$NHCH$_3$ (**40**) are potent inhibitors of penicillopepsin, a fungal aspartyl protease, having K_i values of 10×10^{-9} and 1×10^{-9} M, respectively [127], Each inhibitor was co-crystallized with the enzyme and binding interactions were

Boc-Phe-Phe-X-Leu-Phe-NH$_2$

Compound	X =	K_i (nM)

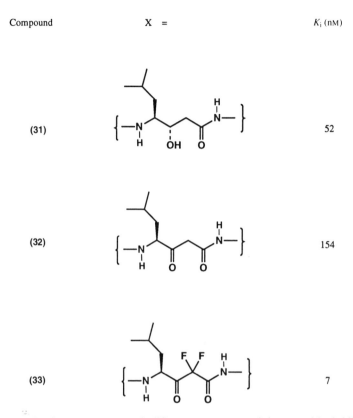

(31) 52

(32) 154

(33) 7

Figure 11.14 Statine-, statone- and difluorostatone-containing peptide inhibitors of renin [124]

analyzed by X-ray crystallography. The (R)-3-hydroxyl group in the StaF$_2$ analogue was found to make hydrogen bonding contacts with the active site carboxyl groups as Asp[33] and Asp[213]. The ketone of StoF$_2$ binds as the hydrate, with the *gem*-diol located between the two aspartic carboxyl groups. In this configuration, it resembles the tetrahedral intermediate expected during hydrolysis of a peptide bond. The additional OH group of the hydrated ketone appears to account for the tenfold greater binding of the StoF$_2$ inhibitor relative to the StaF$_2$ inhibitor. The stereochemistry of the StoF$_2$ group suggests that, during formation of the tetrahedral intermediate during peptide bond hydrolysis, Asp[213] functions as the general base, Asp[33] acts as the general acid, and the nucleophile is a water molecule bound between the two carboxyl groups. This is consistent with earlier proposals [128].

Boc-Phe-His-X-Ile-AMP

Compound   X = IC_{50} (nm)

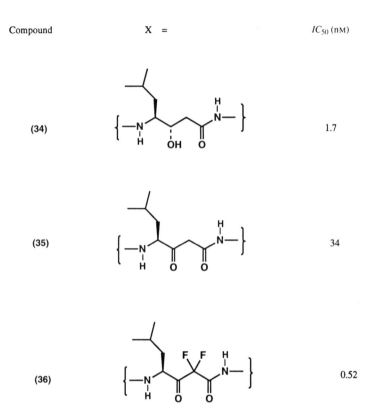

(34)   1.7

(35)   34

(36)   0.52

Figure 11.15 Example of the design of potent and selective renin inhibitors based on the incorporation of difluorostatone [125]

11.5.7 METALLO PROTEASES

The potential of renin and pepsin inhibitors as antihypertensive agents through blockade of angiotensin I production was discussed above. The conversion of angiotensin I to angiotensin II is another obvious target in the design of antihypertensives. Angiotensin-converting enzyme (ACE) is a zinc metalloprotease that cleaves the terminal His–Leu dipeptide from the biologically inactive angiotensin I to produce the powerful vasoconstrictor, angiotensin II. ACE also is responsible for inactivation of the vasodilator and natriuretic nonapeptide bradykinin. A snake venom pentapeptide, PGlu-Lys-Trp-Ala-Pro (BPP$_{5a}$), is a potent inhibitor. Enzyme-binding affinities increase in the order angiotensin I, bradykinin and BPP$_{5a}$, and many inhibitors have been designed based on the Ala–Pro and Gly–Pro dipeptide sequence. Trifluoromethyl ketone analogues incorporating Pro or a benzoproline have been

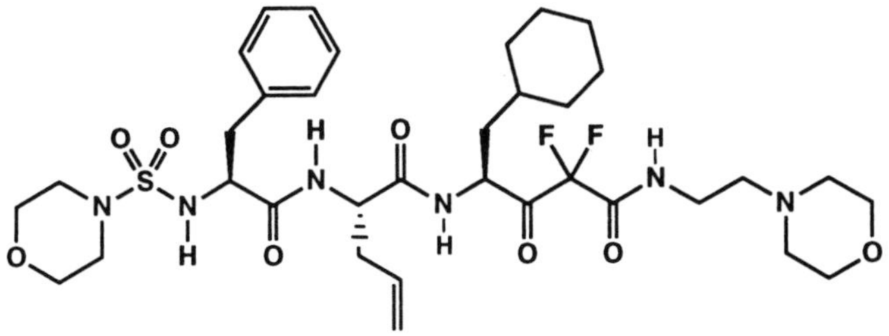

(37)

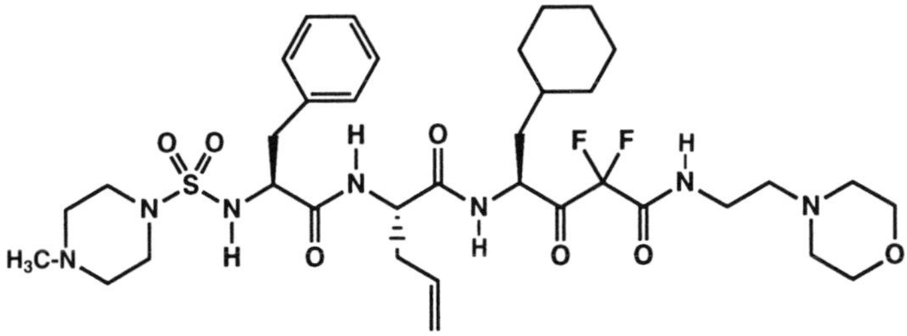

(38)

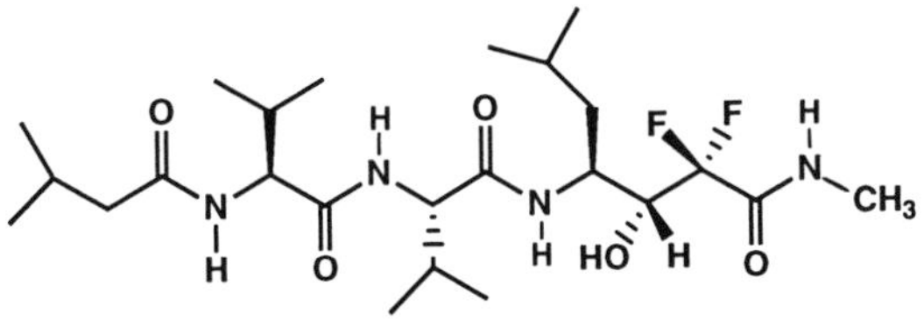

(39)

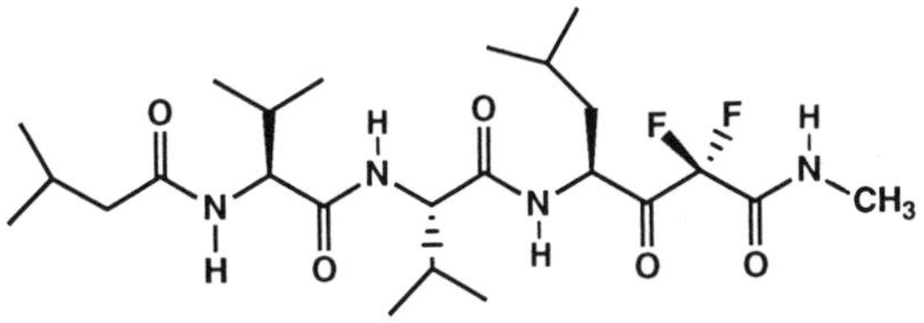

(40)

Table 11.5 Trifluoromethyl ketone analogues of Gly-Pro, Ala-Pro and Ala-benzoPro as inhibitors of angiotensin-converting enzyme [115]

Compound	K_i (μM)
$CH_3COCH_2CH_2CO$-Pro (42)	4500
$CF_3COCH_2CH_2CO$-Pro (41)	15
$CF_3COCH_2CH(CH_3)CO$-Pro	0.2
$CF_3COCH_2CH(CH_3)CO$-benzoPro (43)	0.012

shown to be potent inhibitors of ACE. The trifluoromethyl ketone analogue **41** is 300 times more potent than the methyl ketone **42**, illustrating the importance of fluorine substitution. The benzoproline analogue **43**, modeled after an Ala–Trp inhibitor, is the most potent of the series (Table 11.5) [115].

Captopril (**44**) is an early generation potent inhibitor of ACE. Ojima and Jameison [129] prepared (*R,S*)- and (*S,S*)-captopril-f$_3$ (**45a,b**) by the route shown in Figure 11.16 (*R,S*)-Captopril-f$_3$ (**45a**) ($IC_{50} = 2.9 \times 10^{-10}$ M) was markedly more potent than (*S,S*)-captopril ($IC_{50} = 3.6 \times 10^{-9}$ M). The improved potency was ascribed to increased lipophilicity and stereoelectronic effects.

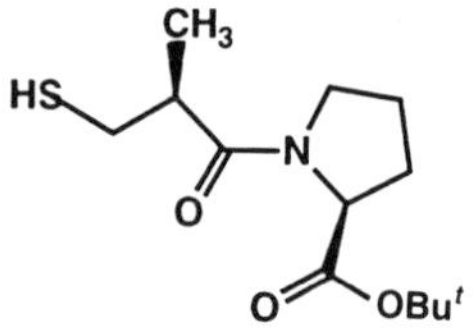

(44)

11.5.8 HIV PROTEASE

Viral encoded HIV protease cleaves a Phe–Pro amide linkage in precursor *gag* and *gag-pol* proteins to produce the mature proteins required for viral replication. Much recent research has been directed toward the development of potent and selective inhibitors of HIV protease as potential antiviral agents for the treatment of AIDS. Included in a variety of transition-state peptidomimetic inhibitors of HIV protease are the pseudo-symmetric fluorine-containing Phe–Pro cleavage site mimics **46** and **47**. The key intermediate, (2*S*,5*S*,4*R*)-2,5-diamino-3,3-difluoro-1,6-diphenylhydroxyhexane (**48**), was prepared from oxazolidinone **49** as shown in Figure 11.17. Coupling of the diamine **48** with Z-L-Val produced **46**, a potent inhibitor of HIV-1 protease ($K_i = 1.0$ nM). Oxidation of **46** to the difluoro ketone **47** gave an even more potent inhibitor ($K_i = 0.1$ nM) [130].

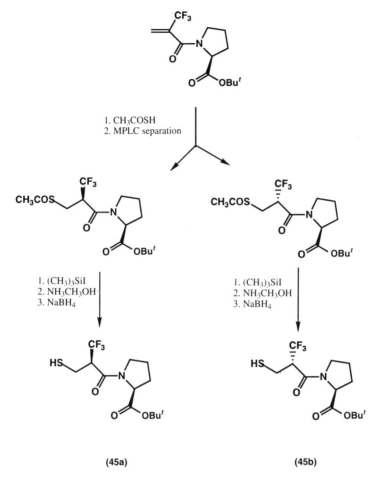

1. CH₃COSH
2. MPLC separation

1. (CH₃)₃SiI
2. NH₃CH₃OH
3. NaBH₄

1. (CH₃)₃SiI
2. NH₃CH₃OH
3. NaBH₄

(45a)   (45b)

Figure 11.16 Synthesis of (R,S)- and (S,S)-captopril-f₃ [129]

11.5.9 FLUORO ALKENE PEPTIDE ISOSTERES

A major problem with the use of peptides as therapeutic agents is their rapid degradation by proteases. Several non-hydrolyzable isosteres of amides have been developed in approaches to overcome this drawback. The use of α,α-difluoro ketones as isosteric replacements for tetrahedral intermediates formed by scissile amide bonds is an obvious important example of this strategy. Another approach has been the use of a *trans*-alkene **50** as a mimic of the most stable, transoid conformation of the amide bond (Figure 11.18). However, results of calculation of molecular profiles of *N*-methylacetamide, *trans*-but-1-ene and 2-fluoro-(*Z*)but-2-ene have suggested that the fluoroalkene **51** is a better amide bond mimic than the *trans*-alkene. Not only are steric features

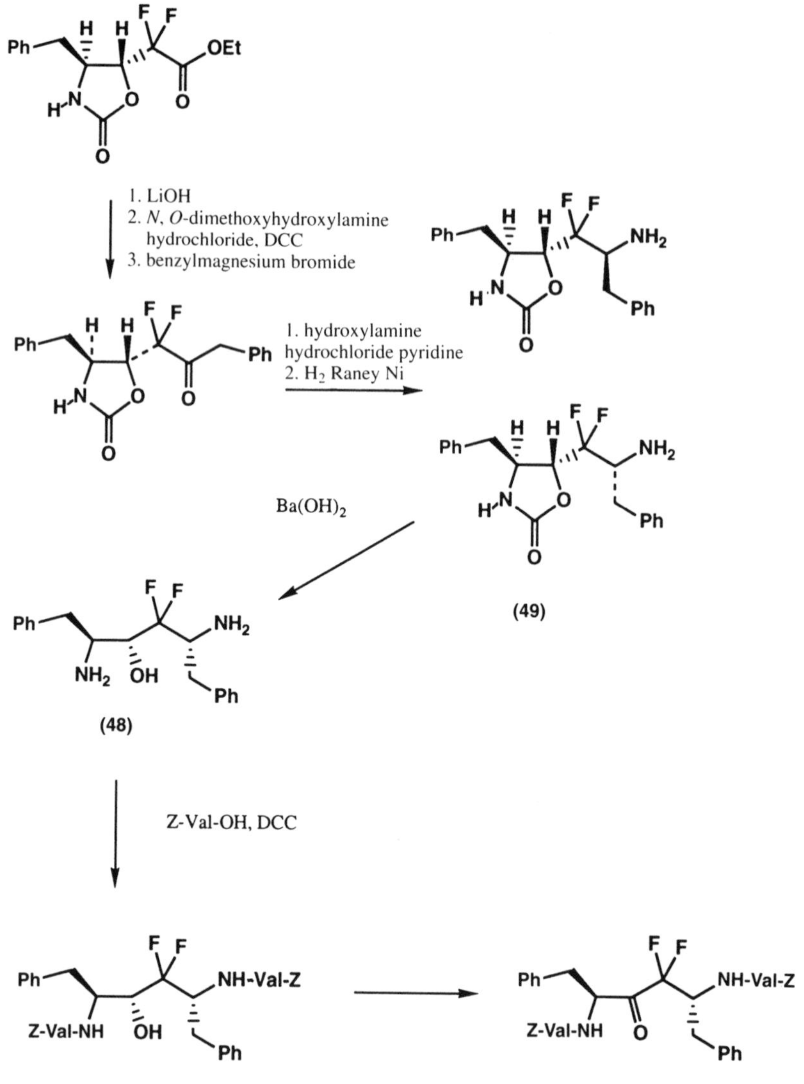

Figure 11.17 Synthesis of fluorine-containing pseudosymmetric inhibitors of HIV protease [130]

comparable, but also electronic properties are more closely mimicked. All-mendinger and co-workers [131] have developed syntheses of GlyΨ(CF=CH)Gly (**52**), PheΨ(CF=CH)Gly (**53**), PhΨ(CF=CH)Pro (**54**), and PheΨ(CF=CH)Phe (**55**) dipeptide isosteres (Figure 11.18). In general, the

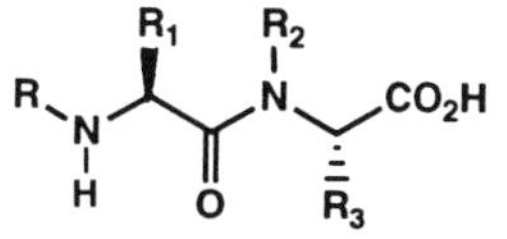

dipeptide

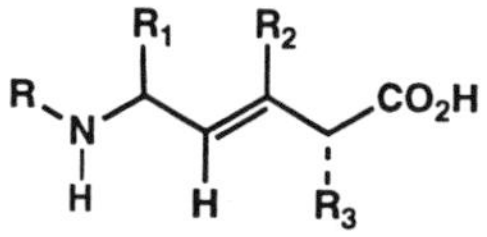

(50)

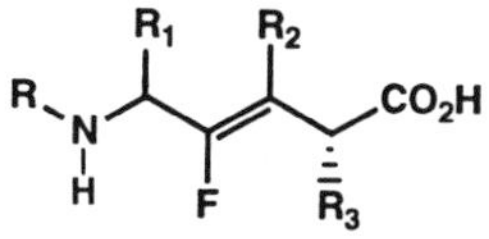

(51)

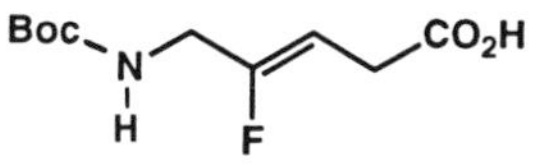

(52)

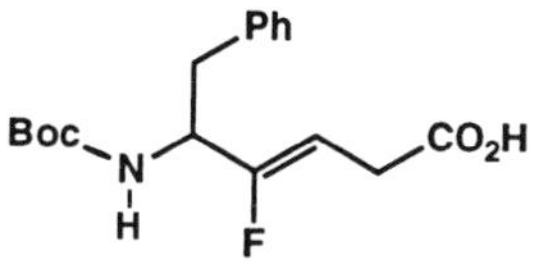

(53) (Z)

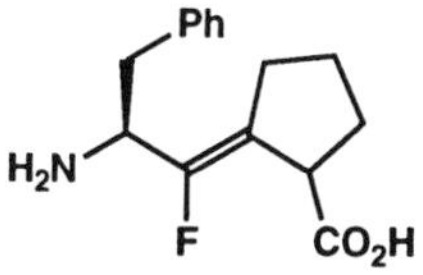

(54)

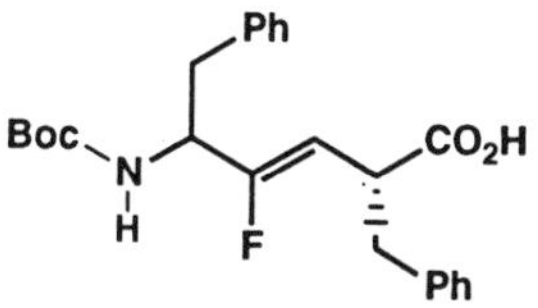

(55)

Figure 11.18 Fluoroalkene peptide isosteres [131]

synthetic procedures developed in this work permit the preparation of dipeptide combinations of amino acids having no other functional group, e.g. Gly, Ala, Val, Phe and Pro.

FMOC protected (−) and (+)-PheΨ(CF=CH)Gly analogues were incorporated into the full sequence of substance P by the solid-phase FMOC procedure to give the peptide analogue **56**, and into the C-terminal hexapeptide of substance P to give **57**. Substance P receptor binding studies confirmed that the fluoroalkene isosteres were effective replacements for the amide bond (IC_{50} = 1.3 and 2 nM for substance P and **56**, respectively). The C-terminal hexapeptide **58** having the *trans*-alkene isosteric replacement had lower affinity (IC_{50} > 10 μM) than the fluoroalkene analogue (IC_{50} = 0.8 μM). Further, unlike the double bonds of the non-fluorinated alkenes which readily isomerize to give α,β-unsaturated amides,

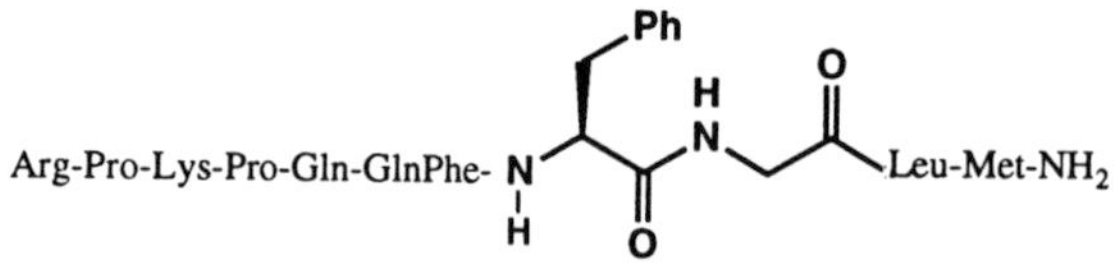
Ph
O
H
N
Arg-Pro-Lys-Pro-Gln-GlnPhe- N
H
O
Leu-Met-NH2

Substance P

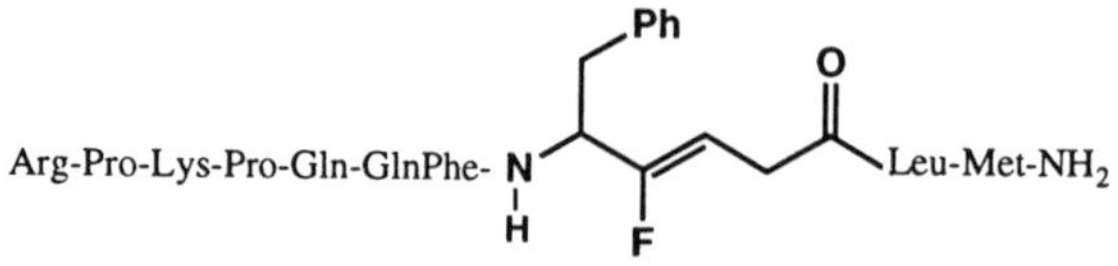
Ph
O
Arg-Pro-Lys-Pro-Gln-GlnPhe- N
H
F
Leu-Met-NH2

(56)

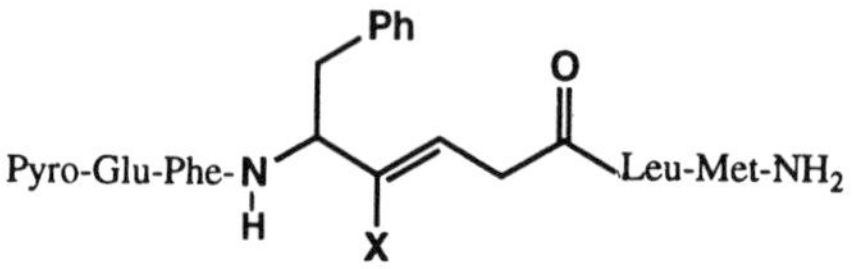
Ph
O
Pyro-Glu-Phe- N
H
X
Leu-Met-NH2

(57) X = F
(58) X = H

the double bonds of the fluoroalkenes were resistant to isomerization under all basic or acidic conditions tried [132].

Welch *et al.* [133] have initiated studies of fluoroalkene peptide isosteres as potential inhibitors of peptidyl prolyl isomerases (PPIase), enzymes that catalyze the *cis–trans* isomerization about the peptidyl prolyl amide bond. The tetrapeptide **59** has been prepared as a potential competitive inhibitor of PPIase [133].

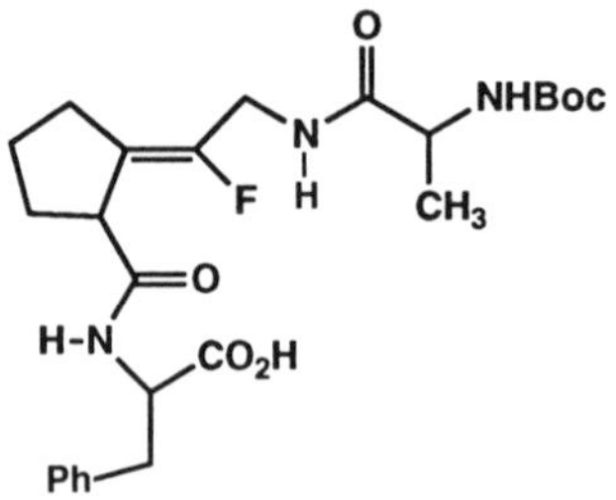
O
NHBoc
N
F
H
CH3
O
H-N
CO2H
Ph

(59)

11.6 REFERENCES

1. For reviews, see (a) J. T. Gerig, in L. S. Berliner and J. Reuben (Eds), *Biological Magnetic Resonance*, Vol. I, Plenum Press, New York, 1978, pp. 139–203; (b) J. T. Gerig, *Methods Enzymol.*, **177**, 3 (1989); (c) D. H. Gregory and J. T. Gerig, *Biopolymers*, **31**, 845 (1991).
2. M. H. Richmond, *Bacteriol. Rev.*, **26**, 398 (1962).
3. K. L. Kirk, *Biochemistry of Halogenated Organic Compounds*, Plenum Press, New York, 1991, (a) pp. 296–299; (b) pp. 9–14.
4. For a review of early work, see R. E. Marquis, in O. Eicher, A. Farah, H. Herken and A. D. Welch (Eds), *Handbook of Experimental Pharmacology*, Vol. XX/II, Springer, New York, 1979, pp. 166–192.
5. (a) D. Hebel, K. L. Kirk, L. A. Cohen and V. M. Labroo, *Tetrahedron Lett.*, **31**, 619 (1990); (b) V. M. Labroo, D. Hebel, K. L. Kirk, L. A. Cohen, C. Lemieux and P. W. Schiller, *Int. J. Pep. Protein Res.*, **37**, 430 (1991).
6. V. M. Labroo, R. B. Labroo and L. A. Cohen, *Tetrahedron Lett.*, **31**, 5705 (1990).
7. W. H. Huistis and M. A. Raftery, *Biochemistry*, **10**, 1181 (1971).
8. (a) N. Staudenmayer, M. B. Smith, H. T. Smith, F. K. Spies, Jr, and F. Millett, *Biochemistry*, **15**, 3198 (1976); (b) N. Staudenmayer, S. Ng, M. B. Smith and F. Millett, *Biochemistry*, **16**, 600 (1977); (c) H. T. Smith, N. Staudenmayer and F. Millett, *Biochemistry*, **16**, 4971 (1977).
9. D. Levy and R. A. Paselk, *Biochim. Biophys. Acta*, **310**, 398 (1973).
10. R. A. Paselk and D. Levy, *Biochemistry*, **13**, 3340 (1974).
11. R. A. Paselk and D. Levy, *Biochim. Biophys. Acta*, **359**, 215 (1974).
12. Y. Shai, K. L. Kirk, M. A. Channing, B. B. Dunn, M. A. Lesniak, R. C. Eastman, R. D. Finn, J. Roth and K. A. Jacobson, *Biochemistry*, **28**, 4801 (1989).
13. R. C. Eastman, R. E. Carson, K. A. Jacobson, Y. Shai, M. A. Channing, B. B. Dunn, J. D. Bacher, E. Bass, E. Jones, K. L. Kirk, M. A. Lesniak and J. Roth, *Diabetes*, **41**, 855 (1992).
14. M. Cairi and J. T. Gerig, *J. Am. Chem. Soc.*, **105**, 4793 (1983).
15. M. Cairi and J. T. Gerig, *J. Am. Chem. Soc.*, **106**, 3640 (1984).
16. M. Kairi and J. T. Gerig, *Magn. Reson. Chem.*, **28**, 47 (1990).
17. (a) M. Kairi and J. T. Gerig, *Biochim. Biophys. Acta*, **1039**, 157 (1990); (b) M. Kairi, N. L. Keder and J. T. Gerig, *Arch. Biochem. Biophys.*, **279**, 305 (1990).
18. M. R. Bendall and G. Lowe, *Eur. J. Biochem.*, **65**, 481 (1976); M. R. Bendall and G. Lowe, *Eur. J. Biochem.*, **65**, 493 (1976); M. R. Bendall and G. Lowe, *FEBS Lett.*, **72**, 231 (1976).
19. I. M. Chaiken, M. H. Freedman, J. R. Lyerla, Jr, and J. S. Cohen, *J. Biol. Chem.*, **248**, 884 (1973).
20. B. M. Dunn, C. DiBello, K. L. Kirk, L. A. Cohen and I. M. Chaiken, *J. Biol. Chem.*, **249**, 6295 (1974).
21. H. C. Taylor and I. M. Chaiken, *Fed. Proc., Fed. Am. Soc. Exp. Biol.*, **36**, 864 (1977).
22. P. Marbach and J. Rudinger, *Helv. Chim. Acta*, **57**, 403 (1974).
23. E. D. Nicolaides, M. K. Craft and H. A. DeWald, *J. Med. Chem.*, **6**, 524 (1963).
24. G. H. Fisher, J. W. Ryan and P. Berryer, *Cardiovasc. Med.*, **2**, 1179 (1977).
25. L. Bernardi, G. Bosisio, F. Chillimi, C. De Cara, R. de Castiglione, V. Ersparmer and O. Goffrido, *Experientia*, **22**, 29 (1966).
26. W. H. Vine, D. A. Brueckner, P. Needleman and G. R. Marshall, *Biochemistry*, **12**, 1630 (1973).
27. W. H. Vine, K. -H. Hsieh and G. R. Marshall, *J. Med. Chem.*, **24**, 1043 (1981).
28. K. -H. Hsieh, P. Needleman and G. R. Marshall, *J. Med. Chem.*, **30**, 1097 (1987).

29. (a) V. M. Labroo, K. L. Kirk, L. A. Cohen, D. Delbeke and P. S. Dannies, *Biochem. Biophys. Res. Commun.*, **113**, 581 (1983); (b) V. M. Labroo, G. Feuerestein and L. A. Cohen, in C. M. Deber, V. J. Hruby and K. D. Kopple (Eds), *Peptides, Structure and Function*, Pierce, Rockford, IL, 1985, pp. 703–706; (c) G. Feuerstein, D. Lozovsky, L. A. Cohen, V. M. Labroo, K. L. Kirk, I. J. Kopin and A. I. Faden, *Neuropeptides*, **4**, 303 (1984); (d) A. -L. Siren, G. Feuerstein, V. M. Labroo, L. A. Cohen and D. Lozovsky, *Neuropeptides*, **8**, 63 (1986); (e) S. Vonhof, I. Paakkari, G. Feuerstein, L. A. Cohen and V. M. Labroo, *Eur. J. Pharmacol.*, **164**, 77 (1989).

30. J. M. Walker, D. H. Coy, E. A. Young, G. Baldrighi, S. F. Siegel, W. Bowen and H. Akil, *Peptides*, **8**, 811 (1987).

31. R. M. Kostrzewa, R. Brus, D. H. Coy, H. Criswell, P. S. Coogan and A. J. Kastin, *Pol. J. Pharmacol. Pharm.*, **44**, 109 (1992).

32. M. Maeda, K. Kawasaki, J. Watanabe and H. Kaneto, *Chem. Pharm. Bull.*, **37**, 826 (1989).

33. J. Watanabe, S. Tokuyama, M. Takahashi, H. Kaneto, M. Maeda, K. Kawasaki, T. Taguchi, Y. Kobayashi, Y. Yamamoto and K. Shimokawa, *J. Pharmacobio-Dyn.*, **14**, 101 (1991).

34. Y. Sasaki, H. Kohno, Y. Ohkubo and K. Suzuki, *Chem. Pharm. Bull.*, **38**, 3162 (1990).

35. P. D. Gesellchen, R. C. A. Frederickson, S. Tafur and D. Smiley, in D. H. Rich and E. Gross (Eds), *Peptides. Synthesis–Structure–Function. Proceedings of Seventh American Peptide Symposium*, Pierce, Rockford IL, 1982 (a) pp. 621–624; (b) pp. 637–640.

36. D. H. Coy, A. J. Kastin, M. J. Walker, R. F. McGivern and C. A. Sandman, *Biochem. Biophys. Res. Commun.*, **83**, 977 (1978).

37. P. W. Schiller, T. M. -D. Nguyen, N. N. Chung and C. Lemieux, *J. Med. Chem.*, **32**, 698 (1989), and references cited therein.

38. Y. Sasaki, A. Ambo and K. Suzuki, *Chem. Pharm. Bull.*, **39**, 2316 (1991).

39. H. Tanaka, F. Osakada, S. -I. Ohashi, M. Shiraki and E. Munekata, *Chem. Lett.*, 391 (1986).

40. B. L. Martin, D. Wu, S. Jakes and D. J. Graves, *J. Biol. Chem.*, **265**, 7108 (1990).

41. A. P. Campbell, P. J. Cachia and B. D. Sykes, *Biochem. Cell Biol.*, **69**, 674 (1991).

42. J. J. McGuire and J. K. Coward, *J. Biol. Chem.*, **260**, 6747 (1985).

43. J. Galivan, J. K. Coward and J. J. McGuire, *Biochem. Pharmacol.*, **34**, 2995 (1985).

44. J. Galivan, J. Inglese, J. J. McGuire, Z. Nimec and J. K. Coward, *Proc. Natl. Acad. Sci. USA*, **82**, 2598 (1985).

45. J. J. McGuire, M. Graber, N. Licato, C. Vincenz, J. K. Coward, Z. Nimec and J. Galivan, *Cancer Res.*, **49**, 4517 (1989).

46. J. K. Coward, J. J. McGuire and J. Galivan, in J. T. Welch (Ed.), *Selective Fluorination in Organic and Bioorganic Chemistry*, ACS Symposium Series, No. 456, American Chemical Society, Washington, DC, 1991, pp. 196–204.

47. A. M. Stern, B. M. Foxman, A. H. Tashjian, Jr, and R. H. Abeles, *J. Med. Chem.*, **25**, 544 (1982).

48. G. Hortin, A. M. Stern, B. Miller, R. H. Abeles and I. Boime, *J. Biol. Chem.*, **258**, 4047 (1983).

49. P. K. Rathod, A. H. Tashjian, Jr, and R. H. Abeles, *J. Biol. Chem.*, **261**, 6461 (1986).

50. For a review, see J. W. Payne, *Drugs Exp. Clin. Res.*, **12**, 585 (1986).

51. W. D. Kingsbury, J. C. Boehm, R. J. Mehta and S. F. Grappel, *J. Med. Chem.*, **26**, 1725 (1983).

52. D. H. Young and R. J. Mehta, *Experientia*, **45**, 325 (1989).
53. R. Munier and G. N. Cohen, *C. R. Acad. Sci.*, **248**, 1870 (1959).
54. E. W. Westhead and P. D. Boyer, *Biochim. Biophys. Acta*, **54**, 145 (1961).
55. B. D. Sykes and J. H. Weiner, *Magn. Reson. Biol.*, **1**, 171 (1980).
56. (a) B. D. Sykes, H. I. Weingarten and M. J. Schlesinger, *Proc. Natl. Acad. Sci. USA*, **71**, 469 (1974); (b) D. T. Browne and J. D. Otvos, *Biochem. Biophys. Res. Commun.*, **68**, 907 (1976).
57. E. A. Pratt and C. Ho, *Biochemistry*, **14**, 3035 (1975).
58. B. J. Kimber, D. V. Griffiths, B. Birdsall, R. W. King, P. Scudder, J. Feeney, G. C. K. Roberts and A. S. V. Burgen, *Biochemistry*, **16**, 3492 (1977).
59. D. B. Wacks and H. K. Schachman, *J. Biol. Chem.*, **260**, 11651 (1985).
60. D. B. Wacks and H. K. Schachman, *J. Biol. Chem.*, **260**, 11659 (1985).
61. R. A. Anderson, Y. Nakashima and J. E. Coleman, *Biochemistry*, **14**, 907 (1975).
62. J. E. Coleman, R. A. Anderson, R. G. Ratcliffe and I. M. Armitage, *Biochemistry*, **15**, 5419 (1976).
63. S. D. Hagen, J. H. Weiner and B. D. Sykes, *Biochemistry*, **17**, 3860 (1978).
64. S. D. Hagen, J. H. Weiner and B. D. Sykes, *Biochemistry*, **18**, 2007 (1979).
65. M. L. Wilson and F. W. Dahlquist, *Biochemistry*, **24**, 1920 (1985).
66. J. T. -F. Wong, *Proc. Natl. Acad. Sci. USA*, **80**, 6303 (1983).
67. P. M. Bronskill and J. T. -F. Wong, *Biochem. J.*, **249**, 305 (1988).
68. P. Lu, M. Jarema, K. Mosser and W. E. Daniel, Jr, *Proc. Natl. Acad. Sci. USA*, **73**, 3471 (1976); P. Lu, W. J. Metzler, F. Rastinejad and J. Wsilewski, in F. Y. -H. Wu and C. -W. Wu (Eds), *Structure and Function of Nucleic Acids and Proteins*, Raven Press, New York, 1990, pp. 19–35.
69. M. A. C. Jarema, P. Lu and J. H. Miller, *Proc. Natl. Acad. Sci. USA*, **78**, 2707 (1981).
70. F. Boschelli, M. A. C. Jarema and P. Lu, *J. Biol. Chem.*, **256**, 11595 (1981).
71. M. Ring, I. M. Armitage and R. E. Huber, *Biochem. Biophys. Res. Commun.*, **131**, 675 (1985).
72. Reviewed by C. Ho, S. R. Dowd and J. F. M. Post, *Curr. Top. Bioenerg.*, **14**, 53 (1985).
73. G. S. Rule, E. A. Pratt, V. Simplaceanu and C. Ho, *Biochemistry*, **26**, 549 (1987).
74. O. B. Peersen, E. A. Pratt, H. -T. N. Truong, C. Ho and G. S. Rule, *Biochemistry*, **29**, 3256 (1990).
75. R. Gosh, R. Bachofen and H. Hauser, *FEBS Lett.*, **188**, 107 (1985).
76. D. E. Robertson, P. A. Kroon and C. Ho, *Biochemistry*, **16**, 1443 (1977); J. F. M. Post, P. F. Cottam, V. Simplaceanu and C. Ho, *J. Mol. Biol.*, **179**, 729 (1984).
77. Q. Shen, V. Simplaceanu, P. F. Cottam, J. -L. Wu, J. -S. Hong and C. Ho, *J. Mol. Biol.*, **210**, 859 (1989).
78. E. Li, S. -J. Quian, L. Nader, N. -C. Yang, A. d'Avignon, J. C. Sacchettini and J. I. Gordon, *J. Biol. Chem.*, **264**, 17041 (1989).
79. E. Li, S. -J. Quian, N. -C. C. Yang, A. d'Avignon and J. I. Gordon, *J. Biol. Chem.*, **265**, 11549 (1990).
80. E. Li, S. -J. Quian, N. S. Winter, A. d'Avignon, M. S. Levin and J. I. Gordon, *J. Biol. Chem.*, **266**, 3622 (1991).
81. I. J. Ropson and C. Frieden, *Proc. Natl. Acad. Sci. USA*, **89**, 7222 (1992).
82. (a) F. Sixl, R. W. King, M. Bracken and J. Feeney, *Biochem. J.*, **266**, 545 (1990); (b) M. G. Hinds, R. W. King and J. Feeney, *FEBS Lett.*, **283**, 127 (1991); (c) M. G. Hinds, R. W. King and J. Feeney, *Biochem. J.*, **287**, 627 (1992).
83. L. A. Luck and J. J. Falke, *Biochemistry*, **30**, 4248 (1991).
84. L. A. Luck and J. J. Falke, *Biochemistry*, **30**, 4257 (1991).
85. L. A. Luck and J. J. Falke, *Biochemistry*, **30**, 6484 (1991).
86. K. Brindle, S. -P. Williams and M. Boulton, *FEBS Lett.*, **255**, 121 (1989).

87. D. N. Wheatley and J. Y. Henderson, *Natural (London)*, **247**, 281 (1974).
88. P. S. Sunkara, B. M. Chakroborty, D. A. Wright and P. N. Rao, *Eur. J. Cell Biol.*, **23**, 312 (1981).
89. M. S. Inglis and D. N. Wheatley, *Cell Biol. Int. Rep.*, **3**, 739 (1979).
90. P. F. Torrence, R. M. Friedman, K. L. Kirk, L. A. Cohen and C. R. Creveling, *Biochem. Pharmacol.*, **28**, 1565 (1979).
91. C. R. Creveling, W. L. Padgett, E. T. McNeal, L. A. Cohen and K. L. Kirk, *Life Sci.*, **51**, 1197 (1992), and references cited therein.
92. D. C. Klein, J. L. Weller, K. L. Kirk and R. W. Hartley, *Mol. Pharmacol.*, **13**, 1105 (1977); D. C. Klein and K. L. Kirk, in R. Filler (Ed.), *Biochemistry Involving Carbon–Fluorine Bonds*, ACS Symposium Series, No. 28, American Chemical Society, Washington, DC, 1976, pp. 37–56, and references cited therein.
93. J. Uitto and D. J. Prockop, *Biochim. Biophys. Acta*, **336**, 234 (1974).
94. J. T. Gerig, J. C. Klinkenborg and R. A. Nieman, *Biochemistry*, **22**, 2076 (1983).
95. M. P. Gamcsik, J. T. Gerig and D. H. Gregory, *Biochim. Biophys. Acta*, **912**, 303 (1987).
96. M. P. Gamcsik, J. T. Gerig and R. B. Swenson, *Biochim. Biophys. Acta*, **874**, 372 (1986).
97. M. P. Gamcsik and J. T. Gerig, *FEBS Lett.*, **196**, 71 (1986).
98. S. Schoellmann and E. Shaw, *Biochemistry*, **2**, 252 (1963).
99. For reviews, see J. C. Powers, *Methods Enzymol*, **46**, 197 (1977); J. C. Powers, *Chemistry and Biochemistry of Amino Acids, Peptides and Proteins*, Marcel Dekker, New York, 1977, pp. 65–178.
100. E. Shaw, in P. D. Boyer (Ed.), *The Enzymes*, Vol. I, 3rd edn., Academic Press, New York, 1970, pp. 91–146 and references cited therein.
101. B. Imperiali and R. H. Abeles, *Tetrahedron Lett.*, **27**, 135 (1986).
102. B. Imperiali and R. H. Abeles, *Biochemistry*, **25**, 3760 (1986).
103. P. Rauber, H. Angliker, B. Walker and E. Shaw, *Biochem. J.*, **239**, 633 (1986).
104. H. Angliker, P. Wikstrom, P. Rauber and E. Shaw, *Biochem. J.*, **241**, 871 (1987).
105. B. Imperiali, in *Synthetic Peptides in Biotechnology*, Alan R. Liss, New York, 1988, pp. 97–129.
106. (a) J. -L. Dimicoli, J. Bieth and J. -M. Lhoste, *Biochemistry*, **15**, 2230 (1976); (b) P. Lestienne, J. -L. Dimicoli and J. Bieth, *J. Biol. Chem.*, **253**, 3459 (1978).
107. A. Renaud, J. -L. Dimicoli, P. Lestienne and J. Bieth, *J. Am. Chem. Soc.*, **100**, 1005 (1978).
108. J. -L. Dimicoli, A. Renaud and J. Bieth, *Eur. J. Biochem.*, **107**, 423 (1980).
109. P. Lestienne, J. -L. Dimicoli, A. Renaud and J. G. Bieth, *J. Biol. Chem.*, **254**, 5219 (1979).
110. I. L. de la Sierra, E. Papamichael, C. Sakarellos, J. L. Dimicoli and T. Prangé, *J. Mol. Recogn.*, **3**, 36 (1990).
111. J. O. Westerik and R. Wolfenden, *J. Biol. Chem.*, **247**, 8195 (1972).
112. R. C. Thompson, *Biochemistry*, **12**, 47 (1973).
113. D. O. Shah, K. Lai and D. G. Gorenstein, *J. Am. Chem. Soc.*, **106**, 4272 (1984).
114. R. E. Galardy and Z. P. Kortylewicz, *Biochemistry*, **23**, 2083 (1984).
115. M. H. Gelb, J. P. Svaren and R. H. Asbeles, *Biochemistry*, **24**, 1813 (1985).
116. J. W. Skiles, B. Fuchs, G. Chow and M. Skoog, *Res. Commun. Chem. Pathol. Pharmacol.*, **68**, 365 (1990); J. W. Skiles, V. Fuchs, C. Miao, R. Sorcek, K. G. Grozinger, S. C. Mauldin, J. Vitous, P. W. Mui, S. Jacober, G. Chow, M. Matteo, M. Skoog, S. M. Weldon, G. Possanza, J. Keirns, G. Letts and A. S. Rosenthal, *J. Med. Chem.*, **35**, 641 (1992).
117. S. M. Welden, L. G. Letts, J. Keirns, G. Chow, M. Skoog, J. Skiles, V. Fuchs and G. J. Possanza, *FASEB J.*, **4**, Abstr. 5212 (1990).

118. N. P. Peet, J. P. Burkhart, M. R. Angelastro, E. L. Girous, S. Mehdi, P. Bey, M. Kolb, B. Nieses and D. Schirlin, *J. Med. Chem.*, **33**, 394 (1989).
119. B. Imperiali and R. H. Abeles, *Biochemistry*, **26**, 4474 (1987).
120. J. W. Skiles, C. Miao, R. Sorcek, S. Jacober, P. W. Mui, G. Chow, S. M. Weldon, G. Possanza, M. Skoog, J. Keirns, G. Letts and A. S. Rosenthal, *J. Med. Chem.*, **35**, 4795 (1992).
121. D. Rasnick, *Anal. Biochem.*, **146**, 461 (1985).
122. D. H. Rich, A. S. Boparai and M. S. Bernatowicz, *Biochem. Biophys. Res. Commun.*, **104**, 1127 (1982).
123. D. H. Rich, M. S. Bernatowicz and P. G. Schmidt, *J. Am. Chem. Soc.*, **104**, 3535 (1982).
124. K. Fearon, A. Spaltenstein, P. B. Hopkins and M. H. Gelb, *J. Med. Chem.*, **30**, 1617 (1987).
125. S. Thaisrivongs, D. T. Pals, M. W. Kati, S. R. Turner, L. M. Thomasco and W. Watt, *J. Med. Chem.*, **29**, 2080 (1986).
126. A. M. Doherty, I. Sircar, B. E. Kornberg, J. Quinn, III, R. T. Winters, J. S. Kaltenbronn, M. D. Taylor, B. L. Batley, S. R. Rapundalo, M. J. Ryan and C. A. Painchaud, *J. Med. Chem.*, **35**, 2 (1992).
127. M. N. G. James, A. R. Sielecki, K. Hayakawa and M. H. Gelb, *Biochemistry*, **31**, 3872 (1992).
128. K. Suguna, E. A. Padlan, C. W. Smith, W. D. Carlson and D. R. Davies, *Proc. Natl. Acad. Sci USA*, **84**, 7009 (1987).
129. I. Ojima and F. A. Jameison, *Bioorg. Med. Chem. Lett.*, **1**, 581 (1991).
130. H. L. Sham, N. E. Wideburg, S. G. Spanton, W. E. Kohlbrenner, D. A. Betebenner, D. J. Kemph, D. W. Norbeck, J. J. Plattner and J. W. Erickson, *J. Chem. Soc., Chem. Commun.*, 110 (1991).
131. T. Allmendinger, P. Furet and E. Hungerbühler, *Tetrahedron Lett.*, **31**, 7927 (1990); T. Allmendinger, E. Felder and E. Hungerbuehler, in J. T. Welch (Ed.), *Selective Fluorination in Organic and Bioorganic Chemistry*, ACS Symposium Series, No. 456, American Chemical Society, Washington, DC, 1991, pp. 186–195.
132. T. Allmendinger, E. Felder and E. Hungerbühler, *Tetrahedron Lett.*, **31**, 7301 (1990).
133. J. T. Welch, L. Boros and B. DeCorte, in *Abstracts of the ACS Eleventh Winter Fluorine Conference, St Petersburg, FL, January 25–30, 1993*, American Chemical Society, Washington, DC, 1993, p. 25.

Index

Index compiled by G. Jones